W9-BSF-366

QUANTUM STATISTICS OF LINEAR AND NONLINEAR OPTICAL PHENOMENA

QUANTUM STATISTICS OF LINEAR AND NONLINEAR OPTICAL PHENOMENA

by

JAN PEŘINA

Palacký University, Olomouc, Czechoslovakia

D. REIDEL PUBLISHING COMPANY

A MEMBER OF THE KLUWER ACADEMIC PUBLISHERS GROUP

DORDRECHT / BOSTON / LANCASTER

Library of Congress Cataloging in Publication Data

CIP

Peřina, Jan, 1936—
Quantum statistics of linear and nonlinear optical phenomena.

Bibliography: p.
Includes index.
1. Optics. 2. Quantum statistics. I. Title.
QC355.2.P47 1983 535 83—15988
ISBN 90-277-1512-2 (Reidel)

Published by D. Reidel Publishing Company,
P.O. Box 17, 3300 AA Dordrecht, Holland
In co-edition with SNTL—Publishers of Technical Literature—Prague, Czechoslovakia.

Distributed in Albania, Bulgaria, Chinese People's Republic, Cuba, Czechoslovakia, German Democratic Republic, Hungary, Korean People's Democratic Republic, Mongolia, Poland, Rumania, the U.S.S.R., Vietnam, and Yugoslavia by Artia, Prague.

Sold and distributed in the U.S.A. and Canada
by Kluwer Academic Publishers,
190 Old Derby Street, Hingham, MA 02043, U.S.A.

Sold and distributed in all other countries by
Kluwer Academic Publishers Group,
P.O. Box 322, 3300 AH Dordrecht, Holland.

Printed in Czechoslovakia by SNTL, Prague.

PREFACE

The quantum statistical properties of radiation represent an important branch of modern physics with rapidly increasing applications in spectroscopy, quantum generators of radiation, optical communication, etc. They have also an increasing role in fields other than pure physics, such as biophysics, psychophysics, biology, etc.

The present monograph represents an extension and continuation of the previous monograph of this author entitled *Coherence of Light* (Van Nostrand Reinhold Company, London 1972, translated into Russian in the Publishing House Mir, Moscow 1974) and of a review chapter in Progress in Optics, Vol. 18 (E. Wolf (Ed.), North-Holland Publishing Company, Amsterdam 1980), published just recently. It applies the fundamental tools of the coherent-state technique, as described in *Coherence of Light*, to particular studies of the quantum statistical properties of radiation in its interaction with matter. In particular, nonlinear optical processes are considered, and purely quantum phenomena such as antibunching of photons are discussed.

This book will be useful to research workers in the fields of quantum optics and electronics, quantum generators, optical communication and solid-state physics, as well as to students of physics, optical engineering and opto-electronics.

It is a great pleasure for me to thank sincerely many people who have been engaged in this research or supported it for many years. I wish to thank Professor B. Havelka, the founder of the Laboratory of Optics of Palacký University in Olomouc, for his support, and my co-workers Drs. R. Horák, L. Mišta and V. Peřinová, who contributed very much to the subject. Professors M. C. Teich and P. Diament of Columbia University, Professor G. Lachs of Pennsylvania State University and Professor S. Kielich and Dr. P. Szlachetka of A. Mickiewicz University are acknowledged for their fruitful co-operation which resulted in publication of joint papers. The author is obliged to Professor Emil Wolf of Rochester University for his kind support and to many scientists abroad, particularly to Professor D. F. Walls of the University of Waikato, for regularly sending him preprints and reprints in the field for many years. He is very indebted to Prof. R. M. Sillitto for his careful reading of the manuscript and for its improvement. He also thanks very much Mrs. M. Rozsypalová for her careful preparation of the figures.

Permission to reproduce figures is acknowledged to: The American Physical Society, Fig. 3.7 by B. L. Morgan and L. Mandel, Fig. 5.2 by A. W. Smith and J. A. Armstrong, Fig. 5.5 by F. T. Arecchi, V. Degiorgio and B. Querzola, Fig. 5.3

by R. F. Chang, V. Korenman, C. O. Alley and R. W. Detenbeck, Figs. 9.31 and 9.32 by M. Dagenais and L. Mandel, Fig. 9.30 by N. Nayak and B. K. Mohanty, Fig. 9.29 by M. S. Zubairy and J. J. Yeh; North-Holland Publishing Company, Fig. 3.6 by F. T. Arecchi, E. Gatti and A. Sona, Fig. 6.12 by R. E. Slusher, Fig. 9.11 by K. J. McNeil, P. D. Drummond and D. F. Walls, Fig. 9.12 by P. D. Drummond, K. J. McNeil and D. F. Walls, Fig. 9.26 by N. Tornau and A. Bach, Figs. 7.2 and 7.3 by F. T. Arecchi, A. Politi and L. Ulivi; Optical Society of America, Fig. 3.10 by B. I. Cantor and M. C. Teich, Fig. 5.10 by M. C. Teich and G. Vannucci, Figs. 8.6 and 8.7 by V. Bluemel, L. M. Narducci and R. A. Tuft; Taylor and Francis Publishing House, Fig. 9.9 by S. K. Srinivasan and S. Udayabaskaran, Figs. 9.19, 9.22 and 9.23 by A. Pieczonková, Fig. 9.30 by P. S. Gupta and B. K. Mohanty; The Institute of Physics (UK), Fig. 9.10 by D. F. Walls and C. T. Tindle, Figs. 9.24 and 9.25 by H. D. Simaan; Springer Publishing House, Fig. 9.13 by J. Wagner, P. Kurowski and W. Martienssen, Fig. 9.28 by H. Voigt, A. Bandilla and H. H. Ritze; The Institute of Electrical and Electronics Engineers, Fig. 5.4 by F. T. Arecchi, A. Berné, A. Sona and P. Burlamacchi; J. A. Barth Publishing House, Fig. 9.27 by U. Mohr and H. Paul; Plenum Press, Fig. 8.8 by M. Bertolotti; Macmillan Publishing House, Fig. 9.33 by D. F. Walls; Publishing House Academia, Figs. 9.20 and 9.21 by A. Pieczonková.

Olomouc, March, 1984 J. P.

TABLE OF CONTENTS

PREFACE V

CHAPTER 1. INTRODUCTION 1

CHAPTER 2. QUANTUM THEORY OF THE ELECTROMAGNETIC FIELD 10

2.1 Quantum description of the field 11
2.2 Statistical states 15
2.3 Multimode description 17
2.4 Calculation of commutators of the field operators 17
2.5 Time development of quantum states 19

CHAPTER 3. OPTICAL CORRELATION PHENOMENA 22

3.1 Definition of quantum correlation functions 23
3.2 Properties of quantum correlation functions 27
3.2.1 Analytic properties 27
3.2.2 Spectral properties 28
3.2.3 Wave equations in vacuo 29
3.2.4 Symmetries and inequalities 30
3.2.5 Examples of the second-order degrees of coherence 32
3.3 Quantum coherence 32
3.3.1 Second-order phenomena 32
3.3.2 Higher-order phenomena 34
3.4 Measurements corresponding to antinormally ordered products of field operators – quantum counters 38
3.5 Quantum characteristic functionals 40
3.6 Measurements of mixed-order correlation functions 41
3.7 Photocount distribution and photocount statistics 43
3.8 Determination of the integrated intensity probability distribution from the photocount distribution 49
3.9 Short-time measurements 52
3.10 Bunching and antibunching of photons 57
3.11 Hanbury Brown – Twiss effect – correlation interferometry and correlation spectroscopy 61

CHAPTER 4. COHERENT-STATE DESCRIPTION OF THE ELECTROMAGNETIC FIELD 69

4.1 Coherent states of a harmonic oscillator and of the electromagnetic field 69
4.1.1 Definitions 69
4.1.2 Expansions in terms of coherent states 71
4.1.3 Minimum-uncertainty wave packets 72
4.1.4 Properties of the displacement operator $\hat{D}(\alpha)$ 73
4.1.5 Expectation values of operators in coherent states 74
4.1.6 Generalized coherent states 76
4.1.7 Multimode description 76
4.1.8 Time development of the coherent states 77
4.1.9 Even and odd coherent states 78
4.2 Glauber – Sudarshan representation of the density matrix 78
4.3 The existence of the Glauber – Sudarshan representation 86
4.4 The phase operators 88
4.5 Multimode description 88
4.6 Relation between the quantum and classical descriptions 89
4.6.1 Quantum and classical correlation functions 89
4.6.2 Photon-number and photocount distributions 90
4.7 Stationary conditions for the field 92
4.7.1 Time invariance properties of the correlation functions 92
4.7.2 Stationary conditions in phase space 93
4.8 Ordering of field operators 94
4.8.1 Ω- and s-ordering and general decompositions 94
4.8.2 Connecting relations 97
4.8.3 Multimode description 98
4.9 Interference of independent light beams 101
4.10 Two-photon coherent states, atomic coherent states and coherent states for general potentials 103
4.10.1 Two-photon coherent states 103
4.10.2 Atomic coherent states 106

CHAPTER 5. SPECIAL STATES OF THE ELECTROMAGNETIC FIELD 109

5.1 Chaotic (Gaussian) light 109
5.1.1 Distributions and characteristic functions 109
5.1.2 The second-order correlation function for blackbody radiation 112
5.1.3 Photocount statistics 114
5.2 Laser radiation 118
5.2.1 Ideal laser model 118
5.2.2 Real laser model 122
5.3 Superposition of coherent and chaotic fields 129
5.3.1 One-mode field 130

5.3.2 Multimode field – characteristic generating function 131
5.3.3 Integrated intensity probability distribution 133
5.3.4 The photocount distribution 134
5.3.5 Factorial moments 140
5.3.6 Factorial cumulants 146
5.3.7 Accuracy of approximate M-mode formulae 147

CHAPTER 6. REVIEW OF NONLINEAR OPTICAL PHENOMENA 152

6.1 General classical description 152
6.2 The second-order phenomena 153
6.3 The third- and higher-order phenomena 157
6.4 Transient coherent optical effects 163
6.4.1 Self-induced transparency 164
6.4.2 Photon echo 165
6.4.3 Superradiance 165

CHAPTER 7. HEISENBERG–LANGEVIN AND MASTER EQUATIONS APPROACHES TO THE STATISTICAL PROPERTIES OF RADIATION INTERACTING WITH MATTER 167

7.1 The Heisenberg – Langevin approach 167
7.2 The master equation and generalized Fokker – Planck equation approaches 172
7.3 The interaction of radiation with the atomic system of a nonlinear medium 174

CHAPTER 8. QUANTUM STATISTICS OF RADIATION IN RANDOM MEDIA 178

8.1 Phenomenological description of propagation of radiation through turbulent atmosphere and Gaussian media 178
8.2 The hamiltonian for radiation interacting with a random medium 185
8.3 Heisenberg – Langevin equations and the generalized Fokker – Planck equation 187
8.4 Solutions of the generalized Fokker – Planck equation and the Heisenberg – Langevin equations 189
8.5 Photocount statistics 193
8.6 Diament – Teich and Tatarskii descriptions 196
8.7 Comparison of the quantum and phenomenological descriptions 200
8.8 Speckle phenomenon 204

CHAPTER 9. QUANTUM STATISTICS OF RADIATION IN NONLINEAR MEDIA 205

9.1 Optical parametric processes with classical pumping 205
9.1.1 Degenerate case 205

9.1.2 Non-degenerate case 210
9.2 Interaction of three one-mode boson quantum fields 219
9.3 Second and higher harmonic and subharmonic generation 226
9.4 Raman, Brillouin and hyper-Raman scattering 234
9.4.1 Reservoir phonon system 234
9.4.2 Dynamics of photon and phonon modes 245
9.4.3 Completely quantum description 252
9.4.4 Hyper-Raman scattering 259
9.5 Multiphoton absorption 261
9.6 Multiphoton emission 266
9.7 Resonance fluorescence 269
9.8 Other interesting nonlinear phenomena 273
9.8.1 Coherent γ-emission by stimulated annihilation of electron–positron pairs 273
9.8.2 A solvable model for light scattering 274
9.9 Phase-transition analogies 278

CHAPTER 10. CONCLUSIONS 281

REFERENCES 283

INDEX 307

CHAPTER 1

INTRODUCTION

At present, the role of the statistical and coherence properties of light fields in optics is well recognized, from both the physical point of view and applications. The statistical properties of optical fields manifest themselves in the form of their coherence properties in the optical region, where the classical wave behaviour of the long-wave region overlaps with the chaotic particle behaviour of the short-wave region. Many papers and other works have appeared which are devoted to the classical aspects of the problem, originally related to the second-order theory of coherence in the context of interference, diffraction and polarization phenomena. In this respect, we can recommend several works for reading: Born and Wolf (1965) Françon (1966), Françon and Slansky (1965), O'Neil (1963), Beran and Parrent (1964), Peřina (1972) and Potechin and Tatarinov (1978). At present, many papers are available in which these general methods are applied to various problems, mostly in instrumental optics. This may be seen, for instance, in publications by Dainty (1975) and Baltes (1978), where problems such as speckle phenomenon [Goodman (1975), Elbaum and Diament (1976), Jakeman et al. (1976), Dainty (1975, 1976), Welford (1977), Ohtsubo and Asakura (1977a, b), May (1977), Zardecki and Delisle (1977), Peřina (1977), Zardecki (1978), De Santis et al. (1978), Ross and Fiddy (1978), Saleh and Irshid (1979), Barakat (1981), Peřina and Horák (1981)]; super-resolution [Frieden (1971), Peřina, Peřinová and Braunerová (1977), Schmidt–Weinmar (1978), Schmidt–Weinmar et al. (1978/79)]; the phase problem [Wolf (1962), Roman and Marathay (1963), Dialetis (1967), Nussenzveig (1967), Kohler and Mandel (1970, 1973), Misell (1973), Saxton (1974), Peřina (1972), Burge et al. (1974, 1976), Spence (1974), Hoenders (1975), Chopra and Dudeja (1977), Ross et al. (1977), Agrawal (1978), Ferwerda (1978), Nieto–Vesperinas and Hignette (1979), Taylor (1980), Ablekov, Babaev et al. (1980), Ablekov, Avdyrevskii et al. (1980), Nieto–Vesperinas (1980), Barakat (1980), Kiedroń (1980, 1981), Walker (1981)]; the connection of partial coherence and radiometry [Marchand and Wolf (1972, 1974a, b), Wolf and Carter (1975, 1976), Carter and Wolf (1975, 1977, 1981a, b), Baltes, Steinle and Antes (1976), Steinle and Baltes (1977), Wolf (1978), Friberg (1978a, b, 1979a, b, 1981a, b), Baltes, Geist and Walther (1978), Carter and Bertolotti (1978), Collet and Wolf (1979, 1980), Farina et al. (1980), Jannson (1980), Gori (1980)]; and relations between the object and its image [Peřina and Peřinová (1969), Martinéz–Herrero (1979, 1981), Martinéz–Herrero and Durán (1981), Baltes et al. (1981), Martinéz–Herrero and Mejías (1981)] have been discussed. These areas just represent the most fruitful applications of the second-order classical coherence theory of free optical fields. Moreover, applications to the X-ray region have been dealt with [Holý (1980)]. It should be noted also that the correlation tensor theory, originally proposed and developed by Roman and Wolf (1960) and reviewed by Beran and Parrent (1964) and Peřina (1972), is being applied [Carter (1980), Ross and Nieto – Vesperinas (1981)].

A more complete description of the field, however requires the definition of correlation functions of all orders, introduced by Wolf (1963, 1964, 1965, 1966) on the basis of classical stochastic functions. Only such mathematical and physical tools have been able to explain the pioneering experiments of Hanbury Brown and Twiss (1956a, b, c, 1957a, b, 1958) which concerned the correlation of intensities with the help of two quadratic detectors – photodetectors. Also laser light, being non-Gaussian (non-chaotic) light, requires a complete statistical description by means of all-order moments. In this connection it was important to derive the relation between the statistics of stochastic photons incident on a photocathode and the statistics of the emitted photoelectrons in the form of the photodetection equation [Mandel (1958, 1959, 1963a)], so that one can determine the statistical properties of light from photoelectric measurements [Peřina (1970)]. Intensity correlation measurements or photon coincidence measurements provide information on the spatial, temporal, spectral and polarization properties of radiation. These questions are discussed in greater detail by Saleh (1978), Mehta (1970) and Barakat and Blake (1980).

A fruitful development of the theory of the statistical and coherence properties and its applications was realized when Glauber (1963a, b, c, 1964, 1965, 1966a, b, 1967, 1969, 1970, 1972) used the coherent states for the harmonic oscillator to describe coherence as a cooperative boson phenomenon. The coherent-state technique enabled the relationship to be established between the quantum and classical descriptions of optical fields: these descriptions have been found to be formally equivalent if the field is registered by the absorption of a photon from the field, i.e. by photodetectors [Sudarshan (1963a, b), Mandel and Wolf (1965, 1966), Klauder and Sudarshan (1968), Troup (1967), Vinson (1971), Peřina (1972), Picinbono and Rousseau (1977)]. The coherent-state technique has proven to be extremely useful for describing also the interaction of optical fields with matter [Scully and Whitney (1972), Agarwal (1973), Haake (1973), Davies (1976)], particularly in the laser [Risken (1968, 1970), Lax (1967, 1968a, b), Haken (1967, 1970a, b, 1972), Lax and Zwanziger (1970, 1973), Louisell (1970, 1973), Sargent and Scully (1972), Sargent et al. (1974)], in nonlinear optical processes [Shen (1967), Loudon (1973), Nussenzveig (1973), Graham (1973, 1974), Schubert and Wilhelmi (1978, 1980), Peřina (1980a, b)] and in scattering processes [Crosignani et al. (1975)]. Many experimental results have been reported in the literature [Armstrong and Smith (1967), Pike (1969, 1970), Pike and Jakeman (1974), Arecchi and Degiorgio (1972)].

The coherent states of the harmonic oscillator have now been generalized in many areas. Arecchi, Courtens, Gilmore and Thomas (1973) have introduced the atomic coherent states as eigenstates of the angular-momentum operators and Yuen (1975,

1976) has introduced the so-called two-photon coherent states to describe two-photon stimulated emission, provided that atomic variables are treated classically. A general theory of the generalized coherent states has been developed [Malkin et al. (1969), Mikhailov (1971, 1976), Trifonov (1974), Narducci et al. (1975), Nieto and Simmons (1978, 1979), Nieto et al. (1981), Lugiato et al. (1977), Perelomov (1977), Shustov (1978), Malkin and Man'ko (1979), Jannussis et al. (1979), Nikolov and Trifonov (1980), Rowe and Ryman (1980)] and has been used to study other co-operative phenomena such as superconductivity, superfluidity and ferromagnetism [Carruthers and Dy (1966), López (1967), Carruthers and Nieto (1968), Langer (1968, 1969), Rezende and Zagury (1969), Man'ko (1972), Sargent et al. (1974), Haken (1978)]. There have also been applications in other fields of solid-state physics, and in elementary particle physics and quantum field theories. Such an approach has been found to be fruitful also in studying propagation in optical waveguides [Krivoshlykov and Sissakian (1979, 1980a, b)]. Another aspect of coherent states has been developed in the context of free electron lasers [Dattoli, Renieri, Romanelli and Bonifacio (1980), Dattoli, Renieri and Romanelli (1980), Bonifacio (1980)].

The basic property of coherent states is that they have a completely known phase but an uncertain number of particles, whereas the Fock occupation number states have an uncertain phase but a definite number of particles. The coherent states are as close to classical deterministic states (possessing no fluctuations), as quantum states can be, i.e. they include only incoherent vacuum fluctuations. So the Fock states are appropriate in describing the interaction of several particles in quantum electrodynamics, whereas the coherent states are able to describe cooperative phenomena, such as coherence when photons in the coherent state are "seeing" each other.

The importance of investigating the statistical properties of optical fields is emphasized by the qualitative difference between natural sources of light and artificial sources such as lasers. Natural sources operate in thermal equilibrium, so chaotic photons from such sources are governed in one mode by the Bose – Einstein statistics, in agreement with the application of the central limit theorem to macroscopic field produced by many independent elementary sources (atoms). On the other hand, the laser is a nonlinear device not in thermal equilibrium, which operates only if the density of photons is so high that the photons see each other and, consequently, they are governed by the probability distribution of classical particles, i.e. by the Poisson distribution. As a result, photons from natural sources create clusters and are bunched when arriving at the photodetector; this correlation of photons in the beam is balanced out by the correlation of atoms in the laser source so that the photons of a laser beam arrive independently at the photocathode.

The monograph *Coherence of Light* was mainly devoted to the statistical and coherence properties of free optical fields. The purpose of this monograph is to apply the methods described there, based on the coherent-state technique, to the investigation of the quantum statistical properties, including the measurable photocount statistics, of radiation interacting with random and nonlinear media. To do this we use both the Heisenberg approach, based on the solution of the Heisenberg–Langevin equations of motion for the field operators, and the calculation of the quantum characteristic functions and the corresponding quasi-distributions, and the Schrödinger approach, based on the solution of the master or generalized Fokker–Planck equations.

It should be noted that these methods are of increasing importance, not only in solving various problems of interferometry, spectroscopy, imaging theory with photons and electrons, the theory of masers and lasers, optical communication and ranging, and the propagation of radiation through random and nonlinear media, but also in cross-disciplinary sciences such as biophysics and psychophysics.

Chapters 2–5 of this monograph present a modified and extended treatment of the corresponding chapters of *Coherence of Light* in order that this monograph is self-contained; of course, the most recent results are included. Chapter 6 reviews the basic properties of certain nonlinear optical processes which are discussed from a completely quantum statistical point of view in Chapters 8 and 9. More complete treatments of these basic properties are available in the monographs and reviews by Bloembergen (1965), Baldwin (1969), Yariv (1967, 1975), Akhmanov et al. (1972), Kleinman (1972), Courtens (1972), Akhmanov and Tchirkin (1971), Zernike and Midwinter (1973), Paul (1973), Loudon (1973), Nussenzveig (1973), Svelto (1974), Sargent et al. (1974), Allen and Eberly (1975), Shen (1976), Schubert and Wilhelmi (1978), Klyshko (1980) and Kielich (1981). Chapter 7 is devoted to the introduction of coherent-state methods for interacting systems; these are explained in greater detail in, for instance, the monograph by Louisell (1973).

Let us make now several historical remarks relating to the main subjects of this book which are presented in Chapters 5, 8 and 9.

The quantum statistical properties of the superposition of coherent and chaotic fields (coherent and chaotic fields can be considered as special cases) have been discussed by a number of authors from 1965 onwards in connection with the properties of laser radiation above the threshold, as well as with the properties of scattered radiation. Such a model can, however, be used everywhere if a signal is embedded in noise, particularly in optical communication, but also, for instance, in psychophysics and neural counting [Teich and McGill (1976), Teich et al. (1978, 1982)]. The superposition of coherent and chaotic fields was investigated in terms of coherent states for the first time by Lachs (1965), Glauber (1966a) and Morawitz (1965, 1966). Multimode formulae have been obtained by Peřina (1967b, 1968a, 1969b, 1970) and independently by McGill (1967) in neural counting [cf. also Teich and McGill (1976)] and further work has been reported by Peřina and Horák (1969a) and Horák et al. (1971a, b). Contributions to this field have been summarized in a table by Jakeman and Pike (1969a), Pike and Jakeman (1974) and Lax and Zwanziger (1973), including purely coherent and chaotic fields as special cases. More recent references have been summarized, too [Peřina (1975, 1980b), also Peřina (1974) and references added therein by the translators, and Saleh (1978)]. Particular mention should be made of results relating to arbitrary detection-time intervals and photocathode areas, as well as to arbitrary temporal and spatial spectra of radiation with arbitrary polarization [Jaiswal and Mehta (1970), Mehta and Jaiswal (1970), Peřina et al. (1971, 1972), Mišta et al. (1973)]; modulated fields [Troup and Lyons (1969), Lyons and Troup (1970a, b), Diament and Teich (1970a), Bendjaballah and Perrot (1971, 1973), Teich and Vannucci (1978), Mišta (1973), Prucnal and Teich (1978), Koňák et al. (1982)]; fields propagating through the turbulent atmosphere [Diament and Teich (1970b), Bertolotti et al. (1970), Peřina and Peřinová (1972), Lachs and Laxpati (1973)]; and the photocount statistics of multiphoton processes [Teich and Diament (1969), Jaiswal and Agarwal (1969), Barashev (1970a, 1976), Mišta and Peřina (1971), Mišta (1971), Peřina et al. (1972), Delone and Masalov (1980), Dixit and Lambropoulos (1980)]. In addition, the superposition of a coherent component and a chaotic component composed of two or more spectral lines has been investigated [Tornau and Echtermeyer (1973), Mehta and Gupta (1975), Mišta and Peřina (1977a)]. Making use of the exact recursion formulae derived by Laxpati and Lachs (1972), it was possible to estimate the accuracy of the approximate multimode formulae [Peřina, Peřinová, Lachs and Braunerová (1973)] and to find that it is quite good in usual situations, particularly in optical communication channels. Further applications have been worked out, in connection with definitions of coherence time, coherence area and coherence volume [Peřina and Mišta (1974)] and the statistics of speckles including a coherent background [Peřina (1977)].

Recently great attention has been paid to systematic research in the statistical properties of radiation in random media [see for instance Rousseau (1971), Chu (1974), Crosignani et al. (1975)]. Radiation of arbitrary statistical behaviour (e.g. chaotic, coherent, the superposition of the two, etc.) propagating through a random medium is modified by fluctuations in the random medium. Such a modification must be taken into account for the purposes of optical communication, particularly if a small number of photons is incident on a photocathode [Helstrom (1972, 1976, 1981), Sheremetyev (1971), Yen et al. (1972), Kuriksha (1973), Teich and Rosenberg (1973), Rosenberg and Teich (1973a, b), Saleh (1978), Teich and Cantor (1978), Prucnal and Teich (1978)]. If highly stabilized laser radiation is used, then the statistical properties of scattered radiation directly reflect fluctuations of the scattering medium and they can serve as a source of information about the medium. The propagation equations for the coherence tensors of light scattered by a random medium have been discussed by Ross and Nieto–Vesperinas (1981).

Diament and Teich (1970b, 1971) have applied Mandel's photodetection equation to determine the photocount statistics of coherent radiation, chaotic radiation and their narrow-band superposition, propagating through the turbulent atmosphere, under the assumption that the detection time-interval is much shorter than the coherence time. The effect of fluctuations of the medium is then introduced through a modulation of the instantaneous intensity or the integrated intensity. The same approach has been applied by Bertolotti et al. (1970) [cf. also Bertolotti (1974)] in connection with the photocount statistics in Gaussian media. The calculations of these authors have been extended to include the arbitrary spectral width of the radiation and arbitrary detection time-intervals [Peřina and Peřinová (1972)]. Similar investigations of the photocount statistics in the turbulent atmosphere have been performed by Lachs and Laxpati (1973) who made use of the recursion formulae for the superposition of coherent and chaotic fields.

Independently, an alternative description of the statistical properties of radiation propagating through random media has been suggested by Tatarskii (1971) and Zavorotnyi and Tatarskii (1973), which is based on a modulation of the photon number rather than of the classical integrated intensity. Tatarskii calculated the second moment, while the present author (1972a) continued his calculations to obtain the moments of all orders, the characteristic function and the so-called modified photodetection equation. However, both these approaches have been found to differ from one another only by shot noise in the vacuum state, while for non-vacuum states and with the incident field in a Fock state, or in a coherent state, or with chaotic radiation, or with a multimode superposition of coherent and chaotic radiation, they are practically identical [Peřina, Peřinová, Teich and Diament (1973)].

In the next step the quantum dynamical description has been developed [Crosignani, Di Porto and Solimeno (1971), Peřina, Peřinová and Horák (1973a, b), Peřina, Peřinová, Mišta and Horák (1974), Peřina, Peřinová and Mišta (1974), Peřina and Peřinová (1975, 1976a, b)], based on the use of the coherent-state technique and the Heisenberg and Schrödinger pictures (which have been shown to be equivalent to each other by explicit calculations). Since in the interaction problems the weighting functional of the Glauber – Sudarshan diagonal representation of the density matrix does not usually exist as an ordinary functional but is rather a generalized functional (ultradistribution), a special procedure has been suggested; it employs the quasi-distribution related to antinormal ordering of field operators and it provides the measurable photocount statistics. This approach makes it possible to describe propagation through passive as well as active random media (lossy, lossless, as well as amplifying) with slow or fast, weak or strong fluctuations, taking into account the coupling of modes. In the case of lossless media this quantum dynamical description has been found [Peřina et al. (1975)] to be in good agreement with the above mentioned descriptions, based on the modulation of the integrated intensity or the number of photons, under the assumption of energy conservation. A typical feature of this quantum approach is that it includes the self-radiation of the medium.

The theoretical results stimulated a number of experimental investigations, but they are still not numerous. The statistical properties of radiation scattered by a rotating ground glass disk have been experimentally investigated by Bluemel et al. (1972) and Parry et al. (1978) and elegant experiments on the statistical properties of radiation scattered by liquid crystals have been performed by Bertolotti et al. (1973), Bartolino et al. (1973), Scudieri and Bertolotti (1974), Scudieri et al. (1974), Bertolotti (1974). Regarding the turbulent atmosphere we may mention investigations by Gurvich and Tatarskii (1973), Churnside and McIntyre (1978a, b) and recent experimental results by Phillips and Andrews (1981) and by Parry (1981) for the higher-order reduced factorial moments and the photocount distribution of laser radiation in the saturation regime of the turbulent atmosphere over a 1.125 km path. Further Davidson and Gonzales-Del-Valle (1975) have observed the photocount statistics of radiation propagating through a mixture of hot and cold water.

Chapter 9 of this book deals with the statistical properties of radiation associated with the nonlinear optical processes reviewed in Chapter 6. We consider mainly optical parametric processes with classical pumping [Mollow and Glauber (1967a, b), Mishkin and Walls (1969), Mišta (1969), Tucker and Walls (1969), Raiford (1970), Stoler (1974), Smithers and Lu (1974), Mišta et al. (1977), Trung and Schütte (1977), Kryszewski and Chrostowski (1977), Mišta and Peřina (1978), Mielniczuk (1979), Srinivasan and Udayabaskaran (1979), Bandilla and Ritze (1980a), Peřinová (1981), Peřinová and Peřina (1981), Mielniczuk and Chrostowski (1981)], as well as with quantum pumping [Graham and Haken (1968), Graham (1968b, 1970, 1973), Walls and Barakat (1970), Walls and Tindle (1972), Agrawal and Mehta (1974), Dewael (1975), Peřina (1976), Stoljarov (1976), Mišta and Peřina (1977b, c), Peřinová et al. (1977), Peřina, Peřinová and Knesel (1977), Kozierowski and Tanaś (1977), Nayak and Mohanty (1977), Gambini (1977), Chmela (1977, 1978, 1979a–e, 1981a, b), Kielich, Kozierowski and Tanaś (1978), Trung and Schütte (1978), Peřinová and Peřina (1978a–c), Neumann and Haug (1979), Orszag (1979), Gorbatchev and Zanadvorov (1980), Oliver and Bendjaballah (1980), Drummond, McNeil and Walls (1980b, 1981)]; we consider non-degenerate processes (sum- and difference-frequency generation, frequency conversion, parametric amplification and generation, etc.) as well as degenerate ones (higher harmonic and subharmonic generation). In addition we treat Brillouin, Raman and hyper-Raman scattering [Walls (1970, 1973), Simaan (1975, 1978), Loudon (1973), Szlachetka, Kielich, Peřina and Peřinová (1979, 1980a, b), Szlachetka, Kielich, Peřinová and Peřina (1980), Peřinová et al. (1979a, b), Gupta and Mohanty (1980, 1981), Mohanty and Gupta (1980, 1981a–c), Orszag (1981), Peřina (1981a, b), Pieczonková and Peřina (1981), DosReis and Sharma (1982)]. Reviews of the statistical properties of these and other nonlinear optical processes, such as multiphoton absorption [Tornau and Bach (1974), Simaan and Loudon (1975, 1978), Every (1975), Paul et al. (1976), Bandilla and Ritze (1976a, b), Bandilla (1977), Chaturvedi et al. (1977), Chrostowski and Karczewski (1977), Mohr and Paul (1978), Voigt et al. (1980), Hildred (1980), Chrostowski (1980), Voigt and Bandilla (1981)] and two-photon spontaneous and stimulated emission [McNeil and Walls (1974, 1975a–c), Yuen (1976), Nayak and Mohanty (1979), Zubairy and Yeh (1980)] have been published recently [Peřina (1980a, b), Schubert and Wilhelmi (1980), Kozierowski (1981), Kielich (1981), Akhmanov et al. (1981)]. It should be noted that results obtained for the propagation of radiation through random media, taking into account the self-radiation, can be directly applied to second subharmonic generation employing the classical description of the pumping laser light [Stoler (1974), Mišta et al. (1977)]; further they can serve to introduce the two-photon coherent states, originally thought to describe radiation generated by two-photon stimulated emission [Stoler (1970, 1971, 1972, 1975), Yuen (1976), Narducci et al. (1977), Yuen and Shapiro (1978a, 1979), Helstrom (1979), Shapiro et al. (1979)], in the same way as one-photon coherent states describe the radiation generated by one-photon stimulated emission. Moreover such a model can serve as a prototype for an amplification process [Yuen (1976)].

The most interesting property of these nonlinear processes is that, under certain conditions, they provide radiation exhibiting the so-called anticorrelation effect or antibunching of photons, i.e. they are able to generate fields having no classical analogues [Kozierowski and Tanaś (1977), Mišta and Peřina (1977b, c), Chandra and Prakash (1970), Anisimov and Sotskii (1977a), Hildred and Hall (1978), Bandilla and Ritze (1979, 1980a, b), Ritze and Bandilla (1979), Loudon (1980), Walls (1979)]. Such fields may have photocount distributions which are narrower than the Poisson distribution (which corresponds to the coherent state), i.e. they have less uncertainty in the number of photons than the coherent state, and the variance of their intensity is negative. On the other hand, they are able to produce fields whose fluctuations are enhanced compared to chaotic fields.

A great success was achieved recently in this field of physics when these new features of quantum-optical systems were observed experimentally in pioneering measurements by Mandel and his coworkers. Kimble et al. (1977, 1978) and Dagenais and Mandel (1978) have observed antibunching of photons obtained by means of the resonance fluorescence of sodium atoms [Carmichael and Walls (1976), Kimble and Mandel (1975, 1976, 1977), Jakeman et al. (1977), Carmichael et al. (1978, 1980), Sobolewska and Sobolewski (1978), Süsse et al. (1979, 1980a, b), Smirnov and Troshin (1979), Schubert et al. (1980), Cook (1980), Lenstra (1982)]. Similar measurements for resonance fluorescence have been performed by Leuchs, Rateike and Walther [see Walls (1979), Cresser et al. (1982)]. At present proposals have been made to measure antibunching in nonlinear optical processes [Paul and Brunner (1980), Chmela et al. (1981)], and some simulation experiments have been performed [Wagner et al. (1979)]. Scattering processes seem to be particularly convenient for generating radiation exhibiting these peculiar properties. Sub-Poissonian behaviour has been observed in radiation from one atom by Short and Mandel (1983) and the Franck – Hertz experiment has been suggested to produce sub-Poissonian light [Teich, Saleh and Stoler (1983), Teich, Saleh and Peřina (1983)].

Anticorrelation has been observed also in light scattering by non-spherical particles [Griffin and Pusey (1979)], and in ring lasers, in which one mode is coherent and the other chaotic above the threshold and there is anticorrelation between them [M-Tehrani and Mandel (1977, 1978a, b)]. Anticorrelations in intensity fluctuations of orthogonally polarized components of amplified spontaneous emission have been observed also [Abraham and Smith (1981)]. Further investigation of ring lasers, including the effect of periodic modulation of the spatial population (Bragg grating), has been done, too [Kühlke and Horák (1979, 1981)].

Substantial effort has been devoted to studying the statistical properties of the laser near threshold [Davidson and Mandel (1967), Meltzer et al. (1970), Jakeman, Oliver, Pike, Lax and Zwanziger (1970), Cantrell (1969), Vrbová (1970), Meltzer and Mandel (1971), Arecchi and Degiorgio (1971), Cantrell and Smith (1971), Chopra and Mandel (1972, 1973), Zardecki, Bures and Delisle (1972), Corti et al. (1973), Cantrell et al. (1973), Seybold and Risken (1974), Corti and Degiorgio (1974, 1976a, b), Chopra and Dudeja (1976), Arecchi and Ricca (1977), Zubairy (1979)], and to ring and dye lasers [Davidson et al. (1974), Abate et al. (1976), Kruglik (1978), Kühlke and Horák (1979), Hioe et al. (1979), Roy and Mandel (1977, 1980), Singh (1981), Kaminishi et al. (1981)] and Raman lasers [Steÿn – Ross and Walls (1981)]. Spatial coherence of the laser output far below threshold has been considered by Kimble and Mandel (1973).

Much progress has been made in the phase-transition studies of the laser and of nonlinear optical processes, and in the study of optical bistability [see e.g. Graham (1973), Dembinski and Kossakowski (1976), Haken (1978), Roy and Mandel (1977), Drummond and Walls (1980, 1981), Brand et al. (1980), Lugiato (1981)].

The coherent-state technique proved fruitful in a formulation of phase-operator algebra [Carruthers and Nieto (1968), Paul (1974, 1976)].

Finally, we mention some works on coherent resonant phenomena, such as ultrashort pulse propagation, self-induced transparency, photon echo, superradiance, etc. [Sargent et al. (1974), Arecchi et al. (1969), Courtens (1972), Allen and Eberly (1975), Slusher (1974), Greenhow and Schmidt (1974), Shapiro (1977), Schubert and Wilhelmi (1978), Kujawski and Eberly (1978), Eljutin et al. (1981), Richter (1981), Haake, Haus, King et al. (1981)].

CHAPTER 2

QUANTUM THEORY OF THE ELECTROMAGNETIC FIELD

The development of improved and new measuring techniques (e.g. coincidence techniques) created new possibilities of detecting single photons in the visible region, thus allowing the measurement of, for example, the statistics of photons. This was in contrast to earlier investigations where the detecting devices were able to respond only to intense currents of photons. Thus the particle character of the electromagnetic field may now be studied directly. Under such circumstances it is obviously advantageous to describe the coherence phenomena in the field by the methods of quantum electrodynamics, where a field operator corresponds to the classical field and the quanta of the field correspond to the classical wave field. As is well-known this correspondence is realized by the second quantization of the field.

The standard treatment of quantum electrodynamics is formulated on the basis of perturbation theory and such a formulation is appropriate to the discussion of processes involving a few particles. The treatment of many-particle processes in this way is complex and consequently the classical limit of quantum electrodynamics has never been fully developed using the traditional approach. In contrast, a discussion of phenomena such as coherence involves a considerable number of photons and their mutual relationships in space and time. It is only this way that the operation of a laser source – a typical nonlinear device functioning only at sufficiently high field intensities – can be explained. Consequently it is necessary to develop quantum electrodynamics in a manner which exhibits clearly the classical limit. To this end we will not use the Fock states, the basis of the usual formulation, but instead use the coherent states. These are linear combinations of Fock states and have the advantage that they retain information about the phase of the field – information lost when Fock states are used. To conserve the phase information, which is so important for cooperative phenomena such as coherence, one must determine an infinite number of off-diagonal Fock matrix elements, whereas all the information about the cooperative properties of the system is contained in one matrix element in the coherent state representation. With these coherent states it is possible to obtain a close formal correspondence between the quantum and classical descriptions of light beams. Of course, this does not imply the physical identity of these descriptions. In general, one can say that the classical description fails for measurements which are sensitive to fluctuations of the physical vacuum. Such measurements are described by functions of the field operators in orderings other than normal ordering. In the normal order all creation operators stand to the left of all the annihilation operators and the vacuum expectation value is zero. In other orderings the vacuum expectation values are non zero. The principles of the quantum theory of optical coherence employing the coherent states were given by Glauber (1963a, b, 1964, 1965, 1966a, 1969, 1970, 1972).

2.1 Quantum description of the field

In this chapter we summarize only basic ideas and results of the second quantization of the electromagnetic field in vacuo. For a fuller treatment we refer the reader to the texts by Akhiezer and Berestetsky (1965), Bogolyubov and Shirkov (1959), Dirac (1958), Schweber (1961), Messiah (1961, 1962); texts with particular attention to quantum optics are those by Louisell (1964, 1973), Yariv (1975), Klauder and Sudarshan (1968) and Loudon (1973).

In the quantum theory of the electromagnetic field the electric and magnetic intensities, $\mathbf{E}$ and $\mathbf{H}$, are regarded as operators in the space of states which describe the field. These quantities satisfy Maxwell's equations in vacuum

$$\begin{aligned}
\nabla \times \hat{\mathbf{E}} &= -\frac{1}{c}\frac{\partial \hat{\mathbf{H}}}{\partial t}, \\
\nabla \times \hat{\mathbf{H}} &= \frac{1}{c}\frac{\partial \hat{\mathbf{E}}}{\partial t}, \\
\nabla \cdot \hat{\mathbf{E}} &= 0, \\
\nabla \cdot \hat{\mathbf{H}} &= 0,
\end{aligned} \tag{2.1}$$

∇ being the nabla operator, c the light velocity in the vacuum and $\times$ and . denote the vectorial and scalar products. The operators $\hat{\mathbf{E}}$ and $\hat{\mathbf{B}}$ can be derived from the vector potential operator $\hat{\mathbf{A}}$ according to the relations

$$\hat{\mathbf{E}} = -\frac{1}{c}\frac{\partial \hat{\mathbf{A}}}{\partial t}, \tag{2.2a}$$

$$\hat{\mathbf{H}} = \nabla \times \hat{\mathbf{A}}, \tag{2.2b}$$

$\hat{\mathbf{A}}$ obeys the wave equation

$$\Delta \hat{\mathbf{A}} - \frac{1}{c^2}\frac{\partial^2 \hat{\mathbf{A}}}{\partial t^2} = 0, \tag{2.3}$$

with the calibration condition $\nabla \cdot \hat{\mathbf{A}} = 0$. Here Δ is the Laplace operator.

Decomposing the vector potential operator $\hat{\mathbf{A}}(\mathbf{x}, t)$ in terms of real mode functions $\mathbf{u}_l(\mathbf{x})$, we have

$$\hat{\mathbf{A}}(x) = \sum_l \left(\frac{\hbar}{2\omega_l}\right)^{1/2} \mathbf{u}_l(\mathbf{x})\,[\hat{a}_l(t) + \hat{a}_l^+(t)], \tag{2.4}$$

where $x \equiv (\mathbf{x}, t)$ is a space-time point and the mode functions $\mathbf{u}_l(\mathbf{x})$ satisfy the Helmholtz equation

$$\Delta \mathbf{u}_l(\mathbf{x}) + \frac{\omega_l^2}{c^2}\mathbf{u}_l(\mathbf{x}) = 0, \tag{2.5}$$

and the annihilation and creation operators $\hat{a}_l(t)$ and $\hat{a}_l^+(t)$ satisfy the equations of motion

$$\frac{\mathrm{d}\hat{a}_l(t)}{\mathrm{d}t} = -\mathrm{i}\omega_l\hat{a}_l(t), \qquad \frac{\mathrm{d}\hat{a}_l^+(t)}{\mathrm{d}t} = \mathrm{i}\omega_l\hat{a}_l^+(t). \tag{2.6}$$

The mode functions $\boldsymbol{u}_l(\boldsymbol{x})$ are orthogonal as follows:

$$\int_{L^3} \boldsymbol{u}_l(\boldsymbol{x}) \,.\, \boldsymbol{u}_m(\boldsymbol{x})\, \mathrm{d}^3x = 4\pi c^2 \delta_{lm}, \tag{2.7}$$

where L^3 is the volume of the electromagnetic field.

Consider now a field in a normalization volume L^3 with periodic boundary conditions, so that we can decompose the vector potential operator $\hat{\mathbf{A}}$ in terms of plane waves in the form

$$\hat{\mathbf{A}}(x) = \frac{(2\pi\hbar c)^{1/2}}{L^{3/2}} \sum_{\mathbf{k}} \sum_{s=1}^{2} k^{-1/2}\{\mathbf{e}^{(s)}(\mathbf{k})\, \hat{a}_{\mathbf{k}s} \exp\left[\mathrm{i}(\mathbf{k}\,.\,\mathbf{x} - ckt)\right] + \\ + \mathbf{e}^{(s)*}(\mathbf{k})\, \hat{a}_{\mathbf{k}s}^+ \exp\left[-\mathrm{i}(\mathbf{k}\,.\,\mathbf{x} - ckt)\right]\}, \tag{2.8}$$

where $\hat{a}_{\mathbf{k}s}$ is the annihilation operator for a photon of momentum $\hbar\mathbf{k}$ and polarization s, and the hermitian conjugate to $\hat{a}_{\mathbf{k}s}$ is the creation operator $\hat{a}_{\mathbf{k}s}^+$. These operators obey the following commutation rules

$$[\hat{a}_\lambda, \hat{a}_{\lambda'}^+] = \delta_{\lambda\lambda'}\hat{1}, [\hat{a}_\lambda, \hat{a}_{\lambda'}] = [\hat{a}_\lambda^+, \hat{a}_{\lambda'}^+] = \hat{0}. \tag{2.9}$$

Here $\lambda \equiv (\mathbf{k}, s)$ is the mode index, $\hat{1}$ and $\hat{0}$ are the identity and zero operators, $[\hat{A}, \hat{B}] = \hat{A}\hat{B} - \hat{B}\hat{A}$ stands for the commutator of operators $\hat{A}$ and $\hat{B}$, and $\delta_{\lambda\lambda'} = \delta_{\mathbf{k}\mathbf{k}'}\delta_{ss'}$. The unit polarization vector satisfies

$$\mathbf{e}^{(s)*}(\mathbf{k})\,.\,\mathbf{e}^{(s')}(\mathbf{k}) = \delta_{ss'},\ \mathbf{k}\,.\,\mathbf{e}^{(s)}(\mathbf{k}) = 0, \tag{2.10}$$

so that the vectors $\mathbf{e}^{(1)}$, $\mathbf{e}^{(2)}$ and $\mathbf{k}/k$ ($k = |\,\mathbf{k}\,|$) form an orthogonal system

$$\sum_{s=1}^{2} e_i^{(s)*}(\mathbf{k})\, e_j^{(s)}(\mathbf{k}) + \frac{k_i k_j}{k^2} = \delta_{ij}, \qquad i, j = 1, 2, 3. \tag{2.11}$$

In (2.8) only transverse polarizations are present ($s = 1, 2$) since the longitudinal and scalar photons are eliminated by the Lorentz condition $\sum_{\mu=1}^{4} \partial\hat{A}_\mu/\partial x_\mu = 0$ ($x_4 = ict$).

We define the number operator $\hat{n} = \hat{a}^+\hat{a}$ (omitting the mode index λ for simplicity), for which, with the use of the commutation rules (2.9),

$$[\hat{a}, \hat{n}] = \hat{a}, \tag{2.12a}$$

$$[\hat{n}, \hat{a}^+] = \hat{a}^+. \tag{2.12b}$$

Repeated use of these relations gives ($m = 0, 1, 2, \ldots$)

$$\hat{n}\hat{a}^m = \hat{a}^m(\hat{n} - m), \tag{2.13a}$$

$$\hat{n}\hat{a}^{+m} = \hat{a}^{+m}(\hat{n} + m) \tag{2.13b}$$

and also

$$\hat{a}^{+m}\hat{a}^{m} = \hat{n}(\hat{n} - 1) \ldots (\hat{n} - m + 1), \tag{2.14a}$$

$$\hat{a}^{m}\hat{a}^{+m} = (\hat{n} + 1)(\hat{n} + 2) \ldots (\hat{n} + m); \tag{2.14b}$$

one can see from (2.14a) that the normal order $\hat{a}^{+m}\hat{a}^{m}$ of the field operators is related to successive annihilation of a photon, whereas the antinormal order $\hat{a}^{m}\hat{a}^{+m}$ in (2.14b) is related to successive creation of a photon. Therefore (2.14a, b) describe absorption and emission, respectively.

Assume that there exists an eigenstate $|n\rangle$ of the number operator $\hat{n}$ such that

$$\hat{n}|n\rangle = n\,|n\rangle, \tag{2.15}$$

where n is a real number. From this equation, $\langle n|\hat{n}|n\rangle = n\langle n|n\rangle$, where $\langle n|$ is hermitian conjugate to $|n\rangle$. Thus $n = \langle n|\hat{a}^{+}\hat{a}|n\rangle/\langle n|n\rangle \geqq 0$. Further, if $|n\rangle$ is an eigenstate of $\hat{n}$, then $\hat{a}^{+}\hat{a}(\hat{a}^{l}|n\rangle) = (n - l)(\hat{a}^{l}|n\rangle)$ from (2.13a), i.e. the state $(\hat{a}^{l}|n\rangle)$ is also an eigenstate of $\hat{n}$ with eigenvalues $(n - l)$, $l = 0, 1, 2, \ldots$ As $n \geqq 0$, n must be a positive integer or zero to terminate this sequence and it can be seen from (2.13b) that all integers are eigenvalues of $\hat{n}$ since $\hat{n}(\hat{a}^{+l}|n\rangle) = (n + l)(\hat{a}^{+l}|n\rangle)$, $l = 0, 1, 2, \ldots$ The ground state (the physical vacuum state) $|0\rangle$ can be defined as the state with $n = 0$, that is $\hat{n}(\hat{a}^{+l}|0\rangle) = l(\hat{a}^{+l}|0\rangle)$, so the state $\hat{a}^{+l}|0\rangle$ must be proportional to the state $|l\rangle$. From (2.14b) $\langle 0|\hat{a}^{l}\hat{a}^{+l}|0\rangle = l!$ and we can define the normalized states

$$|n\rangle = (n!)^{-1/2}\hat{a}^{+n}|0\rangle, \tag{2.16}$$

which are called the Fock states or occupation number states. It is obvious that these states are orthonormal, $\langle n|m\rangle = \delta_{nm}$, and

$$\hat{a}^{+}|n\rangle = (n + 1)^{1/2}|n + 1\rangle, \tag{2.17a}$$

$$\hat{a}|n\rangle = n^{1/2}|n - 1\rangle. \tag{2.17b}$$

The vacuum stability condition demands that $\hat{a}|0\rangle = 0$. Equations (2.17a, b) show that the operators $\hat{a}^{+}$ and $\hat{a}$ can indeed be called the creation and annihilation operators of a photon in the mode of the field.

If we write the hamiltonian of the field in the form

$$\hat{H} = \frac{1}{8\pi}\int_{L^3}(\hat{\mathbf{E}}^2 + \hat{\mathbf{H}}^2)\,\mathrm{d}^3x, \tag{2.18}$$

we obtain by using (2.4) or (2.8) and (2.2a, b)

$$\hat{H} = \sum_{\lambda}\left(\hat{n}_{\lambda} + \frac{1}{2}\right)\hbar\omega_{\lambda}, \tag{2.19a}$$

where $\omega_{\lambda} = kc$. This operator is, apart from the additive constant $\sum_{\lambda}(\hbar\omega_{\lambda}/2)$ (the energy of the physical vacuum), the hamiltonian of a set of dynamically independent harmonic oscillators with energies $n_{\lambda}\hbar\omega_{\lambda}$. Similarly, for the total momentum operator of the field we obtain

$$\hat{\mathbf{P}} = \sum_\lambda \hbar \mathbf{k}_\lambda \hat{n}_\lambda \quad (\sum_\lambda \mathbf{k}_\lambda = 0). \tag{2.19b}$$

Introducing the canonical variables $\hat{q}$ and $\hat{p}$ by the relations

$$\hat{a} = (2\hbar\omega)^{-1/2}(\omega\hat{q} + \mathrm{i}\hat{p}),$$
$$\hat{a}^+ = (2\hbar\omega)^{-1/2}(\omega\hat{q} - \mathrm{i}\hat{p}), \tag{2.20}$$

so that

$$\hat{q} = \left(\frac{\hbar}{2\omega}\right)^{1/2}(\hat{a} + \hat{a}^+),$$
$$\hat{p} = -\mathrm{i}\left(\frac{\hbar\omega}{2}\right)^{1/2}(\hat{a} - \hat{a}^+), \tag{2.21}$$

the commutation rule becomes

$$[\hat{q}, \hat{p}] = \mathrm{i}\hbar\hat{1} \tag{2.22}$$

and (2.19) gives us

$$\hat{H} = \sum_\lambda \frac{1}{2}(\hat{p}_\lambda^2 + \omega_\lambda^2\hat{q}_\lambda^2). \tag{2.23}$$

This shows clearly that the free electromagnetic field is equivalent to an infinite set of harmonic oscillators. With this result Dirac began the development of quantum electrodynamics.

The states $|\,n\rangle$ for all n form a complete orthonormal system, which is expressed by the completeness condition (condition of resolution of unity)

$$\sum_{n=0}^{\infty} |\,n\rangle\langle n\,| = \hat{1}. \tag{2.24}$$

Thus every quantum state $|\,\rangle$ may be decomposed in terms of the states $|\,n\rangle$ in the form

$$|\,\rangle = \sum_l c_l\,|\,l\rangle, \tag{2.25}$$

where

$$c_l = \langle l\,|\,\rangle. \tag{2.26}$$

The space they span is the familiar Fock space of quantum field theory, and nearly all treatments of quantum electrodynamics have been formulated using these states. However, as mentioned above, such formulations do not retain any information about the phase of the field; in early applications of quantum electrodynamics this was unimportant since such information played a minimal role, but in discussing coherence the phase information plays a considerable role. For this the coherent states are extremely convenient, representing an infinite superposition of the Fock states. These states make it possible to consider the classical limit of quantum electrodynamics when the fields are strong and n is large and uncertain.

The vector potential $\hat{\boldsymbol{A}}$ is Hermitian ($\hat{\boldsymbol{A}}^+ = \hat{\boldsymbol{A}}$) since the classical field is real and we see from (2.4) and (2.8) that $\hat{\boldsymbol{A}}$ (and also $\hat{\boldsymbol{E}}$ and $\hat{\boldsymbol{H}}$) can be decomposed into the sum of a positive frequency part $\hat{\boldsymbol{A}}^{(+)}$ and a negative frequency part $\hat{\boldsymbol{A}}^{(-)}$ as follows:

$$\hat{\boldsymbol{A}} = \hat{\boldsymbol{A}}^{(+)} + \hat{\boldsymbol{A}}^{(-)}, \tag{2.27}$$

where

$$\hat{\boldsymbol{A}}^{(+)}(x) = \sum_l \left(\frac{\hbar}{2\omega_l}\right)^{1/2} \boldsymbol{u}_l(\boldsymbol{x})\, \hat{a}_l(t), \tag{2.28a}$$

$$= \frac{(2\pi\hbar c)^{1/2}}{L^{3/2}} \sum_{\boldsymbol{k},s} k^{-1/2} \boldsymbol{e}^{(s)}(\boldsymbol{k})\, \hat{a}_{\boldsymbol{k}s} \exp\left[\mathrm{i}(\boldsymbol{k}\,.\,\boldsymbol{x} - ckt)\right] \tag{2.28b}$$

and $\hat{\boldsymbol{A}}^{(-)} = (\hat{\boldsymbol{A}}^{(+)})^+$. Hence the operators $\hat{\boldsymbol{A}}^{(+)}(x)$ and $\hat{\boldsymbol{A}}^{(-)}(x)$ represent respectively the annihilation and creation operators at a point $\boldsymbol{x}$ and time t. They correspond to some classical fields $\boldsymbol{V}$ and $\boldsymbol{V}^*$ in the strong-field limit and play an important role in the process of detection of fields. For this reason the decomposition of the field operator $\hat{\boldsymbol{A}}$ into two Hermitian conjugate quantities has a basic significance in quantum theory, whereas in the classical theory it is largely a matter of mathematical convenience. In the classical theory, where energy quantities with values comparable with the field quantum $\hbar\omega$ are neglected, one cannot distinguish between absorption and emission and only the real part of the complex field $\hat{\boldsymbol{A}}^{(+)}$ can be measured. In the quantum theory on the other hand, atoms as detectors are able to measure the annihilation operator $\hat{\boldsymbol{A}}^{(+)}$ if they are in ground states and the creation operator $\hat{\boldsymbol{A}}^{(-)}$ if they are in excited states.

To study the process of the detection of the electromagnetic field we introduce the so-called detection operator (Mandel (1964c))

$$\hat{\mathrm{A}}(x) = L^{-3/2} \sum_{\boldsymbol{k},s} \boldsymbol{e}^{(s)}(\boldsymbol{k})\, \hat{a}_{\boldsymbol{k}s} \exp\left[\mathrm{i}(\boldsymbol{k}\,.\,\boldsymbol{x} - kct)\right], \tag{2.29}$$

which is closely related to the positive frequency part of the vector potential operator. The photon number operator in a finite volume V (the linear dimensions of which are large compared with the wavelength) is (Mandel (1964c), Cook (1982))

$$\hat{n}_{Vt} = \int_V \hat{\mathrm{A}}^+(x)\,.\,\hat{\mathrm{A}}(x)\, \mathrm{d}^3x, \tag{2.30a}$$

for $V \equiv L^3$, and making use of (2.10), this gives

$$\hat{n} = \sum_\lambda \hat{a}_\lambda^+ \hat{a}_\lambda. \tag{2.30b}$$

For quasi-monochromatic fields, the most usual case in practice, all the operators $\hat{\boldsymbol{E}}^{(+)}$, $\hat{\boldsymbol{H}}^{(+)}$, $\hat{\boldsymbol{A}}^{(+)}$ and $\hat{\mathrm{A}}$ are proportional to each other.

2.2 Statistical states

From the quantum-mechanical point of view all information about the quantum behaviour of a system, including the quantum statistics, is contained in the density matrix (see e.g. Landau and Lifshitz (1959)). It can be decomposed in terms of the

Fock states in the form

$$\hat{\varrho} = \sum_{n,m} \varrho(n, m) \mid n\rangle \langle m \mid, \tag{2.31}$$

where $\varrho(n, m) = \langle n \mid \hat{\varrho} \mid m\rangle$, since $\langle n \mid m\rangle = \delta_{nm}$. The normalization condition reads

$$\mathrm{Tr}\, \hat{\varrho} = 1, \qquad \text{or} \qquad \sum_{n} \varrho(n, n) = 1. \tag{2.32}$$

The density matrix is clearly Hermitian, $\hat{\varrho}^+ = \hat{\varrho}$ $(\varrho^*(n, m) = \varrho(m, n))$. A pure Fock state of the field is described by $\hat{\varrho} = \mid n\rangle \langle n \mid$. For any pure state $\hat{\varrho}^2 = \hat{\varrho}$, i.e. $\mathrm{Tr}\, \hat{\varrho}^2 = 1$, for mixed states, in general, $\mathrm{Tr}\, \hat{\varrho}^2 < 1$. This is evident when we use an orthonormal system of states in which $\hat{\varrho}$ is diagonal,

$$\hat{\varrho} = \sum_{l} p_l \mid \psi_l\rangle \langle \psi_l \mid = \{\mid \psi_l\rangle \langle \psi_l \mid\}_{\text{average}}, \tag{2.33}$$

giving $\mathrm{Tr}\, \hat{\varrho}^2 = \sum_{l} p_l^2 \leqq 1 = (\mathrm{Tr}\, \hat{\varrho})^2$, $0 \leqq p_l \leqq 1$ being the probabilities for the system to be in the state $\mid \psi_l\rangle$.

Further details about the density matrix can be found, for instance, in texts by Landau and Lifshitz (1959) and by Louisell (1964, 1973).

As an example we can discuss thermal radiation in thermal equilibrium at temperature T. Writing $\Theta = \hbar\omega/KT$, where K is the Boltzmann constant, we have

$$\hat{\varrho} = \sum_{n} \varrho(n) \mid n\rangle \langle n \mid, \tag{2.34}$$

i.e. (2.31) is diagonal and

$$\varrho(n) \equiv \varrho(n, n) = [1 - \exp(-\Theta)] \exp(-n\Theta). \tag{2.35}$$

Hence

$$\begin{aligned} \hat{\varrho} &= \sum_{n=0}^{\infty} [1 - \exp(-\Theta)] \exp(-\Theta \hat{a}^+ \hat{a}) \mid n\rangle \langle n \mid = \\ &= [1 - \exp(-\Theta)] \exp(-\Theta \hat{H}/\hbar\omega) = \frac{\exp(-\Theta \hat{H}/\hbar\omega)}{\mathrm{Tr}\{\exp(-\Theta \hat{H}/\hbar\omega)\}}, \end{aligned} \tag{2.36}$$

where (2.24) has been used and $\hat{H}$ is the renormalized hamiltonian (the vacuum energy is subtracted). The expectation value of $\hat{n}$ is equal to

$$\begin{aligned} \langle \hat{n} \rangle &= \mathrm{Tr}\{\varrho \hat{a}^+ \hat{a}\} = \sum_{n=0}^{\infty} n \exp(-n\Theta) [1 - \exp(-\Theta)] = \\ &= -[1 - \exp(-\Theta)] \frac{\mathrm{d}}{\mathrm{d}\Theta} \frac{1}{1 - \exp(-\Theta)} = \frac{1}{\exp(\Theta) - 1} = \langle n \rangle, \end{aligned} \tag{2.37}$$

so that the mean energy $\langle U \rangle$ is equal to

$$\langle U \rangle = \hbar\omega \langle \hat{n} \rangle = \frac{\hbar\omega}{\exp(\Theta) - 1}. \tag{2.38}$$

From (2.35) and (2.37) one has for the matrix elements of the density matrix

$$\varrho(n) = \frac{\langle n \rangle^n}{(1 + \langle n \rangle)^{1+n}}, \tag{2.39}$$

which is the well-known Bose–Einstein distribution. The quantum photon-number characteristic function becomes

$$\mathrm{Tr}\,\{\hat{\varrho} \exp(\mathrm{i}s\hat{n})\} = \sum_{n=0}^{\infty} \frac{1}{1 + \langle n \rangle} \left[\frac{\langle n \rangle \exp(\mathrm{i}s)}{1 + \langle n \rangle} \right]^n =$$

$$= \frac{1}{1 + \langle n \rangle} \frac{1}{1 - \dfrac{\langle n \rangle}{1 + \langle n \rangle} \exp(\mathrm{i}s)} = \frac{1}{1 - \langle n \rangle (\exp(\mathrm{i}s) - 1)}, \tag{2.40}$$

is being a parameter of the characteristic function. These results are also correct for a superposition of a chaotic field $\hat{a}_{ch}$ and a coherent field with the complex amplitude β_c, since for the resulting operator $\hat{a} = \hat{a}_{ch} + \beta_c$, $[\hat{a}, \hat{a}^+] = [\hat{a}_{ch}, \hat{a}_{ch}^+] = \hat{1}$.

The Poisson distribution

$$\varrho(n) = \frac{\langle n \rangle^n}{n!} \exp(-\langle n \rangle) \tag{2.41}$$

can serve as another example, where $\mathrm{Tr}\,\{\hat{\varrho}\hat{n}\} = \langle n \rangle$ again. The photon-number characteristic function reads in this case

$$\mathrm{Tr}\,\{\hat{\varrho} \exp(\mathrm{i}s\hat{n})\} = \exp[(\mathrm{e}^{\mathrm{i}s} - 1) \langle n \rangle]. \tag{2.42}$$

2.3 Multimode description

To describe multimode fields we introduce the global Fock states

$$|\{n_\lambda\}\rangle = \prod_\lambda |n_\lambda\rangle \tag{2.43}$$

for which the completeness condition reads

$$\sum_{\{n_\lambda\}} |\{n_\lambda\}\rangle \langle\{n_\lambda\}| = \hat{1}. \tag{2.44}$$

The density matrix is expressed in the form

$$\hat{\varrho} = \sum_{\{n_\lambda\}} \sum_{\{m_\lambda\}} \varrho(\{n_\lambda\}, \{m_\lambda\}) \,|\{n_\lambda\}\rangle \langle\{m_\lambda\}|, \tag{2.45}$$

where $\varrho(\{n_\lambda\}, \{m_\lambda\}) = \langle\{n_\lambda\}|\hat{\varrho}|\{m_\lambda\}\rangle$. A pure Fock state is described by the density matrix $\hat{\varrho} = |\{n_\lambda\}\rangle\langle\{n_\lambda\}|$ $(\hat{\varrho}^2 = \hat{\varrho})$.

2.4 Calculation of commutators of the field operators

The correct correspondence between classical optical fields and their operators is given by the commutation rules (2.9). However, the commutation rules for space-time field operators, such as $\hat{\mathbf{A}}(x)$, $\hat{\mathbf{E}}(x)$, $\hat{\mathbf{H}}(x)$, etc., are sometimes needed and therefore we derive some of them here, for use in later calculations. It will be sufficient for us to derive the commutation rules for $\hat{\boldsymbol{A}}$ and $\hat{\mathbf{A}}$ since the commutation rules for $\hat{\boldsymbol{E}}$ and $\hat{\boldsymbol{H}}$ follow using (2.2a, b). Making use of (2.8) and (2.9) we obtain

$$[\hat{A}_i(x), \hat{A}_j(x')] = \hat{1}\,\frac{2\pi\hbar c}{L^3} \sum_{\mathbf{k}, s} k^{-1} e_i^{(s)}(\mathbf{k})\, e_j^{(s)*}(\mathbf{k}) \times$$
$$\times \{\exp[ik(x - x')] - \exp[-ik(x - x')]\} = \tag{2.46}$$
$$= \hat{1}\,\frac{2\pi\hbar c}{L^3} \sum_{\mathbf{k}} k^{-1}\left(\delta_{ij} - \frac{k_i k_j}{k^2}\right)\{\exp[ik(x - x')] - \exp[-ik(x - x')]\},$$

where (2.11) has been used and $kx = \mathbf{k}\,.\,\mathbf{x} - kct$ and

$$[\hat{A}_i^{(+)}(x), \hat{A}_j^{(-)}(x')] = \hat{1}\,\frac{2\pi\hbar c}{L^3} \sum_{\mathbf{k}, s} k^{-1} e_i^{(s)}(\mathbf{k})\, e_j^{(s)*}(\mathbf{k}) \times \tag{2.47}$$
$$\times \exp[ik(x - x')] = \hat{1}\,\frac{2\pi\hbar c}{L^3} \sum_{\mathbf{k}} k^{-1}\left(\delta_{ij} - \frac{k_i k_j}{k^2}\right)\exp[ik(x - x')].$$

The commutation rule for the detection operator (2.29) becomes

$$[\hat{\mathbf{A}}_l(x), \hat{\mathbf{A}}_m^+(x')] = \hat{1}L^{-3} \sum_{\mathbf{k}}\left(\delta_{lm} - \frac{k_l k_m}{k^2}\right)\exp[ik(x - x')] \tag{2.48}$$

and for $t = t'$ this reduces to

$$[\hat{\mathbf{A}}_l(\mathbf{x}, t), \hat{\mathbf{A}}_m^+(\mathbf{x}', t)] =$$
$$= \hat{1}\delta_{lm}\delta(\mathbf{x} - \mathbf{x}') - \frac{\hat{1}}{(2\pi)^3} \int \frac{k_l k_m}{k^2} \exp[i\mathbf{k}\,.\,(\mathbf{x} - \mathbf{x}')]\,\mathrm{d}^3k. \tag{2.49}$$

Here we have replaced the sum $\sum_{\mathbf{k}}$ by the integral $(L/2\pi)^3 \int \ldots \mathrm{d}^3k$. It is easily seen that

$$[\hat{A}_l^{(+)}(x), \hat{A}_m^{(+)}(x')] = [\hat{A}_l^{(-)}(x), \hat{A}_m^{(-)}(x')] =$$
$$= [\hat{\mathbf{A}}_l(x), \hat{\mathbf{A}}_m(x')] = [\hat{\mathbf{A}}_l^+(x), \hat{\mathbf{A}}_m^+(x')] = 0. \tag{2.50}$$

The functions on the right-hand sides of (2.46)–(2.49) are singular functions which are zero outside the light cone, that is, when $(\mathbf{x} - \mathbf{x}')^2 > c^2(t - t')^2$. This is a reflection of the fact that measurements of fields at such pairs of points can be performed with arbitrary accuracy and they cannot influence one another. We remember that if two operators satisfy the commutation rule $[\hat{A}, \hat{B}] = \hat{1}C$, C being a c-number, then, defining $\Delta\hat{A} = \hat{A} - \langle\hat{A}\rangle$, $\Delta\hat{B} = \hat{B} - \langle\hat{B}\rangle$, we find that $|C|/2 = |\langle[\hat{A}, \hat{B}]\rangle|/2 = |\langle[\Delta\hat{A}, \Delta\hat{B}]\rangle|/2 \leqq |\langle\Delta\hat{A}\Delta\hat{B}\rangle| \leqq [\langle(\Delta\hat{A})^2\rangle\langle(\Delta\hat{B})^2\rangle]^{1/2}$, and consequently $\langle(\Delta\hat{A})^2\rangle\langle(\Delta\hat{B})^2\rangle \geqq |C|^2/4$. Writing $\hat{a} = \exp(i\hat{\varphi})\,\hat{n}^{1/2}$, $\hat{a}^+ = \hat{n}^{1/2} \times \exp(-i\hat{\varphi})$, we can introduce the phase operator $\hat{\varphi}$ and it holds that $[\exp(i\hat{\varphi}), \hat{n}] = \exp(i\hat{\varphi})$, which may be reduced to $[\hat{n}, \hat{\varphi}] = i\hat{1}$, giving the uncertainty relation $\langle(\Delta\hat{n})^2\rangle\langle(\Delta\hat{\varphi})^2\rangle \geqq 1/4$. However, this conclusion is not generally correct – it does not hold in the physical vacuum. These questions will be discussed in Sec. 4.4. We see that the commutation rules (2.46) express the principle of causality in quantum electrodynamics (signals cannot propagate more rapidly than with the velocity c of light in vacuo) (Rosenfeld (1958)).

By analogy with the commutation rules (2.12a, b), one can derive the following space-time commutation rules (Mandel (1966a))

$$[\hat{\mathbf{A}}(\mathbf{x}, t), \hat{n}_{Vt'}] = \begin{cases} \hat{\mathbf{A}}(\mathbf{x}, t), & \text{if } (\mathbf{x}, t) \text{ is conjoint with } (V, t'), \\ 0, & \text{if } (\mathbf{x}, t) \text{ is disjoint with } (V, t'), \end{cases} \tag{2.51a}$$

and

$$[\hat{n}_{Vt'}, \hat{\mathbf{A}}^{+}(\mathbf{x}, t)] = \begin{cases} \hat{\mathbf{A}}^{+}(\mathbf{x}, t), & \text{if } (\mathbf{x}, t) \text{ is conjoint with } (V, t'), \\ 0, & \text{if } (\mathbf{x}, t) \text{ is disjoint with } (V, t'). \end{cases} \tag{2.51b}$$

The terms conjoint and disjoint are defined in the following way. We define the function $U(\mathbf{x}, V) = 1$ if $\mathbf{x} \in V$ and $U = 0$ if $\mathbf{x} \notin V$; then $(\mathbf{x}, t)$ is conjoint with (V, t') if $U(\mathbf{x} - \mathbf{k}c(t - t')/k, V) = 1$ and disjoint if $U(\mathbf{x} - \mathbf{k}c(t - t')/k, V) = 0$.

Making use of (2.51a) and (2.30a), we obtain, by analogy with (2.14a),

$$\langle \mathcal{N} \hat{n}_{Vt}^{k} \rangle \equiv \langle \hat{n}_{Vt}^{k} \rangle_{\mathcal{N}} = \langle \hat{n}_{Vt}(\hat{n}_{Vt} - 1) \ldots (\hat{n}_{Vt} - k + 1) \rangle, \tag{2.51c}$$

$\mathcal{N}$ being the normal ordering operator (all the creation operators are to the left of all the annihilation operators, without the use of the commutation rules). From here we obtain the quantum normal characteristic function as follows

$$\langle \mathcal{N} \exp(\mathrm{i}s\hat{n}_{Vt}) \rangle \equiv \langle \exp(\mathrm{i}s\hat{n}_{Vt}) \rangle_{\mathcal{N}} = \sum_{r=0}^{\infty} \frac{(\mathrm{i}s)^r}{r!} \langle \hat{n}_{Vt}^{r} \rangle_{\mathcal{N}} =$$
$$= \langle (1 + \mathrm{i}s)^{\hat{n}_{Vt}} \rangle, \tag{2.51d}$$

and by substituting $1 + \mathrm{i}s = \exp(\mathrm{i}y)$ we arrive at the relation between the photon-number characteristic function and the normal characteristic function,

$$\langle \exp(\mathrm{i}y\hat{n}_{Vt}) \rangle = \langle \exp[\hat{n}_{Vt}(\mathrm{e}^{\mathrm{i}y} - 1)] \rangle_{\mathcal{N}}. \tag{2.51e}$$

2.5 Time development of quantum states

It is well known that in quantum electrodynamics a state of the field is described by a vector in the Hilbert space of states. The time development of a state can be interpreted either as the time development of the vector in the space if the system of coordinates is fixed (the Schrödinger picture) or as the time development of the system of coordinates if the state vector is fixed (the Heisenberg picture).

In the Schrödinger picture the Schrödinger equation holds for the state vector $|\psi_S(t)\rangle$

$$\mathrm{i}\hbar \frac{\partial}{\partial t} |\psi_S(t)\rangle = \hat{H} |\psi_S(t)\rangle, \tag{2.52a}$$

where $\hat{H}$ is the hamiltonian, independent of time; and for an operator $\hat{M}_S$,

$$-\mathrm{i}\hbar \frac{\mathrm{d}\hat{M}_S}{\mathrm{d}t} = \hat{0}. \tag{2.52b}$$

Performing the canonical transformations

$$\hat{M}_H(t) = \exp\left(\mathrm{i}\frac{\hat{H}(t-t_0)}{\hbar}\right)\hat{M}_S(t_0)\exp\left(-\mathrm{i}\frac{\hat{H}(t-t_0)}{\hbar}\right), \tag{2.53a}$$

$$|\psi_S(t)\rangle = \exp\left(-\mathrm{i}\frac{\hat{H}(t-t_0)}{\hbar}\right)|\psi_H(t_0)\rangle, \tag{2.53b}$$

which provide the correspondence between quantities in the Schrödinger and Heisenberg pictures, we obtain in the Heisenberg picture

$$\mathrm{i}\hbar\frac{\mathrm{d}}{\mathrm{d}t}\hat{M}_H(t) = [\hat{M}_H(t), \hat{H}], \tag{2.54a}$$

$$\mathrm{i}\hbar\frac{\partial|\psi_H\rangle}{\partial t} = \hat{0}. \tag{2.54b}$$

For example, in the Heisenberg picture we have for $\hat{a}_\lambda(t)$ with $\hat{H}$ given by (2.19a),

$$\frac{\mathrm{d}\hat{a}_\lambda(t)}{\mathrm{d}t} = -\mathrm{i}\omega_\lambda\hat{a}_\lambda(t), \tag{2.54c}$$

that is

$$\hat{a}_\lambda(t) = \hat{a}_\lambda(0)\exp(-\mathrm{i}\omega_\lambda t), \tag{2.54d}$$

in agreement with (2.6).

For the density matrix $\hat{\varrho} = \{|\psi\rangle\langle\psi|\}_{\text{average}}$ we obtain from (2.52a) in the Schrödinger picture

$$\mathrm{i}\hbar\frac{\partial\hat{\varrho}_S(t)}{\partial t} = [\hat{H}, \hat{\varrho}_S(t)], \tag{2.55a}$$

whereas

$$\mathrm{i}\hbar\frac{\partial\hat{\varrho}_H}{\partial t} = \hat{0} \tag{2.55b}$$

in the Heisenberg picture and

$$\hat{\varrho}_S(t) = \exp\left(-\mathrm{i}\frac{\hat{H}(t-t_0)}{\hbar}\right)\hat{\varrho}_H(t_0)\exp\left(\mathrm{i}\frac{\hat{H}(t-t_0)}{\hbar}\right). \tag{2.56}$$

In calculations of the expectation values of an operator $\hat{M}(t)$, it is convenient to use the following rule:

$$\begin{aligned}\mathrm{Tr}\{\hat{\varrho}_H(t_0)\hat{M}_H(t)\} &= \mathrm{Tr}\left\{\hat{\varrho}_H(t_0)\exp\left(\mathrm{i}\frac{\hat{H}(t-t_0)}{\hbar}\right)\hat{M}_S(t_0)\times\right.\\ &\left.\times\exp\left(-\mathrm{i}\frac{\hat{H}(t-t_0)}{\hbar}\right)\right\} = \mathrm{Tr}\left\{\exp\left(-\mathrm{i}\frac{\hat{H}(t-t_0)}{\hbar}\right)\hat{\varrho}_H(t_0)\times\right.\\ &\left.\times\exp\left(\mathrm{i}\frac{\hat{H}(t-t_0)}{\hbar}\right)\hat{M}_S(t_0)\right\} = \mathrm{Tr}\{\hat{\varrho}_S(t)\hat{M}_S(t_0)\}.\end{aligned} \tag{2.57}$$

Sometimes it is convenient to introduce the interaction picture. If $\hat{H} = \hat{H}_0 + \hat{H}_i$, where $\hat{H}_0$ and $\hat{H}_i$ are the free and interaction hamiltonians, it follows that

$$i\hbar \frac{\partial |\varphi(t)\rangle}{\partial t} = \mathscr{H}_i |\varphi(t)\rangle, \tag{2.58a}$$

$$i\hbar \frac{\partial \mathscr{M}(t)}{\partial t} = [\mathscr{M}(t), \hat{H}_0], \tag{2.58b}$$

where

$$|\varphi(t)\rangle = \exp\left(i\frac{\hat{H}_0 t}{\hbar}\right) |\psi(t)\rangle, \tag{2.59a}$$

$$\mathscr{H}_i = \exp\left(i\frac{\hat{H}_0 t}{\hbar}\right) \hat{H}_i \exp\left(-i\frac{\hat{H}_0 t}{\hbar}\right), \tag{2.59b}$$

$$\mathscr{M}(t) = \exp\left(i\frac{\hat{H}_0 t}{\hbar}\right) \hat{M} \exp\left(-i\frac{\hat{H}_0 t}{\hbar}\right). \tag{2.59c}$$

If $\hat{H}$ in the Schrödinger equation (2.52a) is time dependent, it can be solved in the form

$$|\psi(t)\rangle = \hat{U}(t, t_0) |\psi(t_0)\rangle, \tag{2.60}$$

where $\hat{U}(t, t_0)$ is the time development unitary operator obeying the Schrödinger equation

$$i\hbar \frac{\partial \hat{U}(t, t_0)}{\partial t} = \hat{H}(t) \hat{U}(t, t_0). \tag{2.61}$$

Solving this equation by the standard iterative method with the initial condition $\hat{U}(t_0, t_0) = \hat{1}$, we obtain

$$\hat{U}(t, t_0) = \mathscr{T} \exp\left[-\frac{i}{\hbar}\int_{t_0}^{t} \hat{H}(t')\,dt'\right] =$$

$$= \sum_{n=0}^{\infty} \frac{(-i/\hbar)^n}{n!} \int_{t_0}^{t} \ldots \int_{t_0}^{t} \mathscr{T}\{\hat{H}(t_1) \ldots \hat{H}(t_n)\}\, dt_1 \ldots dt_n, \tag{2.62}$$

where $\mathscr{T}$ is the time-ordering operator. This operator formally reorders the terms so that all operator products are taken with the factors in an increasing time-order sequence from the right to the left. In this case, the functions $\exp(-i\hat{H}(t - t_0)/\hbar)$ and $\exp(i\hat{H}(t - t_0)/\hbar)$ in the above equations are to be replaced by $\hat{U}(t, t_0)$ and $\hat{U}^+(t, t_0)$, respectively.

CHAPTER 3

OPTICAL CORRELATION PHENOMENA

Having prepared the basic quantum formalism we can start a quantum-mechanical treatment of correlation and coherence phenomena in the electromagnetic field, and show relations between various kinds of experiments and the corresponding quantum correlation functions. We consider the detection of a field using the interaction of this field with matter. If the detection device, composed of atoms, is assumed to be in its ground state, then photons are registered by absorption. Such detectors of light are photocells, photographic plates, etc.

In this chapter we discuss unconventional techniques based on photoelectric correlation measurements and on photocount measurements, through which the statistical properties of optical fields, including effects of arbitrary order, can be determined in principle. In the first case one determines the correlation functions of arbitrary order, while in the second case one uses a photoelectric detector and measures the distribution of emitted photoelectrons, whose statistics reflect the statistics of the absorbed photons.

As is well known the second-order coherence effects, related to interference and diffraction of light, are characterized by the second-order correlation function, which has the physical dimension of intensity (in optical correlators, e.g. interferometers, two complex amplitudes of light are multiplied and averaged). The fourth- and higher-order correlation functions can be realized as correlations of intensities (coincidences of photons or photoelectrons) of the field at various space-time points. Such fourth-order measurements were first performed in 1956 by Hanbury Brown and Twiss (1956a–c, 1957a, b, 1958), who used fast photoelectric detectors to measure the correlation of intensity fluctuations. A number of experiments of this kind were performed both by correlation and photoelectric coincidence techniques [Twiss et al. (1957), Rebka and Pound (1957), Brannen et al. (1958), Twiss and Little (1959), Harwit (1960)]. The sixth-order correlation functions were measured by Davidson and Mandel (1968), Davidson (1969) and Corti and Degiorgio (1974, 1976a, b). Another means of gaining information about higher-order moments (correlation functions) is based on measuring the photocount distribution. Sometimes this approach is more suitable [Arecchi (1965), Johnson, McLean and Pike (1966), Arecchi, Berné and Burlamacchi (1966), Arecchi, Berné, Sona and Burlamacchi (1966), Arecchi, Berné and Sona (1966) (for a review see Arecchi (1969), Arecchi and Degiorgio (1972)), Freed and Haus (1965, 1966), Johnson, Jones, McLean and Pike (1966), Fray et al. (1967) (for a review see Pike (1969, 1970), Pike and Jakeman (1974)), Martienssen and Spiller (1966a, b), Smith and Armstrong (1966a, b) (for a review see Armstrong and Smith (1967)), Chang et al. (1969)].

A new branch of interferometry, the so-called correlation interferometry, and higher-order high resolution correlation spectroscopy, are based on fourth- and higher-order correlation measurements [Degiorgio and Lastovka (1971), Cummins and Swinney (1970), Cummins and Pike (1974)]. By this technique the angular diameters of stars can be obtained, with better resolution than in measurements using the Michelson stellar interferometer [Hanbury Brown and Twiss (1956b, 1958), Twiss (1969)]. Correlation measurements also yield information about the spectral properties of light beams [Wolf (1962, 1965)] and about the polarization properties of light [Wolf (1960), Mandel and Wolf (1961), Mandel (1963a, b)]. A theoretical explanation of the fourth-order correlation effects, based on the classical stochastic or the semiclassical descriptions of light beams, was first given by Purcell (1956). The ideas of this paper have been continued by Wolf (1957), Janossy (1957, 1959), Mandel (1958, 1959, 1963a), Kahn (1958), Mandel and Wolf (1961) and Mandel et al. (1964). The problem was investigated in a fully quantum manner by Dicke (1964), Senitzky (1958, 1962, 1978), Fano (1961), Glauber (1963a–c, 1964, 1965, 1969, 1970, 1972), Kelley and Kleiner (1964), Goldberger and Watson (1964, 1965), Holliday and Sage (1964), Peřina (1967a, 1970), Lehmberg (1968), Klauder and Sudarshan (1968), Mandel and Meltzer (1969), Lax (1968a), Lax and Zwanziger (1973) and Selloni (1980).

3.1 Definition of quantum correlation functions

First consider an ideal detector which is quite small and whose sensitivity is independent of the photon frequency. One atom in the ground state may be considered as such a detector. Suppose the initial state of the field is characterized by the state vector $|i\rangle$ and the atom is initially in its ground state. If $|f\rangle$ is a final state of the field after the absorption of a photon described by the annihilation operator $\hat{A}^{(+)}(x)$ (a linearly polarized field is assumed for simplicity), then the transition amplitude of the process is equal to $\langle f|\hat{A}^{(+)}(x)|i\rangle$. The transition probability per unit time from the state $|i\rangle$ to the state $|f\rangle$ of the radiation field due to the absorption of a photon at a space-time point $x \equiv (\mathbf{x}, t)$ is then proportional to $|\langle f|\hat{A}^{(+)}(x)|i\rangle|^2$. But the final state $|f\rangle$ of the field usually remains unobserved so that we must sum over all possible states $|f\rangle$, which gives

$$\sum_f \langle i|\hat{A}^{(-)}(x)|f\rangle\langle f|\hat{A}^{(+)}(x)|i\rangle = \langle i|\hat{A}^{(-)}(x)\hat{A}^{(+)}(x)|i\rangle, \qquad (3.1)$$

because of the completeness condition $\sum_f |f\rangle\langle f| = \hat{1}$. Then the initial states $|i\rangle$ of the field can depend on some random or unknown parameters, if they are not pure states, and consequently all field measurements must be carried out as averages over the ensemble of ways in which the initial field can be prepared. We know that all the information about the quantum statistics of the field as a statistical dynamic system is contained in the density matrix $\hat{\varrho} = \{|i\rangle\langle i|\}_{\text{average}}$. Thus we arrive at

$$\{\langle i|\hat{A}^{(-)}(x)\hat{A}^{(+)}(x)|i\rangle\}_{\text{average}} = \mathrm{Tr}\,\{\hat{\varrho}\hat{A}^{(-)}(x)\hat{A}^{(+)}(x)\}, \qquad (3.2)$$

which represents the quantum mean intensity at x. This quantum intensity is a particular value at $x' = x$ of the correlation function

$$\Gamma_{\mathcal{N}}^{(1,1)}(x, x') = \mathrm{Tr}\,\{\hat{\varrho}\hat{A}^{(-)}(x)\,\hat{A}^{(+)}(x')\}. \tag{3.3}$$

If the vectorial properties of the field are to be taken into account, the vectorial indices have to be added to $\hat{A}^{(+)}$ and $\hat{A}^{(-)}$ and we obtain the second-rank tensor $\Gamma_{\mathcal{N},ij}^{(1,1)}$. For simplicity we use x to denote $(\mathbf{x}, t, j)$. Sometimes one considers fields obtained using polarizing filters which select photons of a polarization $\mathbf{e}$; then the use of only the scalar field operator $\hat{A} = \mathbf{e}\,.\,\hat{\mathbf{A}}$ is sufficient.

More generally, we can define the transition probability per unit (time)^n that a photon of a polarization j_1 is absorbed at $(\mathbf{x}_1, t_1) \equiv x_1$, a photon of a polarization j_2 at $(\mathbf{x}_2, t_2) \equiv x_2$, etc., and a photon of a polarization j_n at $(\mathbf{x}_n, t_n) \equiv x_n$. In the same way as before we conclude that the probability of this process is proportional to the correlation function

$$\begin{aligned} &\Gamma_{\mathcal{N}}^{(n,n)}(x_1, \ldots, x_n, x_n, \ldots, x_1) = \\ &= \{\sum_f \langle f \,|\, \hat{A}^{(+)}(x_1) \ldots \hat{A}^{(+)}(x_n) \,|\, i\rangle\,|^2\}_{\text{average}} = \\ &= \mathrm{Tr}\,\{\hat{\varrho}\hat{A}^{(-)}(x_1) \ldots \hat{A}^{(-)}(x_n)\,\hat{A}^{(+)}(x_n) \ldots \hat{A}^{(+)}(x_1)\}, \end{aligned} \tag{3.4}$$

which is a particular value of the general correlation function (tensor) $\Gamma_{\mathcal{N}}^{(m,n)}(x_1, \ldots, x_{m+n})$ defined by

$$\begin{aligned} &\Gamma_{\mathcal{N}}^{(m,n)}(x_1, \ldots, x_{m+n}) = \\ &= \mathrm{Tr}\,\{\hat{\varrho}\hat{A}^{(-)}(x_1) \ldots \hat{A}^{(-)}(x_m)\,\hat{A}^{(+)}(x_{m+1}) \ldots \hat{A}^{(+)}(x_{m+n})\}, \end{aligned} \tag{3.5a}$$

if $m = n$ and $x_j = x_{n+j}$, $j = 1, 2, \ldots, n$. Note that all the $\hat{A}^{(-)}$ or $\hat{A}^{(+)}$ can be interchanged since the commutation rules (2.50) hold. These correlation functions of the $(m + n)$th order correspond to classical correlation functions via a correspondence $\hat{A}^{(+)} \leftrightarrow V$, $\hat{A}^{(-)} \leftrightarrow V^*$, $\hat{\varrho} \leftrightarrow P$, V being some classical field and P a classical probability density function. Such a correspondence for normal correlation functions (3.5a) (i.e. all annihilation operators $\hat{A}^{(+)}$ stand to the right of all creation operators $\hat{A}^{(-)}$) is obtained with the use of the coherent states $|\,\{\alpha_\lambda\}\rangle$ (Chapter 4), which are eigenstates of the annihilation operator $\hat{A}^{(+)}(x)$, $\hat{A}^{(+)}(x)\,|\,\{\alpha_\lambda\}\rangle = V(x)\,|\,\{\alpha_\lambda\}\rangle$, $\langle\{\alpha_\lambda\}\,|\,\hat{A}^{(-)}(x) = V^*(x)\,\langle\{\alpha_\lambda\}\,|$ $(\hat{a}_\lambda\,|\,\{\alpha_\lambda\}\rangle = \alpha_\lambda\,|\,\{\alpha_\lambda\}\rangle$, $\langle\{\alpha_\lambda\}\,|\,\hat{a}_\lambda^+ = \alpha_\lambda^*\langle\{\alpha_\lambda\}\,|$, ($\alpha_\lambda$ being a complex number), which provides the equivalent "classical" correlation function

$$\begin{aligned} &\Gamma^{(m,n)}(x_1, \ldots, x_{m+n}) = \\ &= \langle V^*(x_1) \ldots V^*(x_m)\,V(x_{m+1}) \ldots V(x_{m+n})\rangle, \end{aligned} \tag{3.5b}$$

where the angled brackets mean the average over the classical complex amplitudes $\{\alpha_\lambda\}$ with some quasi-probability weighting functional. Further details will be discussed in the next chapter. Of course, the operators $\hat{A}^{(+)}$ and $\hat{A}^{(-)}$ do not commute in general (cf. (2.47)) so that this correspondence of the classical and quantum correlation functions is not unique. The Hanbury Brown – Twiss effect (Sec. 3.11) described by the correlation of intensities is just related to the correlation function $\Gamma^{(2,2)}_{\mathcal{N}}(x_1, x_2, x_2, x_1)$.

An important property of the correlation functions (3.5a) is that the operators $\hat{A}^{(+)}$ and $\hat{A}^{(-)}$ are, in their normal order, related to the absorption of photons. A consequence of this, if the vacuum stability condition $\hat{A}^{(+)} | 0\rangle = 0$ is applied, is that the vacuum expectation value of these normally ordered products is zero

$$\Gamma^{(n,n)}_{\mathcal{N},\,\mathrm{vac}}(x_1, \ldots, x_n, x_n, \ldots, x_1) = \\ = \langle 0 | \prod_{j=1}^{n} \hat{A}^{(-)}(x_j) \prod_{k=1}^{n} \hat{A}^{(+)}(x_k) | 0\rangle = 0, \tag{3.6}$$

and the physical vacuum gives no contribution to the quantities characterizing the photoelectric correlation measurements; but it does give a contribution to measurements which are related to other orderings of field operators. As an example we will discuss in Sec. 3.4 some devices (quantum counters), whose operation is connected with the antinormally ordered products of field operators ($\hat{A}^{(+)}$ and $\hat{A}^{(-)}$ are interchanged). These devices detect the field by means of stimulated emission rather than by absorption. In this case $\langle 0 | \hat{A}^{(+)}(x) \hat{A}^{(-)}(x') | 0\rangle = [\hat{A}^{(+)}(x), \hat{A}^{(-)}(x')]$, where the c-number commutator is given in (2.47) and the physical vacuum contributes. Just these circumstances are responsible for the fact that the probability distributions of absorbed photons and emitted photoelectrons have a similar form whereas the probability distribution of photons and the probability distribution of counts of a quantum counter differ from one another (as a consequence of the contribution of the physical vacuum).

From the correlation functions defined in terms of $\hat{\mathbf{A}}$, we can derive, using (2.2a, b), the correlation functions (tensors) defined in terms of $\hat{\mathbf{E}}$,

$$\mathrm{Tr}\,\{\hat{\varrho} \prod_{j=1}^{m} \hat{E}^{(-)}(x_j) \prod_{k=m+1}^{m+n} \hat{E}^{(+)}(x_k)\}, \tag{3.7}$$

or in terms of $\mathbf{H}$, or various mixed forms. In practice quasi-monochromatic light is mostly used and then all these quantities are proportional to one another. Further kinds of correlation tensors, based on the electromagnetic antisymmetric tensor $\hat{F}_{\mu\nu}(x) = \partial \hat{A}_\nu / \partial x_\mu - \partial \hat{A}_\mu / \partial x_\nu$, may also be introduced (Glauber (1963a)) and some of their properties have been investigated in a relativistic manner (Kujawski (1966), Dialetis (1969b)).

A more precise treatment of the detection of an electromagnetic field can be based on the time-dependent perturbation theory. This was done making electric-dipole approximation by Glauber (1965), the interaction hamiltonian being

$$\hat{H}_{\mathrm{int}} = -e \sum_j \hat{\mathbf{q}}_j \cdot \hat{\mathbf{E}}(\mathbf{x}, t), \tag{3.8a}$$

where $-e$ is the charge of an electron, $\hat{\mathbf{q}}_j$ is the spatial coordinate operator of the jth electron of the atom relative to its nucleus, assumed to be located at $\mathbf{x}$. More generally, the interaction hamiltonian can be written, in terms of the vector potential operator $\hat{\mathbf{A}}$, as

$$\hat{H}_{\text{int}} = -\frac{e}{m}\sum_j \hat{\mathbf{p}}_j \cdot \hat{\mathbf{A}}(\mathbf{x}, t), \tag{3.8b}$$

where $\hat{\mathbf{p}}_j$ is the momentum operator and m the mass of an electron.

Consider N atoms as detectors placed at points $\mathbf{x}_1, \dots, \mathbf{x}_n$ and apply the formalism for the time development contained in Sec. 2.5. We can solve the Schrödinger equation (2.52a) with $\hat{H} = \hat{H}_{\text{int}}$ (the interaction picture) in the form (2.60), where the time development operator $\hat{U}$ is given by (2.62). Using the Nth-order approximation of perturbation theory (the Nth term in (2.62)), we obtain for the probability that the first atom has undergone a transition in the time interval $(0, T_1)$, the second one in $(0, T_2)$, etc., and the Nth one in $(0, T_N)$ (Glauber (1965))

$$p^{(N)}(T_1, \dots, T_N) = \int_0^{T_1} \dots \int_0^{T_N} \prod_{j=1}^{N} \mathscr{S}(t'_j - t''_j) \times$$
$$\times \Gamma_{\mathscr{N}}^{(N,N)}(\mathbf{x}_1, \dots, \mathbf{x}_N, \mathbf{x}_N, \dots, \mathbf{x}_1, t'_1, \dots, t'_N, t''_N, \dots, t''_1) \prod_{k=1}^{N} \mathrm{d}t'_k \, \mathrm{d}t''_k, \tag{3.9}$$

where $\mathscr{S}$ is a weight function (a response function of a detector), including the spectral properties of the detector.

For detectors assumed to be broadband (i.e. they do not distinguish between frequencies)

$$\mathscr{S}(t' - t'') = \eta\delta(t' - t''), \tag{3.10}$$

where η is the sensitivity factor (the photoelectron efficiency) and we have from (3.9)

$$p^{(N)}(T_1, \dots, T_N) = \tag{3.11a}$$
$$= \eta^N \int_0^{T_1} \dots \int_0^{T_N} \Gamma_{\mathscr{N}}^{(N,N)}(\mathbf{x}_1, \dots, \mathbf{x}_N, \mathbf{x}_N, \dots, \mathbf{x}_1, t'_1, \dots, t'_N, t'_N, \dots, t'_1) \prod_{j=1}^{N} \mathrm{d}t'_j.$$

This probability must equal the expectation value $\langle \hat{n}_1 \dots \hat{n}_N \rangle$, which is simply the probability that all the detectors $1, 2, \dots, N$ have contributed counts in $(0, T_1)$, $(0, T_2), \dots, (0, T_N)$ respectively. If all the T_j and $\mathbf{x}_j$ are the same and if we suppress the spatial variable,

$$p^{(N)}(T, \dots, T) = \langle \mathscr{N} \hat{n}^N \rangle =$$
$$= \eta^N \int_0^{T} \dots \int_0^{T} \Gamma_{\mathscr{N}}^{(N,N)}(t'_1, \dots, t'_N, t'_N, \dots, t'_1) \prod_{j=1}^{N} \mathrm{d}t'_j. \tag{3.11b}$$

The joint counting rate corresponding to times $T_1, \dots, T_N$ is equal to

$$w(T_1, \dots, T_N) = \frac{\partial^N}{\partial T_1 \dots \partial T_N} p^{(N)}(T_1, \dots, T_N) =$$
$$= \eta^N \Gamma_{\mathscr{N}}^{(N,N)}(\mathbf{x}_1, \dots, \mathbf{x}_N, \mathbf{x}_N, \dots, \mathbf{x}_1, T_1, \dots, T_N, T_N, \dots, T_1), \tag{3.12}$$

so it is determined by the $2N$th-order correlation function.

Equation (3.9) may be interpreted as a result for point detectors which are not sensitive to the instantaneous value of the field but are sensitive rather to the average of the field over a time interval. More generally, we can consider detectors which are not sensitive even to the field at points $\mathbf{x}_1, \ldots, \mathbf{x}_N$, but to the averaged fields over spatial regions $V_1, \ldots, V_N$ and we obtain

$$p^{(N)} = \int_0^{T_1} \cdots \int_0^{T_N} \int_{V_1} \cdots \int_{V_N} \prod_{j=1}^{N} \mathscr{S}(x_j' - x_j'') \times$$
$$\times \Gamma_{\mathcal{N}}^{(N,N)}(x_1', \ldots, x_N', x_N'', \ldots, x_1'') \prod_{k=1}^{N} \mathrm{d}^4 x_k' \, \mathrm{d}^4 x_k''. \tag{3.13}$$

Equation (3.9) is then obtained if $\mathscr{S}(x_j' - x_j'') = \delta(\mathbf{x}_j' - \mathbf{x}_j'')\,\delta(\mathbf{x}_j' - \mathbf{x}_j)\,\mathscr{S}(t_j' - t_j'')$. In (2.13) x generally denotes $(\mathbf{x}, t, j)$ and the integrals over x imply in addition a summation over the polarization indices j.

3.2 Properties of quantum correlation functions

Here we briefly outline a derivation of some inequalities for the quantum correlation functions and summarize their properties.

3.2.1 Analytic properties

It is clear from (2.28b) that the positive-frequency part $\hat{A}^{(+)}$ $(kc > 0)$ of the vector potentional operator is an analytic function in the lower half of the complex t-plane, whereas the negative-frequency part $\hat{A}^{(-)}$ is an analytic function in the upper half of the complex t-plane. Therefore, from the definition (3.5a), the correlation function $\Gamma_{\mathcal{N}}^{(m,n)}$ is an analytic function in the upper half of the complex t-plane in variables $t_1, \ldots, t_m$ and in the lower half-plane in variables $t_{m+1}, \ldots, t_{m+n}$ and consequently the following dispersion relations hold

$$\operatorname{Im} \Gamma_{\mathcal{N}}^{(m,n)}(t_1, \ldots, t_{m+n}) =$$
$$= -\varepsilon_k \frac{P}{\pi} \int_{-\infty}^{+\infty} \frac{\operatorname{Re} \Gamma_{\mathcal{N}}^{(m,n)}(t_1, \ldots, t_k', \ldots, t_{m+n})}{t_k' - t_k} \, \mathrm{d}t_k', \tag{3.14a}$$

$$\operatorname{Re} \Gamma_{\mathcal{N}}^{(m,n)}(t_1, \ldots, t_{m+n}) =$$
$$= \varepsilon_k \frac{P}{\pi} \int_{-\infty}^{+\infty} \frac{\operatorname{Im} \Gamma_{\mathcal{N}}^{(m,n)}(t_1, \ldots, t_k', \ldots, t_{m+n})}{t_k' - t_k} \, \mathrm{d}t_k', \tag{3.14b}$$

where $\varepsilon_k = 1$ for $k = 1, \ldots, m$ and $\varepsilon_k = -1$ for $k = m + 1, \ldots, m + n$, Re and Im denote the real and imaginary parts respectively, and P denotes the Cauchy principal value of the integral at $t_k' = t_k$. We have suppressed the space variables for simplicity here.

3.2.2 Spectral properties

Performing the spectral decomposition of the annihilation operator in the form of the Fourier integral,

$$\hat{A}^{(+)}(\mathbf{x}, t) = \int_0^\infty \hat{A}^{(+)}(\mathbf{x}, \nu) \exp(-\mathrm{i}\, 2\pi\nu t)\, \mathrm{d}\nu, \tag{3.15}$$

the correlation function can be spectrally decomposed as follows

$$\Gamma_{\mathcal{N}}^{(m,n)}(\mathbf{x}_1, \ldots, \mathbf{x}_{m+n}, \mathbf{t}) = \\ = \int_0^\infty \ldots \int_0^\infty G_{\mathcal{N}}^{(m,n)}(\mathbf{x}_1, \ldots, \mathbf{x}_{m+n}, \boldsymbol{\nu}) \exp[\mathrm{i}\, 2\pi(\boldsymbol{\nu}, \mathbf{t})]\, \mathrm{d}^{m+n}\nu, \tag{3.16}$$

where the spectral correlation function reads

$$G_{\mathcal{N}}^{(m,n)}(\mathbf{x}_1, \ldots, \mathbf{x}_{m+n}, \boldsymbol{\nu}) = \mathrm{Tr}\,\{\hat{\varrho} \prod_{j=1}^{m} \hat{A}^{(-)}(\mathbf{x}_j, \nu_j) \prod_{k=m+1}^{m+n} \hat{A}^{(+)}(\mathbf{x}_k, \nu_k)\} = \\ = \int_0^\infty \ldots \int_0^\infty \Gamma_{\mathcal{N}}^{(m,n)}(\mathbf{x}_1, \ldots, \mathbf{x}_{m+n}, \mathbf{t}) \exp[-\mathrm{i}\, 2\pi(\boldsymbol{\nu}, \mathbf{t})]\, \mathrm{d}^{m+n}t,\ \nu_j \geqq 0. \tag{3.17}$$

Here $\mathbf{t} \equiv (t_1, \ldots, t_{m+n})$, $\boldsymbol{\nu} \equiv (\nu_1, \ldots, \nu_m, -\nu_{m+1}, \ldots, -\nu_{m+n})$, $\mathrm{d}^{m+n}t \equiv \prod_{j=1}^{m+n} \mathrm{d}t_j$, $\mathrm{d}^{m+n}\nu \equiv \prod_{j=1}^{m+n} \mathrm{d}\nu_j$ and $(\boldsymbol{\nu}, \mathbf{t}) = \sum_{j=1}^{m+n} \varepsilon_j \nu_j t_j$, ε_j being defined above. The equation (3.16) is called the generalized Wiener–Khintchine theorem.

If the optical field is stationary, then the correlation functions $\Gamma_{\mathcal{N}}^{(m,n)}$ are invariant with respect to the translation of the origin, i.e.

$$\Gamma_{\mathcal{N}}^{(m,n)}(t_1 + \tau, t_2 + \tau, \ldots, t_{m+n} + \tau) = \Gamma_{\mathcal{N}}^{(m,n)}(t_1, \ldots, t_{m+n}). \tag{3.18}$$

This means, referring to (3.16), that the resonance frequency condition

$$\sum_{j=1}^{m+n} \varepsilon_j \nu_j = 0 \tag{3.19}$$

must hold. This is the necessary and sufficient condition for the frequency components $\hat{A}^{(+)}(\nu_1), \ldots, \hat{A}^{(+)}(\nu_{m+n})$ of a stationary field to be correlated. Thus (3.18) depends on $(m + n - 1)$ time variables $\boldsymbol{\tau} \equiv (\tau_2 = t_2 - t_1, \ldots, \tau_{m+n} = t_{m+n} - t_1)$. From the quantum-mechanical point of view this means that for the density matrix $\hat{\varrho}$, $\mathrm{i}\hbar\, \partial\hat{\varrho}/\partial t = [\hat{H}, \hat{\varrho}] = \hat{0}$.

In the particular case of the second-order correlation function (also called the mutual coherence function) $\Gamma_{\mathcal{N}}(\mathbf{x}_1, \mathbf{x}_2, \tau) \equiv \Gamma_{\mathcal{N}}^{(1,1)}(\mathbf{x}_1, \mathbf{x}_2, 0, \tau)$ for a stationary field, we have the standard Wiener–Khintchine theorem

$$\Gamma_{\mathcal{N}}(\mathbf{x}_1, \mathbf{x}_2, \tau) = \int_0^\infty G_{\mathcal{N}}(\mathbf{x}_1, \mathbf{x}_2, \nu) \exp(-\mathrm{i}\, 2\pi\nu\tau)\, \mathrm{d}\nu, \tag{3.20a}$$

where

$$G_{\mathcal{N}}(\mathbf{x}_1, \mathbf{x}_2, \nu) \equiv G_{\mathcal{N}}^{(1,1)}(\mathbf{x}_1, \mathbf{x}_2, \nu, -\nu) =$$
$$= \mathrm{Tr}\,\{\hat{\varrho}\hat{A}^{(-)}(\mathbf{x}_1, \nu)\,\hat{A}^{(+)}(\mathbf{x}_2, \nu)\} =$$
$$= \int_{-\infty}^{+\infty} \Gamma_{\mathcal{N}}(\mathbf{x}_1, \mathbf{x}_2, \tau) \exp(\mathrm{i}\, 2\pi\nu\tau)\, \mathrm{d}\tau, \quad \nu \geqq 0, \tag{3.20b}$$
$$= 0, \ \nu < 0;$$

$G_{\mathcal{N}}$ is called the normal cross-spectral density or the mutual spectral density. The dispersion relations (3.14a, b) are clearly valid for (3.20a) in the complex τ-plane. The Wiener–Khintchine theorem allows the spectral properties to be determined from the correlation measurements.

The generalized Wiener–Khintchine theorem makes it possible to show for stationary quasi-monochromatic fields ($\Delta\nu \ll \bar{\nu}$, $\Delta\nu$ being the spectral half-width and $\bar{\nu}$ the mean frequency) that the correlation functions can be written in the form

$$\Gamma_{\mathcal{N}}^{(n,n)}(\mathbf{x}_1, \ldots, \mathbf{x}_{2n}, \boldsymbol{\tau}) =$$
$$= |\,\Gamma_{\mathcal{N}}^{(n,n)}(\mathbf{x}_1, \ldots, \mathbf{x}_{2n}, \boldsymbol{\tau})\,| \exp\Big[\mathrm{i}\alpha(\mathbf{x}_1, \ldots, \mathbf{x}_{2n}, \boldsymbol{\tau}) + \mathrm{i}2\pi\bar{\nu}\sum_{j=2}^{2n} \varepsilon_j\tau_j\Big], \tag{3.21}$$

where $|\,\Gamma_{\mathcal{N}}^{(n,n)}(\mathbf{x}_1, \ldots, \mathbf{x}_{2n}, \boldsymbol{\tau})\,|$ and $\alpha(\mathbf{x}_1, \ldots, \mathbf{x}_{2n}, \boldsymbol{\tau})$ are amplitude and phase functions whose variations with τ_j are slow compared to the variation of $\exp(\mathrm{i}2\pi\bar{\nu}\varepsilon_j\tau_j)$. If in addition

$$|\,\tau_j\,| \ll 1/\Delta\nu, \tag{3.22}$$

then

$$\Gamma_{\mathcal{N}}^{(n,n)}(\mathbf{x}_1, \ldots, \mathbf{x}_{2n}, \boldsymbol{\tau}) = \Gamma_{\mathcal{N}}^{(n,n)}(\mathbf{x}_1, \ldots, \mathbf{x}_{2n}, \hat{0}) \exp\Big(\mathrm{i}2\pi\bar{\nu}\sum_{j=2}^{2n} \varepsilon_j\tau_j\Big). \tag{3.23}$$

Further discussions of non-stationary spectra have been given by Eberly and Wódkiewicz (1977) and Ponath and Schubert (1980) (for a review, see Cresser (1983)).

3.2.3 Wave equations in vacuo

As the operators $\hat{A}^{(+)}(\mathbf{x}, t)$ and $\hat{A}^{(-)}(\mathbf{x}, t)$ obey the wave equations in vacuo, we easily obtain the propagation equations for the correlation functions in vacuo:

$$\left(\Delta_j - \frac{1}{c^2}\frac{\partial^2}{\partial t_j^2}\right)\Gamma_{\mathcal{N}}^{(m,n)}(\mathbf{x}_1, \ldots, \mathbf{x}_{m+n}, \mathbf{t}) = 0, \tag{3.24a}$$

$$\left[\Delta_j + \left(\frac{2\pi\nu_j}{c}\right)^2\right] G_{\mathcal{N}}^{(m,n)}(\mathbf{x}_1, \ldots, \mathbf{x}_{m+n}, \boldsymbol{\nu}) = 0, \ j = 1, \ldots, m+n, \tag{3.24b}$$

where Δ_j is the Laplace operator with respect to the jth spatial variable.

In the case of stationary fields we obtain similarly

$$\left[\Delta_1 - \frac{1}{c^2}\left(\frac{\partial}{\partial\tau_2} + \frac{\partial}{\partial\tau_3} + \ldots + \frac{\partial}{\partial\tau_{m+n}}\right)^2\right]\Gamma_{\mathcal{N}}^{(m,n)}(\mathbf{x}_1, \ldots, \mathbf{x}_{m+n}, \boldsymbol{\tau}) = 0,$$

$$\left[\Delta_j - \frac{1}{c^2}\frac{\partial^2}{\partial\tau_j^2}\right]\Gamma_{\mathcal{N}}^{(m,n)}(\mathbf{x}_1, \ldots, \mathbf{x}_{m+n}, \tau) = 0, \qquad j = 2, 3, \ldots, m+n, \tag{3.25a}$$

and

$$\left[\Delta_j + \left(\frac{2\pi\nu_j}{c}\right)^2\right]G_{\mathcal{N}}^{(m,n)}(\mathbf{x}_1, \ldots, \mathbf{x}_{m+n}, \nu) = 0, \qquad j = 1, 2, \ldots, m+n; \tag{3.25b}$$

in (3.25b) the frequency condition (3.19) holds.

The wave equations for the mutual coherence function $\Gamma(\mathbf{x}_1, \mathbf{x}_2, \tau)$ were first derived by Wolf (1955). This function is used in the description of classical interference and diffraction phenomena.

3.2.4 Symmetries and inequalities

From the definition (3.5a) the cross-symmetry condition immediately follows:

$$[\Gamma_{\mathcal{N}}^{(m,n)}(x_1, \ldots, x_{m+n})]^* = \Gamma_{\mathcal{N}}^{(n,m)}(x_{m+n}, \ldots, x_1), \tag{3.26a}$$

and as a particular case

$$\Gamma_{\mathcal{N}}^*(\mathbf{x}_1, \mathbf{x}_2, \tau) = \Gamma_{\mathcal{N}}(\mathbf{x}_2, \mathbf{x}_1, -\tau), \tag{3.26b}$$

or

$$G_{\mathcal{N}}^*(\mathbf{x}_1, \mathbf{x}_2, \nu) = G_{\mathcal{N}}(\mathbf{x}_2, \mathbf{x}_1, \nu). \tag{3.26c}$$

A further interesting property of the correlation functions is that an exchange of arguments $x_1, \ldots, x_m$, and also of arguments $x_{m+1}, \ldots, x_{m+n}$ does not change the value of the correlation function $\Gamma_{\mathcal{N}}^{(m,n)}$, because the operators $\hat{A}^{(-)}(x_1), \ldots, \hat{A}^{(-)}(x_m)$, or the operators $\hat{A}^{(+)}(x_{m+1}), \ldots, \hat{A}^{(+)}(x_{m+n})$ mutually commute (see (2.50)). Another property is that $\Gamma_{\mathcal{N}}^{(n,n)}$ is zero for states with a finite number of photons, say N, in the field, when $n > N$; this is because $\hat{A}^{(+)}$ represents the annihilation operator of a photon and the vacuum stability condition holds.

As for any operator $\hat{B}$ it is the case that

$$\mathrm{Tr}\,\{\hat{\varrho}\hat{B}^+\hat{B}\} \geqq 0, \tag{3.27}$$

we obtain, choosing $\hat{B} = \hat{A}^{(+)}(x_1) \ldots \hat{A}^{(+)}(x_n)$, that

$$\Gamma_{\mathcal{N}}^{(n,n)}(x_1, \ldots, x_n, x_n, \ldots, x_1) \geqq 0; \tag{3.28}$$

choosing

$$\hat{B} = \sum_{j=1}^{n} c_j \hat{A}^{(+)}(x_j) \tag{3.29}$$

with some coefficients c_j, we obtain the non-negative definiteness condition [Mehta et al. (1966)]

$$\sum_{j,k}^{n} c_j^* c_k \Gamma_{\mathcal{N}}^{(1,1)}(x_j, x_k) \geqq 0, \tag{3.30a}$$

leading to

$$\operatorname{Det}\{\Gamma_{\mathcal{N}}^{(1,1)}(x_j, x_k)\} \geqq 0. \tag{3.30b}$$

For $n = 1$ we have for the mean intensity $\Gamma_{\mathcal{N}}^{(1,1)}(x, x) \geqq 0$, for $n = 2$

$$|\Gamma_{\mathcal{N}}^{(1,1)}(x_1, x_2)|^2 \leqq \Gamma_{\mathcal{N}}^{(1,1)}(x_1, x_1)\,\Gamma_{\mathcal{N}}^{(1,1)}(x_2, x_2); \tag{3.31}$$

hence the degree of coherence

$$\gamma_{\mathcal{N}}^{(1,1)}(x_1, x_2) = \frac{\Gamma_{\mathcal{N}}^{(1,1)}(x_1, x_2)}{[\Gamma_{\mathcal{N}}^{(1,1)}(x_1, x_1)\,\Gamma_{\mathcal{N}}^{(1,1)}(x_2, x_2)]^{1/2}} \tag{3.32a}$$

satisfies

$$0 \leqq |\gamma_{\mathcal{N}}^{(1,1)}(x_1, x_2)| \leqq 1. \tag{3.33}$$

It is evident that the same inequality holds for the normalized spectral correlation function (3.20b), defined by

$$g_{\mathcal{N}}(\mathbf{x}_1, \mathbf{x}_2, \nu) = \frac{G_{\mathcal{N}}(\mathbf{x}_1, \mathbf{x}_2, \nu)}{[G_{\mathcal{N}}(\mathbf{x}_1, \mathbf{x}_1, \nu)\, G_{\mathcal{N}}(\mathbf{x}_2, \mathbf{x}_2, \nu)]^{1/2}}, \tag{3.32b}$$

as a result of the non-negative definiteness condition in the spectral domain similar to that given in (3.30a) [Mandel and Wolf (1976, 1981)].

Choosing

$$\hat{B} = c_1 \hat{A}^{(+)}(x_1) \ldots \hat{A}^{(+)}(x_n) + c_2 \hat{A}^{(+)}(x_{n+1}) \ldots \hat{A}^{(+)}(x_{2n}), \tag{3.34}$$

we arrive at

$$|\Gamma_{\mathcal{N}}^{(n,n)}(x_1, \ldots, x_{2n})|^2 \leqq \Gamma_{\mathcal{N}}^{(n,n)}(x_1, \ldots, x_n, x_n, \ldots, x_1) \times \\ \times \Gamma_{\mathcal{N}}^{(n,n)}(x_{n+1}, \ldots, x_{2n}, x_{2n}, \ldots, x_{n+1}), \tag{3.35}$$

enabling us to introduce the higher-order degree of coherence [Klauder and Sudarshan (1968)]

$$^{(S)}\gamma_{\mathcal{N}}^{(n,n)}(x_1, \ldots, x_{2n}) = \\ = \frac{\Gamma_{\mathcal{N}}^{(n,n)}(x_1, \ldots, x_{2n})}{[\Gamma_{\mathcal{N}}^{(n,n)}(x_1, \ldots, x_n, x_n, \ldots, x_1)\,\Gamma_{\mathcal{N}}^{(n,n)}(x_{n+1}, \ldots, x_{2n}, x_{2n}, \ldots, x_{n+1})]^{1/2}} \tag{3.36}$$

which satisfies the inequality

$$0 \leqq |{}^{(S)}\gamma_{\mathcal{N}}^{(n,n)}(x_1, \ldots, x_{2n})| \leqq 1. \tag{3.37}$$

Other definitions can be given for the higher-order degree of coherence, e.g. [Peřina and Peřinová (1965), Mehta (1966)]

$$\gamma_{\mathcal{N}}^{(n,n)}(x_1, \ldots, x_{2n}) = \frac{\Gamma_{\mathcal{N}}^{(n,n)}(x_1, \ldots, x_{2n})}{[\prod_{j=1}^{2n} \Gamma_{\mathcal{N}}^{(n,n)}(x_j, \ldots, x_j)]^{1/2n}}, \tag{3.38}$$

or, after Glauber (1963a, 1964, 1965)

$$^{(G)}\gamma_{\mathcal{N}}^{(n,n)}(x_1, \ldots, x_{2n}) = \frac{\Gamma_{\mathcal{N}}^{(n,n)}(x_1, \ldots, x_{2n})}{[\prod_{j=1}^{2n} \Gamma_{\mathcal{N}}^{(1,1)}(x_j, x_j)]^{1/2}}. \tag{3.39}$$

Of course, for $n = 1$ the definitions (3.36), (3.38) and (3.39) reduce to (3.32a). It is obvious that the degrees of coherence (3.36) and (3.38) are only defined if there are at least n photons in the field, and to define the whole sequence an infinite number of photons must be present in the field. For fields having classical analogues, the modulus of (3.38) is bounded by unity, whereas (3.39) is unbounded ($| {}^{(G)}\gamma_{\mathcal{N}}^{(n,n)} | = 1$ for coherent fields and $| {}^{(G)}\gamma_{\mathcal{N}}^{(n,n)} | = n!$ for chaotic fields).

3.2.5 Examples of the second-order degrees of coherence

Considering temporal coherence ($\mathbf{x}_1 = \mathbf{x}_2$) and suppressing the spatial variable, we obtain from (3.20a) the normalized form of the Wiener–Khintchine theorem

$$\gamma_{\mathcal{N}}(\tau) = \int_0^\infty g_{\mathcal{N}}(\nu) \exp(-\mathrm{i}2\pi\nu\tau)\,\mathrm{d}\nu; \tag{3.20c}$$

the normalized spectrum is given by $g_{\mathcal{N}}(\nu) = G_{\mathcal{N}}(\nu)/\int_0^\infty G_{\mathcal{N}}(\nu)\,\mathrm{d}\nu = G_{\mathcal{N}}(\nu)/\langle I\rangle_{\mathcal{N}}$. For the Lorentzian spectral profile, defined by

$$g_{\mathcal{N}}(\nu) = \frac{2\Delta\nu}{4\pi^2(\nu - \bar{\nu})^2 + (\Delta\nu)^2}, \tag{3.20d}$$

we have

$$\gamma_{\mathcal{N}}(\tau) = \exp(-\mathrm{i}2\pi\bar{\nu}\tau - \Delta\nu\,|\,\tau\,|); \tag{3.20e}$$

for the Gaussian spectral profile, defined by

$$g_{\mathcal{N}}(\nu) = \frac{2\pi^{1/2}}{\Delta\nu} \exp\left[-\frac{4\pi^2}{(\Delta\nu)^2}(\nu - \bar{\nu})^2\right], \tag{3.20f}$$

we obtain

$$\gamma_{\mathcal{N}}(\tau) = \exp\left(-\mathrm{i}2\pi\bar{\nu}\tau - \frac{(\Delta\nu)^2\tau^2}{4}\right); \tag{3.20g}$$

and for the rectangular spectral profile given as

$$\begin{aligned} g_{\mathcal{N}}(\nu) &= \frac{\pi}{\Delta\nu}, \qquad |\,\nu - \bar{\nu}\,| \leqq \Delta\nu/2\pi, \\ &= 0, \qquad |\,\nu - \bar{\nu}\,| > \Delta\nu/2\pi, \end{aligned} \tag{3.20h}$$

the corresponding result is

$$\gamma_{\mathcal{N}}(\tau) = \exp(-\mathrm{i}2\pi\bar{\nu}\tau)\frac{\sin(\Delta\nu\tau)}{\Delta\nu\tau}. \tag{3.20i}$$

3.3 Quantum coherence

3.3.1 Second-order phenomena

Second-order coherence can be discussed on the basis of the well-known Young two-slit experiment (Fig. 3.1). This experiment can be described in quantum terms as well as in classical terms [Born and Wolf (1965), Beran and Parrent (1964), Peřina

(1972)]. For the quantum mean intensity at a point $\mathbf{x}$ of the screen $\mathscr{B}$ we obtain

$$I(x) \equiv \Gamma_{\mathcal{N}}^{(1,1)}(x, x) = \mathrm{Tr}\,\{\hat{\varrho}[a_1^*\hat{A}^{(-)}(x_1) + a_2^*\hat{A}^{(-)}(x_2)]\times \times[a_1\hat{A}^{(+)}(x_1) + a_2\hat{A}^{(+)}(x_2)]\} = |a_1|^2\,\Gamma_{\mathcal{N}}^{(1,1)}(x_1, x_1) + + |a_2|^2\,\Gamma_{\mathcal{N}}^{(1,1)}(x_2, x_2) + a_1^*a_2\Gamma_{\mathcal{N}}^{(1,1)}(x_1, x_2) + + a_1a_2^*\Gamma_{\mathcal{N}}^{(1,1)}(x_2, x_1), \tag{3.40}$$

where the a_j are propagators between P_j and Q $(j = 1, 2)$ considered at the mean frequency, and x_j are related to the positions of the pinholes P_j at times t_j. The interference terms depending on $\Gamma_{\mathcal{N}}^{(1,1)}(x_1, x_2)$ and $\Gamma_{\mathcal{N}}^{(1,1)}(x_2, x_1)$ in (3.40) express the fact that one cannot distinguish from which pinhole a photon came (interference can be observed only if photons are within one cell of the phase space or within the coherence volume). Full second-order coherence then implies maximum visibility of the interference fringes, and

$$|\Gamma_{\mathcal{N}}^{(1,1)}(x_1, x_2)|^2 = \Gamma_{\mathcal{N}}^{(1,1)}(x_1, x_1)\,\Gamma_{\mathcal{N}}^{(1,1)}(x_2, x_2). \tag{3.41}$$

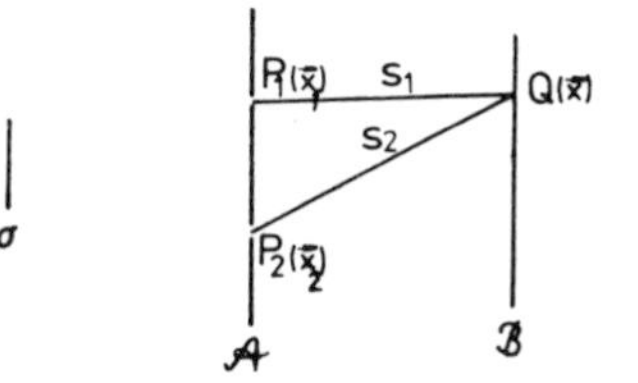

Fig. 3.1 — Significance of second-order coherence in a two-beam interference experiment, σ is an extended quasi-monochromatic source, $\mathscr{A}$ is a screen with two small pinholes $P_1(\mathbf{x}_1)$ and $P_2(\mathbf{x}_2)$ and $\mathscr{B}$ is the screen of observation with a typical point $Q(\mathbf{x})$.

This is the boundary value of (3.31) corresponding, according to the definition (3.32a) of the degree of coherence, to

$$|\gamma_{\mathcal{N}}^{(1,1)}(x_1, x_2)| = 1 \tag{3.42}$$

for all x_1 and x_2; in general (3.33) holds, implying states of partial coherence.

Writing (3.41) in the form

$$\Gamma_{\mathcal{N}}^{(1,1)}(x_1, x_2) = A(x_1)\,B(x_2), \tag{3.43}$$

we have from the cross-symmetry condition (3.26a)

$$A(x_1)\,B(x_2) = A^*(x_2)\,B^*(x_1), \tag{3.44}$$

and so

$$\frac{A(x_1)}{B^*(x_1)} = \frac{A^*(x_2)}{B(x_2)} = k, \tag{3.45}$$

k being a real constant. Therefore $A(x) = kB^*(x)$, and if we define the function $V(x) = k^{1/2}B(x)$, we can rewrite (3.43) in the form

$$\Gamma_{\mathcal{N}}^{(1,1)}(x_1, x_2) = V^*(x_1)\,V(x_2). \tag{3.46}$$

From (3.46) clearly (3.42) follows. Therefore the factorization condition (3.46) is the necessary and sufficient condition for second-order coherence.

We can ask in what manner the higher-order correlation functions are restricted by the factorization condition (3.46). We will follow the method given by Titulaer and Glauber (1965). If (3.46) holds, then (3.41) is also fulfilled and this implies that

$$\operatorname{Tr}\{\hat{\varrho}\hat{B}^{+}\hat{B}\} = 0, \tag{3.47}$$

where

$$\hat{B} = \hat{A}^{(+)}(x) - \frac{\Gamma_{\mathcal{N}}^{(1,1)}(x_0, x)}{\Gamma_{\mathcal{N}}^{(1,1)}(x_0, x_0)} \hat{A}^{(+)}(x_0); \tag{3.48}$$

here x_0 is an arbitrary space-time point. Therefore $\hat{\varrho}\hat{B}^{+} = \hat{B}\hat{\varrho} = \hat{0}$ and we obtain

$$\hat{A}^{(+)}(x)\,\hat{\varrho} = \frac{\Gamma_{\mathcal{N}}^{(1,1)}(x_0, x)}{\Gamma_{\mathcal{N}}^{(1,1)}(x_0, x_0)} \hat{A}^{(+)}(x_0)\,\hat{\varrho}, \tag{3.49a}$$

$$\hat{\varrho}\hat{A}^{(-)}(x) = \frac{\Gamma_{\mathcal{N}}^{(1,1)}(x, x_0)}{\Gamma_{\mathcal{N}}^{(1,1)}(x_0, x_0)} \hat{\varrho}\hat{A}^{(-)}(x_0). \tag{3.49b}$$

Considering identities (3.49a, b) at x_2 and x_1 respectively, we obtain

$$\operatorname{Tr}\{\hat{\varrho}\hat{A}^{(-)}(x_1)\,\hat{A}^{(+)}(x_2)\} = \frac{\Gamma_{\mathcal{N}}^{(1,1)}(x_1, x_0)\,\Gamma_{\mathcal{N}}^{(1,1)}(x_0, x_2)}{[\Gamma_{\mathcal{N}}^{(1,1)}(x_0, x_0)]^2} \times$$
$$\times \operatorname{Tr}\{\hat{\varrho}\hat{A}^{(-)}(x_0)\,\hat{A}^{(+)}(x_0)\}, \tag{3.50a}$$

or

$$\Gamma_{\mathcal{N}}^{(1,1)}(x_1, x_2) = \frac{\Gamma_{\mathcal{N}}^{(1,1)}(x_1, x_0)\,\Gamma_{\mathcal{N}}^{(1,1)}(x_0, x_2)}{\Gamma_{\mathcal{N}}^{(1,1)}(x_0, x_0)} = V^*(x_1)\,V(x_2), \tag{3.50b}$$

in agreement with (3.46), where $V(x) = \Gamma_{\mathcal{N}}^{(1,1)}(x_0, x)/[\Gamma_{\mathcal{N}}^{(1,1)}(x_0, x_0)]^{1/2}$.

3.3.2 Higher-order phenomena

For the $(m + n)$th-order correlation function we obtain in the same way

$$\Gamma_{\mathcal{N}}^{(m,n)}(x_1, x_2, \ldots, x_{m+n}) = \gamma^{(m,n)} \prod_{j=1}^{m} V^*(x_j) \prod_{k=m+1}^{m+n} V(x_k), \tag{3.51}$$

where

$$\gamma^{(m,n)} = \frac{\Gamma_{\mathcal{N}}^{(m,n)}(x_0, \ldots, x_0)}{[\Gamma_{\mathcal{N}}^{(1,1)}(x_0, x_0)]^{(m+n)/2}}. \tag{3.52}$$

However, x_0 is an arbitrary point so that (3.52) must be independent of x_0. Consequently the factorization condition (3.46), expressing the second-order coherence, leads to the factorization of all correlation functions into the form (3.51).

By analogy with the factorization (3.46) for second-order coherence, we may introduce the following set of factorizations

$$\Gamma_{\mathcal{N}}^{(m,n)}(x_1, \ldots, x_{m+n}) = \prod_{j=1}^{m} V^*(x_j) \prod_{k=m+1}^{m+n} V(x_k), \tag{3.53}$$

in which V is independent of m and n. Thus we can speak of $2N$th-order coherence if (3.53) holds for $m, n \leqq N$. If (3.53) holds for all m, n (in practice for all $m = n$ since the phase of an optical field is usually random), then the field possesses full coherence. It is clear from a classical point of view that fully coherent fields are noiseless fields whose distribution function is the Dirac δ-function. In the sense of the even-order factorization, they may possess phase fluctuations. This illustrates a close relation between the noiselessness of fields and full coherence. In the quantum theory, also, there exist states of fields for which the factorization (3.53) holds. These are called coherent states, and their properties will be investigated in the next chapter. It is clear that the field $V(x, t)$ in (3.53) may have any spectral composition, and full coherence does not require monochromaticity for general fields. Only for stationary fields, for which $\Gamma_{\mathcal{N}}^{(1,1)}(t_1, t_2) = \Gamma_{\mathcal{N}}^{(1,1)}(t_1 - t_2) = V^*(t_1)\, V(t_2)$, which is a functional equation satisfied only by an exponential function $V(t) \sim \exp(-i2\pi\nu_0 t)$, is monochromaticity the necessary and sufficient condition for the field to be coherent. The connection between fluctuations of the field and its coherence can be demonstrated as follows. Let the field be uniformly fluctuating, $V(x) = c V_{\mathrm{det}}(x)$, c being a random complex variable and V_{det} a non-fluctuating deterministic field. Second-order coherence demands $\langle | c |^2 \rangle = \int P(c) \, | c |^2 \, d^2 c = 1$ ($P(c)$ is a probability distribution in c, and the integration is taken over the whole complex plane). Hence, some phase and amplitude fluctuations are admissible in this case. Similarly, from the fourth-order coherence $\langle | c |^4 \rangle = 1 = \langle | c |^2 \rangle^2$, from which $| c |^2 = 1$, so only phase fluctuations are admissible, etc.

A further discussion of coherent optical fields can be found in Bojcov et al. (1973), while the less familiar topic of entropy considerations has been treated by Anisimov and Sotskii (1977b), and Kikuchi and Soffer (1977).

Recently, a new approach to the theory of coherence has been developed [Mandel and Wolf (1976, 1981), Bastiaans (1977, 1981), Wolf (1981a, b)] in the space-frequency domain. Full coherence in second order is then defined, in analogy to (3.46), by the factorization $G_{\mathcal{N}}(\mathbf{x}_1, \mathbf{x}_2, \nu) = U^*(\mathbf{x}_1, \nu)\, U(\mathbf{x}_2, \nu)$, where $U(\mathbf{x}, \nu)$ satisfies the Helmholtz equation, and $| g_{\mathcal{N}}(\mathbf{x}_1, \mathbf{x}_2, \nu) | \equiv 1$ from (3.32b).

Comparing (3.53) with (3.51), we see that these equations differ from one another by the factor $\gamma^{(m,n)}$. Consequently, second-order coherence does not in general imply higher-order coherence, for which

$$\gamma^{(m,n)} = 1. \tag{3.54}$$

If a finite number of photons, say M, are present in the field, then $\gamma^{(n,n)} = 0$ for $n > M$ and such a field cannot possess coherence to all orders. Considering a one-mode field only, we have from (2.14a)

$$\gamma^{(k,k)} = \frac{\langle \hat{a}^{+k} \hat{a}^k \rangle}{\langle \hat{a}^+ \hat{a} \rangle^k} = \frac{\langle \hat{n}(\hat{n} - 1) \ldots (\hat{n} - k + 1) \rangle}{\langle \hat{n} \rangle^k}. \tag{3.55}$$

For a Fock state $|n\rangle$ we have $\varrho = |n\rangle\langle n|$ and so

$$\gamma^{(k,k)} = \frac{n!}{(n-k)!\,n^k}, \tag{3.56}$$

which is zero for $k > n$. An interesting property of the Fock state is that the normal variance is negative, $\langle(\Delta\hat{n})^2\rangle_{\mathcal{N}} = \langle\hat{a}^{+2}\hat{a}^2\rangle - \langle\hat{a}^+\hat{a}\rangle^2 = n(n-1) - n^2 = -n < 0$, so the variance of $\hat{n}$ is $\langle(\Delta\hat{n})^2\rangle = \langle(\hat{a}^+\hat{a})^2\rangle - \langle\hat{a}^+\hat{a}\rangle^2 = \langle\hat{n}\rangle + \langle(\Delta\hat{n})^2\rangle_{\mathcal{N}} = 0$. This expresses the antibunching (anticorrelation) effect for these states (Chapter 9) and the Fock states have no classical analogue [Miller and Mishkin (1967a), Bertrand and Mishkin (1967)]. [Note that for the coherent states (Chapter 4) $\langle\hat{a}^{+2}\hat{a}^2\rangle - \langle\hat{a}^+\hat{a}\rangle^2 = 0$.]

For thermal radiation $\hat{\varrho}$ is given by (2.36) and

$$\gamma^{(k,k)} = [\mathrm{Tr}\{\exp(-\Theta\hat{n})\}]^{k-1} \frac{\mathrm{Tr}\left\{\exp(-\Theta\hat{n})\dfrac{\hat{n}!}{(\hat{n}-k)!}\right\}}{[\mathrm{Tr}\{\exp(-\Theta\hat{n})\,\hat{n}\}]^k}. \tag{3.57}$$

Further

$$\begin{aligned}
\mathrm{Tr}\left\{\exp(-\Theta\hat{n})\frac{\hat{n}!}{(\hat{n}-k)!}\right\} &= \sum_{n=0}^{\infty}\exp(-\Theta n)\frac{n!}{(n-k)!} = \exp(-\Theta k)\times \\
&\times \frac{\mathrm{d}^k}{\mathrm{d}(\exp(-\Theta))^k}\sum_{n=0}^{\infty}\exp(-\Theta n) = \\
&= \exp(-\Theta k)\frac{\mathrm{d}^k}{\mathrm{d}(\exp(-\Theta))^k}\frac{1}{1-\exp(-\Theta)} = \\
&= \exp(-\Theta k)\frac{k!}{(1-\exp(-\Theta))^{k+1}}.
\end{aligned} \tag{3.58}$$

Using this in (3.57) we obtain

$$\begin{aligned}
\gamma^{(k,k)} &= \frac{1}{(1-\exp(-\Theta))^{k-1}}\frac{\exp(-\Theta k)\,k!}{(1-\exp(-\Theta))^{k+1}} \times \\
&\times \frac{(1-\exp(-\Theta))^{2k}}{\exp(-\Theta k)} = k!,
\end{aligned} \tag{3.59}$$

and $\gamma^{(k,l)} = k!\,\delta_{kl}$. Thus if the definition in (3.39) is adopted, chaotic radiation cannot possess coherence of higher than second order ($m = n = 1$).

The conditions (3.54) and (3.59) (i.e. the validity of the factorization conditions) for laser and chaotic light were experimentally tested by Jakeman et al. (1968b). Very good agreement was found up to twelfth order ($m = n$).

Taking into account that for a fully coherent field

$$\mathrm{Tr}\{\hat{\varrho}\hat{A}^{(-)}(x)\,\hat{A}^{(+)}(x)\} - |\,\mathrm{Tr}\{\hat{\varrho}\hat{A}^{(+)}(x)\}\,|^2 = 0, \tag{3.60}$$

we arrive at

$$V(x) = \mathrm{Tr}\{\hat{\varrho}\hat{A}^{(+)}(x)\} = \Gamma_{\mathcal{N}}^{(0,1)}(x). \tag{3.61}$$

We have seen that the idea of second-order coherence can be derived from the Young experiment and that it has a clear physical significance – the interference fringes have the maximum visibility. We can also ask for the physical significance of higher-order coherence conditions. A natural experiment to investigate this is the photon-coincidence counting experiment.

If the definition (3.39) of the degree of coherence is adopted, equation (3.53) implies that $|{}^{(G)}\gamma_{\mathcal{N}}^{(n,n)}(x_1, \ldots, x_{2n})| \equiv 1$, i.e. ${}^{(G)}\gamma_{\mathcal{N}}^{(n,n)}(x_1, \ldots, x_n, x_n, \ldots, x_1) \equiv 1$ and consequently

$$\Gamma_{\mathcal{N}}^{(n,n)}(x_1, \ldots, x_n, x_n, \ldots, x_1) = \prod_{j=1}^{n} \Gamma_{\mathcal{N}}^{(1,1)}(x_j, x_j). \tag{3.62}$$

This means, according to our earlier results, that the N-fold joint counting rate defined by (3.12) is equal to the product of the counting rates which would be measured by each of the N counters in the absence of all the others; the responses of the counters are statistically independent of one another. If the average intensity of the field is independent of time (the field is stationary), the counters detect no tendency toward any sort of correlation in the arrival times of photons. In this case there is no Hanbury Brown – Twiss effect, i.e. no correlation of intensity fluctuations will occur (Sec. 3.11). Fields for which the Hanbury Brown – Twiss effect occurs cannot be coherent in the fourth- and higher orders. A typical example is the chaotic field, for which there is a tendency for photons to arrive in pairs (Sec. 3.10).

The full coherence conditions are equivalent to the fact that the photon statistics are Poissonian. Since $\langle \hat{a}^{+n}\hat{a}^n \rangle = \langle \hat{a}^+\hat{a} \rangle^n$ for a fully coherent field, the normally ordered characteristic function is equal to

$$\langle \exp(\mathrm{i}s\hat{n}) \rangle_{\mathcal{N}} = \sum_{n=0}^{\infty} \frac{(\mathrm{i}s)^n}{n!} \langle \hat{a}^{+n}\hat{a}^n \rangle = \sum_{n=0}^{\infty} \frac{(\mathrm{i}s)^n}{n!} \langle \hat{a}^+\hat{a} \rangle^n = \exp(\mathrm{i}s\langle \hat{n} \rangle). \tag{3.63}$$

Then, making use of (2.14a), we find that

$$\langle \exp(\mathrm{i}s\hat{n}) \rangle_{\mathcal{N}} = \sum_{n=0}^{\infty} \varrho(n, n)(1 + \mathrm{i}s)^n. \tag{3.64}$$

Hence

$$\varrho(n, n) = \frac{1}{n!} \frac{\mathrm{d}^n}{\mathrm{d}(\mathrm{i}s)^n} \langle \exp(\mathrm{i}s\hat{n}) \rangle_{\mathcal{N}} \Big|_{\mathrm{i}s=-1} = \frac{\langle n \rangle^n}{n!} \exp(-\langle n \rangle), \tag{3.65}$$

which is the Poissonian distribution.

While the degrees of coherence $\gamma_{\mathcal{N}}$, ${}^{(G)}\gamma_{\mathcal{N}}$ and ${}^{(S)}\gamma_{\mathcal{N}}$, defined by (3.38), (3.39) and (3.36), are equivalent if the factorization condition (3.53) is fulfilled and $|\gamma_{\mathcal{N}}| = |{}^{(G)}\gamma_{\mathcal{N}}| = |{}^{(S)}\gamma_{\mathcal{N}}| = 1$, they are in general different (although they are identical for $m = n = 1$). As we have seen, the degree of coherence ${}^{(G)}\gamma_{\mathcal{N}}$ is suitable for analyzing photon correlation experiments, but it cannot be used as a measure of partial coherence in higher orders, since it has no bound. For this purpose the degrees of coherence $\gamma_{\mathcal{N}}$ and ${}^{(S)}\gamma_{\mathcal{N}}$ can be used, since the inequalities such as (3.37) hold under the assumption that $\gamma_{\mathcal{N}}$ and ${}^{(S)}\gamma_{\mathcal{N}}$ exist. Adopting, for example, the definition (3.38), we can speak of partially coherent fields in higher orders if $0 < |\gamma_{\mathcal{N}}^{(n,n)}| < 1$ (in analogy with the case $n = 1$). From the quantal definition of the degree of coherence it follows that this quantity can be defined so long as the number of photons in the field is larger than n. To define the whole sequence an infinite number of photons must be present in the field. Note that the degree of coherence (3.38) is useful in connection with imaging problems [Peřina and Peřinová (1965), Peřina (1969a, 1972)], whereas the degree of coherence (3.36) can serve to characterize the visibility in the two-slit experiment, if the resulting measured quantity is the n-fold intensity [Klauder and Sudarshan (1968)]. It may be said that the various definitions of the degree of coherence correspond to various types of experiments.

Sometimes one speaks of weakly coherent fields if the $\gamma^{(1,1)}$ of eqn. (3.52) is equal to unity, and of strongly coherent fields if $\gamma^{(n,n)} = 1$ for every n. It is obvious that if for a weakly coherent field $|\gamma_{\mathcal{N}}^{(1,1)}(x_1, x_2)| = 1$, then $|\gamma_{\mathcal{N}}^{(n,n)}(x_1, \ldots, x_{2n})| = 1$ in the sense of (3.38) and vice versa; the same is true in the sense of (3.36), as can be easily verified by using (3.51). This means that relation (3.42) completely determines a weakly coherent field. Note that for strongly coherent fields, considered as quantum fields having classical analogues, a sufficient specification is $\gamma^{(1,1)} = \gamma^{(2,2)} = 1$ [Titulaer and Glauber (1965), Picinbono (1967)]. Terms for quasi-stationary and pseudo-Gaussian fields have been introduced and discussed by Picinbono (1969), Picinbono and Rousseau (1970) and Barashev (1971).

3.4 Measurements corresponding to antinormally ordered products of field operators—quantum counters

In this section we show that in principle there exists an interesting method of measuring the statistics of optical fields, in which the photoelectric detectors are replaced by atomic counting devices, called quantum counters [Mandel (1966b)]. Such detectors operate by stimulated emission rather than by absorption of photons in an external field and consequently correlation measurements by means of quantum counters correspond to antinormally ordered products of field operators. Although this device may be useful to measure only highly degenerate fields (it is scarcely sensitive to fields with low degeneracy, i.e. with low mean number of photons per mode), it is interesting to compare it with the photoelectric detector, as to how the two devices measure fields.

The operation of a quantum counter, as proposed by Mandel (1966b) can be explained in terms of the energy level scheme in Fig. 3.2. In this figure a represents a terminal energy level and c is a metastable level which is radiatively coupled to a broad energy band d, corresponding to a very short-lived state. The system will make spontaneous radiative transitions from d to a. We assume that this system is prepared in the state c by optical pumping from a to the broad energy level band b, from which it makes a non-radiative transition to c. Further, suppose the interval $(E_c - E_d)/\hbar c$, defined by the energy levels E_c and E_d, is of the same order as the wave number of a typical mode of the external field, and that $(E_d - E_a) > (E_c - E_d)$.

Under the interaction of the system placed at a space point $\mathbf{x}$ at time t with the external field, the system can make a stimulated transition from c to d with the emission of a photon. As level d is very short lived, the system decays spontaneously from d to a with the further emission of a photon. The latter photon can be distinguished from the former since $(E_d - E_a) > (E_c - E_d)$ and a photodetector placed in the neighbourhood of $\mathbf{x}$ with sufficiently high photoelectric threshold will register the second photon alone. The combination of the photodetector with a large number of such atomic systems acts as a quantum counter of the external field operating by stimulated emission of radiation. Note that two photons are emitted into the field by the system but only one is absorbed by the photodetector;

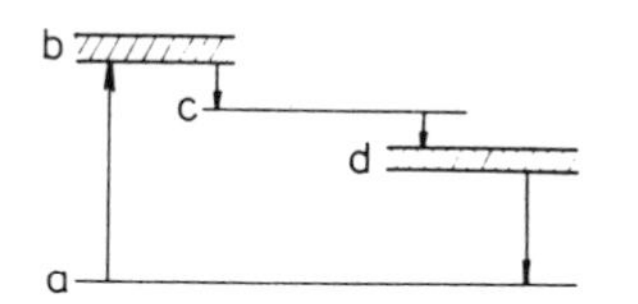

Fig. 3.2 — Energy levels for the quantum counter operating by stimulated emission.

the photodetector plays an auxiliary role only, the field is actually measured by means of the first induced transition.

Making use of arguments similar to those used in connection with the theory of the measurement of the field by means of photoelectric detectors, we find that the probability that one count is registered at $(\mathbf{x}, t)$ is proportional to [Mandel (1966b), Peřina (1968b)],

$$\begin{aligned}
&\{\sum_f | \langle f | \hat{A}^{(-)}(x) | i \rangle |^2\}_{\text{average}} = \\
&= \{\sum_f \langle i | \hat{A}^{(+)}(x) | f \rangle \langle f | \hat{A}^{(-)}(x) | i \rangle\}_{\text{average}} = \\
&= \mathrm{Tr}\,\{\hat{\varrho}\hat{A}^{(+)}(x)\,\hat{A}^{(-)}(x)\} = \Gamma_{\mathscr{A}}^{(1,1)}(x, x),
\end{aligned} \tag{3.66}$$

so this probability is proportional to the antinormally ordered correlation function $\Gamma_{\mathscr{A}}^{(1,1)}(x, x)$. Similarly, the joint probability that counts are registered by N quantum counters at space-time points $x_1, \ldots, x_N$ is proportional to

$$\begin{aligned}
&\Gamma_{\mathscr{A}}^{(N,N)}(x_1, \ldots, x_N, x_N, \ldots, x_1) = \\
&= \mathrm{Tr}\,\{\varrho\hat{A}^{(+)}(x_1) \ldots \hat{A}^{(+)}(x_N)\,\hat{A}^{(-)}(x_N) \ldots \hat{A}^{(-)}(x_1)\},
\end{aligned} \tag{3.67}$$

which is the $2N$th-order antinormal correlation function.

An alternating (symmetrical) ordering is typical for scattering processes [Loudon (1973)]. In such a process a photon of the frequency $\omega_1(\mathbf{k}_1)$ is absorbed, while a photon of the frequency $\omega_2(\mathbf{k}_2)$ is emitted; the field makes a transition from the initial state $| i \rangle$ to the final state $| f \rangle$. The transition probability is proportional to the following correlation function with the alternating order,

$$\mathrm{Tr}\,\{\hat{\varrho}\hat{A}_1^{(-)}\hat{A}_2^{(+)}\hat{A}_2^{(-)}\hat{A}_1^{(+)}\} \approx \langle \hat{n}_1 \rangle (\langle \hat{n}_2 \rangle + 1). \tag{3.68}$$

Spontaneous scattering takes place if $\langle \hat{n}_2 \rangle = 0$, stimulated scattering if $\langle \hat{n}_2 \rangle \gg 0$s.

3.5 Quantum characteristic functionals

We can define the normal characteristic functional

$$C_{\mathcal{N}}\{y(x)\} = \mathrm{Tr}\,\{\hat{\varrho}\exp\left[\int y(x)\hat{A}^{(-)}(x)\,\mathrm{d}^4x\right]\times \times\exp\left[-\int y^*(x)\hat{A}^{(+)}(x)\,\mathrm{d}^4x\right]\}, \tag{3.69}$$

with the parametric function $y(x)$. The complete set of normally ordered correlation functions can be derived from (3.69) with the help of functional differentiation:

$$\mathrm{Tr}\,\{\hat{\varrho}\hat{A}^{(-)}(x_1)\ldots\hat{A}^{(-)}(x_m)\,\hat{A}^{(+)}(x_{m+1})\ldots\hat{A}^{(+)}(x_{m+n})\} = = \prod_{j=1}^{m}\frac{\delta}{\delta y(x_j)}\prod_{k=m+1}^{m+n}\frac{\delta}{\delta(-y^*(x_k))}C_{\mathcal{N}}\{y(x)\}\bigg|_{y(x)\equiv 0}; \tag{3.70}$$

here $x \equiv (\mathbf{x}, t, j)$, so that the integrals over x include the summation over the polarization indices j.

Decomposing $A^{(+)}(x)$ in terms of orthogonal functions $\{\varphi_l(x)\}$, for which

$$\int \varphi_l^*(x)\,\varphi_m(x)\,\mathrm{d}^4x = \delta_{lm}, \tag{3.71}$$

we have

$$\hat{A}^{(+)}(x) = \sum_l \hat{a}_l\varphi_l(x),\ \hat{a}_l = \int \hat{A}^{(+)}(x)\,\varphi_l^*(x)\,\mathrm{d}^4x. \tag{3.72}$$

Introducing parameters

$$\beta_l = \int y(x)\,\varphi_l^*(x)\,\mathrm{d}^4x, \tag{3.73}$$

we can rewrite (3.69) in the form

$$C_{\mathcal{N}}(\{\beta_l\}) = \mathrm{Tr}\,\{\hat{\varrho}\prod_l\exp(\beta_l\hat{a}_l^+)\prod_k\exp(-\beta_k^*\hat{a}_k)\}. \tag{3.74}$$

The multimode correlation function is then obtained as

$$\mathrm{Tr}\,\{\hat{\varrho}\hat{a}_1^+\ldots\hat{a}_m^+\hat{a}_{m+1}\ldots\hat{a}_{m+n}\} = = \prod_{j=1}^{m}\frac{\partial}{\partial\beta_j}\prod_{k=m+1}^{m+n}\frac{\partial}{\partial(-\beta_k^*)}C_{\mathcal{N}}(\{\beta_l\})\bigg|_{\{\beta_l\}=\{\beta_l^*\}=0}. \tag{3.75}$$

Also characteristic functionals related to symmetric (or other) ordering can be introduced,

$$C(\{\beta_l\}) = \mathrm{Tr}\,\{\hat{\varrho}\prod_l\exp(\beta_l\hat{a}_l^+ - \beta_l^*\hat{a}_l)\}, \tag{3.76a}$$

$$C\{y(x)\} = \mathrm{Tr}\,\{\hat{\varrho}\exp\left[\int\left[y(x)\hat{A}^{(-)}(x) - y^*(x)\hat{A}^{(+)}(x)\right]\mathrm{d}^4x\right]\}, \tag{3.76b}$$

which are quantum expectation values of the unitary operators.

The functional formalism can be found in an excellent book by Volterra (1959). A systematic functional approach to the coherence theory and to photocount statistics has been developed by Zardecki (1969, 1971). Applications to laser theory have been suggested by Arecchi, Asdente and Ricca (1976).

3.6 Measurements of mixed-order correlation functions

Quadratic detectors allow only the even-order correlation functions of the type $\Gamma_{\mathcal{N}}^{(n,n)}$ to be measured; however some further experiments are possible which permit, at least in principle, the measurement of mixed-order correlation functions of the type $\Gamma_{\mathcal{N}}^{(m,n)}$ for $m \neq n$, which contain additional phase information. We will follow considerations suggested by Glauber (1970), based on nonlinear optics.

Considering a stationary field and returning to the frequency stationary condition (3.19), which is necessary if the $\Gamma_{\mathcal{N}}^{(m,n)}$ are to be non-vanishing, we see that it is easily satisfied if $m = n$ and $\nu_j = \nu_{n+j}$, as is usually the case. In this case one time dependence cancels the other and it is easy to have a non-vanishing function. A more complicated situation will occur if $m \neq n$, because in this case a very special set of frequencies must be occupied in the field. For example for a non-vanishing $\Gamma_{\mathcal{N}}^{(1,2)}$ it is necessary that $\nu_1 = \nu_2 + \nu_3$, a condition valid in the parametric amplification process or in Raman scattering (Chapter 9). This condition is necessary in order that the correlation function is nonvanishing, but a statistical dependence of modes must also exist, for if modes are independent then the density matrix factorizes into the product of the mode density matrices, and again $\Gamma_{\mathcal{N}}^{(1,2)} = 0$. In the parametric amplifier and in Raman scattering this dependence exists and $\Gamma_{\mathcal{N}}^{(1,2)} \neq 0$. For second harmonic generation in a nonlinear medium we also have $2\nu = \nu + \nu$, since in this case pairs of "red" photons are joined together to form "blue" photons. In the outgoing field there are two components, one of which is precisely twice the frequency of the other, that is

$$\hat{A}^{(+)}(t) = \hat{B}^{(+)}(t) + \hat{B}^{(+)2}(t) = \hat{U}^{(+)}(t) \exp(-i\omega t) + \\ + \hat{U}^{(+)2}(t) \exp(-2i\omega t), \tag{3.77}$$

where $\hat{B}^{(+)}(t) = \hat{U}^{(+)}(t) \exp(-i\omega t)$ is the incident field, and $\omega = 2\pi\nu$. As there are phase relations between these two fields, $\Gamma_{\mathcal{N}}^{(1,2)} \neq 0$.

Measuring the correlation functions with $m \neq n$ implies a measurement of the high frequency field correlated with the low frequency field. Assume we have a photodetector (an atom) with a threshold for photon detection which lies higher in frequency than the red frequency used. A single red photon will not give a photoelectron while two red photons will and so two red photons can be observed simultaneously. The photoelectric effect will be proportional to $\hat{A}_{\text{red}}^{(+)2}$. Of course, a blue photon will also cause a photoelectric transition and these two processes interfere with one another, giving the odd-order correlation function

$$\Gamma_{\mathcal{N}}^{(1,2)} = \mathrm{Tr}\,\{\hat{\varrho}\hat{A}_{\text{blue}}^{(-)}\hat{A}_{\text{red}}^{(+)}\hat{A}_{\text{red}}^{(+)}\}. \tag{3.78}$$

Another possibility of measuring $\Gamma_{\mathcal{N}}^{(1,2)}$ (or $\Gamma_{\mathcal{N}}^{(2,1)}$) is to use two frequency converters C_1 and C_2 (Fig. 3.3) formed from nonlinear dielectrics. The first converter C_1 is illuminated by red photons, some of which are converted to blue photons and some of which remain red. Some of the remaining red photons are converted to blue ones by the second converter C_2 and the remaining red photons are filtered by a filter F.

The detector D detects the function $\Gamma_{\mathcal{N}}^{(1,1)}$. Denoting the detected field by $\hat{A}_{\text{blue}}^{(+)}$ we have

$$\hat{A}_{\text{blue}}^{(+)} = \hat{B}_{\text{blue}}^{(+)} + \hat{B}_{\text{red}}^{(+)2}. \tag{3.79}$$

The photodetector measures the correlation function

$$\begin{aligned}\operatorname{Tr}\{\hat{\varrho}\hat{A}_{\text{blue}}^{(-)}\hat{A}_{\text{blue}}^{(+)}\} &= \operatorname{Tr}\{\hat{\varrho}(\hat{B}_{\text{blue}}^{(-)} + \hat{B}_{\text{red}}^{(-)2})(\hat{B}_{\text{blue}}^{(+)} + \hat{B}_{\text{red}}^{(+)2})\} = \\ &= \operatorname{Tr}\{\hat{\varrho}\hat{B}_{\text{blue}}^{(-)}\hat{B}_{\text{blue}}^{(+)}\} + \operatorname{Tr}\{\hat{\varrho}\hat{B}_{\text{red}}^{(-)2}\hat{B}_{\text{blue}}^{(+)}\} + \\ &+ \operatorname{Tr}\{\hat{\varrho}\hat{B}_{\text{blue}}^{(-)}\hat{B}_{\text{red}}^{(+)2}\} + \operatorname{Tr}\{\hat{\varrho}\hat{B}_{\text{red}}^{(-)2}\hat{B}_{\text{red}}^{(+)2}\},\end{aligned} \tag{3.80}$$

which involves also the correlation functions $\Gamma_{\mathcal{N}}^{(1,2)}$ and $\Gamma_{\mathcal{N}}^{(2,1)}$.

However, in usual experiments with stationary fields the correlation functions with $m \neq n$ are practically zero [Mehta and Mandel (1967), Mandel (1964a)] and only in the circumstances of nonlinear optics is $\Gamma_{\mathcal{N}}^{(m,n)} \neq 0$ for $m \neq n$.

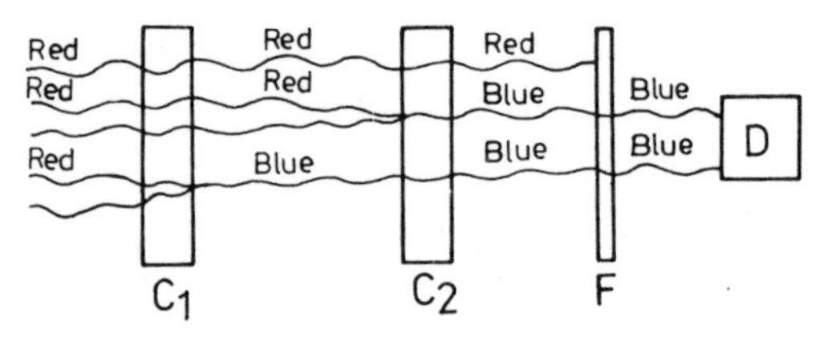

Fig. 3.3 — A scheme for measuring $\Gamma_{\mathcal{N}}^{(1,2)}$ and $\Gamma_{\mathcal{N}}^{(2,1)}$; C_1 and C_2 are nonlinear frequency converters, F is a filter transmitting only the second harmonic wave and D is a detector.

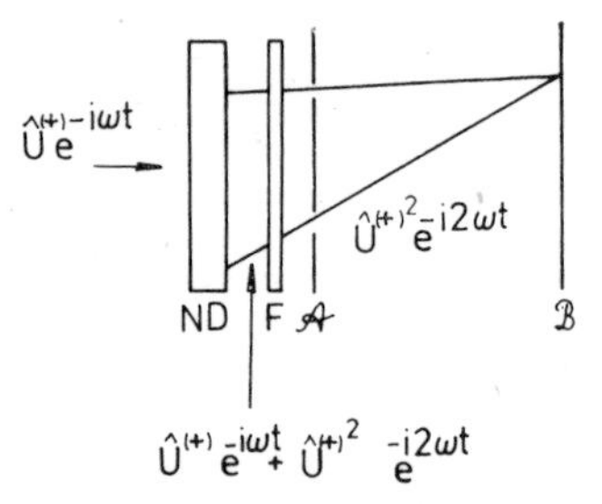

Fig. 3.4 — An outline of combination of the nonlinear dielectric ND with the Young's arrangement for measuring $\Gamma_{\mathcal{N}}^{(2,2)}$; F is a filter transmitting only the second harmonic radiation.

Using nonlinear dielectrics we can also measure the fourth-order correlation function $\Gamma_{\mathcal{N}}^{(2,2)}$ (an analogue of the Hanbury Brown–Twiss intensity correlation measurements, Sec. 3.11), as suggested by Beran, De Velis and Parrent (1967). In Fig. 3.4 we see a scheme of the experimental arrangement, where a nonlinear dielectric ND is combined with the Young two-slit arrangement. If the incident field has the complex amplitude $\hat{U}^{(+)}(t)\exp(-i\omega t)$, then, according to (3.77), the outgoing field has the frequency components ω and 2ω. If the fundamental frequency component is filtered by the filter F, then one can observe interference fringes, described by the fourth-order correlation function $\Gamma_{\mathcal{N}}^{(2,2)} = \operatorname{Tr}\{\hat{\varrho}\hat{U}^{(-)2}\hat{U}^{(+)2}\}$, on the screen $\mathscr{B}$.

3.7 Photocount distribution and photocount statistics

We can start our treatment of photocount statistics by a simple semiclassical stochastic model of the photodetection process, considering the act of absorption of a photon as a quantum process and the radiation field as a classical system. Applying the perturbation theory of quantum mechanics, Mandel, Sudarshan and Wolf (1964) arrived at the probability $p(t)\,dt$ of emission (absorption) of a photoelectron (a photon) within the time interval $(t, t + dt)$ caused by quasi-monochromatic light of the classical intensity $I(t)$ which is incident on the photocathode,

$$p(t)\,dt = \eta I(t)\,dt = \eta V^*(t)\,V(t)\,dt, \tag{3.81}$$

where η is the photoefficiency of the photodetector and linearly polarized light is assumed. This is in correspondence with the equation (3.1) or (3.2) (for incident coherent radiation). Consequently, when we obtain the photocount distribution in classical stochastic terms, it will be easy to obtain it in quantum terms by adopting the normal ordering.

We can determine the probability $p(n, T, t)$ of emission of n photoelectrons, or n photoelectron counts (the probability of detection of n photons) by the plane photocathode on which light is normally incident for a time interval $(t, t + T)$, where T is a fixed time. First we assume that the light intensity $I(t)$ is deterministic and has no fluctuations. Following Mandel (1963a), we divide the interval $(t, t + T)$ into sub-intervals with points $t + i\Delta T = t_i$, $i = 0, 1, \ldots, T/\Delta T$ and we can calculate the probability $p(n, T, t)$ of n counts occurring in the interval $(t, t + T)$. This is the sum over all possible sequences of counts of the product of probabilities of obtaining a count at time t_{r_1}, a count at t_{r_2}, ..., and a count at t_{r_n}, multiplied by probabilities of obtaining no counts in the remaining $T/\Delta T - n$ intervals,

$$\begin{aligned} p(n, T, t) &= \lim_{\Delta T \to 0} \sum_{r_1=0}^{T/\Delta T} \ldots \sum_{r_n=0}^{T/\Delta T} \frac{1}{n!} \eta^n I(t_{r_1}) \ldots I(t_{r_n}) (\Delta T)^n \times \\ &\quad \times \frac{\prod_{i=0}^{T/\Delta T} [1 - \eta I(t_i)\,\Delta T]}{\prod_{j=1}^{n} [1 - \eta I(t_{r_j})\,\Delta T]} = \\ &= \frac{1}{n!} \Big[\eta \int_t^{t+T} I(t')\,dt'\Big]^n \exp\Big[-\eta \int_t^{t+T} I(t')\,dt'\Big], \end{aligned} \tag{3.82}$$

which is a Poisson distribution for the number of counts n with the mean value of counts $\langle n \rangle = \eta \int_t^{t+T} I(t')\,dt'$.

If the intensity $I(t)$ is a stochastic function, (3.82) must be averaged over all possible realizations of $I(t)$ and we arrive at the photocount distribution

$$p(n, T, t) = \frac{1}{n!}\langle(\eta W)^n \exp(-\eta W)\rangle =$$
$$= \frac{1}{n!}\int_0^\infty (\eta W)^n \exp(-\eta W)\, P(W)\, \mathrm{d}W, \tag{3.83}$$

where

$$W = \int_t^{t+T} I(t')\, \mathrm{d}t' \tag{3.84}$$

and $P(W)$ is the probability distribution of the integrated intensity W. The photodetection equation (3.83) was first derived by Mandel (1958, 1959, 1963a).

For stationary fields $p(n, T, t)$ is independent of t and $p(n, T, t) \equiv p(n, T)$.

In general the probability distribution of counts will depart from the classical Poisson distribution, but an ideal laser source, perfectly stabilized so that the intensity does not fluctuate, leads to the Poisson distribution of counts and will behave like a source of classical particles.

Now we can generalize this classical approach to the quantum one. Making use of the relation between the quantum normal correlation functions (3.5a) and the classical ones given by (3.5b), established by the coherent states, we can write the quantum photodetection equation in the form

$$p(n, T, t) = \mathrm{Tr}\left\{\hat{\varrho}\mathcal{N}\frac{(\eta\hat{W})^n}{n!}\exp(-\eta\hat{W})\right\} =$$
$$= \int_0^\infty \frac{(\eta W)^n}{n!}\exp(-\eta W)\, P_{\mathcal{N}}(W)\, \mathrm{d}W, \tag{3.85}$$

which is just the Fourier transform of the relation (2.51e) between the photon-number and normal characteristic functions. In (3.85) $\mathcal{N}$ represents again the normal ordering operator (without the use of the commutation rules), $P_{\mathcal{N}}(\mathrm{W})$ is the probability distribution related to normal ordering, and the operator of the integrated intensity is

$$\hat{W} = \int_t^{t+T} \hat{A}^{(-)}(t)\hat{A}^{(+)}(t)\, \mathrm{d}t. \tag{3.86}$$

More generally, for interacting fields, $\mathcal{N}$ also includes the time-ordering operator [Lax (1968a), Lax and Zwanziger (1973)]. In fact the photodetection equation (3.85) is a consequence of the definition of the number operator and the relation of the normal correlation functions to the photodetection process.

Following Glauber (1965, 1966b, 1972) we can rederive the quantum photodetection equation (3.85) in the following way. Assume that the sensitive element of a photodetector consists of many atoms, say N, any of which may undergo a photodetection process and give rise to a detected photon count. Let $\hat{n}(T) = \sum_{j=1}^{N} \hat{n}_j(T)$ be the number of counts in the time interval $(0, T)$, where $\hat{n}_j(T)$ represent the contributions of the individual atoms. The eigenvalues of $\hat{n}_j$ are one or zero; one for final states to which the jth atom has contributed and zero for states to which it has not contributed. Therefore we can write for the generating function [cf. (3.64)]

$$(1+\mathrm{i}s)^{\hat{n}(T)} = (1+\mathrm{i}s)^{\sum_{j=1}^{N}\hat{n}_j(T)} = \prod_{j=1}^{N}(1+\mathrm{i}s)^{\hat{n}_j(T)} = \prod_{j=1}^{N}[1+\mathrm{i}s\hat{n}_j(T)]. \quad (3.87)$$

Expanding (3.87), expressing $\langle \hat{n}_1(T) \ldots \hat{n}_m(T)\rangle = p^{(m)}$ from (3.13) and approximating the combination number $N!/[(N-m)!\,m!]$, appearing in the expansion of (3.87), by $1/m!$ (N is large so that $p^{(m)}$ are non zero only for $m \ll N$), we obtain

$$\langle (1+\mathrm{i}s)^{\hat{n}(T)}\rangle = 1 + \sum_{m=1}^{\infty} \frac{(\mathrm{i}s)^m}{m!} F_m, \quad (3.88)$$

where

$$F_m = \int_0^T \cdots \int_0^T \underbrace{\int \cdots \int}_{\text{Volume of detector}} \Gamma_{\mathcal{N}}^{(m,m)}(x'_1, \ldots, x'_m, x''_m, \ldots, x''_1) \times \quad (3.89)$$

$$\times \prod_{j=1}^{m} \mathscr{S}(x'_j - x''_j)\, \mathrm{d}^4x'_j\, \mathrm{d}^4x''_j.$$

Using the definitions (3.5a, b) of the correlation functions we arrive at

$$\langle (1+\mathrm{i}s)^{\hat{n}(T)}\rangle = \mathrm{Tr}\,\{\hat{\varrho}_{\mathcal{N}} \exp(\mathrm{i}s\hat{W}_\eta)\} = \langle \exp(\mathrm{i}sW_\eta)\rangle_{\mathcal{N}}, \quad (3.90)$$

where

$$\hat{W}_\eta = \iint_0^T \underbrace{\iint}_{\text{Volume of detector}} \mathscr{S}(x'-x'')\,\hat{A}^{(-)}(x')\,\hat{A}^{(+)}(x'')\,\mathrm{d}^4x'\,\mathrm{d}^4x'' \quad (3.91a)$$

and

$$W_\eta = \iint_0^T \underbrace{\iint}_{\text{Volume of detector}} \mathscr{S}(x'-x'')\,V^*(x')\,V(x'')\,\mathrm{d}^4x'\,\mathrm{d}^4x'', \quad (3.91b)$$

which is the eigenvalue of (3.91a) in the coherent state. Equation (3.90) corresponds to equation (2.51d). In general, the integrals also imply the summation over the polarization indices. Equations (3.91a, b) are generalizations of the expressions (3.86) and (3.84), which are obtained, apart from η, for broad-band detectors and plane quasi-monochromatic waves normally incident in the z-direction on a plane photocathode of area S, as $\mathscr{S}(x'-x'') = (\eta/ScT)\,\delta(x''-x')$ and

$$W_\eta = \frac{\eta}{ScT} \int_0^T \mathrm{d}t' \int_0^{cT} \mathrm{d}z' \int_S I(x')\,\mathrm{d}^2x' = \eta \int_0^T I(t')\,\mathrm{d}t' = \eta W, \quad (3.91c)$$

where $I(x) = \mathbf{V}^*(x)\,.\,\mathbf{V}(x) = \sum_j V_j^*(x)\,V_j(x)$ is explicitly independent of x, y, z.

Thus the photocount distribution (3.85) for the detection interval $(0, T)$ is obtained from (3.90) as

$$p(n,T) = \frac{1}{n!}\frac{\mathrm{d}^n}{\mathrm{d}(\mathrm{i}s)^n}\langle (1+\mathrm{i}s)^{\hat{n}}\rangle\bigg|_{\mathrm{i}s=-1} = \frac{1}{2\pi}\oint \frac{\langle \exp(\mathrm{i}sn)\rangle_{\mathcal{N}}}{(1+\mathrm{i}s)^{n+1}}\,\mathrm{d}s =$$

$$= \mathrm{Tr}\left\{\hat{\varrho}_{\mathcal{N}} \frac{\hat{W}_\eta^n}{n!}\exp(-\hat{W}_\eta)\right\} = \int_0^\infty \frac{W_\eta^n}{n!}\exp(-W_\eta)\,P_{\mathcal{N}}(W_\eta)\,\mathrm{d}W_\eta. \quad (3.92)$$

The factorial moments of $p(n, T)$ are defined as

$$\left\langle \frac{\hat{n}!}{(\hat{n}-k)!} \right\rangle = \frac{d^k}{d(is)^k} \langle (1+is)^{\hat{n}} \rangle \bigg|_{is=0} = \int_0^T \ldots \int_0^T \int_{\substack{\text{Volume} \\ \text{of detector}}} \ldots \int \prod_{j=1}^{k} \mathscr{S}(x'_j - x''_j) \times$$

$$\times \Gamma_{\mathscr{N}}^{(k,k)}(x'_1, \ldots, x'_k, x''_k, \ldots, x''_1) \prod_{l=1}^{k} d^4x'_l \, d^4x''_l, \tag{3.93a}$$

or, in correspondence with (3.91c),

$$\left\langle \frac{\hat{n}!}{(\hat{n}-k)!} \right\rangle = \eta^k \int_0^T \ldots \int_0^T \Gamma_{\mathscr{N}}^{(k,k)}(t'_1, \ldots, t'_k, t'_k, \ldots, t'_1) \prod_{j=1}^{k} dt'_j \tag{3.93b}$$

[in close analogy to (3.11b)].

It can be verified that the photocount distribution $p(n, T)$, given by (3.85), and the photon-number probability distribution $p(n) \equiv \varrho(n, n)$ are related by the Bernoulli distribution

$$p(n, T) = \sum_{m=n}^{\infty} \binom{m}{n} (\eta T)^n (1 - \eta T)^{m-n} \, p(m). \tag{3.94}$$

For closed systems [Mandel (1981d)] and in the one-mode case, $\eta W(t) = \eta \, |\alpha|^2 \, T$ in the photodetection equation ($I = |\alpha|^2$) is replaced by $(1 - \exp(-\eta T)) \, |\alpha|^2$ [Mollow (1968a), Scully and Lamb (1969), Arecchi and Degiorgio (1972), Sargent et al. (1974), Selloni et al. (1978a), Shepherd (1981), Srinivas and Davies (1981)].

We note that also the photocount statistics of multiphoton absorption processes have been investigated. The perturbation theory of quantum mechanics leads to the result [Lambropoulos et al. (1966), Carusotto et al. (1967, 1968)] that the probability of the transition for k-quantum absorption is proportional to I^k, i.e. all the results concerning the statistics of the photodetection are correct if $I(t)$ in the corresponding expressions is replaced by $I^k(t)$. The relation between the two-quantum photocount distributions and intensity fluctuations and the comparison with the one-photon case have been discussed by Teich and Diament (1969). Further investigations of the statistics of the two- and multiphoton absorption processes have been performed by a number of authors [Teich and Wolga (1966), Lambropoulos (1968), Mollow (1968b), Diament and Teich (1969), Millet and Varnier (1969), Jaiswal and Agarwal (1969), Agarwal (1970), Barashev (1970a, b, 1976), Millet and Usselio-La-Verna (1970), Tunkin and Tchirkin (1970), Peřina, Peřinová and Mišta (1971), Mišta and Peřina (1971), Mišta (1971), Delone and Masalov (1980)]. Some experimental results have been obtained by Shiga and Inamura (1967), Clark, Estes and Narducci (1970), Lyons and Troup (1970b) and Teich, Abrams and Gandrud (1970), particularly in relation to the dependence of the two-photon absorption process on the statistics of radiation (the k-photon process is $k!$ more effective for chaotic radiation than for coherent radiation).

We deal now with some properties of the photocount distribution. First we calculate the variance of the number of counts n. From (3.83)

$$\langle \hat{n} \rangle = \langle n \rangle = \sum_{n=0}^{\infty} p(n, T, t)\, n = \eta \langle W \rangle_{\mathcal{N}} = \eta \langle \hat{W} \rangle, \tag{3.95a}$$

$$\langle \hat{n}^2 \rangle = \langle n^2 \rangle = \sum_{n=0}^{\infty} p(n, T, t)\, n^2 = \eta \langle W \rangle_{\mathcal{N}} + \eta^2 \langle W^2 \rangle_{\mathcal{N}} = \\ = \eta \langle \hat{W} \rangle + \eta^2 \langle \mathcal{N} \hat{W}^2 \rangle. \tag{3.95b}$$

In the same way we obtain

$$\langle \hat{n}^3 \rangle = \eta \langle \hat{W} \rangle + 3\eta^2 \langle \mathcal{N} \hat{W}^2 \rangle + \eta^3 \langle \mathcal{N} \hat{W}^3 \rangle, \tag{3.95c}$$

$$\langle \hat{n}^4 \rangle = \eta \langle \hat{W} \rangle + 7\eta^2 \langle \mathcal{N} \hat{W}^2 \rangle + 6\eta^3 \langle \mathcal{N} \hat{W}^3 \rangle + \eta^4 \langle \mathcal{N} \hat{W}^4 \rangle, \tag{3.95d}$$

etc. The variance of n is obtained from (3.95a, b):

$$\langle (\Delta n)^2 \rangle = \langle n^2 \rangle - \langle n \rangle^2 = \eta \langle W \rangle_{\mathcal{N}} + \eta^2 \langle (\Delta W)^2 \rangle_{\mathcal{N}} = \\ = \langle n \rangle + \eta^2 \langle (\Delta W)^2 \rangle_{\mathcal{N}}, \tag{3.96}$$

where

$$\langle (\Delta W)^2 \rangle_{\mathcal{N}} = \langle W^2 \rangle_{\mathcal{N}} - \langle W \rangle_{\mathcal{N}}^2 \tag{3.97}$$

is the normal variance of the integrated intensity W. The formula (3.96) has a very simple physical interpretation. It shows that the variance of fluctuations in the number of ejected photoelectrons is the sum of the fluctuations in the number of classical particles obeying the Poisson distribution (the term $\langle n \rangle$) and of the fluctuations in a classical wave field (the wave interference term $\eta^2 \langle (\Delta W)^2 \rangle_{\mathcal{N}}$). This result generalizes the earlier results by Einstein (1909), Bothe (1927) and Fürth (1928a, b) and is correct for any light beam. Although the result refers to the fluctuations of photoelectric counts, it may be regarded as reflecting the fluctuation properties of the light itself.

The relations (3.95) between the photoelectron number moments and the normal integrated intensity moments are a direct consequence of the definitions of the number operator and the normally ordered moments. For instance, in the one-mode case we have $\langle \hat{n}^2 \rangle = \langle \hat{a}^+ \hat{a} \hat{a}^+ \hat{a} \rangle = \langle \hat{a}^{+2} \hat{a}^2 \rangle + \langle \hat{a}^+ \hat{a} \rangle = \langle W^2 \rangle_{\mathcal{N}} + \langle W \rangle_{\mathcal{N}}$; for general fields the same follows from (2.51a, b) or (2.51c). From this point of view the term $\langle n \rangle = \langle W \rangle_{\mathcal{N}}$ $(\eta = 1)$ in (3.96) is the quantum term arising from the commutator. Associating the number operator $\hat{m}$ with W we obtain the quantum analogue of the Burgess variance theorem $\langle (\Delta n)^2 \rangle = \eta(1 - \eta) \langle m \rangle + \eta^2 \langle (\Delta m)^2 \rangle$ or $\langle (\Delta n)^2 \rangle / \langle n \rangle - 1 = \eta(\langle (\Delta m)^2 \rangle / \langle m \rangle - 1)$ [Peřina, Saleh and Teich (1983)]. If m-particles are sub-Poissonian ($\langle (\Delta m)^2 \rangle < \langle m \rangle$), also n-particles are sub-Poissonian [Teich and Saleh (1982)]. This is a basis for the proposal to use the Franck–Hertz experiment to produce sub-Poissonian light (Sec. 3.10) [Teich, Saleh and Stoler (1983), Teich, Saleh and Peřina (1983)].

The photocount characteristic function is obtained from the photodetection equation (3.85),

$$\langle \exp(\mathrm{i}s\hat{n}) \rangle = \sum_{n=0}^{\infty} p(n, T, t) \exp(\mathrm{i}sn) = \int_0^{\infty} P_{\mathcal{N}}(W) \exp[\eta(\mathrm{e}^{\mathrm{i}s} - 1) W]\, \mathrm{d}W = \\ = \langle \exp[\eta(\mathrm{e}^{\mathrm{i}s} - 1) W] \rangle_{\mathcal{N}}, \tag{3.98}$$

in agreement with (2.51e); the function $\langle \exp(\mathrm{i}sW) \rangle_{\mathscr{N}}$ is the normal characteristic function of the integrated intensity probability distribution $P_{\mathscr{N}}(W)$ related to the normal ordering. The photocount distribution $p(n, T, t)$ may be calculated from the characteristic function $\langle \exp(\mathrm{i}s\hat{n}) \rangle$ by means of a Fourier transform,

$$p(n, T, t) = \frac{1}{2\pi} \int_0^{2\pi} \exp(-\mathrm{i}sn) \langle \exp(\mathrm{i}s\hat{n}) \rangle \, \mathrm{d}s. \tag{3.99}$$

The moments $\langle \hat{n}^k \rangle$, $k = 0, 1, \ldots$, may be calculated as

$$\langle \hat{n}^k \rangle = \frac{\mathrm{d}^k}{\mathrm{d}(\mathrm{i}s)^k} \langle \exp(\mathrm{i}s\hat{n}) \rangle \bigg|_{\mathrm{i}s=0}. \tag{3.100}$$

Substituting $\exp(\mathrm{i}s) - 1 \to \mathrm{i}s$ in (3.98) we obtain for the normal characteristic function

$$\langle \exp(\mathrm{i}s\eta W) \rangle_{\mathscr{N}} = \langle (1 + \mathrm{i}s)^n \rangle, \tag{3.101}$$

in agreement with (3.90), (3.64) and (2.51d). Hence $p(n, T, t)$ is obtained by means of the derivatives

$$p(n, T, t) = \frac{1}{n!} \frac{\mathrm{d}^n}{\mathrm{d}(\mathrm{i}s)^n} \langle \exp(\mathrm{i}s\eta W) \rangle_{\mathscr{N}} \bigg|_{\mathrm{i}s=-1}, \tag{3.102}$$

in agreement with (3.65) and (3.92). The factorial moments (3.93a, b) are equal to

$$\left\langle \frac{n!}{(n-k)!} \right\rangle = \frac{\mathrm{d}^k}{\mathrm{d}(\mathrm{i}s)^k} \langle (1 + \mathrm{i}s)^n \rangle \bigg|_{\mathrm{i}s=0} =$$

$$= \frac{\mathrm{d}^k}{\mathrm{d}(\mathrm{i}s)^k} \langle \exp(\mathrm{i}s\eta W) \rangle_{\mathscr{N}} \bigg|_{\mathrm{i}s=0} = \eta^k \langle W^k \rangle_{\mathscr{N}}. \tag{3.103}$$

One can define the cumulants $\varkappa_j$ as

$$\varkappa_j^{(n)} = \frac{\mathrm{d}^j}{\mathrm{d}(\mathrm{i}s)^j} \log \langle \exp(\mathrm{i}s\hat{n}) \rangle \bigg|_{\mathrm{i}s=0}, \qquad j = 1, 2, \ldots, \tag{3.104}$$

($\varkappa_0^{(n)} = 0$) so that the characteristic function can be written as

$$\langle \exp(\mathrm{i}s\hat{n}) \rangle = \exp\left[\sum_{j=1}^{\infty} \varkappa_j^{(n)} \frac{(\mathrm{i}s)^j}{j!} \right]. \tag{3.105}$$

In the same way the factorial cumulants $\varkappa_j^{(W)}$ are defined by means of $\langle \exp(\mathrm{i}s\eta W) \rangle_{\mathscr{N}}$. The relation between the cumulants $\varkappa_j^{(n)}$ and $\varkappa_k^{(W)}$ is obtained with the help of (3.98)

$$\sum_{j=1}^{\infty} \varkappa_j^{(n)} \frac{(\mathrm{i}s)^j}{j!} = \sum_{j=1}^{\infty} \varkappa_j^{(W)} \frac{\eta^j (\mathrm{e}^{\mathrm{i}s} - 1)^j}{j!}. \tag{3.106}$$

Comparing coefficients of the powers $(\mathrm{i}s)^j$ in the expansion of (3.106) we arrive at the same relations for the cumulants as for the moments, given in (3.95) [this may be seen by expanding the exponential functions in (3.98), which generates for $\langle \hat{n}^j \rangle$ and

$\langle W^k \rangle_{\mathcal{N}}$ the same relation as (3.106)],

$$\begin{aligned} \varkappa_1^{(n)} &= \eta \varkappa_1^{(W)}, \\ \varkappa_2^{(n)} &= \eta \varkappa_1^{(W)} + \eta^2 \varkappa_2^{(W)}, \end{aligned} \tag{3.107}$$

etc. The significance of the cumulants is as follows:

$$\varkappa_1^{(n)} = \langle \hat{n} \rangle, \tag{3.108a}$$

$$\varkappa_2^{(n)} = \langle \hat{n}^2 \rangle - \langle \hat{n} \rangle^2 = \langle (\Delta \hat{n})^2 \rangle, \tag{3.108b}$$

$$\varkappa_3^{(n)} = \langle \hat{n}^3 \rangle - 3\langle \hat{n}^2 \rangle \langle \hat{n} \rangle + 2\langle \hat{n} \rangle^3, \tag{3.108c}$$

etc. The same relations hold for $\varkappa_j^{(W)}$.

From (3.106) we obtain the general relation

$$\varkappa_k^{(n)} = \sum_{j=1}^{k} \varkappa_j^{(W)} \eta^j \sum_{l=0}^{j} \frac{(-1)^{j+l} l^k}{l!(j-l)!}, \tag{3.109}$$

where the identity $\sum_{l=0}^{j} (-1)^l l^k / l!\,(j-l)! = 0$ for $j > k \geqq 0$ has been used.

When the radiation is very weak, the first term in (3.107) or (3.95) is dominant and the distribution $p(n, T, t)$ becomes Poissonian. On the other hand, in strong fields, the last term is dominant and the probability distribution of n tends towards the probability distribution of ηW (the output of the photodetector can be regarded as a continuous signal).

Further discussion of cumulants can be found in a paper by Cantrell (1970) and the monograph by Saleh (1978).

Measurements of the photocount distribution can be realized in the following way. Assuming a stationary field, we compute the number of emitted photoelectrons within a time interval of length T in many realizations and determine the number of intervals for the same n; the normalization then provides the photocount distribution $p(n, T)$. The measurement may be performed using a capacitor which is charged by the emitted photoelectrons, and whose voltage is proportional to the number of photoelectrons emitted within the interval T. Or a multichannel analyzer may be used automatically registering the number of intervals with the same n, after standardization of the photoelectron pulses [see e.g. Arecchi and Degiorgio (1972)].

3.8 Determination of the integrated intensity probability distribution from the photocount distribution

Knowing $p(n, T)$ we can calculate the moments $\langle \hat{n}^k \rangle$ as well as the factorial moments $\langle \hat{n}!/(\hat{n} - k)! \rangle$; hence the statistics of the emitted photoelectrons can be determined. The problem now arises of how information about the statistics of the field [i.e. about the distribution $P_{\mathcal{N}}(W)$] can be obtained from the photocount distribution $p(n, T) \equiv \equiv p(n)$. This implies inverting the photodetection equation (3.85).

Denoting $n!\,p(n)$ by M_n and $P_{\mathcal{N}}(W) \exp(-W)$ by $Q(W)$ in (3.85) (we also put $\eta = 1$ for simplicity), we find that the inversion problem reduces to the moment problem, i.e. to determine the function $Q(W)$ if the moment sequence

$$\int_0^\infty Q(W)\, W^k \,\mathrm{d}W = \langle W^k \rangle = M_k, \qquad k = 0, 1, \ldots \tag{3.110}$$

is given.

One can easily see that a formal solution of (3.85) is

$$P_{\mathcal{N}}(W) = \exp(W) \sum_{n=0}^{\infty} (-1)^n\, p(n)\, \delta^{(n)}(W), \tag{3.111}$$

where $\delta^{(n)}(W)$ is the nth derivative of the Dirac δ-function.

Although such a mathematical quantity with an infinite number of terms may be meaningless sometimes, it is well defined here because of the analyticity of the characteristic function

$$\langle \exp(\mathrm{i}sW) \rangle_{\mathcal{N}} = \int_0^\infty P_{\mathcal{N}}(W) \exp(\mathrm{i}sW)\,\mathrm{d}W; \tag{3.112}$$

it is analytic in the upper half of the complex s-plane, since $P_{\mathcal{N}}(W) = 0$ for $W < 0$. Expressing the $\delta^{(n)}$-function in (3.111) by means of the Fourier integral,

$$\delta^{(n)}(W) = \frac{1}{2\pi} \int_{-\infty}^{+\infty} (-\mathrm{i}s)^n \exp(-\mathrm{i}sW)\,\mathrm{d}s, \tag{3.113}$$

we obtain from (3.111) [Wolf and Mehta (1964)]

$$P_{\mathcal{N}}(W) = \frac{1}{2\pi} \exp(W) \int_{-\infty}^{+\infty} C(\mathrm{i}s) \exp(-\mathrm{i}sW)\,\mathrm{d}s, \tag{3.114}$$

where

$$C(\mathrm{i}s) = \sum_{n=0}^{\infty} (\mathrm{i}s)^n\, p(n), \tag{3.115}$$

which is the characteristic function of $P_{\mathcal{N}}(W) \exp(-W)$. From (3.112) and (3.114) $C(\mathrm{i}s) = \langle \exp[(\mathrm{i}s - 1)\, W] \rangle_{\mathcal{N}}$. The analyticity of $C(\mathrm{i}s)$ makes it possible to obtain the values of $C(\mathrm{i}s)$ for all s by analytic continuation if the series (3.115) has a finite radius of convergence. Thus $P_{\mathcal{N}}(W)$, determined in this way, is unique.

An alternative solution of the moment problem can be obtained in terms of the Laguerre polynomials [Bédard (1967a), Piovoso and Bolgiano (1967), Morse and Feshbach (1953), Klauder and Sudarshan (1968), Peřina and Mišta (1968a, 1969)], which may be used to construct $P_{\mathcal{N}}(W)$ when a finite number of $p(n)$ are obtained from an experiment.

Assuming that $P_{\mathcal{N}}(W)$ is square integrable, we may look for a solution of the inverse problem in the form

$$P_{\mathcal{N}}(W) = \exp[-(\zeta - 1)\, W] \sum_{j=0}^{\infty} c_j L_j^0(\zeta W), \tag{3.116}$$

where L_j^0 are the Laguerre polynomials defined by [Morse and Feshbach (1953), Chapter 6]

$$L_j^k(x) = [\Gamma(j+k+1)]^2 \sum_{s=0}^{j} \frac{(-x)^s}{s!(j-s)!\,\Gamma(s+k+1)}, \tag{3.117}$$

obeying the orthogonality condition

$$\int_0^\infty x^k \exp(-x)\, L_j^k(x)\, L_l^k(x)\, dx = \delta_{jl} \frac{[\Gamma(j+k+1)]^3}{\Gamma(j+1)}; \tag{3.118}$$

$\zeta \geqq 1$ is a real number and the gamma function is $\Gamma(k) = \int_0^\infty x^{k-1} \exp(-x)\, dx$. Multiplying (3.116) by $L_k^0(\zeta W) \exp(-W)$, integrating over ζW and using (3.118) with $k = 0$, we obtain

$$c_k = \frac{\zeta}{(k!)^2} \int_0^\infty P_{\mathcal{N}}(W) \exp(-W)\, L_k^0(\zeta W)\, dW =$$

$$= \zeta \sum_{s=0}^{k} p(s) \frac{(-\zeta)^s}{(k-s)!\, s!}, \tag{3.119}$$

where (3.117) and (3.85) have also been used. The accuracy of an approximation in which $p(n)$ is obtained as far as the first N terms, from measurements, is given by

$$\int_0^\infty [P_{\mathcal{N}}(W) - P_{\mathcal{N}}^{(N)}(W)]^2 \exp(-W)\, dW = \sum_{n=N+1}^{\infty} (n!)^2 c_n^2 < \varepsilon, \tag{3.120}$$

where $P_{\mathcal{N}}^{(N)}$ denotes a function constructed as the Nth partial sum of (3.116) and ε is an arbitrarily small number.

Sometimes it is more suitable for calculations to use a decomposition showing the explicit dependence on the number of degrees of freedom (modes) M,

$$P_{\mathcal{N}}(W) = W^{M-1} \sum_{j=0}^{\infty} c_j L_j^{M-1}(W), \tag{3.121}$$

where

$$c_j = \frac{j!}{\Gamma(j+M)} \sum_{s=0}^{j} \frac{(-1)^s p(s)}{(j-s)!\, \Gamma(s+M)}. \tag{3.122}$$

We can use some theorems from the moment theory [Akhiezer (1970), Berezansky (1968)] to make clear the conditions under which the function (3.116) represents either an ordinary non-negative function or a generalized function with the support composed of a finite number of points [Peřina and Mišta (1969)].

We define the moment sequence

$$M_n = \int_0^\infty W^n\, dF(W), \qquad n = 0, 1, 2, \ldots, \tag{3.123}$$

where $dF(W) = F'(W)\, dW = P_{\mathcal{N}}(W)\, dW$ [in the same way one can consider the sequence $\{p(n)\}$ denoting $M_n = p(n)\, n!$ and $F'(W) = \exp(-W) P_{\mathcal{N}}(W)$]. We introduce the following system of quadratic forms [Akhiezer (1970)]:

$$q_m = \sum_{i,k=0}^{m} M_{i+k} u_i u_k = \int_0^\infty \left(\sum_{i=0}^{m} W^i u_i\right)^2 dF(W), \qquad m = 0, 1, \ldots, \tag{3.124}$$

where $\{u_i\}$ is an arbitrary non-trivial real vector. A necessary and sufficient condition for the measure $F(W)$ to be non-negative is

$$q_m \geqq 0 \tag{3.125}$$

for all m and an arbitrary non-trivial vector $\{u_i\}$. The positiveness of the quadratic forms q_m is also necessary and sufficient for the existence of a non-decreasing function $F(W)$ $(0 \leqq W < \infty)$ satisfying (3.123) and having an infinite number of points at which the function increases [Akhiezer (1970)]. Thus if $q_m > 0$ for every m and for an arbitrary non-trivial $\{u_i\}$, it is possible to construct the distribution $F'(W) = P_{\mathcal{N}}(W)$ from the given sequence M_n in a unique way. Another case occurs if for some m and non-trivial u_i the quadratic forms (3.124) equal zero. It can be shown [Brezansky (1968)] in this case that there exists a unique distribution $P_{\mathcal{N}}(W)$ the support of which is composed of a finite number of points equal to the minimal number of the m's for which $q_m = 0$.

These conclusions are illustrated by the following examples. Let us consider system of determinants

$$\mathscr{D}_m = \begin{vmatrix} M_0 & M_1 & \dots & M_m \\ M_1 & M_2 & \dots & M_{m+1} \\ \dots & \dots & \dots & \dots \\ M_m & M_{m+1} & \dots & M_{2m} \end{vmatrix}, \qquad m = 0, 1, \dots . \tag{3.126}$$

If $\mathscr{D}_m > 0$, then $q_m > 0$ $(m = 0, 1, \dots)$. These conditions can easily be shown to be valid for chaotic light, for which $M_n = n!\,\langle n \rangle^n$, where $\langle n \rangle$ is the mean number of photons in the field. Thus the function $P_{\mathcal{N}}(W)$ must exist as a unique ordinary function; indeed $P_{\mathcal{N}}(W) = \langle n \rangle^{-1} \exp(-W/\langle n \rangle)$ (see next section). For coherent light $M_n = \langle n \rangle^n$ and so $\mathscr{D}_m = 0$ for $m \geqq 1$ and $\mathscr{D}_0 = 1$. Thus $q_0 > 0$ and $q_m = 0$ for $m \geqq 1$. Therefore $P_{\mathcal{N}}(W)$ is a generalized function having one-point support, $P_{\mathcal{N}}(W) = \delta(W - \langle n \rangle)$ (see next section).

3.9 Short-time measurements

We begin with a simple description of the chaotic field which is the most usual state of the field in nature (more detailed discussion will be given in Chapter 5). Such a field is generated by a thermal source composed of many independent atomic radiators and consists of superpositions of waves of many different frequencies lying within some continuous range. These elementary waves can be regarded as independent waves with random phases. Using the central limit theorem of mathematical statistics, we can conclude that the field represents a Gaussian random process for the amplitude with zero mean value. This was explicitly proved by Van Cittert (1934, 1939) and Janossy (1957, 1959) by direct calculations. The quantum analogue of the central limit theorem has been discussed by Klauder and Sudarshan (1968). Of course, this cannot be applied to laser radiation. The following random walk considerations have been generalized to an arbitrary number of dimensions by Jakeman (1980).

Consider first a linearly polarized chaotic field. Since Re $V(t)$ is a Gaussian process with zero mean, Im $V(t)$ also is a Gaussian process with zero mean, and they are uncorrelated. Denoting the variances of these processes as $2\sigma^2$, it follows that the joint probability distribution of Re V and Im V is

$$P_{\mathcal{N}}(\text{Re } V, \text{Im } V) = \frac{1}{2\pi\sigma^2} \exp\left[-\frac{(\text{Re } V)^2 + (\text{Im } V)^2}{2\sigma^2} \right]. \tag{3.127}$$

However $(\text{Re } V)^2 + (\text{Im } V)^2$ is the intensity I of the field and (3.127) can be written in the form

$$P_{\mathcal{N}}(I^{1/2}, \arg V) = \frac{1}{2\pi\sigma^2} \exp\left(-\frac{I}{2\sigma^2} \right) I^{1/2}, \tag{3.128}$$

with $\mathrm{d}(\text{Re } V)\,\mathrm{d}(\text{Im } V) = I^{1/2}\,\mathrm{d}I^{1/2}\,\mathrm{d}(\arg V)$. This probability density is independent of the phase of V, i.e. all phase angles in the range $0 \leqq \arg V \leqq 2\pi$ are equally probable. Integrating over the phase and taking into account that $\langle I \rangle_{\mathcal{N}} = 2\sigma^2$, we finally arrive at

$$P_{\mathcal{N}}(I) = \frac{1}{\langle I \rangle_{\mathcal{N}}} \exp\left(-\frac{I}{\langle I \rangle_{\mathcal{N}}} \right), \tag{3.129}$$

which is a negative exponential probability distribution of I. The kth moment is

$$\langle I^k \rangle_{\mathcal{N}} = k! \langle I \rangle_{\mathcal{N}}^k \tag{3.130}$$

and the variance

$$\langle (\Delta I)^2 \rangle_{\mathcal{N}} = \langle I^2 \rangle_{\mathcal{N}} - \langle I \rangle_{\mathcal{N}}^2 = \langle I \rangle_{\mathcal{N}}^2. \tag{3.131}$$

Consider now a resolving time T of the detector much smaller than the coherence time $1/\Delta\nu$. The integrated intensity W is equal to IT since the intensity $I(t)$ may be regarded as practically constant over such a time interval and the photodetection equation becomes

$$p(n, T) = \int_0^\infty \frac{(\eta I T)^n}{n!} \exp(-\eta I T)\, P_{\mathcal{N}}(I)\,\mathrm{d}I. \tag{3.132}$$

Such short-time measurements provide a deeper insight into the physical problem since the intensity has a simple physical meaning whereas the integrated intensity is a more complicated quantity. The distribution $p(n, T)$ for chaotic light can be obtained by substituting (3.129) into (3.132), leading to

$$p(n, T) = \frac{\langle n \rangle^n}{(1 + \langle n \rangle)^{1+n}}, \tag{3.133}$$

where $\langle n \rangle = \eta \langle I \rangle_{\mathcal{N}} T$. This is the well-known Bose–Einstein distribution for n identical particles in one quantum state, or in one cell of the phase space. This can be understood as follows. In the direction of the beam propagation, photons cannot be distinguished in the linear distance cT ($T \approx 1/\Delta\nu = \tau_c$) as a consequence of the uncertainty principle and so they occupy one cell of the phase space, that is they behave like Bose–Einstein particles. The variance (3.96) can be calculated as

$$\begin{aligned}\langle(\Delta n)^2\rangle &= \eta\langle I\rangle_{\mathcal{N}}T + \eta^2\langle(\Delta I)^2\rangle_{\mathcal{N}}T^2 = \\ &= \eta\langle I\rangle_{\mathcal{N}}T + \eta^2\langle I\rangle_{\mathcal{N}}^2T^2 = \langle n\rangle(1 + \langle n\rangle),\end{aligned} \tag{3.134}$$

where (3.131) has been used. This variance exceeds the value of the variance for the Poisson distribution, $\langle n\rangle$.

If the N-photon absorption process is considered, it is sufficient to replace I by I^N in the integrated intensity. Then $\langle I_N^k\rangle_{\mathcal{N}} = \langle I^{kN}\rangle_{\mathcal{N}}$, in terms of the one-photon moment. For chaotic light $\langle I_N^2\rangle_{\mathcal{N}}/\langle I_N\rangle_{\mathcal{N}}^2 = (2N)!/(N!)^2$ and for $N = 2$ this quantity equals $4!/4 = 6$, whereas for the one-photon process it equals 2. These relations have been experimentally verified by Shiga and Inamura (1967) and Teich, Abrams and Gandrud (1970).

The characteristic function is

$$\langle\exp(\mathrm{i}s\eta W)\rangle_{\mathcal{N}} = \int_0^\infty \frac{\exp\left(\mathrm{i}s\eta IT - \dfrac{I}{\langle I\rangle_{\mathcal{N}}}\right)}{\langle I\rangle_{\mathcal{N}}}\,\mathrm{d}I = \frac{1}{1 - \mathrm{i}s\langle n\rangle}. \tag{3.135}$$

These results may be generalized to fields with M degrees of freedom and to partially polarized fields.

Assume $W = \sum_j W_j$ is a stochastic quantity, the W_j (with probability distributions $P_{\mathcal{N},j}(W_j)$) being statistically independent. Then the probability distribution $P_{\mathcal{N}}(W)$ of the resulting quantity W is equal to the convolution of $P_{\mathcal{N},j}(W_j)$, namely

$$P_{\mathcal{N}}(W) = \int_0^\infty \dots \int_0^\infty \delta(W - \sum_j W_j)\prod_j P_{\mathcal{N},j}(W_j)\,\mathrm{d}W_j. \tag{3.136}$$

Consequently the corresponding characteristic function of $P_{\mathcal{N}}(W)$ equals the product of the characteristic functions of $P_{\mathcal{N},j}(W_j)$,

$$\langle\exp(\mathrm{i}sW)\rangle_{\mathcal{N}} = \prod_j \langle\exp(\mathrm{i}sW_j)\rangle_{\mathcal{N}}. \tag{3.137}$$

Applying this result to a system with M independent degrees of freedom with the same mean numbers per mode, $\langle n\rangle/M$, we have

$$\langle\exp(\mathrm{i}s\eta W)\rangle_{\mathcal{N}} = \left(1 - \mathrm{i}s\frac{\langle n\rangle}{M}\right)^{-M}. \tag{3.138}$$

Hence (Mandel (1959), Troup (1965a)), applying the inverse transformation to (3.112), (3.102) and (3.103),

$$P_{\mathcal{N}}(W) = \left(\frac{\eta M}{\langle n\rangle}\right)^M \frac{W^{M-1}}{\Gamma(M)}\exp\left(-\frac{\eta WM}{\langle n\rangle}\right), \tag{3.139}$$

$$p(n, T) = \frac{\Gamma(n + M)}{n!\Gamma(M)}\left(1 + \frac{M}{\langle n \rangle}\right)^{-n}\left(1 + \frac{\langle n \rangle}{M}\right)^{-M}, \tag{3.140}$$

$$\langle W^k \rangle_{\mathcal{N}} = \left(\frac{\langle n \rangle}{\eta}\right)^k \frac{\Gamma(k + M)}{\Gamma(M) M^k}, \tag{3.141a}$$

$$\langle (\Delta n)^2 \rangle = \langle n \rangle \left(1 + \frac{\langle n \rangle}{M}\right). \tag{3.141b}$$

The number $\langle n \rangle / M$ represents the so-called degeneracy parameter.

Consider next partially polarized chaotic light whose degree of polarization

$$P = \left[1 - \frac{4 \operatorname{Det} \hat{J}}{(\operatorname{Tr} \hat{J})^2}\right]^{1/2} = \left[1 - \frac{4 \operatorname{Det} \hat{\sigma}}{(\operatorname{Tr} \hat{\sigma})^2}\right]^{1/2}, \tag{3.142}$$

where $\hat{J} = (J_{ij})$, $J_{ij} = \langle E_i^*(t)\, E_j(t) \rangle_{\mathcal{N}}$ is the coherence matrix and $\hat{\sigma} = (\sigma_{ij})$, $\sigma_{ij} = \langle \Delta I_i(t)\, \Delta I_j(t) \rangle_{\mathcal{N}}^{1/2}$, $\Delta I_j = I_j - \langle I_j \rangle_{\mathcal{N}}$, is the correlation matrix (Born and Wolf (1965), Beran and Parrent (1964), Peřina (1972), Carter and Wolf (1973), Barakat (1977)). If I_1 and I_2 are the intensities of two linearly independent modes of polarization of Gaussian light (in which case the mutual correlations are zero), we obtain for the characteristic function:

$$\langle \exp(\mathrm{i}s\eta W) \rangle_{\mathcal{N}} = \frac{1}{(1 - \mathrm{i}s\langle n_1 \rangle)(1 - \mathrm{i}s\langle n_2 \rangle)}, \tag{3.143}$$

where

$$\langle n_j \rangle = \eta \langle I_j \rangle_{\mathcal{N}} T, \qquad j = 1, 2, \tag{3.144}$$

and

$$\langle n_1 \rangle = \frac{1}{2}(1 + P)\langle n \rangle, \tag{3.145a}$$

$$\langle n_2 \rangle = \frac{1}{2}(1 - P)\langle n \rangle, \tag{3.145b}$$

$\langle n \rangle = \langle n_1 \rangle + \langle n_2 \rangle$. Assuming $\langle I_1 \rangle_{\mathcal{N}} \geqq \langle I_2 \rangle_{\mathcal{N}}$ (which does not entail any generality), we have from (3.143)

$$\langle \exp(\mathrm{i}s\eta W) \rangle_{\mathcal{N}} = \frac{1}{\langle n_1 \rangle - \langle n_2 \rangle}\left[\frac{\langle n_1 \rangle}{1 - \mathrm{i}s\langle n_1 \rangle} - \frac{\langle n_2 \rangle}{1 - \mathrm{i}s\langle n_2 \rangle}\right] \tag{3.146}$$

and (Mandel (1963b))

$$P_{\mathcal{N}}(I) = \frac{1}{P\langle I \rangle_{\mathcal{N}}}\left[\exp\left(-\frac{2I}{(1 + P)\langle I \rangle_{\mathcal{N}}}\right) - \exp\left(-\frac{2I}{(1 - P)\langle I \rangle_{\mathcal{N}}}\right)\right], \tag{3.147}$$

$$p(n, T) = \frac{1}{P\langle n \rangle}\left[\left(1 + \frac{2}{1 + P\langle n \rangle}\right)^{-n-1} - \left(1 + \frac{2}{1 - P\langle n \rangle}\right)^{-n-1}\right]. \tag{3.148}$$

The variance of n is equal to

$$\langle(\Delta n)^2\rangle = \langle n\rangle\left(1 + \frac{1}{2}(1 + P^2)\,\langle n\rangle\right) \tag{3.149}$$

so that $M = 2/(1 + P^2)$. If the light is completely polarized, $P = 1$ and we have (3.134); if it is unpolarized, $P = 0$ and

$$\langle(\Delta n)^2\rangle = \langle n\rangle\left(1 + \frac{1}{2}\langle n\rangle\right) \tag{3.150}$$

and $M = 2$.

A qualitatively different situation occurs for an ideal laser whose intensity can be regarded as perfectly stabilized and whose intensity probability distribution is the δ-function,

$$P_{\mathcal{N}}(I) = \delta(I - \langle I\rangle_{\mathcal{N}}). \tag{3.151}$$

The photodetection equation (3.132) gives in this case

$$p(n, T) = \frac{\langle n\rangle^n}{n!}\exp(-\langle n\rangle), \tag{3.152}$$

which is the Poisson distribution, in agreement with (3.65). The variance $\langle(\Delta n)^2\rangle = \langle n\rangle$, since $\langle(\Delta W)^2\rangle_{\mathcal{N}} = 0$ in (3.96).

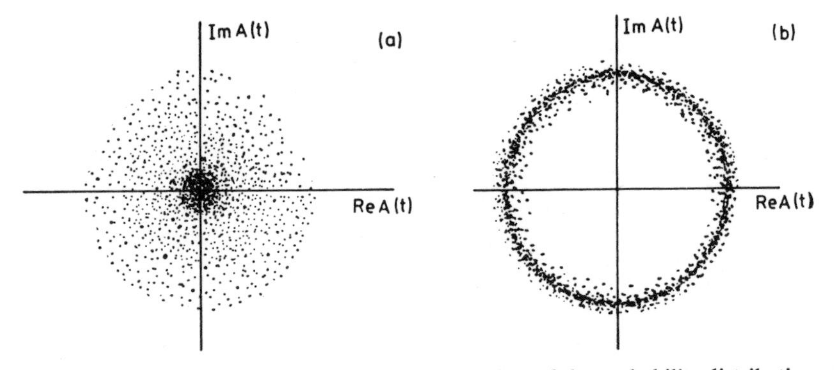

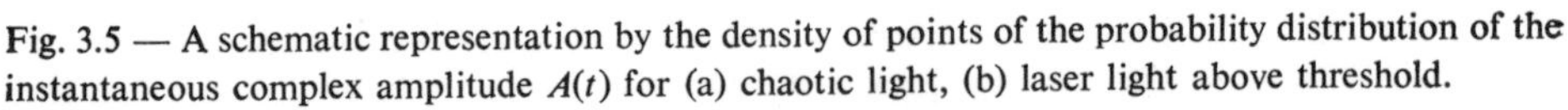
Fig. 3.5 — A schematic representation by the density of points of the probability distribution of the instantaneous complex amplitude $A(t)$ for (a) chaotic light, (b) laser light above threshold.

The difference between the statistics of the instantaneous intensities for chaotic light and for laser light is shown in Fig. 3.5a, b (Loudon (1973)). For a real laser source, the distribution, with a random phase, is a smoothed out δ-function in the form of a Gaussian distribution (the comparison of the photocount distributions for chaotic and laser light can be seen in Fig. 5.7a).

With respect to the radiation states discussed above, we can conclude that ($\eta = 1$)

$$\frac{\langle(\Delta W)^2\rangle_{\mathcal{N}}}{\langle W\rangle_{\mathcal{N}}} = \begin{cases}\langle n\rangle, & \text{for chaotic light,}\\ 0, & \text{for coherent light,}\\ -1, & \text{for the Fock state.}\end{cases} \tag{3.153}$$

Experimental verification of the applicability of the distr'butions (3.133) and (3.152) to chaotic and laser light, respectively, was carried out by Arecchi (1965) [also Arecchi, Berné and Burlamacchi (1966), Arecchi, Berné, Sona and Burlamacchi (1966)] and others (for further details, see Chapter 5). In Arecchi's experiments chaotic light was obtained by sending the light of an amplitude-stabilized single-mode He–Ne laser onto a moving ground-glass disk. Such light, called pseudothermal light, was first utilized by Martienssen and Spiller (1964). The photocount distribution for a chaotic field with M degrees of freedom given by (3.140), and for polarized and unpolarized light given by (3.148), was experimentally tested by Martienssen and Spiller (1966a), who obtained very good agreement with the theory. Experimental verification of the validity of the Bose–Einstein distribution for a laser operating below threshold was performed by Freed and Haus (1965).

For N-photon processes the photocount distribution is expressed by the Poisson distribution (3.152) with $\langle n \rangle = \eta \langle I^N \rangle_{\mathcal{N}} T$, provided that the field is initially coherent; this was experimentally verified in the course of second-harmonic generation by Clark et al. (1970). The pohotocount statistics for non-Gaussian light scattered from polydispersive suspensions have been discussed by Barakat and Blake (1976). Finite-aperture corrections to the photocount statistics have been investigated by Bark and Smith (1977) and the effects of spatial coherence and polarization have been taken into account by Jakeman, Oliver and Pike (1970), Srinivasan and Sukavanam (1972, 1978) and Zardecki and Delisle (1973). Some generalized photocount distributions have been suggested by Teich and Saleh (1981a, b), Teich (1981), Saleh et al. (1981), Teich et al. (1982), Saleh and Teich (1982) and Saleh et al. (1983).

3.10 Bunching and antibunching of photons

We have noted that the photocount distribution will depart, in general, from the Poisson distribution. If the variance of the number of counts is in excess of that given by the Poisson distribution, photons do not arrive at random, but are bunched. On the other hand, if the variance is less, the photons exhibit antibunching.

To describe the bunching effect of chaotic photons, as well as the fourth-order correlation effect first observed by Hanbury Brown and Twiss, we need to use the fourth-order correlation function, which, for linearly polarized light, is

$$\begin{aligned} \Gamma_{\mathcal{N}}^{(2,2)}(\mathbf{x}_1, \mathbf{x}_2, \mathbf{x}_2, \mathbf{x}_1, 0, \tau, \tau, 0) &= \langle I_1(t)\, I_2(t+\tau) \rangle_{\mathcal{N}} = \\ &= \langle V_1^*(t)\, V_1(t)\, V_2^*(t+\tau)\, V_2(t+\tau) \rangle_{\mathcal{N}}, \end{aligned} \tag{3.154}$$

with $V_j(t) \equiv V(\mathbf{x}_j, t)$; this expresses the intensity correlation. For chaotic light we must calculate the fourth-order correlation function for the Gaussian distribution function. It was shown by Wang and Uhlenbeck (1945), Reed (1962) and Glauber (1963b, 1965) (see Sec. 5.1) that

$$\Gamma_{\mathcal{N}}^{(m,n)}(x_1, \ldots, x_{m+n}) = \delta_{mn} \sum_{\pi} \Gamma_{\mathcal{N}}^{(1,1)}(x_1, x_{n+1}) \ldots \Gamma_{\mathcal{N}}^{(1,1)}(x_n, x_{2n}), \tag{3.155}$$

where $\sum_{\pi}$ stands for the sum of $n!$ possible permutations of the indices $n+1, \ldots, 2n$ (or $1, \ldots, n$). Expression (3.155) shows that the $2n$th-order correlation function for the chaotic field is determined completely by the second-order correlation function. Putting $x_1 = x_2 = \ldots = x_n$, we have

$$\Gamma_{\mathcal{N}}^{(n,n)}(x, \ldots, x) = \langle I^n(x)\rangle_{\mathcal{N}} = n!\, \langle I(x)\rangle_{\mathcal{N}}^n, \tag{3.156}$$

which agrees with (3.130). For the fourth-order correlation (3.155) can be derived easily. Decomposing $V(x_j)$ in the form of a Fourier series, we have

$$\langle V^*(x_1)\, V^*(x_2)\, V(x_3)\, V(x_4)\rangle_{\mathcal{N}} = \sum_{k_1} \cdots \sum_{k_4} \langle V_{k_1}^* V_{k_2}^* V_{k_3} V_{k_4}\rangle_{\mathcal{N}} \times$$
$$\times \exp(-ik_1x_1 - ik_2x_2 + ik_3x_3 + ik_4x_4). \tag{3.157}$$

As the Gaussian probability distribution is phase independent,

$$\langle V^*(x_1)\, V^*(x_2)\, V(x_3)\, V(x_4)\rangle_{\mathcal{N}} = \sum_{k_1,k_2} \langle |V_{k_1}|^2\rangle_{\mathcal{N}} \langle |V_{k_2}|^2\rangle_{\mathcal{N}} \times$$
$$\times \exp[ik_1(x_3 - x_1) + ik_2(x_4 - x_2)] + \sum_{k_1,k_2} \langle |V_{k_1}|^2\rangle_{\mathcal{N}} \langle |V_{k_2}|^2\rangle_{\mathcal{N}} \times$$
$$\times \exp[ik_1(x_4 - x_1) + ik_2(x_3 - x_2)] =$$
$$= \langle V^*(x_1)\, V(x_3)\rangle_{\mathcal{N}} \langle V^*(x_2)\, V(x_4)\rangle_{\mathcal{N}} +$$
$$+ \langle V^*(x_1)\, V(x_4)\rangle_{\mathcal{N}} \langle V^*(x_2)\, V(x_3)\rangle_{\mathcal{N}}. \tag{3.158}$$

Applying (3.158) to (3.154) we obtain (Wolf (1957))

$$\langle I_1(t)\, I_2(t+\tau)\rangle_{\mathcal{N}} = \Gamma_{\mathcal{N},11}(0)\, \Gamma_{\mathcal{N},22}(0) + |\Gamma_{\mathcal{N},12}(\tau)|^2 =$$
$$= \langle I_1\rangle_{\mathcal{N}} \langle I_2\rangle_{\mathcal{N}} (1 + |\gamma_{\mathcal{N},12}|^2), \tag{3.159}$$

where $\Gamma_{\mathcal{N},ij}(\tau) \equiv \Gamma_{\mathcal{N}}^{(1,1)}(\mathbf{x}_i, \mathbf{x}_j, 0, \tau)$, $\gamma_{\mathcal{N},12}(\tau) \equiv \gamma_{\mathcal{N}}^{(1,1)}(\mathbf{x}_1, \mathbf{x}_2, \tau)$, $\langle I_j\rangle_{\mathcal{N}} \equiv \Gamma_{\mathcal{N},jj}(0)$ Thus the correlation of the fluctuations $\Delta I_j = I_j - \langle I_j\rangle_{\mathcal{N}}$ is

$$\langle \Delta I_1(t)\, \Delta I_2(t+\tau)\rangle_{\mathcal{N}} = \langle I_1(t)\, I_2(t+\tau)\rangle_{\mathcal{N}} - \langle I_1\rangle_{\mathcal{N}} \langle I_2\rangle_{\mathcal{N}} =$$
$$= \langle I_1\rangle_{\mathcal{N}} \langle I_2\rangle_{\mathcal{N}} |\gamma_{\mathcal{N},12}(\tau)|^2. \tag{3.160}$$

This shows that the modulus of the degree of coherence may be determined by measuring the correlation of the intensity fluctuations for chaotic light, and this is the basis of correlation interferometry and spectroscopy.

If partially polarized light is assumed, we can obtain (in correspondence to (3.149))

$$\langle I_1(t)\, I_2(t+\tau)\rangle_{\mathcal{N}} = \frac{1}{2} \langle I_1\rangle_{\mathcal{N}} \langle I_2\rangle_{\mathcal{N}} (1 + P^2)\, |\gamma_{\mathcal{N},12}(\tau)|^2. \tag{3.161}$$

This shows that intensity correlation measurements may also provide information about the degree of polarization of light beams.

Now introduce the conditional probability $p_c(t \mid t+\tau)\,\Delta\tau$ as the probability that a photoelectric count will be registered in the time interval $(t+\tau, t+\tau+\Delta\tau)$ if a count has been registered at a time t (Mandel (1963a)). The probability of observing a count at both times t and $t+\tau$ within $\mathrm{d}t$ and $\mathrm{d}\tau$ is obviously $\eta^2 I(t)\, I(t+\tau)\, \mathrm{d}t\, \mathrm{d}\tau$

and the probability of finding two counts separated by the interval τ is $\eta^2 \langle I(t)\, I(t+\tau)\rangle_{\mathcal{N}}\, dt\, d\tau$. If this is divided by the ensemble average of the probability $\eta I(t)\, dt$ of finding one count at t within dt, we arrive at the conditional probability $p_c(\tau)$ of obtaining a second count τ seconds after the first one,

$$p_c(\tau)\, d\tau = \frac{\eta^2 \langle I(t)\, I(t+\tau)\rangle_{\mathcal{N}}\, dt\, d\tau}{\langle I(t)\rangle_{\mathcal{N}}\, dt} =$$

$$= \eta \langle I(t)\rangle_{\mathcal{N}} \left[1 + \frac{1}{2}(1 + P^2)\, |\gamma_{\mathcal{N}}(\tau)|^2\right] d\tau, \tag{3.162}$$

where (3.161) has been used. Assuming completely polarized light we have, putting $P = 1$,

$$p_c(\tau) = \eta \langle I(t)\rangle_{\mathcal{N}}\, (1 + |\gamma_{\mathcal{N}}(\tau)|^2). \tag{3.163}$$

Since $|\gamma_{\mathcal{N}}(\tau)| \approx 1$ for $\tau \ll \tau_c = 1/\Delta\nu$ and $|\gamma_{\mathcal{N}}(\tau)| \approx 0$ for $\tau \gg \tau_c$, $p_c(\tau)$ starts at the value $2\eta \langle I(t)\rangle_{\mathcal{N}}$ for small τ and tends to $\eta \langle I(t)\rangle_{\mathcal{N}}$ for large τ. The degree of temporal coherence for the Lorentzian spectrum is given in (3.20e) (Sec. 3.2.5). These results illustrate the bunching properties of a chaotic photon beam. Their first experimental investigation was by Arecchi, Gatti and Sona (1966) and Morgan and Mandel (1966), whose results are given in Figs. 3.6 and 3.7. The experimental results are in very good agreement with the theoretical predictions.

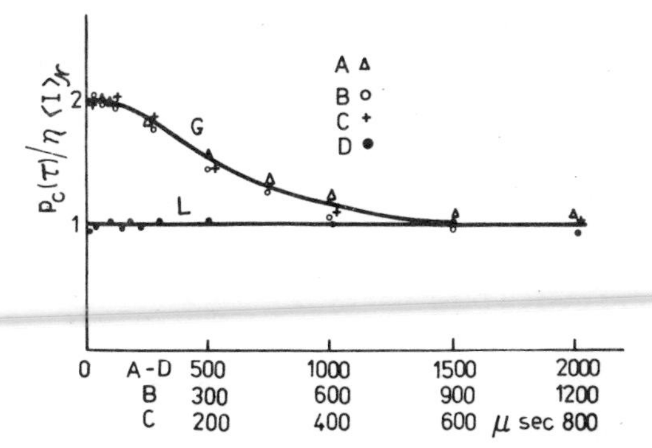

Fig. 3.6 — The conditional probability $p_c(\tau)$ of a second count occurring at a time τ after a first count has occurred at time $\tau = 0$. Experimental results apply to a laser (L, ●) and to an artificially synthesized chaotic source (G) created by passing the laser radiation through ground glass which is rotated at speeds $v = 1.25$ cm/s (△), 2.09 cm/s (○) and 3.14 cm/s (+). Theoretical predictions are shown by the full curves (after Arecchi, Gatti and Sona, 1966, Phys. Lett. **20**, 27).

In these experiments, light from a chaotic source was incident on a photomultiplier with a fast response. The output of the photomultiplier reaches a coincidence counter via two different paths, one of which contains a time-delay line so that only pulses in the output of the photomultiplier which are separated by the time delay τ can produce an output from the coincidence counter. The conditional probability $p_c(\tau)$ is then determined by the normalization of the distribution of coincidences as a function of τ. While in the experiment of Morgan and Mandel light from a low pressure ^{198}Hg light source was used, in the experiment of Arecchi, Gatti and Sona pseudothermal light was used, synthesized from laser light by a rotating ground glass disk [Parry et al. (1978)], or more conveniently, liquid crystals with an applied voltage can be used [Bertolotti (1974)]. A measurement of the distribution of time intervals between the photoelectrons emitted by a photodetector [Glauber (1967)] illuminated by a one-mode laser and a pseudothermal source was reported by Bendjaballah (1969). All these experiments are in very good agreement with the theory.

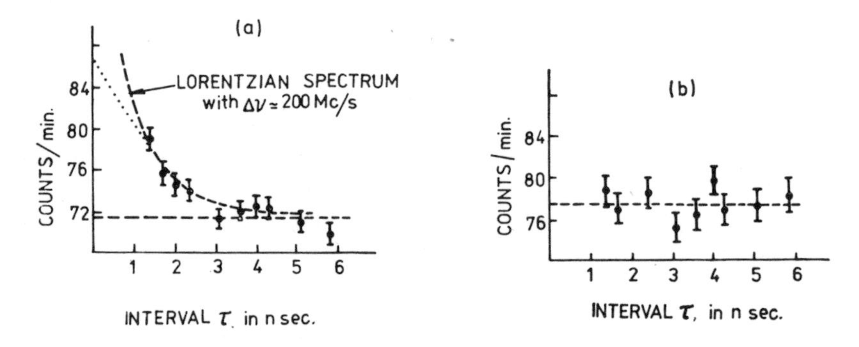

Fig. 3.7 — Counting rates illustrating the phenomenon of photon bunching with light from (a) ^{198}Hg source, (b) tungsten lamp; in the latter case the wide frequency spectrum leads to intensity correlations in an unmeasurably short time interval (after Morgan and Mandel, 1966, Phys. Rev. Lett. **16**, 1012).

For laser light, which is non-Gaussian light, $\langle I_1(t)\, I_2(t+\tau)\rangle_{\mathcal{N}} \approx \langle I_1\rangle_{\mathcal{N}} \langle I_2\rangle_{\mathcal{N}}$ and $\langle \Delta I_1(t)\, \Delta I_2(t+\tau)\rangle_{\mathcal{N}}$ is practically zero. Consequently no bunching of laser photons occurs, as was indeed observed by Arecchi, Gatti and Sona (1966) (Fig. 3.6). Very illustrative demonstrations of the bunching effect for pseudothermal radiation and of the absence of this effect for laser radiation have been obtained by Bendjaballah (1971). The temperature dependence of photon bunching in thermal radiation has been investigated by Dudeja and Chopra (1977).

We can conclude that the second term in (3.163), typical of chaotic light, represents a positive departure from the Poisson statistics which is responsible for the photon bunching effect.

On the other hand, optical fields having $\langle (\Delta I)^2\rangle_{\mathcal{N}} < 0$ [cf. (3.153) for the Fock states] will exhibit photon antibunching and sub-Poissonian behaviour [for their relations, see Mandel (1982a), Teich and Saleh (1982), Singh (1983), Teich, Saleh and Stoler (1983)]. The possibility of generating such fields, which have no classical analogues, will be discussed in Chapter 9.

Note that a relationship between stimulated and spontaneous emission and photon correlation has been considered by Abraham and Smith (1977) and Abraham et al.

(1981). Correlations between intervals in a stationary random point process [Bendjaballah (1975)] and time-interval photoelectron statistics [Blake and Barakat (1972, 1977)] have been investigated, too.

3.11 Hanbury Brown-Twiss effect—correlation interferometry and correlation spectroscopy

Historically the correlation of photoelectrons in the output from two photodetectors was demonstrated first by Hanbury Brown and Twiss (1956a) in the form of a correlation between two photocurrents treated as continuous signals. The scheme of their apparatus is shown in Fig. 3.8. Light from a thermal source (mercury lamp) S was split into two beams by a half-silvered mirror M and fell on two photocells P_1 and P_2, whose outputs were sent through band-limited amplifiers A_1 and A_2 to a correlator C. The averaged value of the product of the intensity fluctuations was recorded by an integrating motor Mo. This experiment was repeated by Martienssen and Spiller (1964) with a pseudothermal source, among others.

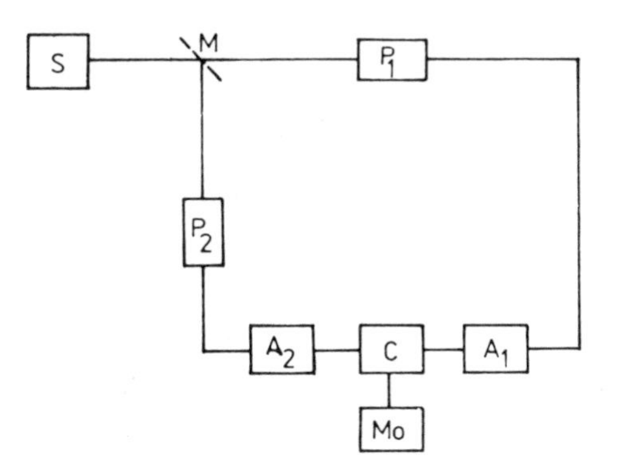

Fig. 3.8 — An outline of the apparatus for demonstrating the correlation between intensity fluctuations; S is a chaotic source, M is a half-silvered mirror, P_1 and P_2 are photomultipliers, A_1 and A_2 are amplifiers, C is an electric correlater and Mo is an integrating device.

To explain this experiment we must look more deeply into a physical model working with the field intensity rather than with the integrated intensity. We follow a treatment by Mandel (1963a).

First consider the variance of the number of photoelectrons in one photodetector and for simplicity assume linearly polarized light. Using the photodetection equation (3.85), we obtain

$$\langle n^2 \rangle = \langle n \rangle + \eta^2 \iint_{t}^{t+T} \langle I(t')\, I(t'') \rangle_{\mathcal{N}}\, \mathrm{d}t'\, \mathrm{d}t'' =$$

$$= \langle n \rangle + \eta^2 \iint_{0}^{T} \langle I(t + t' - t'')\, I(t) \rangle_{\mathcal{N}}\, \mathrm{d}t'\, \mathrm{d}t'', \tag{3.164}$$

where $\langle n \rangle = \eta \langle I \rangle_{\mathcal{N}} T$ and the stationary condition is assumed. Writing [Mandel (1958, 1959, 1963a)]

$$T\xi(T) = \iint_0^T |\gamma_{\mathcal{N}}(t' - t'')|^2 \, dt' \, dt'' = 2 \int_0^T (T - t') |\gamma_{\mathcal{N}}(t')|^2 \, dt', \tag{3.165}$$

we obtain from (3.164)

$$\langle n^2 \rangle = \langle n \rangle^2 + \langle n \rangle \left(1 + \langle n \rangle \frac{\xi(T)}{T}\right), \tag{3.166}$$

where (3.159) has also been used. Since $|\gamma_{\mathcal{N}}(t' - t'')| \leqq 1$, then $\xi(T) \leqq T$ and if $T \ll 1/\Delta\nu$, then $|\gamma_{\mathcal{N}}(\tau)| \approx 1$ and $\xi(T) \approx T$. Thus the variance following from (3.166) agrees with (3.134). However, if $T \gg 1/\Delta\nu$, then, since $|\gamma_{\mathcal{N}}(\tau)|$ is non-vanishing only within an interval of length of a few times $1/\Delta\nu$,

$$\xi(T) \approx \xi(\infty) = \int_{-\infty}^{+\infty} |\gamma_{\mathcal{N}}(t)|^2 \, dt; \tag{3.167}$$

this quantity can be adopted as the coherence time τ_c for chaotic radiation [Mandel (1959, 1963a)].

From (3.166) we obtain

$$\langle (\Delta n)^2 \rangle = \langle n \rangle \left(1 + \langle n \rangle \frac{\tau_c}{T}\right), \tag{3.168}$$

which is the variance of the number of bosons divided into T/τ_c cells of phase space – the length of the cell being $c\tau_c$ [see (3.141b)]. The photocount distribution (3.140) becomes, for $M = T/\tau_c$

$$p(n, T) = \frac{\Gamma(n + T/\tau_c)}{n!\Gamma(T/\tau_c)} \left(1 + \frac{T}{\tau_c \langle n \rangle}\right)^{-n} \left(1 + \frac{\langle n \rangle \tau_c}{T}\right)^{-T/\tau_c}. \tag{3.169}$$

It has been shown by Bédard et al. (1967) that this formula is valid not only for $T \gg \tau_c$, but represents a very good approximation for all T, if M is appropriately chosen. Similar results have been obtained for the superposition of coherent and chaotic fields [Horák et al. (1971a, b), Peřina, Peřinová and Mišta (1971), Mišta and Peřina (1971)]. These questions are discussed in greater detail in Chapter 5.

Departures from Poisson counting statistics depend on the degeneracy parameter

$$\delta = \frac{\langle n \rangle \tau_c}{T} = \eta \langle I \rangle_{\mathcal{N}} \tau_c, \tag{3.170}$$

which gives the mean number of counts per cell of the phase space; $\delta \ll 1$ for thermal light and $\delta \gg 1$ for laser light. The fact that $\delta \ll 1$ for thermal light implies that a direct measurement of the excess photon noise is not easy. The conditions are much better for pseudothermal light.

The correlation between the counts in two photodetectors may be obtained in the same way,

$$\langle n_1 n_2 \rangle = \eta_1 \eta_2 \iint_t^{t+T} \langle I_1(t')\, I_2(t'') \rangle_{\mathcal{N}}\, \mathrm{d}t'\, \mathrm{d}t'' =$$
$$= \eta_1 \eta_2 \iint_0^T \langle I_1(t + t' - t'')\, I_2(t) \rangle_{\mathcal{N}}\, \mathrm{d}t'\, \mathrm{d}t''. \tag{3.171}$$

If we assume the validity of the cross-spectral purity condition $\gamma_{\mathcal{N},12}(\tau) = \gamma_{\mathcal{N},12}(0)\gamma_{\mathcal{N},11}(\tau)$ [Mandel and Wolf (1961)] (independence of spatial and temporal coherence), we obtain

$$\langle \Delta n_1 \Delta n_2 \rangle = \langle n_1 \rangle \langle n_2 \rangle \,|\, \gamma_{\mathcal{N},12}(0) \,|^2 \frac{\tau_c}{T}. \tag{3.172}$$

The normalized correlation of the fluctuations is equal to

$$\frac{\langle \Delta n_1 \Delta n_2 \rangle}{[\langle (\Delta n_1)^2 \rangle \langle (\Delta n_2)^2 \rangle]^{1/2}} = \frac{\delta}{1 + \delta} \,|\, \gamma_{\mathcal{N},12}(0) \,|^2, \tag{3.173}$$

where the degeneracy parameter δ is given by (3.170) and $\langle n_1 \rangle = \langle n_2 \rangle$. As $\delta \approx 10^{-3}$ for a thermal source, the effect is difficult to observe. It is easy for pseudothermal light, since then $\delta \approx 10^{12}$.

For partially polarized fields one obtains, using (3.161),

$$\langle \Delta n_1 \Delta n_2 \rangle = \frac{1}{2}(1 + P^2) \langle n_1 \rangle \langle n_2 \rangle \,|\, \gamma_{\mathcal{N},12}(0) \,|^2 \frac{\tau_c}{T}. \tag{3.174}$$

Thus, in general, the correlation fluctuation measurements provide information on the spectral properties (temporal coherence), the spatial coherence, and the polarization properties.

If non-Gaussian light is considered now one can write for the intensity correlation

$$\langle I_1(t)\, I_2(t + \tau) \rangle_{\mathcal{N}} = \langle I_1 \rangle_{\mathcal{N}} \langle I_2 \rangle_{\mathcal{N}} + \langle \Delta I_1(t)\, \Delta I_2(t + \tau) \rangle_{\mathcal{N}} =$$
$$= \langle I_1 \rangle_{\mathcal{N}} \langle I_2 \rangle_{\mathcal{N}} (1 + |\, \zeta_{12}(\tau) \,|^2), \tag{3.175}$$

where

$$|\, \zeta_{12}(\tau) \,|^2 = \frac{\langle \Delta I_1(t)\, \Delta I_2(t + \tau) \rangle_{\mathcal{N}}}{\langle I_1 \rangle_{\mathcal{N}} \langle I_2 \rangle_{\mathcal{N}}}. \tag{3.176}$$

For chaotic light $|\, \zeta_{12}(\tau) \,| = |\, \gamma_{\mathcal{N},12}(\tau) \,|$. The photocount correlation is then

$$\langle n_1 n_2 \rangle = \langle n_1 \rangle \langle n_2 \rangle + \eta_1 \eta_2 \iint_0^T \langle \Delta I_1(t')\, \Delta I_2(t'') \rangle_{\mathcal{N}}\, \mathrm{d}t'\, \mathrm{d}t'' \tag{3.177}$$

and the fluctuation correlation is equal to

$$\langle \Delta n_1 \Delta n_2 \rangle = \langle n_1 \rangle \langle n_2 \rangle \frac{\bar{\xi}(T)}{T}, \tag{3.178}$$

where

$$\bar{\xi}(T) = \frac{1}{T \langle I_1 \rangle_{\mathcal{N}} \langle I_2 \rangle_{\mathcal{N}}} \iint_0^T \langle \Delta I_1(t')\, \Delta I_2(t'') \rangle_{\mathcal{N}}\, \mathrm{d}t'\, \mathrm{d}t'' =$$
$$= \frac{2}{T \langle I_1 \rangle_{\mathcal{N}} \langle I_2 \rangle_{\mathcal{N}}} \int_0^T (T - \tau) \langle \Delta I_1(\tau)\, \Delta I_2(0) \rangle_{\mathcal{N}}\, \mathrm{d}\tau, \tag{3.179}$$

again the stationary condition has been assumed. For the ideal laser, $\Delta I \approx 0$ and $\langle \Delta n_1 \Delta n_2 \rangle \approx 0$ and no Hanbury Brown–Twiss effect occurs.

These expressions can serve to define the coherence time for arbitrary statistics of the field [Klauder and Sudarshan (1968), Peřina (1980b), Mandel (1981a)]. With the help of (3.176) we can define $\tau_c = \int_{-\infty}^{+\infty} | \zeta(\tau)/\zeta(0) |^2 \, d\tau$, which reduces to (3.167) for chaotic fields. Some corrections to the coherence time, coherence area and coherence volume, based on the photocount statistics for the superposition of coherent and chaotic fields, have been also found [Peřina and Mišta (1974)] [see equation (5.121)].

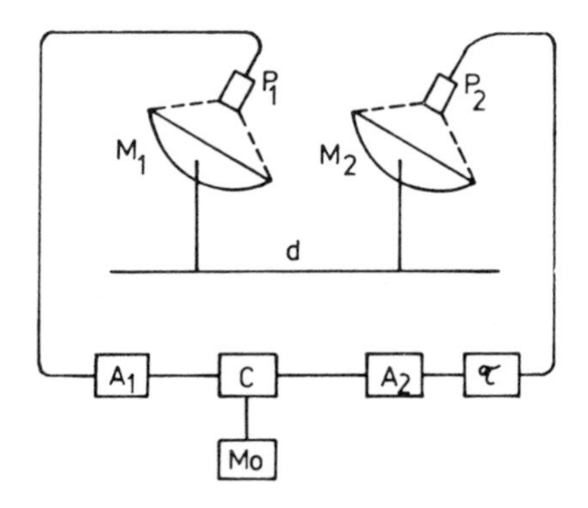

Fig. 3.9 — A scheme of a stellar correlation interferometer; M_1 and M_2 are mirrors, P_1 and P_2 are photodetectors, A_1 and A_2 are amplifiers, τ is a delay line, C is a correlater and Mo is an integrating device.

The correlation technique has been applied to measurements of angular diameters of stars by Hanbury Brown and Twiss (1956b, 1958) and forms the basis of correlation interferometry. A scheme of the correlation interferometer is shown in Fig. 3.9. According to (3.174) such correlation measurements allow $| \gamma_{\mathcal{N},12}(0) |$ to be determined when the separation d of the mirrors is changed. As is well known, the modulus of the degree of spatial coherence in the Fraunhofer approximation for a source of radius ϱ at distance r is [Born and Wolf (1965), Beran and Parrent (1964), Peřina (1972)]

$$| \gamma_{\mathcal{N},12}(0) | = \frac{| 2J_1(\bar{k}\varrho d/r) |}{\bar{k}\varrho d/r} . \tag{3.180}$$

where d is the distance between the points considered, J_1 is the Bessel function and $\bar{k}$ the mean wave number. Thus for $d = d_0$, where the Bessel function has the first zero point, we have for the angular radius ϱ/r of a star, $\varrho/r = 3.83/\bar{k}d_0 \approx 0.61\bar{\lambda}/d_0$, $\bar{\lambda}$ being the mean wavelength. A large stellar correlation interferometer was built at Narrabri (Australia) [Hanbury Brown (1964), Wolf (1966)] and further applications of intensity correlation interferometry in physics and astronomy have been reviewed by Twiss (1969).

There are advantages in the correlation technique for measuring stellar diameters compared with the use of the Michelson stellar interferometer. The correlation measurements involve the intensity $I = |V|^2$ which varies slowly in comparison with the rapidly varying amplitude V, and also the phase of V is absent. Consequently phase distortions due to turbulence of the air, changes in the refraction index of the atmosphere, etc., do not affect the measurements, whereas they may be intolerable for the classical Michelson stellar interferometer. Also much larger separation d can be used in the correlation interferometer, giving an increase of the resolving power.

On the other hand, the phase information about the degree of coherence is definitely lost in correlation measurements and other methods may have to be adopted to recover the phase. This phase problem has been intensively discussed, and relevant references have been given in the Introduction to this book.

Spectroscopic correlation measurements employ the sensitivity of the photocount distribution and its factorial moments to the spectral parameters, such as the mean frequency, the half-width of the spectrum, etc., as will be demonstrated in Chapter 5. They may be determined by fitting the experimental results to the theoretical model. One such a powerful method is heterodyne detection, when the spectrally unknown chaotic field is superimposed, before detection, on a known coherent component.

Corresponding to (3.168), the correlation of the photocurrent from a photodetector can be written in the form [Cummins and Swinney (1970)]

$$C(\tau) = e\langle i\rangle\,\delta(\tau) + \langle i\rangle^2\,\gamma_{\mathcal{N}}^{(2,2)}(\tau), \tag{3.181}$$

where $\gamma_{\mathcal{N}}^{(2,2)}(\tau) = \Gamma_{\mathcal{N}}^{(2,2)}(\tau)/[\Gamma_{\mathcal{N}}^{(1,1)}(0)]^2$, i being the photoelectric current and e the electron charge. The first term in (3.181) corresponds to the quantum shot noise and the second one describes the intensity fluctuation correlation. For chaotic light, one applies (3.159) to (3.181), which give

$$C(\tau) = e\langle i\rangle\,\delta(\tau) + \langle i\rangle^2 + \langle i\rangle^2\,|\gamma_{\mathcal{N}}(\tau)|^2, \tag{3.182a}$$

and the Fourier transformation provides the photocurrent spectrum

$$P(\omega) = \frac{1}{2\pi}\int_{-\infty}^{+\infty} C(\tau)\exp(i\omega\tau)\,d\tau =$$

$$= \frac{1}{2\pi}e\langle i\rangle + \langle i\rangle^2\,\delta(\omega) + \langle i\rangle^2\,h(\omega), \tag{3.182b}$$

where $h(\omega)$, for the degree of coherence given by (3.20e) (Sec. 3.2.5), has the Lorentzian spectrum centered at $\omega = 0$,

$$h(\omega) = \frac{1}{2\pi}\int_{-\infty}^{+\infty} |\gamma_{\mathcal{N}}(\tau)|^2\exp(i\omega\tau)\,d\tau = \frac{1}{\pi}\frac{2\Delta\nu}{4(\Delta\nu)^2 + \omega^2}. \tag{3.183}$$

The first spectral term in (3.182b) corresponds to the shot noise in (3.182a), the second to the d.c. component, and the third term represents the correlation fluctuation spectrum, having the Lorentzian shape. Thus, the measurement of the low-frequency photocount spectrum permits an accurate determination of the width of the optical spectral line. The resolving power of high-resolution correlation spectroscopy is given by $1/T$ and lies within $1-10^8$ Hz. The quantum statistics of homodyne and heterodyne detection have been discussed by Yuen and Shapiro (1978b).

If the correlation measurements are digitized the correlation function of counts can be written as

$$\Gamma_s = \frac{1}{N} \sum_{j=1}^{N} n_j n_{j+s}, \tag{3.184}$$

where N is the number of subintervals in $(0, T)$ and n_j is the digitized signal. For rapid computations of these correlations the clipping technique is used. This means that the signal is replaced, before the correlation stage, by a series of ones and zeros according as it lies above or below a specified clipping level. For the digital signals in optical measurements, it is convenient to use single-channel clipping [Jakeman and Pike (1969b)]. Here one leaves the n_{j+s} unchanged and introduces a positive integer clipping level K for the number of counts collected in the interval T at $t_j = j\Delta t$, $\Delta t = T/N$, $j = 0, 1, \ldots, N$. The clipped count $n_j^{(K)}$ is then

$$n_j^{(K)} = \begin{cases} 1, & n_j > K, \\ 0, & n_j \leqq K. \end{cases} \tag{3.185}$$

The clipped correlation function $\Gamma_s^{(K)} = \langle n_j^{(K)} n_{j+s} \rangle$ is then easily measured with a single electronic circuitry. Double clipping also is possible, and is appropriate in delayed coincidence measurements (the clipping level is $K = 0$). In this case all measurements give $n_j, n_{j+s} = 0$ or 1 and the multiplication is equivalent to coincidence measurement. The method can be extended to higher-order correlation measurements. A number of papers have addressed this topic [Bendjaballah (1973, 1980), Blake and Barakat (1973), Hughes et al. (1973), Jakeman (1970), Jakeman and Pike (1969b), Jakeman, Oliver and Pike (1971a), Jakeman, Pike and Swain (1970, 1971), Saleh and Hendrix (1975)].

Further details concerning correlation spectroscopy can be found in excellent reviews by Cummins and Swinney (1970), Arecchi and Degiorgio (1972), Jakeman (1974), Pike and Jakeman (1974) and Saleh (1978). The spectroscopic correlation technique for the investigation of turbulent liquid flow has been developed and applied by Crosignani and Di Porto (1974) and Crosignani, Di Porto and Bertolotti (1975).

In measurements of photocount statistics the so-called dead-time effect occurs. This means that for a counter there exists a time interval τ_D after each registration, during which no photoemission can be registered. Consequently the number of events registered during the counting interval T will be smaller than the actual number of events and the measured photocount distributions are narrower (as if antibunching occurred) and a correction must be applied. This effect is shown in Fig. 3.10. The corrected distribution can be simply expressed in terms of the uncorrected distribution. Even small values of the ratio τ_D/T ($\sim 1\%$) alter the photocount distribution markedly. The corrected distribution is equal to

$$p(n, T, \tau_D/T) = \sum_{k=0}^{n} p_0(k, T'_n) - \sum_{k=0}^{n-1} p_0(k, T'_{n-1}),\ T'_n = T(1 - n\tau_D/T), \tag{3.186a}$$

p_0 being the uncorrected distribution. The first-order correction modifies the photodetection equation to:

$$p(n, T, \tau_D/T) = \left\langle \frac{(\eta W)^n}{n!} \exp(-\eta W)\left[1 + n(\eta W - n + 1)\frac{\tau_D}{T}\right]\right\rangle_{\mathcal{N}}. \tag{3.186b}$$

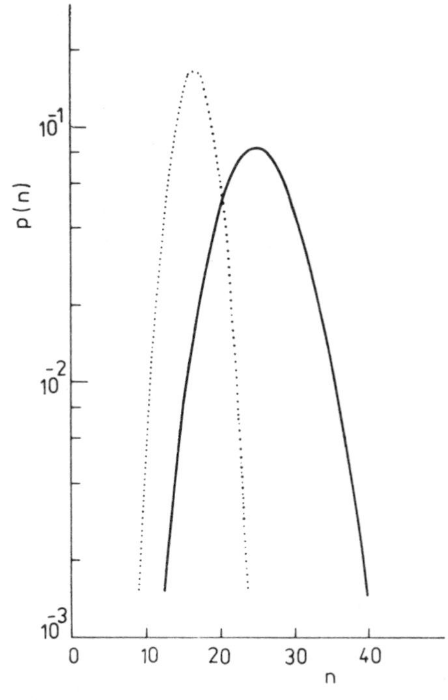

Fig. 3.10 — The Poisson photocount distribution with $\langle n \rangle = \langle(\Delta n)^2\rangle = 25$ (full curve) is reduced by the effects of dead time ($\tau_D/T = 0.025$) to the photocount distribution exhibiting "antibunching" with $\langle n \rangle = 16.3$ and $\langle(\Delta n)^2\rangle = 5.6$ (dotted curve) (after Cantor and Teich, 1975, J. Opt. Soc. Am. **65**, 786).

The problem of inverting the photodetection equation incorporated the dead-time effect has been considered by Mehta (1970) and Mandel (1980). More detail about the dead-time effect is available in the literature [Johnson, Jones, McLean and Pike (1966), Bédard (1967c), Cantor and Teich (1975), Srinivasan (1974, 1978), Teich and McGill (1976), Saleh (1978), Vannucci and Teich (1981)].

Finally, note that an interesting use of the intensity correlation technique in the problem of scattering and of determining the phase of the scattering matrix has been suggested by Goldberger, Lewis and Watson (1963) and Goldberger and Watson (1964, 1965). Spectroscopic applications have been discussed by Goldberger, Lewis and Watson (1966). Intensity correlation linewidth measurements have been carried out, for instance, by Phillips, Kleiman and Davis (1967), while a determination of the linewidth using photocount statistics has been described by Jakeman, Oliver and Pike (1968a).

Some effort has been devoted to the development of multifold counting statistics, in a strict analogy to the above one-fold treatment [for reviews see Peřina (1972) and Saleh (1978)]. Applications of the N-fold formalism have been discussed by Bédard (1967d), Arecchi, Berné and Sona (1966), Fillmore (1969), Cantrell (1970), Blake and Barakat (1973, 1976), Aoki (1977), Bendjaballah (1979) and Srinivasan and Gururajan (1981), among others. Sixth-order correlation functions have been experimentally investigated by Davidson and Mandel (1968), Davidson (1969), and Corti and Degiorgio (1974, 1976a, b).

Temperature effects in photodetection have been investigated by Selloni et al. (1978b) and some other questions related to photocount statistics have been discussed by Astafunov and Glazov (1980) and Durnin et al. (1981). Arguments against neoclassical theories of radiation based on photocount statistics have been reviewed by Mandel (1976a) and Paul (1980, 1981).

CHAPTER 4

COHERENT-STATE DESCRIPTION OF THE ELECTROMAGNETIC FIELD

In this chapter we show that in the quantum theory of the electromagnetic field there exist quantum coherent states, in which the quantum correlation functions factorize in the form (3.53), i.e. a field in such a state satisfies the full coherence conditions. We discuss the basic properties of these states. Although the coherent states were first introduced by Schrödinger (1927) and were studied by others [e.g. Klauder (1960)], their full utilization in connection with quantum optics is due to Glauber (1963a–c, 1964, 1965, 1966a, b, 1969, 1970, 1972), who used them to study the quantum coherence of optical fields. Generalizations of this concept have been discussed in the Introduction.

4.1 Coherent states of a harmonic oscillator and of the electromagnetic field

4.1.1 Definitions

Let us introduce displaced vacuum states

$$|\alpha\rangle = \hat{D}(\alpha)|0\rangle, \tag{4.1}$$

where

$$\hat{D}(\alpha) = \exp(\alpha\hat{a}^{+} - \alpha^{*}\hat{a}) \tag{4.2}$$

is a displacement operator. The states $|\alpha\rangle$ are coherent states, α is a complex number. Using the Baker–Hausdorff identity [e.g. Louisell (1964)]

$$\exp(\hat{A} + \hat{B}) = \exp(\hat{A})\exp(\hat{B})\exp\left(-\frac{1}{2}[\hat{A}, \hat{B}]\right), \tag{4.3}$$

where $\hat{A}$ and $\hat{B}$ are operators for which the commutator $[\hat{A}, \hat{B}]$ is a c-number ($[[\hat{A}, \hat{B}], \hat{A}] = [[\hat{A}, \hat{B}], \hat{B}] = 0$), we can rewrite (4.1) in the form

$$|\alpha\rangle = \exp\left(-\frac{1}{2}|\alpha|^2\right)\exp(\alpha\hat{a}^{+})\exp(-\alpha^{*}\hat{a})|0\rangle = \\ = \exp\left(-\frac{1}{2}|\alpha|^2\right)\sum_{n=0}^{\infty}\frac{\alpha^n}{(n!)^{1/2}}|n\rangle, \tag{4.4}$$

where we have expanded $\exp(\alpha\hat{a}^{+})$ and $\exp(-\alpha^{*}\hat{a})$ in Taylor series and used the vacuum stability condition $\hat{a}|0\rangle = 0$ together with the definition of the Fock state, given in (2.16). Making use of the relation (2.17b) we can verify that

$$\hat{a} \mid \alpha\rangle = \alpha \mid \alpha\rangle, \tag{4.5a}$$

$$\langle\alpha \mid \hat{a}^+ = \langle\alpha \mid \alpha^*, \tag{4.5b}$$

i.e. the coherent state $|\alpha\rangle$ is an eigenstate of the annihilation operator $\hat{a}$ with eigenvalue α, which is in general a complex number since $\hat{a}$ is not a Hermitian operator. Equation (4.5a) shows that $|\alpha\rangle$ must contain an uncertain number of photons since the annihilation operator $\hat{a}$ does not change this state.

One can prove [Miller and Mishkin (1966)] that it is not possible to construct states $\| \alpha\rangle\rangle$ with a finite α, which are eigenstates of the creation operator $\hat{a}^+$,

$$\hat{a}^+ \| \alpha\rangle\rangle = \alpha \| \alpha\rangle\rangle, \tag{4.6a}$$

$$\langle\langle \alpha \| \hat{a} = \langle\langle\alpha \| \alpha^*. \tag{4.6b}$$

Writing

$$\|\alpha\rangle\rangle = \sum_{n=0}^{\infty} |n\rangle \langle n \| \alpha\rangle\rangle, \tag{4.7}$$

we obtain a recurrence relation for the coefficient $\langle n \| \alpha\rangle\rangle$: using (4.6a) and making repeated use of the Hermitian adjoint to (2.17b),

$$\alpha\langle n \| \alpha\rangle\rangle = n^{1/2}\langle n - 1 \| \alpha\rangle\rangle, \tag{4.8}$$

we obtain

$$\langle n \| \alpha\rangle\rangle = \frac{(n!)^{1/2}}{\alpha^n} \langle 0 \|\alpha\rangle\rangle. \tag{4.9}$$

Substituting (4.9) into (4.7) ,we have

$$\| \alpha\rangle\rangle = \langle 0 \| \alpha\rangle\rangle \sum_{n=0}^{\infty} \frac{(n!)^{1/2}}{\alpha^n} | n\rangle, \tag{4.10}$$

and consequently we have for the squared norm of $\| \alpha\rangle\rangle$

$$\langle\langle\alpha \| \alpha\rangle\rangle = | \langle 0 \| \alpha\rangle\rangle |^2 \sum_{n=0}^{\infty} \frac{n!}{|\alpha|^{2n}}, \tag{4.11}$$

recalling that $\langle n \mid m\rangle = \delta_{nm}$. This is divergent for all finite complex amplitudes α, and the states $\| \alpha\rangle\rangle$ cannot be regarded as physically admissible states of the radiation field. The states $|n\rangle$ and $|\alpha\rangle$ are the most useful bases for representations of the radiation field.

From (4.4) we see that the coherent states are normalized, $\langle\alpha \mid \alpha\rangle = 1$ and

$$\langle n \mid \alpha\rangle = \exp\left(-\frac{1}{2} |\alpha|^2\right) \frac{\alpha^n}{(n!)^{1/2}}, \tag{4.12a}$$

which provides

$$|\langle n \mid \alpha\rangle|^2 = \frac{|\alpha|^{2n}}{n!} \exp(-|\alpha|^2); \tag{4.12b}$$

this is a Poisson distribution with $\langle n \rangle = |\alpha|^2$ so the probability that there are n photons in the coherent state is given by the Poisson distribution, in agreement with (3.65) and (3.152).

The scalar product of two coherent states is

$$\langle \alpha | \beta \rangle = \exp\left(-\frac{1}{2} |\alpha|^2 - \frac{1}{2} |\beta|^2 + \alpha^* \beta \right), \tag{4.13a}$$

and so

$$|\langle \alpha | \beta \rangle|^2 = \exp(-|\alpha - \beta|^2). \tag{4.13b}$$

This shows that the coherent states are not orthogonal (they can be regarded as approximately orthogonal if $|\alpha - \beta| \gg 1$).

4.1.2 Expansions in terms of coherent states

The coherent states form an overcomplete system of states, enabling us to expand any vector or operator (particularly the density operator) in their basis.

We first note that the coherent states are mutually dependent, which is the most characteristic property of an overcomplete set of vectors. On multiplying (4.4) by α^k and integrating over the whole complex α-plane we have

$$\int \alpha^k |\alpha\rangle \, d^2\alpha = 0, \qquad k = 1, 2, \ldots,$$

where $\alpha = r \exp(i\varphi)$, r and φ being polar coordinates and $d^2\alpha = d(\mathrm{Re}\,\alpha)\, d(\mathrm{Im}\,\alpha) = r\, dr\, d\varphi$. The Fock states $|n\rangle$ can be expanded in terms of $|\alpha\rangle$ by multiplying (4.4) by $\pi^{-1}(n!)^{-1/2}\alpha^{*n} \exp(-|\alpha|^2/2)$ and integrating over α:

$$|n\rangle = \frac{1}{\pi} \int \exp\left(-\frac{|\alpha|^2}{2} \right) \frac{\alpha^{*n}}{(n!)^{1/2}} |\alpha\rangle \, d^2\alpha. \tag{4.14}$$

Substituting (4.14) in the completeness condition (2.24) we obtain

$$\frac{1}{\pi^2} \iint \exp\left(-\frac{|\alpha|^2}{2} - \frac{|\beta|^2}{2} + \alpha\beta^* \right) |\beta\rangle \langle \alpha| \, d^2\alpha \, d^2\beta = \hat{1}. \tag{4.15}$$

The integral over β becomes

$$\frac{1}{\pi} \int |\beta\rangle \exp\left(\frac{|\beta|^2}{2} \right) \exp(-|\beta|^2 + \alpha\beta^*) \, d^2\beta = |\alpha\rangle \exp\left(\frac{|\alpha|^2}{2} \right), \tag{4.16}$$

where we have expressed $|\beta\rangle \exp(|\beta|^2/2)$ using (4.4), have written $\exp(\alpha\beta^*)$ in the form of a series, and have used the integral

$$\int \beta^n \beta^{*m} \exp(-s|\beta|^2) \, d^2\beta = \frac{\pi n!}{s^{n+1}} \delta_{nm}, \qquad \mathrm{Re}\, s > 0 \tag{4.17}$$

for $s = 1$. In this way we arrive at the "resolution of unity" in terms of coherent states

$$\frac{1}{\pi}\int|\alpha\rangle\langle\alpha|\,\mathrm{d}^2\alpha \doteq \hat{1}. \tag{4.18}$$

For an entire function $f(\alpha)$, in analogy with (4.16),

$$\frac{1}{\pi}\int f(\beta)\exp(-|\beta|^2+\alpha\beta^*)\,\mathrm{d}^2\beta = f(\alpha). \tag{4.19}$$

The identity (4.18) makes it possible to express an arbitrary vector $|\,\rangle$ as

$$|\,\rangle = \frac{1}{\pi}\int|\alpha\rangle\langle\alpha|\,\rangle\,\mathrm{d}^2\alpha \tag{4.20}$$

and an arbitrary operator $\hat{M}$ as

$$\hat{M} = \frac{1}{\pi^2}\iint|\alpha\rangle\langle\alpha|\hat{M}|\beta\rangle\langle\beta|\,\mathrm{d}^2\alpha\,\mathrm{d}^2\beta. \tag{4.21}$$

The coefficient $\langle\alpha|\,\rangle$ in the decomposition (4.20) can be written, expanding the state $|\,\rangle$ in terms of the Fock states, as

$$\langle\alpha|\sum_n f_n|n\rangle = \exp\left(-\frac{1}{2}|\alpha|^2\right)\sum_n\frac{\alpha^{*n}}{(n!)^{1/2}}f_n = \\ = f(\alpha^*)\exp\left(-\frac{|\alpha|^2}{2}\right). \tag{4.22}$$

For a normalized vector $|\,\rangle$, $\langle\,|\,\rangle = \sum_n |f_n|^2 = 1$ and the series $f(\alpha^*) = \sum_n \alpha^{*n}f_n/(n!)^{1/2}$ converges for all values of α in the finite plane and therefore $f(\alpha^*)$ is an entire function.

4.1.3 Minimum-uncertainty wave packets

Defining $\Delta\hat{q}$ and $\Delta\hat{p}$ by

$$\Delta\hat{q} = \hat{q} - \langle\hat{q}\rangle, \qquad \Delta\hat{p} = \hat{p} - \langle\hat{p}\rangle, \tag{4.23}$$

we have from the commutation rule (2.22)

$$\frac{\hbar}{2} = \frac{1}{2}|\langle[\hat{q},\hat{p}]\rangle| = \frac{1}{2}|\langle[\Delta\hat{q},\Delta\hat{p}]\rangle| \leqq |\langle\Delta\hat{q}\Delta\hat{p}\rangle| \leqq \\ \leqq [\langle(\Delta\hat{q})^2\rangle\langle(\Delta\hat{p})^2\rangle]^{1/2}. \tag{4.24}$$

Making use of (4.5a, b), we obtain for the expectation values of (2.21) in the coherent states

$$\langle\hat{q}\rangle = \left(\frac{\hbar}{2\omega}\right)^{1/2}(\alpha+\alpha^*), \qquad \langle\hat{p}\rangle = -\mathrm{i}\left(\frac{\hbar\omega}{2}\right)^{1/2}(\alpha-\alpha^*), \tag{4,25a}$$

$$\langle(\Delta\hat{q})^2\rangle = \frac{\hbar}{2\omega}, \qquad \langle(\Delta\hat{p})^2\rangle = \frac{\hbar\omega}{2}. \tag{4.25}$$

Consequently for the coherent states the inequality (4.24) reduces to the equality

$$\langle(\Delta\hat{q})^2\rangle\,\langle(\Delta\hat{p})^2\rangle = \frac{\hbar^2}{4}. \tag{4.26}$$

Thus the coherent states are the closest to the classical states that quantum theory allows.

Following Picard and Willis (1965), we can use the arithmetic mean – geometric mean inequality and $\langle(\Delta\hat{p})^2\rangle = \omega^2\langle(\Delta\hat{q})^2\rangle$ (for the coherent states) to obtain

$$\frac{1}{2}[\langle(\Delta\hat{p})^2\rangle + \omega^2\langle(\Delta\hat{q})^2\rangle] \geqq \omega[\langle(\Delta\hat{q})^2\rangle\,\langle(\Delta\hat{p})^2\rangle]^{1/2} = \frac{\hbar\omega}{2}, \tag{4.27}$$

which reduces to the equality in the coherent states. The quantity

$$\frac{1}{2}[\langle(\Delta\hat{p})^2\rangle + \omega^2\langle(\Delta\hat{q})^2\rangle] =$$

$$= \frac{1}{2}[\langle\hat{p}^2\rangle + \omega^2\langle\hat{q}^2\rangle] - \frac{1}{2}[\langle\hat{p}\rangle^2 + \omega^2\langle\hat{q}\rangle^2] \tag{4.28}$$

represents the difference between the total energy of the field and its coherent energy, i.e. it represents the incoherent energy of the field. Therefore it may be said that in the coherent state the incoherent energy takes on its minimum value $\hbar\omega/2$ representing only the energy of the zero-point fluctuations. This is demonstrated in Fig. 4.1

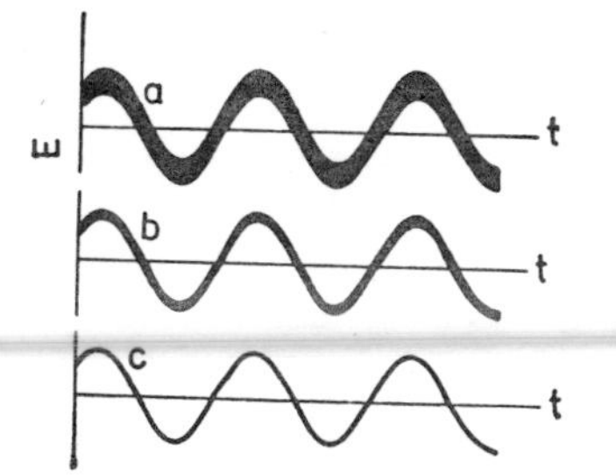

Fig. 4.1 — An outline of the qualitative time behaviour of the electric field in the coherent state with (a) $\langle n\rangle = |\alpha|^2 = 1$, (b) 10 and (c) 100; the black area represents the uncertainty arising from vacuum fluctuations.

showing the dependence of the electric intensity on time for various values of $\langle n\rangle = |\alpha|^2$ of the coherent state. The black area represents the physical vacuum fluctuations.

4.1.4 Properties of the displacement operator $\hat{D}(\alpha)$

From the definition of the displacement operator (4.2)

$$\hat{D}^+(\alpha) = \hat{D}^{-1}(\alpha) = \hat{D}(-\alpha). \tag{4.29}$$

Applying (4.3),

$$\hat{a}\hat{D}(\alpha) = \exp\left(-\frac{1}{2}|\alpha|^2\right)\hat{a}\exp(\alpha\hat{a}^+)\exp(-\alpha^*\hat{a}) =$$
$$= \exp\left(-\frac{1}{2}|\alpha|^2\right)\sum_{n=0}^{\infty}\frac{\alpha^n}{n!}\hat{a}\hat{a}^{+n}\exp(-\alpha^*\hat{a}). \tag{4.30}$$

Since from the commutation rule $\hat{a}\hat{a}^{+n} = \hat{a}^{+n}\hat{a} + n\hat{a}^{+n-1}$, we have from (4.30)

$$\hat{a}\hat{D}(\alpha) = \hat{D}(\alpha)\,\hat{a} + \alpha\hat{D}(\alpha). \tag{4.31}$$

In the same way we can prove for any entire operator $\hat{M}(\hat{a}^+, \hat{a})$ that

$$[\hat{a}, \hat{M}] = \frac{\partial\hat{M}}{\partial\hat{a}^+}, \qquad [\hat{M}, \hat{a}^+] = \frac{\partial\hat{M}}{\partial\hat{a}}, \tag{4.32}$$

giving the standard commutation rules for $\hat{M} = \hat{a}^+$ and $\hat{a}$. Therefore $\hat{D}(\alpha)$ represents a displacement operator,

$$\hat{D}^{-1}(\alpha)\,\hat{a}\hat{D}(\alpha) = \hat{a} + \alpha,$$
$$\hat{D}^{-1}(\alpha)\,\hat{a}^+\hat{D}(\alpha) = \hat{a}^+ + \alpha^*. \tag{4.33}$$

Or more generally for any entire operator function $\hat{M}(\hat{a}^+, \hat{a})$

$$\hat{D}^{-1}(\alpha)\,\hat{M}(\hat{a}^+, \hat{a})\,\hat{D}(\alpha) = \hat{M}(\hat{a}^+ + \alpha^*, \hat{a} + \alpha). \tag{4.34}$$

The product of the two displacement operators is equal to

$$\hat{D}(\alpha)\,\hat{D}(\beta) = \hat{D}(\alpha + \beta)\exp\left(\frac{1}{2}(\alpha\beta^* - \alpha^*\beta)\right), \tag{4.35}$$

where (4.3) has been used. Taking the trace of $\hat{D}$ using (4.18) we arrive at

$$\operatorname{Tr}\hat{D}(\alpha) = \frac{1}{\pi}\int\langle\beta\,|\,\hat{D}(\alpha)\,|\,\beta\rangle\,\mathrm{d}^2\beta = \pi\delta(\alpha), \tag{4.36}$$

where (4.3) and (4.5a, b) have been applied. This gives for (4.35)

$$\operatorname{Tr}\{\hat{D}^+(\alpha)\,\hat{D}(\beta)\} = \pi\delta(\alpha - \beta). \tag{4.37}$$

The complex representation of the $\delta(\alpha)$-function is expressed in the form of the Fourier integral,

$$\delta(\alpha) = \frac{1}{\pi^2}\int\exp(\alpha\beta^* - \alpha^*\beta)\,\mathrm{d}^2\beta. \tag{4.39}$$

4.1.5 Expectation values of operators in coherent states

Assume an operator $\hat{M}(\hat{a}^+, \hat{a})$ to be in normal form. Then it holds that, applying (4.5a, b),

$$\langle\alpha\,|\,\hat{M}(\hat{a}^+, \hat{a})\,|\,\alpha'\rangle = \langle\alpha\,|\,\hat{M}^{(\mathcal{N})}(\hat{a}^+, \hat{a})\,|\,\alpha'\rangle =$$
$$= M^{(\mathcal{N})}(\alpha, \alpha')\,\langle\alpha\,|\,\alpha'\rangle, \tag{4.39}$$

and in particular

$$\langle \alpha | \hat{M}^{(\mathcal{N})}(\hat{a}^+, \hat{a}) | \alpha \rangle = M^{(\mathcal{N})}(\alpha^*, \alpha), \tag{4.40}$$

i.e. the expectation value of a normally ordered operator $\hat{M}^{(\mathcal{N})}$ is obtained by the substitutions $\hat{a} \to \alpha$, $\hat{a}^+ \to \alpha^*$ in the operator function. If $\hat{M}^{(\mathcal{N})}(\hat{a}^+, \hat{a}) = \hat{a}^{+k}\hat{a}^l$ (k, l are integers), then these q-c-number substitutions provide $\langle \alpha | \hat{a}^{+k}\hat{a}^l | \alpha \rangle = \alpha^{*k}\alpha^l$.

Further we can prove, using (4.4), that

$$\hat{a}^+ | \alpha \rangle \langle \alpha | = \left(\alpha^* + \frac{\partial}{\partial \alpha} \right) | \alpha \rangle \langle \alpha |,$$
$$| \alpha \rangle \langle \alpha | \hat{a} = \left(\alpha + \frac{\partial}{\partial \alpha^*} \right) | \alpha \rangle \langle \alpha |. \tag{4.41}$$

For an operator $\hat{M}$ we have

$$\hat{M}(\hat{a}^+, \hat{a}) | \alpha \rangle \langle \alpha | = M\left(\alpha^* + \frac{\partial}{\partial \alpha}, \alpha \right) | \alpha \rangle \langle \alpha |,$$
$$| \alpha \rangle \langle \alpha | \hat{M}(\hat{a}^+, \hat{a}) = M\left(\alpha^*, \alpha + \frac{\partial}{\partial \alpha^*} \right) | \alpha \rangle \langle \alpha | \tag{4.42}$$

and multiplying this by $| \alpha \rangle$ from the right and by $\langle \alpha |$ from the left, we arrive at

$$\langle \alpha | \hat{M}(\hat{a}^+, \hat{a}) | \alpha \rangle = M\left(\alpha^* + \frac{\partial}{\partial \alpha}, \alpha \right) = M\left(\alpha^*, \alpha + \frac{\partial}{\partial \alpha^*} \right). \tag{4.43}$$

These rules are useful for solving operator equations of motion. As an example we can consider the Schrödinger equation for an operator $\hat{U}$,

$$i\hbar \frac{\partial \hat{U}(\hat{a}^+, \hat{a}, t)}{\partial t} = \hat{H}(\hat{a}^+, \hat{a})\, \hat{U}(\hat{a}^+, \hat{a}, t). \tag{4.44}$$

If we multiply (4.44) by $| \alpha \rangle$ from the right and by $\langle \alpha |$ from the left and if we use (4.43), we obtain

$$i\hbar \frac{\partial U\left(\alpha^* + \frac{\partial}{\partial \alpha}, \alpha, t \right)}{\partial t} = H\left(\alpha^* + \frac{\partial}{\partial \alpha}, \alpha \right) U\left(\alpha^* + \frac{\partial}{\partial \alpha}, \alpha, t \right), \tag{4.45}$$

which is a classical equation. Solving this equation we have

$$U\left(\alpha^* + \frac{\partial}{\partial \alpha}, \alpha, t \right) = U^{(\mathcal{N})}(\alpha^*, \alpha, t) = \langle \alpha |\ \hat{U}^{(\mathcal{N})}(\hat{a}^+, \hat{a}, t) | \alpha \rangle. \tag{4.46}$$

This procedure leads directly to the normal form of $\hat{U}$.

4.1.6 Generalized coherent states

A generalization of the coherent states has been introduced by Titulaer and Glauber (1966). Such generalized coherent states can be written in the form

$$| \alpha, \{\varphi_n\}\rangle = \exp\left(-\frac{1}{2}|\alpha|^2\right)\sum_{n=0}^{\infty} \frac{\alpha^n \exp(i\varphi_n)}{(n!)^{1/2}} | n\rangle, \tag{4.47}$$

where φ_n is a sequence of real numbers. These states may be regarded as the most general pure fully coherent states in the sense of $m = n$-factorization for a harmonic oscillator. Some of their properties were investigated by Titulaer and Glauber (1966), Crosignani, Di Porto and Solimeno (1968a) and Bialynicka–Birula (1968). Some properties of subsets of the coherent states were discussed by Campagnoli and Zambotti (1968). Putting $\varphi_n = n\varphi$, we see that $\hat{a} | \alpha, \{\varphi_n\}\rangle = \alpha \exp(i\varphi) | \alpha, \{\varphi_n\}\rangle$ and so $\langle\alpha, \{\varphi_n\} | \hat{a}^{+k}\hat{a}^k | \alpha, \{\varphi_n\}\rangle = \alpha^{*k}\alpha^k$ independently of the phase φ. This shows that the field in the generalized coherent state, which is a mixture of coherent fields with various phases, also fulfils the even-order full coherence factorization conditions. However, the full set of factorization conditions for all m and n can be satisfied only by the coherent state $| \alpha\rangle$.

4.1.7 Multimode description

The coherent-state formalism, just developed for one mode, can be applied to the electromagnetic field in a finite volume having a finite or a countably infinite number of degrees of freedom. (A generalization to fields in infinite volume, having a non-denumerable number of degrees of freedom, is also possible if (continuous) integrals are adopted.) For this purpose we introduce the global coherent state

$$| \{\alpha_\lambda\}\rangle \equiv \prod_\lambda | \alpha_\lambda\rangle, \tag{4.48}$$

so that

$$| \{\alpha_\lambda\}\rangle = \sum_{\{n_\lambda\}=0}^{\infty} \left\{\prod_{\lambda'} \frac{\alpha_{\lambda'}^{n_{\lambda'}}}{(n_{\lambda'}!)^{1/2}} \exp\left(-\frac{|\alpha_{\lambda'}|^2}{2}\right) | \{n_\lambda\}\rangle\right\}. \tag{4.49}$$

The completeness condition reads

$$\int | \{\alpha_\lambda\}\rangle \langle\{\alpha_\lambda\} | \mathrm{d}\mu(\{\alpha_\lambda\}) = \hat{1}, \tag{4.50}$$

where

$$\mathrm{d}\mu(\{\alpha_\lambda\}) \equiv \prod_\lambda \frac{\mathrm{d}^2\alpha_\lambda}{\pi} \tag{4.51}$$

and $| \{n_\lambda\}\rangle$ is given by (2.43).

If (2.28b) and (4.5a) are used, we obtain the following eigen-property of the global coherent state

$$\hat{\mathbf{A}}^{(+)}(x) | \{\alpha_\lambda\}\rangle = \mathbf{V}(x) | \{\alpha_\lambda\}\rangle \tag{4.52a}$$

and its Hermitian adjoint is

$$\langle\{\alpha_\lambda\}\mid \hat{\mathbf{A}}^{(-)}(x) = \langle\{\alpha_\lambda\}\mid \mathbf{V}^*(x), \tag{4.52b}$$

where

$$\mathbf{V}(x) = \frac{(2\pi\hbar c)^{1/2}}{L^{3/2}} \sum_{\mathbf{k},s} k^{-1/2}\mathbf{e}^{(s)}(\mathbf{k})\,\alpha_{\mathbf{k}s} \exp\left[\mathrm{i}(\mathbf{k}\,.\,\mathbf{x} - ckt)\right]. \tag{4.53}$$

Hence, the quantum and classical correlation functions (3.5a, b) are equivalent. Choosing the density matrix in the form $\hat{\varrho} = \mid\{\alpha_\lambda\}\rangle\langle\{\alpha_\lambda\}\mid (\hat{\varrho}^2 = \hat{\varrho})$, these correlation functions are factorized in the form (3.53) and the field in the coherent state fulfils the full coherence conditions.

4.1.8 Time development of the coherent states

Considering a one-mode field for simplicity with the free renormalized hamiltonian $\hbar\omega\hat{a}^+\hat{a}$, the time dependence of the coherent state $\mid\alpha(t)\rangle$ is given by the Schrödinger equation

$$\mathrm{i}\hbar\frac{\partial\mid\alpha(t)\rangle}{\partial t} = \hat{H}\mid\alpha(t)\rangle. \tag{4.54}$$

If the initial condition is $\mid\alpha(0)\rangle = \mid\alpha\rangle$, the solution can be written as

$$\mid\alpha(t)\rangle = \exp\left(-\mathrm{i}\frac{\hat{H}t}{\hbar}\right)\mid\alpha\rangle = \mid\alpha\exp(-\mathrm{i}\omega t)\rangle. \tag{4.55}$$

Therefore for this hamiltonian the coherent state remains coherent at all times. The complex amplitude $\alpha\exp(-\mathrm{i}\omega t)$ of the coherent state describes circles in the complex plane.

It can be shown more generally that the coherent state remains coherent at all times when the hamiltonian takes the form

$$\hat{H} = \sum_{j,k} f_{jk}(t)\,\hat{a}_j^+\hat{a}_k + \sum_k \left[g_k(t)\,\hat{a}_k^+ + \text{h.c.}\right] + h(t),$$

where $f_{jk} = f_{kj}$, $h^*(t) = h(t)$ and $g_k(t)$ are arbitrary functions of time [Glauber (1966b), Mehta and Sudarshan (1966), Mehta et al. (1967)]. This hamiltonian describes the free energy, the exchange of energies among modes and the interaction of classical currents with the radiation field. This means that a coherent state is generated by non-random prescribed currents [Glauber (1963b, 1965)]. A hamiltonian including third-order nonlinear terms yet preserving special coherent states has been discussed by Mišta (1967). Hamiltonians for quantum harmonic oscillators with time-dependent and random frequencies have been considered, in connection with the coherent states, by Crosignani, Di Porto and Solimeno (1968b, 1969) and Solimeno et al. (1969). Further questions of the time evolution of the coherent states have been discussed by Kano (1976), Trifonov and Ivanov (1977), Trias (1977), Chand (1979) and Malkin and Man'ko (1979).

4.1.9 Even and odd coherent states

One can introduce the even and odd coherent states as follows [Malkin and Man'ko (1979)]

$$|\alpha_+\rangle = \frac{1}{2}(|\alpha\rangle + |-\alpha\rangle) = \exp\left(-\frac{|\alpha|^2}{2}\right)\sum_{n=0}^{\infty}\frac{\alpha^{2n}}{((2n)!)^{1/2}}|2n\rangle, \tag{4.57a}$$

$$|\alpha_-\rangle = \frac{1}{2}(|\alpha\rangle - |-\alpha\rangle) = $$
$$= \exp\left(-\frac{|\alpha|^2}{2}\right)\sum_{n=0}^{\infty}\frac{\alpha^{2n+1}}{((2n+1)!)^{1/2}}|2n+1\rangle, \tag{4.57b}$$

corresponding to the displacement operators

$$\frac{1}{2}[\hat{D}(\alpha) + \hat{D}(-\alpha)] = \mathrm{ch}\,(\alpha\hat{a}^+ - \alpha^*\hat{a}), \tag{4.58a}$$

$$\frac{1}{2}[\hat{D}(\alpha) - \hat{D}(-\alpha)] = \mathrm{sh}\,(\alpha\hat{a}^+ - \alpha^*\hat{a}) \tag{4.58b}$$

and

$$\hat{a}^2\,|\,\alpha_\pm\rangle = \alpha^2\,|\,\alpha_\pm\rangle. \tag{4.59}$$

4.2 Glauber—Sudarshan representation of the density matrix

Mostly we deal with mixed states rather than pure states and we must be able to express the density matrix for such general fields in terms of the coherent states. Using the operator expansion (4.21) for the density matrix we obtain [Glauber (1963b)]

$$\hat{\varrho} = \frac{1}{\pi^2}\iint |\,\alpha\rangle\langle\alpha\,|\,\hat{\varrho}\,|\,\beta\rangle\langle\beta\,|\,\mathrm{d}^2\alpha\,\mathrm{d}^2\beta, \tag{4.60}$$

which is an expansion in terms of dyadic products $|\,\alpha\rangle\langle\beta\,|$. Using the Fock form (2.31) of $\hat{\varrho}$ and (4.12a) we obtain for the weight function $\langle\alpha\,|\,\hat{\varrho}\,|\,\beta\rangle$ in this expansion

$$\langle\alpha\,|\,\varrho\,|\,\beta\rangle = \sum_n\sum_m \varrho(n,m)\frac{\alpha^{*n}\beta^m}{(n!m!)^{1/2}}\exp\left(-\frac{|\alpha|^2}{2} - \frac{|\beta|^2}{2}\right). \tag{4.61}$$

Since $\mathrm{Tr}\,\hat{\varrho}^2 = \sum_{n,m}|\,\varrho(n,m)\,|^2 \leqq 1$ $(|\,\varrho(n,m)\,|^2 \leqq \varrho(n,n)\,\varrho(m,m))$, the series in (4.61) converge for all finite values of α^* and β and so $\langle\alpha\,|\,\hat{\varrho}\,|\,\beta\rangle$ is an entire function in α and β, i.e. it is a well-behaved function.

The advantage of the expansion (4.60) is that the quantum expectation values of normally ordered operators (q-numbers) reduce to expectation values of eigenvalues in the coherent states (c-numbers) with weight function $\langle\alpha\,|\,\hat{\varrho}\,|\,\beta\rangle$ and they may be calculated simply by integration over the complex planes of α and β. The expansion

(4.60) corresponds to the condition of "resolution of unity" in the form

$$\frac{1}{\pi^2} \iint |\alpha\rangle \langle\alpha|\beta\rangle \langle\beta| \, d^2\alpha \, d^2\beta =$$

$$= \frac{1}{\pi^2} \iint \exp\left(-\frac{|\alpha|^2}{2} - \frac{|\beta|^2}{2} + \alpha^*\beta\right) |\alpha\rangle \langle\beta| \, d^2\alpha \, d^2\beta = \hat{1}, \tag{4.62}$$

where (4.13a) has been used. This is just the identity (4.15) which leads to the simpler completeness relation (4.18) using (4.16). We can now ask whether the density matrix $\hat{\varrho}$ can be expanded in the simpler diagonal form [Glauber (1963b, c), Sudarshan (1963a)]

$$\hat{\varrho} = \int \Phi_{\mathcal{N}}(\alpha) |\alpha\rangle \langle\alpha| \, d^2\alpha \tag{4.63}$$

as a mixture of the projection operators $|\alpha\rangle\langle\alpha|$ onto the coherent states $((|\alpha\rangle\langle\alpha|)^2 = |\alpha\rangle\langle\alpha|)$, where $\Phi_{\mathcal{N}}(\alpha)$ is the weight function. We call this diagonal representation the Glauber–Sudarshan representation of the density matrix. Since $\mathrm{Tr}\,\hat{\varrho} = 1$,

$$\int \Phi_{\mathcal{N}}(\alpha) \, d^2\alpha = 1 \tag{4.64}$$

and from $\hat{\varrho}^+ = \hat{\varrho}$ it follows that $[\Phi_{\mathcal{N}}(\alpha)]^* = \Phi_{\mathcal{N}}(\alpha)$, i.e. $\Phi_{\mathcal{N}}(\alpha)$ is a real function of the complex variable α.

The function $\langle\alpha|\hat{\varrho}|\beta\rangle$ can be expressed in terms of $\Phi_{\mathcal{N}}(\alpha)$, using (4.63) and (4.13a), in the form

$$\langle\alpha|\hat{\varrho}|\beta\rangle =$$

$$= \exp\left(-\frac{|\alpha|^2}{2} - \frac{|\beta|^2}{2}\right) \int \Phi_{\mathcal{N}}(\gamma) \exp(-|\gamma|^2 + \alpha^*\gamma + \beta\gamma^*) \, d^2\gamma \tag{4.65a}$$

and for $\alpha = \beta$ we arrive at

$$\langle\alpha|\hat{\varrho}|\alpha\rangle = \int \Phi_{\mathcal{N}}(\beta) \exp(-|\alpha - \beta|^2) \, d^2\beta. \tag{4.65b}$$

An interesting relation between $\Phi_{\mathcal{N}}(\alpha)$ and the Fock matrix elements $\varrho(n, m)$ of the density matrix is obtained, using (4.63),

$$\langle n|\hat{\varrho}|m\rangle = \varrho(n, m) = \int \Phi_{\mathcal{N}}(\alpha) \frac{\alpha^n \alpha^{*m}}{(n!m!)^{1/2}} \exp(-|\alpha|^2) \, d^2\alpha. \tag{4.66}$$

Note that the representation (4.63) can formally be derived from the Fock form (2.31) of the density matrix. Substituting (4.14) into (2.31) and making use of (4.66) we find

$$\varrho = \frac{1}{\pi^2} \iiint \exp\left(-\frac{|\alpha|^2}{2} - \frac{|\beta|^2}{2} - |\gamma|^2 + \alpha^*\gamma + \beta\gamma^*\right) \times$$

$$\times \Phi_{\mathcal{N}}(\gamma) |\alpha\rangle \langle\beta| \, d^2\alpha \, d^2\beta \, d^2\gamma \tag{4.67}$$

and using (4.16) we obtain (4.63).

The diagonal form (4.63) of the density matrix is very convenient for calculations of expectation values of normally ordered operators, i.e. the normal correlation functions. For example, for $\hat{a}^{+k}\hat{a}^{l}$ (k, l are integers) we have

$$\begin{aligned}\mathrm{Tr}\,\{\hat{\varrho}\hat{a}^{+k}\hat{a}^{l}\} &= \mathrm{Tr}\,\{\textstyle\int \Phi_{\mathcal{N}}(\alpha)\,|\,\alpha\rangle\,\langle\alpha\,|\,\mathrm{d}^2\alpha\hat{a}^{+k}\hat{a}^{l}\} = \\ &= \textstyle\int \Phi_{\mathcal{N}}(\alpha)\,\langle\alpha\,|\,\hat{a}^{+k}\hat{a}^{l}\,|\,\alpha\rangle\,\mathrm{d}^2\alpha = \int \Phi_{\mathcal{N}}(\alpha)\,\alpha^{*k}\alpha^{l}\,\mathrm{d}^2\alpha = \\ &= \langle\alpha^{*k}\alpha^{l}\rangle_{\mathcal{N}},\end{aligned} \tag{4.68}$$

where we have used the eigenvalue properties (4.5a, b) of the coherent states. The suffix $\mathcal{N}$ on Φ expresses the fact that this function is related to the averaging of the normally ordered operators. Consequently, the quantum expectation value of normally ordered operators can be expressed as a "classical" expectation value in a generalized phase space with the quasi-distribution $\Phi_{\mathcal{N}}(\alpha)$ if the integration is carried out over the whole complex α-plane. As we have seen, the correlation functions suitable for the description of experiments with photodetectors, yielding information about the coherence properties of light, represent just the expectation values of normally ordered products of the field operators and so they may be expressed, using the Glauber–Sudarshan representation, in a form very close to the classical one. This representation provides a basis for a general formal equivalence between the classical and quantum descriptions of optical coherence.

The field in the coherent state $|\,\beta\rangle$ possesses the weight function $\Phi_{\mathcal{N}}(\alpha)$ in the form

$$\Phi_{\mathcal{N}}(\alpha) = \delta(\alpha - \beta). \tag{4.69}$$

Although $\Phi_{\mathcal{N}}(\alpha)$ has some of the properties of a probability distribution [it is a real-valued function satisfying the normalization (4.64)], it cannot generally be interpreted as a probability distribution since it is not non-negative and it may have singularities stronger than the δ-function. This property is a consequence of the fact that the quantities Re α and Im α, which are mean values of the non-commuting operators $(\hat{a} + \hat{a}^{+})/2$ and $(\hat{a} - \hat{a}^{+})/2\mathrm{i}$ in the coherent state, are not simultaneously measurable and so $\Phi_{\mathcal{N}}(\alpha)$ cannot be measured directly. We call the $\Phi_{\mathcal{N}}$-function the quasi-probability distribution for this reason. However, it represents one of the whole family of quasi-probability functions. Another quasi-distribution function was introduced by Wigner (1932) and $\langle\alpha\,|\,\hat{\varrho}\,|\,\alpha\rangle/\pi$ is also a quasi-probability function, although it has many properties in common with an actual probability density, including positive-definiteness and regularity.

The function $\pi^{-1}\langle\alpha\,|\,\hat{\varrho}\,|\,\alpha\rangle$ plays the same role in the averaging of antinormally ordered products as $\Phi_{\mathcal{N}}(\alpha)$ plays for normally ordered products, since

$$\begin{aligned}\mathrm{Tr}\,\{\hat{\varrho}\hat{a}^{l}\hat{a}^{+k}\} &= \mathrm{Tr}\,\{\hat{a}^{+k}\hat{\varrho}\hat{a}^{l}\} = \mathrm{Tr}\left\{\frac{1}{\pi}\int |\,\alpha\rangle\,\langle\alpha\,|\,\mathrm{d}^2\alpha\hat{a}^{+k}\hat{\varrho}\hat{a}^{l}\right\} = \\ &= \frac{1}{\pi}\int \alpha^{*k}\alpha^{l}\langle\alpha\,|\,\hat{\varrho}\,|\,\alpha\rangle\,\mathrm{d}^2\alpha;\end{aligned} \tag{4.70}$$

here we have carried out a cyclic permutation of the order of operators, which does not change the trace, and introduced the unit operator (4.18). We denote $\pi^{-1}\langle\alpha\,|\,\hat{\varrho}\,|\,\alpha\rangle$ as $\Phi_{\mathcal{A}}(\alpha)$, which emphasizes that this function is related to the average of antinormally ordered operators. From (4.65b) we have

$$\Phi_{\mathscr{A}}(\alpha) = \frac{1}{\pi} \int \Phi_{\mathscr{N}}(\beta) \exp(-|\alpha - \beta|^2) \, d^2\beta. \tag{4.71}$$

Putting $k = l = 0$ in (4.70) we arrive at the normalization

$$\frac{1}{\pi} \int \langle \alpha | \hat{\varrho} | \alpha \rangle \, d^2\alpha = 1. \tag{4.72}$$

Now we introduce the quantum characteristic function [Glauber (1966a)]

$$C(\beta) = \mathrm{Tr}\,\{\hat{\varrho}\hat{D}(\beta)\} = \mathrm{Tr}\,\{\hat{\varrho} \exp(\beta\hat{a}^+ - \beta^*\hat{a})\} \tag{4.73}$$

and making use of (4.3) we have

$$C_{\mathscr{N}}(\beta) = \exp\left(\frac{|\beta|^2}{2}\right) C(\beta) = \exp(|\beta|^2)\, C_{\mathscr{A}}(\beta), \tag{4.74}$$

where the normal characteristic function is

$$\begin{aligned} C_{\mathscr{N}}(\beta) &= \mathrm{Tr}\,\{\hat{\varrho} \exp(\beta\hat{a}^+) \exp(-\beta^*\hat{a})\} = \\ &= \int \Phi_{\mathscr{N}}(\alpha) \exp(\beta\alpha^* - \beta^*\alpha) \, d^2\alpha \end{aligned} \tag{4.75}$$

and the antinormal characteristic function is

$$\begin{aligned} C_{\mathscr{A}}(\beta) &= \mathrm{Tr}\,\{\hat{\varrho} \exp(-\beta^*\hat{a}) \exp(\beta\hat{a}^+)\} = \\ &= \int \Phi_{\mathscr{A}}(\alpha) \exp(\beta\alpha^* - \beta^*\alpha) \, d^2\alpha. \end{aligned} \tag{4.76}$$

If the complex δ-function (4.38) is employed, the inverse Fourier transforms are

$$\Phi_{\mathscr{N}}(\alpha) = \frac{1}{\pi^2} \int C_{\mathscr{N}}(\beta) \exp(\alpha\beta^* - \alpha^*\beta) \, d^2\beta \tag{4.77}$$

and

$$\Phi_{\mathscr{A}}(\alpha) = \frac{1}{\pi^2} \int C_{\mathscr{A}}(\beta) \exp(\alpha\beta^* - \alpha^*\beta) \, d^2\beta. \tag{4.78}$$

For the Wigner function $\Phi_{\mathrm{sym}}(\alpha)$, related to the symmetric ordering, we have

$$\Phi_{\mathrm{sym}}(\alpha) = \frac{1}{\pi^2} \int C(\beta) \exp(\alpha\beta^* - \alpha^*\beta) \, d^2\beta. \tag{4.79}$$

The moments of the Wigner function can be calculated as

$$\int \alpha^{*k}\alpha^l \Phi_{\mathrm{sym}}(\alpha) \, d^2\alpha = \frac{\partial^k}{\partial\beta^k} \frac{\partial^l}{\partial(-\beta^*)^l} C(\beta) \bigg|_{\beta=\beta^*=0} = \langle \alpha^{*k}\alpha^l \rangle_{\mathrm{sym}}, \tag{4.80}$$

where

$$C(\beta) = \int \Phi_{\mathrm{sym}}(\alpha) \exp(\beta\alpha^* - \beta^*\alpha) \, d^2\alpha. \tag{4.81}$$

If $k = l = 0$ in (4.80) we have the normalization

$$\int \Phi_{\mathrm{sym}}(\alpha) \, d^2\alpha = C(0) = 1. \tag{4.82}$$

In the same way the moments of $\Phi_{\mathcal{N}}(\alpha)$ and $\Phi_{\mathcal{A}}(\alpha)$ can be calculated from the normal and antinormal characteristic functions $C_{\mathcal{N}}(\beta)$ and $C_{\mathcal{A}}(\beta)$ respectively. From (4.75) and (4.76) we have

$$\frac{\partial^k}{\partial\beta^k}\frac{\partial^l}{\partial(-\beta^*)^l}C_{\mathcal{N}}(\beta)\bigg|_{\beta=\beta^*=0} = \mathrm{Tr}\,\{\hat{\varrho}\hat{a}^{+k}\hat{a}^l\} = \int \Phi_{\mathcal{N}}(\alpha)\,\alpha^{*k}\alpha^l\,\mathrm{d}^2\alpha = \\ = \langle\alpha^{*k}\alpha^l\rangle_{\mathcal{N}}, \tag{4.83}$$

$$\frac{\partial^k}{\partial\beta^k}\frac{\partial^l}{\partial(-\beta^*)^l}C_{\mathcal{A}}(\beta)\bigg|_{\beta=\beta^*=0} = \mathrm{Tr}\,\{\hat{\varrho}\hat{a}^l\hat{a}^{+k}\} = \int \Phi_{\mathcal{A}}(\alpha)\,\alpha^{*k}\alpha^l\,\mathrm{d}^2\alpha = \\ = \langle\alpha^{*k}\alpha^l\rangle_{\mathcal{A}}. \tag{4.84}$$

If $|\psi_n\rangle$ are normalized eigenstates of $\hat{\varrho}$ with eigenvalues λ_n, then $\hat{\varrho} = \sum_n \lambda_n |\psi_n\rangle\langle\psi_n|$, and since $0 \leqq \lambda_n \leqq 1$ ($\mathrm{Tr}\,\hat{\varrho} = \sum_n \lambda_n = 1$) we obtain for $\Phi_{\mathcal{A}}(\alpha)$ the following important properties:

$$\Phi_{\mathcal{A}}(\alpha) = \frac{1}{\pi}\sum_n \lambda_n |\langle\alpha|\psi_n\rangle|^2 \geqq 0 \tag{4.85a}$$

and

$$\Phi_{\mathcal{A}}(\alpha) \leqq \frac{1}{\pi}\sum_n |\langle\alpha|\psi_n\rangle|^2 = \frac{1}{\pi} \tag{4.85b}$$

(from $|\alpha\rangle = \sum_n |\psi_n\rangle\langle\psi_n|\alpha\rangle$, $\langle\alpha|\alpha\rangle = \sum_n |\langle\alpha|\psi_n\rangle|^2 = 1$). Further, since $\hat{D}(\alpha)$ is a unitary operator ($\hat{D}^+\hat{D} = \hat{1}$), $|C(\beta)| \leqq 1$ and from (4.74)

$$|C_{\mathcal{A}}(\beta)| \leqq \exp\left(-\frac{1}{2}|\beta|^2\right), \tag{4.86a}$$

$$|C_{\mathcal{N}}(\beta)| \leqq \exp\left(\frac{1}{2}|\beta|^2\right). \tag{4.86b}$$

Consequently $|C_{\mathcal{A}}(\beta)|$ always decreases at least as fast as $\exp(-|\beta|^2/2)$ for $|\beta| \to \infty$, while $|C_{\mathcal{N}}(\beta)|$ may diverge as rapidly as $\exp(|\beta|^2/2)$ in that limit.

We can see from (4.61) that $\Phi_{\mathcal{A}}(\alpha) = \langle\alpha|\hat{\varrho}|\alpha\rangle/\pi$ is an entire analytic function,

$$\Phi_{\mathcal{A}}(\alpha) = \frac{1}{\pi}\sum_{n,m} \varrho(n, m)\frac{\alpha^{*n}\alpha^m}{(n!m!)^{1/2}}\exp(-|\alpha|^2). \tag{4.87}$$

The quasi-distribution $\Phi_{\mathcal{A}}(\alpha)$ was first introduced and its properties were studied by Glauber (1965), Kano (1964a, 1965) and Mehta and Sudarshan (1965).

Relations between functions $\Phi_{\mathcal{N}}(\alpha)$, $\Phi_{\mathcal{A}}(\alpha)$ and $\Phi_{\mathrm{sym}}(\alpha)$ can be derived from the relation (4.74) between characteristic functions using the convolution theorem:

$$\Phi_{\mathcal{A}}(\alpha) = \frac{2}{\pi}\int \exp(-2|\alpha-\beta|^2)\,\Phi_{\mathrm{sym}}(\beta)\,\mathrm{d}^2\beta, \tag{4.88a}$$

$$\Phi_{\mathrm{sym}}(\alpha) = \frac{2}{\pi}\int \exp(-2|\alpha-\beta|^2)\,\Phi_{\mathcal{N}}(\beta)\,\mathrm{d}^2\beta \tag{4.88b}$$

and (4.71) relates $\Phi_{\mathscr{A}}(\alpha)$ and $\Phi_{\mathscr{N}}(\alpha)$. In general, the functions $\Phi_{\mathscr{A}}$ and Φ_{sym} on the left-hand side represent averages of the functions Φ_{sym} and $\Phi_{\mathscr{N}}$ on the right-hand side with the exponential weight factors. This averaging process, which takes us from $\Phi_{\mathscr{N}}$ to Φ_{sym} and $\Phi_{\mathscr{A}}$ tends to smooth out any unruly behaviour of $\Phi_{\mathscr{N}}$ and to transform it into a smooth function. This smoothing process makes $\Phi_{\mathscr{A}}$ non-negative and regular. While the quasi-distributions $\Phi_{\mathscr{A}}$ and Φ_{sym} exist for all quantum states, the cases in which $\Phi_{\mathscr{N}}$ cannot be defined in the usual sense are precisely those in which the convolution integrals (4.71) and (4.88b) cannot be inverted.

The quasi-distributions $\Phi_{\mathscr{N}}$ or $\Phi_{\mathscr{A}}$ can be obtained directly from the density matrix, adopting appropriate ordering [Lax (1967), Haken, Risken and Weidlich (1967), Lax and Louisell (1967)]. Suppose that an operator $\hat{M}$ can be written in the normal and antinormal forms with the help of the commutation rules,

$$\hat{M}(\hat{a}^+, \hat{a}) = \hat{M}^{(\mathscr{N})}(\hat{a}^+, \hat{a}) = \hat{M}^{(\mathscr{A})}(\hat{a}^+, \hat{a}) = \sum_{r,s} M_{rs}^{(\mathscr{N})}\hat{a}^{+r}\hat{a}^s = $$
$$= \sum_{r,s} M_{rs}^{(\mathscr{A})}\hat{a}^s\hat{a}^{+r}. \tag{4.89}$$

Assuming similar expansions to be valid for the density matrix

$$\hat{\varrho}(\hat{a}^+, \hat{a}) = \hat{\varrho}^{(\mathscr{N})}(\hat{a}^+, \hat{a}) = \hat{\varrho}^{(\mathscr{A})}(\hat{a}^+, \hat{a}) = \sum_{r,s} G_{rs}^{(\mathscr{N})}\hat{a}^{+r}\hat{a}^s = $$
$$= \sum_{r,s} G_{rs}^{(\mathscr{A})}\hat{a}^s\hat{a}^{+r}, \tag{4.90}$$

we obtain, by inserting the unit operator (4.18), permuting the operators cyclically, and using the eigenvalue properties of the coherent states,

$$\operatorname{Tr}\{\hat{\varrho}\hat{M}\} = \operatorname{Tr}\{\hat{\varrho}^{(\mathscr{A})}\hat{M}^{(\mathscr{N})}\} = \frac{1}{\pi}\int \varrho^{(\mathscr{A})}(\alpha^*, \alpha)\, M^{(\mathscr{N})}(\alpha^*, \alpha)\, \mathrm{d}^2\alpha, \tag{4.91}$$

where

$$\frac{1}{\pi}\varrho^{(\mathscr{A})}(\alpha^*, \alpha) \equiv \Phi_{\mathscr{N}}(\alpha) = \sum_{r,s} G_{rs}^{(\mathscr{A})}\alpha^{*r}\alpha^s \tag{4.92}$$

and

$$M^{(\mathscr{N})}(\alpha^*, \alpha) = \sum_{r,s} M_{rs}^{(\mathscr{N})}\alpha^{*r}\alpha^s. \tag{4.93}$$

Similarly

$$\operatorname{Tr}\{\hat{\varrho}\hat{M}\} = \operatorname{Tr}\{\hat{\varrho}^{(\mathscr{N})}\hat{M}^{(\mathscr{A})}\} = \frac{1}{\pi}\int \varrho^{(\mathscr{N})}(\alpha^*, \alpha)\, M^{(\mathscr{A})}(\alpha^*, \alpha)\, \mathrm{d}^2\alpha, \tag{4.94a}$$

that is

$$\Phi_{\mathscr{A}}(\alpha) = \frac{1}{\pi}\varrho^{(\mathscr{N})}(\alpha^*, \alpha). \tag{4.94b}$$

Hence the quasi-distributions $\Phi_{\mathscr{N}}$ and $\Phi_{\mathscr{A}}$ can be obtained from the antinormal and normal forms of the density matrix respectively (making use of the commutation rules) with the substitutions $\hat{a} \to \alpha$, $\hat{a}^+ \to \alpha^*$. Substituting $\alpha \to \hat{a}$, $\alpha^* \to \hat{a}^+$ in $\Phi_{\mathscr{N}}$ and $\Phi_{\mathscr{A}}$ we obtain the antinormal and normal forms of the density matrix, respectively. These important rules will be applied particularly in Chapters 8 and 9, in connection with a derivation of the generalized Fokker–Planck equation for nonlinear optical processes.

This can be demonstrated by an interesting example of Gaussian light with mean photon-number $\langle n\rangle$ superimposed on a coherent component with complex amplitude β, which is described by the density matrix

$$\hat{\varrho} = \frac{\langle n\rangle^{\hat{b}^+\hat{b}}}{(1+\langle n\rangle)^{\hat{b}^+\hat{b}+1}}, \tag{4.95}$$

where $\hat{b} = \hat{a} - \beta$ and $[\hat{b}, \hat{b}^+] = [\hat{a}, \hat{a}^+] = \hat{1}$. We can write

$$\varrho = \frac{1}{\langle n\rangle}\left(1 + \frac{1}{\langle n\rangle}\right)^{-\hat{b}^+\hat{b}-1} = \frac{1}{\langle n\rangle}\sum_{n=0}^{\infty}\frac{(-1)^n}{n!}\frac{(\hat{b}^+\hat{b}+n)!}{(\hat{b}^+\hat{b})!}\langle n\rangle^{-n}. \tag{4.96}$$

Making use of (2.14b) we obtain

$$\hat{\varrho} = \hat{\varrho}^{(\mathscr{A})} = \frac{1}{\langle n\rangle}\sum_{n=0}^{\infty}\frac{(-1)^n}{n!}\frac{\hat{b}^n\hat{b}^{+n}}{\langle n\rangle^n} = \frac{1}{\langle n\rangle}\mathscr{A}\left\{\exp\left(-\frac{\hat{b}^+\hat{b}}{\langle n\rangle}\right)\right\}, \tag{4.97}$$

which gives, by the substitutions $\hat{b} \to \alpha - \beta$, $\hat{b}^+ \to \alpha^* - \beta^*$

$$\Phi_{\mathscr{N}}(\alpha) = \frac{1}{\pi\langle n\rangle}\exp\left(-\frac{|\alpha-\beta|^2}{\langle n\rangle}\right). \tag{4.98}$$

In (4.97) $\mathscr{A}$ denotes the antinormal-ordering operator. This quasi-distribution is shown in Fig. 4.2 for coherent light ($\langle n\rangle = 0$), chaotic light ($\beta = 0$) and their general superposition.

Similarly

$$\begin{aligned}\hat{\varrho} &= \frac{1}{1+\langle n\rangle}\left(1 - \frac{1}{1+\langle n\rangle}\right)^{\hat{b}^+\hat{b}} = \\ &= \frac{1}{1+\langle n\rangle}\sum_{n=0}^{\infty}(-1)^n\binom{\hat{b}^+\hat{b}}{n}(1+\langle n\rangle)^{-n} = \\ &= \frac{1}{1+\langle n\rangle}\sum_{n=0}^{\infty}\frac{(-1)^n}{n!}\frac{(\hat{b}^+\hat{b})!}{(\hat{b}^+\hat{b}-n)!}(1+\langle n\rangle)^{-n}.\end{aligned} \tag{4.99}$$

Making use of (2.14a) we obtain

$$\begin{aligned}\hat{\varrho} = \hat{\varrho}^{(\mathscr{N})} &= \frac{1}{1+\langle n\rangle}\sum_{n=0}^{\infty}\frac{(-1)^n}{n!}\frac{\hat{b}^{+n}\hat{b}^n}{(1+\langle n\rangle)^n} = \\ &= \frac{1}{1+\langle n\rangle}\mathscr{N}\left\{\exp\left(-\frac{\hat{b}^+\hat{b}}{1+\langle n\rangle}\right)\right\}\end{aligned} \tag{4.100}$$

and so

$$\Phi_{\mathscr{A}}(\alpha) = \frac{1}{\pi(\langle n\rangle + 1)}\exp\left(-\frac{|\alpha-\beta|^2}{\langle n\rangle + 1}\right). \tag{4.101}$$

Substituting the expression (2.37) for $\langle n \rangle$ into these equations we can obtain the corresponding results for light with a superimposed coherent component β in thermal equilibrium at temperature T. Comparing (4.96) with (4.100) and (4.97) we arrive at

$$\hat{\varrho} = (1 - e^{-\Theta}) \exp\left[-\Theta(\hat{a}^+ - \beta^*)(\hat{a} - \beta)\right] =$$
$$= (1 - e^{-\Theta}) \mathcal{N}\{\exp\left[-(1 - e^{-\Theta})(\hat{a}^+ - \beta^*)(\hat{a} - \beta)\right]\} \equiv \hat{\varrho}^{(\mathcal{N})} \quad (4.102)$$
$$= (e^{\Theta} - 1) \mathcal{A}\{\exp\left[-(e^{\Theta} - 1)(\hat{a}^+ - \beta^*)(\hat{a} - \beta)\right]\} = \hat{\varrho}^{(\mathcal{A})}. \quad (4.103)$$

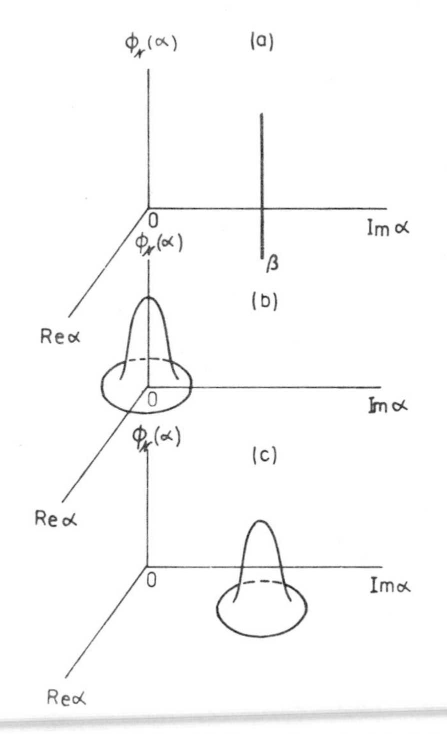

Fig. 4.2 — The quasi-distribution $\Phi_{\mathcal{N}}(\alpha)$ for (a) the coherent state $|\beta\rangle$, (b) single-mode chaotic field with the mean number of photons $\langle n \rangle$, and (c) single-mode superposition of these coherent and chaotic fields.

It is interesting to note that equations such as (4.102) follow simply in the coherent-state algebra ($\beta = 0$ for simplicity):

$$\langle \alpha | \exp(-\Theta \hat{a}^+ \hat{a}) | \alpha \rangle = \sum_n \exp(-\Theta n) | \langle \alpha | n \rangle |^2 =$$
$$= \langle \alpha | \mathcal{N}\{\exp\left[-\hat{a}^+ \hat{a}(1 - e^{-\Theta})\right]\} | \alpha \rangle, \quad (4.104)$$

where we have used (2.24) and (4.12b) [cf. (3.98) and (2.51e)].

The superposition principle, usually assumed in classical optics, is a natural consequence of the properties of $\Phi_{\mathcal{N}}$. Assume the complex amplitude α to be composed of two components β and γ, $\alpha = \beta + \gamma$. Then the density matrix describing fluctuations of β is

$$\hat{\varrho}_1 = \int \Phi_{\mathcal{N}}^{(1)}(\beta) \mid \beta + \gamma\rangle \langle \beta + \gamma \mid \mathrm{d}^2\beta, \tag{4.105}$$

while that describing fluctuations of γ is

$$\hat{\varrho}_2 = \int \Phi_{\mathcal{N}}^{(2)}(\gamma) \mid \beta + \gamma\rangle \langle \beta + \gamma \mid \mathrm{d}^2\gamma. \tag{4.106}$$

The resulting density matrix is equal to

$$\begin{aligned}\hat{\varrho} &= \iint \Phi_{\mathcal{N}}^{(1)}(\beta)\, \Phi_{\mathcal{N}}^{(2)}(\gamma) \mid \beta + \gamma\rangle \langle \beta + \gamma \mid \mathrm{d}^2\beta\, \mathrm{d}^2\gamma = \\ &= \int \Phi_{\mathcal{N}}(\alpha) \mid \alpha\rangle \langle \alpha \mid \mathrm{d}^2\alpha,\end{aligned} \tag{4.107}$$

where

$$\Phi_{\mathcal{N}}(\alpha) = \int \Phi_{\mathcal{N}}^{(1)}(\beta)\, \Phi_{\mathcal{N}}^{(2)}(\gamma)\, \delta(\alpha - \beta - \gamma)\, \mathrm{d}^2\beta\, \mathrm{d}^2\gamma. \tag{4.108}$$

More generally, if $\alpha = \sum_j \alpha_j$,

$$\Phi_{\mathcal{N}}(\alpha) = \int \delta(\alpha - \sum_j \alpha_j) \prod_j \Phi_{\mathcal{N}}^{(j)}(\alpha_j)\, \mathrm{d}^2\alpha_j, \tag{4.109}$$

which is analogous to the convolution law (3.136). As a consequence we have the quantum superposition principle

$$\begin{aligned}\langle\alpha\rangle &= \int \Phi_{\mathcal{N}}(\alpha)\, \alpha\, \mathrm{d}^2\alpha = \iint \Phi_{\mathcal{N}}^{(1)}(\beta)\, \Phi_{\mathcal{N}}^{(2)}(\gamma)\, (\beta + \gamma)\, \mathrm{d}^2\beta\, \mathrm{d}^2\gamma = \\ &= \langle\beta\rangle + \langle\gamma\rangle,\end{aligned} \tag{4.110a}$$

and similarly

$$\langle|\alpha|^2\rangle = \langle|\beta|^2\rangle + \langle|\gamma|^2\rangle + \langle\beta\gamma^*\rangle + \langle\beta^*\gamma\rangle. \tag{4.110b}$$

4.3 The existence of the Glauber—Sudarshan representation

It may be seen from (4.77) that if $C_{\mathcal{N}}(\beta)$ is a square integrable function ($C_{\mathcal{N}} \in L_2$), then $\Phi_{\mathcal{N}}(\alpha)$ is also square integrable. Since (4.86b) holds $C_{\mathcal{N}}(\beta)$ is not in general an L_2-function and consequently $\Phi_{\mathcal{N}}(\alpha)$ need not exist as an ordinary function; nevertheless it may exist as a generalized function (distribution) [Gelfand and Shilov (1964)].

An interesting L_2-criterion for $\Phi_{\mathcal{N}}(\alpha)$ has been suggested by Mehta (1967). Using (4.63) and (4.13a) we have

$$\begin{aligned}\langle -\alpha \mid \hat{\varrho} \mid \alpha\rangle &= \int \Phi_{\mathcal{N}}(\beta) \langle -\alpha \mid \beta\rangle \langle \beta \mid \alpha\rangle\, \mathrm{d}^2\beta = \\ &= \int \Phi_{\mathcal{N}}(\beta) \exp(-|\alpha|^2 - |\beta|^2 - \alpha^*\beta + \alpha\beta^*)\, \mathrm{d}^2\beta\end{aligned} \tag{4.111}$$

and so

$$\Phi_{\mathcal{N}}(\beta) = \frac{\exp(|\beta|^2)}{\pi^2} \int \langle -\alpha \mid \hat{\varrho} \mid \alpha\rangle \exp(|\alpha|^2) \exp(\alpha^*\beta - \alpha\beta^*)\, \mathrm{d}^2\alpha. \tag{4.112}$$

Hence, if $\langle -\alpha \mid \hat{\varrho} \mid \alpha\rangle \exp(|\alpha|^2)$ is an L_2-function, then $\Phi_{\mathcal{N}}(\beta) \exp(-|\beta|^2)$ is also an L_2-function.

Although the original general expression for $\Phi_{\mathcal{N}}(\alpha)$ has been obtained by Sudarshan (1963a, b) in terms of an infinite series of derivatives of the δ-function, a "regularized" form of this function has been suggested later [Peřina and Mišta (1968a, 1969), Lukš (1976)] in terms of the Laguerre polynomials

$$\Phi_{\mathcal{N}}(\alpha) = \exp[-(\zeta - 1)|\alpha|^2] \sum_{j=0}^{\infty} \{\sum_{l=0}^{\infty} c_{jl}(\zeta\alpha)^l L_j^l(\zeta|\alpha|^2) + \\ + \sum_{l=1}^{\infty} c_{jl}^*(\zeta\alpha^*)^l L_j^l(\zeta|\alpha|^2)\}, \tag{4.113}$$

where $\zeta \geqq 1$ is a real number (its choice is related to the convergence of the series) and L_j^l are defined in (3.117) and fulfil the orthogonality condition (3.118). Multiplying (4.113) by $\alpha^{*l'} L_k^{l'}(\zeta|\alpha|^2) \exp(-|\alpha|^2)$, integrating over $\zeta\alpha$ and using (3.118) we obtain for the decomposition coefficients c_{jl} in (4.113) the following expression in terms of the Fock matrix elements $\varrho(n, m)$:

$$c_{jl} = \frac{j!}{\pi[(j+l)!]^3} \int \Phi_{\mathcal{N}}(\alpha) \exp(-|\alpha|^2) \alpha^{*l} L_j^l(\zeta|\alpha|^2) \, d^2(\zeta\alpha) = \\ = \frac{\zeta}{\pi(j+l)!} \sum_{s=0}^{j} (-1)^s \binom{j}{s} \left[\frac{s!}{(s+l)!}\right]^{1/2} \zeta^s \varrho(s, s+l), \tag{4.114a}$$

where we have used (3.117) and (4.66). This relation can be inverted in the form

$$\varrho(s, s+l) = \frac{\pi}{\zeta^{s+1}} \left[\frac{(s+l)!}{s!}\right]^{1/2} \sum_{j=0}^{s} (-1)^j \binom{s}{j} (j+l)! \, c_{jl}. \tag{4.114b}$$

One can verify that the substitution of (4.113) and (4.4) into (4.63) with the use of (4.114b) leads to the Fock form (2.31) of the density matrix.

The moments $\langle\alpha^{*k}\alpha^l\rangle_{\mathcal{N}}$ can be calculated using (4.113) in the form [Holliday and Sage (1965)]

$$\langle\alpha^{*k}\alpha^l\rangle_{\mathcal{N}} = \sum_{r=0}^{\infty} \varrho(r+l, r+k) \frac{[(r+l)!(r+k)!]^{1/2}}{r!}. \tag{4.115}$$

Equations (4.113) and (4.114a) with $l = 0$ correspond to equations (3.116) and (3.119).

It has been shown by Miller and Mishkin (1967b) that $\Phi_{\mathcal{N}}(\alpha)$ may be in general very singular and it is an element of a space of ultradistributions. Further discussions of the existence problems of the quasi-distribution $\Phi_{\mathcal{N}}(\alpha)$ have been given by Cahill (1965, 1969), Holliday and Sage (1965), Bonifacio, Narducci and Montaldi (1966), Mehta and Sudarshan (1965), Klauder, McKenna and Currie (1965), Klauder (1966), Rocca (1967) and Nath (1979) [reviews have been given by Klauder and Sudarshan (1968) (Sec. 8-4) and Peřina (1972) (Sec. 13.3)].

In connection with the quantum statistical properties of nonlinear optical processes Drummond and Gardiner (1980) suggested a generalized positive complex representation, and another procedure avoiding the non-existence of $\Phi_{\mathcal{N}}(\alpha)$ in interaction problems will be explained in Chapters 8 and 9.

4.4 The phase operators

The phase operator $\hat{\varphi}$ has been introduced by Dirac [see Heitler (1954) and Akhiezer and Berestetsky (1965)] in analogy to a classical procedure,

$$\hat{a} = \exp(i\hat{\varphi})\, \hat{n}^{1/2} \qquad \text{(wrong)}. \tag{4.116}$$

Assuming $\hat{\varphi}$ to be Hermitian, so that $\exp(i\hat{\varphi})$ is unitary and

$$[\exp(i\hat{\varphi}), \hat{n}] = \exp(i\hat{\varphi}) \tag{4.117a}$$

or

$$[\hat{\varphi}, \hat{n}] = -i\hat{1} \qquad \text{(wrong)}, \tag{4.117b}$$

it follows that

$$\Delta n \Delta\varphi \geqq \frac{1}{2} \qquad \text{(wrong)}. \tag{4.118}$$

However, one can show [Carruthers and Nieto (1968)] that the operator $\hat{U} = \exp(i\hat{\varphi})$ is not unitary, since $\langle 0 \mid \hat{U}^+\hat{U} \mid 0\rangle = 0$, in contradiction with $\hat{U}^+\hat{U} = \hat{1}$. To avoid this difficulty Carruthers and Nieto (1968) defined the Hermitian operators

$$\hat{C} = \frac{1}{2}(\hat{U} + \hat{U}^+), \qquad \hat{S} = \frac{1}{2i}(\hat{U} - \hat{U}^+), \tag{4.119a}$$

and consequently

$$[\hat{C}, \hat{n}] = i\hat{S}, \qquad [\hat{S}, \hat{n}] = -i\hat{C} \tag{4.119b}$$

and rigorously

$$\Delta n \Delta C \geqq \frac{1}{2} \mid S \mid, \qquad \Delta n \Delta S \geqq \frac{1}{2} \mid C \mid. \tag{4.120}$$

The phase operators in terms of the coherent states have been introduced and studied by Paul (1974, 1976),

$$\hat{\varphi} = \frac{1}{2\pi i} \int (\log \alpha - \log \alpha^*) \mid \alpha\rangle \langle\alpha \mid d^2\alpha, \tag{4.121}$$

making use of (4.18). One can also introduce a set of phase operators as follows

$$\hat{\varphi}_k = \frac{1}{\pi} \int \frac{\alpha^k}{\mid \alpha \mid^k} \mid \alpha\rangle \langle\alpha \mid d^2\alpha, \qquad \hat{\varphi}_{-k} = \frac{1}{\pi} \int \frac{\alpha^{*k}}{\mid \alpha \mid^k} \mid \alpha\rangle \langle\alpha \mid d^2\alpha,$$
$$k = 1, 2, \ldots, \tag{4.122}$$

which also describe phases of higher powers of the field operator.

4.5 Multimode description

All earlier results and relations can be generalized to fields and systems having finite or countably infinite number of degrees of freedom, if the global coherent states defined by (4.48) are adopted. For instance, the Glauber – Sudarshan representation has the form

$$\hat{\varrho} = \int \Phi_{\mathcal{N}}(\{\alpha_\lambda\}) \mid \{\alpha_\lambda\}\rangle \langle\{\alpha_\lambda\} \mid \mathrm{d}^2\{\alpha_\lambda\}, \tag{4.123}$$

where $\Phi_{\mathcal{N}}(\{\alpha_\lambda\}) \equiv \Phi_{\mathcal{N}}(\alpha_1, \alpha_2, \ldots)$ is the weight functional and $\mathrm{d}^2\{\alpha_\lambda\} \equiv \prod_\lambda \mathrm{d}^2\alpha_\lambda$. Instead of (4.66) we have

$$\langle\{n_\lambda\} \mid \hat{\varrho} \mid \{m_\lambda\}\rangle = \varrho(\{n_\lambda\}, \{m_\lambda\}) = \int \Phi_{\mathcal{N}}(\{\alpha_\lambda\}) \prod_\lambda \frac{\alpha_\lambda^{n_\lambda}\alpha_\lambda^{*m_\lambda}}{(n_\lambda m_\lambda!)^{1/2}} \times$$
$$\times \exp(-\mid \alpha_\lambda \mid^2)\, \mathrm{d}^2\alpha_\lambda. \tag{4.124}$$

Further,

$$\Phi_{\mathcal{A}}(\{\alpha_\lambda\}) = \int \Phi_{\mathcal{N}}(\{\beta_\lambda\}) \prod_\lambda \exp(-\mid \alpha_\lambda - \beta_\lambda \mid^2) \frac{\mathrm{d}^2\beta_\lambda}{\pi}, \tag{4.125}$$

which corresponds to (4.71) and

$$\Phi_{\mathcal{N}}(\{\alpha_\lambda\}) = \sum_{\{n_\lambda\}} \sum_{\{m_\lambda\}} \varrho(\{n_\lambda\}, \{m_\lambda\}) \prod_\lambda \frac{\alpha_\lambda^{*n_\lambda}\alpha_\lambda^{m_\lambda}}{(n_\lambda! m_\lambda!)^{1/2}} \exp(-\mid \alpha_\lambda \mid^2), \tag{4.126}$$

which corresponds to (4.87). All other relations can be obtained in an identical way.

Using (4.123) and (4.52a, b) we can write the correlation function (3.5a) in the form

$$\Gamma_{\mathcal{N}}^{(m,n)}(x_1, \ldots, x_{m+n}) = \int \Phi_{\mathcal{N}}(\{\alpha_\lambda\}) \prod_{j=1}^{m} V^*(x_j) \prod_{k=m+1}^{m+n} V(x_k)\, \mathrm{d}^2\{\alpha_\lambda\}, \tag{4.127}$$

where $V(x)$ is given by (4.53); this corresponds to (3.5b).

The superposition of two fields described by $\Phi_{\mathcal{N}}^{(1)}(\{\beta_\lambda\})$ and $\Phi_{\mathcal{N}}^{(2)}(\{\gamma_\lambda\})$ is described, in analogy to (4.108), by the convolution law for fields,

$$\Phi_{\mathcal{N}}(\{\alpha_\lambda\}) = \iint \Phi_{\mathcal{N}}^{(1)}(\{\beta_\lambda\})\, \Phi_{\mathcal{N}}^{(2)}(\{\gamma_\lambda\}) \prod_\lambda \delta(\alpha_\lambda - \beta_\lambda - \gamma_\lambda)\, \mathrm{d}^2\beta_\lambda\, \mathrm{d}^2\gamma_\lambda, \tag{4.128}$$

which expresses the quantum superposition principle for fields.

4.6 Relation between the quantum and classical descriptions

4.6.1 Quantum and classical correlation functions

Introducing the probability distribution $P_n(V_1, \ldots, V_n)$, $V_j = V(x_j)$, by the relation [Mandel (1963c)]

$$P_n(V_1, \ldots, V_n) = \int \Phi_{\mathcal{N}}(\{\alpha'_\lambda\}) \prod_{j=1}^{n} \delta(V_j - V'_j)\, \mathrm{d}^2\{\alpha'_\lambda\} =$$
$$= \langle \prod_{j=1}^{n} \delta(V_j - V'_j) \rangle_{\mathcal{N}}, \tag{4.129}$$

where V_j are given by (4.53) with $\alpha_\lambda \to \alpha'_\lambda$, we can rewrite (4.127) in the form (3.5b),

$$\Gamma^{(m,n)}(x_1, \ldots, x_{m+n}) = \int P_{m+n}(V_1, \ldots, V_{m+n}) \prod_{j=1}^{m} V_j^* \prod_{k=m+1}^{m+n} V_k \prod_{l=1}^{m+n} \mathrm{d}^2 V_l =$$
$$= \langle \prod_{j=1}^{m} V^*(x_j) \prod_{k=m+1}^{m+n} V(x_k) \rangle, \tag{4.130}$$

which provides a classical form of the correlation function. Thus there is a close correspondence between the classical and the normally ordered quantum correlation functions, established by the coherent states. In this correspondence the Glauber–Sudarshan representation serves as a bridge.

4.6.2 Photon-number and photocount distributions

We show now that the photodetection equation (3.85) follows simply from the properties of the quantized electromagnetic field and the definitions of the number operator and normal correlation functions [Ghielmetti (1964), Peřina (1965, 1967a)].

Consider the probability distribution $p(n)$ of the number of photons n within the normalization volume L^3 of the field. Writing $\varrho(\{n_\lambda\}) \equiv \varrho(\{n_\lambda\}, \{n_\lambda\})$, we have

$$p(m) = \sum_{\{n_\lambda\}} \varrho(\{n_\lambda\})\, \delta_{nm}, \tag{4.131}$$

where $\sum_\lambda n_\lambda = n$. For the power moment $\langle \hat{n}^k \rangle$, where $\hat{n}$ is defined in (2.30b), we obtain with the help of the Fock and Glauber–Sudarshan representations of the density matrix

$$\begin{aligned}\langle \hat{n}^k \rangle &= \sum_{\{n_\lambda\}} \varrho(\{n_\lambda\}) \langle \{n_\lambda\} \mid \hat{n}^k \mid \{n_\lambda\} \rangle = \\ &= \int \Phi_{\mathcal{N}}(\{\alpha_\lambda\}) \langle \{\alpha_\lambda\} \mid \hat{n}^k \mid \{\alpha_\lambda\} \rangle \, \mathrm{d}^2\{\alpha_\lambda\};\end{aligned} \tag{4.132}$$

substituting the unit operator (2.44) into the right-hand side of (4.132) and using tne multimode analogue of (4.12b), viz.,

$$|\langle \{n_\lambda\} \mid \{\alpha_\lambda\} \rangle|^2 = \prod_\lambda \frac{|\alpha_\lambda|^{2n_\lambda}}{n_\lambda!} \exp(-|\alpha_\lambda|^2), \tag{4.133}$$

we obtain

$$\sum_{n=0}^{\infty} p(n)\, n^k = \sum_{\{n_\lambda\}} \left\{ \int \Phi_{\mathcal{N}}(\{\alpha_\lambda\}) \prod_\lambda \frac{|\alpha_\lambda|^{2n_\lambda}}{n_\lambda!} \exp(-|\alpha_\lambda|^2)\, \mathrm{d}^2\alpha_\lambda n^k \right\}, \tag{4.134}$$

where $n = \sum_\lambda n_\lambda$ is the eigenvalue of the operator $\hat{n}$ in the Fock state $|\{n_\lambda\}\rangle$ and $p(n)$ is given by (4.131). As (4.134) must hold for every k, we arrive at

$$p(n) = \int \Phi_{\mathcal{N}}(\{\alpha_\lambda\}) \frac{W^n}{n!} \exp(-W)\, \mathrm{d}^2\{\alpha_\lambda\}, \tag{4.135}$$

where the multinomial theorem

$$\sum_{\{n_\lambda\}}' \prod_\lambda \frac{x_\lambda^{n_\lambda}}{n_\lambda!} = \frac{\left(\sum_\lambda x_\lambda\right)^n}{n!} \tag{4.136}$$

has been used and Σ' is evaluated subject to $\sum_\lambda n_\lambda = n$. The integrated intensity W is equal to

$$W = \langle\{\alpha_\lambda\} \,|\, \hat{n} \,|\, \{\alpha_\lambda\}\rangle = \int_{L^3} \mathbf{A}^*(x) \cdot \mathbf{A}(x)\, \mathrm{d}^3x = \sum_\lambda |\alpha_\lambda|^2, \tag{4.137}$$

where $\mathbf{A}(x)$ is the eigenvalue of the detection operator $\hat{\mathbf{A}}(x)$ given by (2.29), in the coherent state. If we consider a detector with a plane cathode of surface S on which plane waves are normally incident, then $L^3 = ScT$ and (4.137) is proportional to (3.84), or to (3.86) averaged in the coherent state. The substitution

$$P_{\mathcal{N}}(W) = \int \Phi_{\mathcal{N}}(\{\alpha'_\lambda\})\, \delta(W - \sum_\lambda |\alpha'_\lambda|^2)\, \mathrm{d}^2\{\alpha'_\lambda\} \tag{4.138}$$

finally gives the photodetection equation (3.85) with $\eta = 1$. With the relation (4.138) the normal characteristic function reads

$$\langle \exp(\mathrm{i}sW)\rangle_{\mathcal{N}} = \int_0^\infty P_{\mathcal{N}}(W) \exp(\mathrm{i}sW)\, \mathrm{d}W =$$
$$= \int \Phi_{\mathcal{N}}(\{\alpha_\lambda\}) \exp(\mathrm{i}s \sum_\lambda |\alpha_\lambda|^2)\, \mathrm{d}^2\{\alpha_\lambda\}. \tag{4.139}$$

Thus we have obtained the same expression for the distribution of the number of photons in the volume L^3 at time t as for the distribution of emitted photoelectrons within the time interval $(t, t + T)$ and it may be said that the statistical properties of photoelectrons reflect the statistical properties of photons. The use of normally ordered correlation functions in the photodetection measurements, ensuring that the contribution of the physical vacuum is zero, is responsible for this result.

Note that the form of the photon-number and photocount distributions is rigorously conserved for fields having the so-called scaling properties, $p(n, \langle n\rangle, \eta T) = p(n, \eta\langle n\rangle T)$ [cf. (3.94)], which means that the characteristic function as a function of $\mathrm{i}s\eta T$ and $\langle n\rangle$ (the mean number of photons/sec) is a function of $\mathrm{i}s$ and $\eta\langle n\rangle T$ and the kth factorial moment depends on η^k [Ghielmetti (1976), Jakeman (1981)]. Then the photocount distribution is obtained by the substitution $\langle n\rangle \to \eta\langle n\rangle T$ in the photon-number distribution. For the Fock states the scaling properties are not satisfied, as follows from (3.94). However, most of the fields encountered in practice possess the scaling properties.

The formal equivalence between the quantum and classical descriptions of optical beams, established by the coherent states, does not mean a real physical equivalence, since the classical description is applicable to strong fields where energies comparable with the photon energy $\hbar\omega$ are neglected, whereas the quantum description is valid even for arbitrarily weak fields. The physical inequivalence of the quantum and classical descriptions reflects itself in the fact that classical probability distributions are well-behaved non-negative functions while the quasi-distribution $\Phi_{\mathcal{N}}(\{\alpha_\lambda\})$ can take on negative values and can be very singular (for example, for the Fock state $|n\rangle$ it is proportional to the nth derivative of the δ-function and the Fock state can have no classical analogue; however, in general it is a very singular ultradistribution). Only in the limit of strong fields where the ordering of the field operators plays no role and $\Phi_{\mathcal{N}} \to \Phi_{\mathcal{A}}$ (which is a well-behaved function), may the complete physical equivalence of the quantum and classical descriptions be reached.

4.7 Stationary conditions for the field

4.7.1 Time invariance properties of the correlation functions

The quantum stationary condition is expressed as

$$[\hat{H}, \hat{\varrho}] = \hat{0}, \tag{4.140}$$

following from (2.55a), since $\partial\hat{\varrho}/\partial t = \hat{0}$ in this case. Here $\hat{H}$ is the hamiltonian of the system. In the Heisenberg picture we can write [Kano (1964b), Glauber (1965, 1969)]

$$\begin{aligned}
\Gamma_{\mathcal{N}}^{(m,n)}(x_1, \ldots, x_{m+n}) &= \operatorname{Tr}\{\hat{\varrho} \prod_{j=1}^{m} \hat{A}^{(-)}(\mathbf{x}_j, t_j) \prod_{k=m+1}^{m+n} \hat{A}^{(+)}(\mathbf{x}_k, t_k)\} = \\
&= \operatorname{Tr}\left\{\exp\left(\mathrm{i}\frac{\hat{H}\tau}{\hbar}\right)\hat{\varrho}\exp\left(-\mathrm{i}\frac{\hat{H}\tau}{\hbar}\right)\exp\left(\mathrm{i}\frac{\hat{H}\tau}{\hbar}\right)\hat{A}^{(-)}(\mathbf{x}_1, t_1)\times\right. \\
&\times\exp\left(-\mathrm{i}\frac{\hat{H}\tau}{\hbar}\right)\exp\left(\mathrm{i}\frac{\hat{H}\tau}{\hbar}\right)\hat{A}^{(-)}(\mathbf{x}_2, t_2)\exp\left(-\mathrm{i}\frac{\hat{H}\tau}{\hbar}\right)\ldots \\
&\left.\ldots\exp\left(\mathrm{i}\frac{\hat{H}\tau}{\hbar}\right)\hat{A}^{(+)}(\mathbf{x}_{m+n}, t_{m+n})\exp\left(-\mathrm{i}\frac{\hat{H}\tau}{\hbar}\right)\right\},
\end{aligned} \tag{4.141}$$

where we have used the cyclic property of the trace and substituted the unit operator $\exp(-\mathrm{i}\hat{H}\tau/\hbar)\exp(\mathrm{i}\hat{H}\tau/\hbar)$. Using (2.53a) and assuming (4.140), we arrive at

$$\begin{aligned}
\Gamma_{\mathcal{N}}^{(m,n)}(x_1, \ldots, x_{m+n}) &= \operatorname{Tr}\{\hat{\varrho} \prod_{j=1}^{m} \hat{A}^{(-)}(\mathbf{x}_j, t_j + \tau) \prod_{k=m+1}^{m+n} \hat{A}^{(+)}(\mathbf{x}_k, t_k + \tau)\} = \\
&= \Gamma_{\mathcal{N}}^{(m,n)}(\mathbf{x}_1, \ldots, \mathbf{x}_{m+n}, t_1 + \tau, \ldots, t_{m+n} + \tau).
\end{aligned} \tag{4.142}$$

Putting $\tau = -t_1$ we have

$$\begin{aligned}
&\Gamma_{\mathcal{N}}^{(m,n)}(x_1, \ldots, x_{m+n}) = \\
&= \Gamma_{\mathcal{N}}^{(m,n)}(\mathbf{x}_1, \ldots, \mathbf{x}_{m+n}, 0, t_2 - t_1, \ldots, t_{m+n} - t_1),
\end{aligned} \tag{4.143}$$

i.e. the correlation function of the stationary field depends on $m + n - 1$ time variables $t_j - t_1$, $j = 2, \ldots, m + n$, as discussed in Sec. 3.2.2 [see equation (3.18)]. Of course, from (4.142) or (4.143) the condition (4.140) also follows.

Another stationary condition for the correlation functions can be obtained in the form [Horák (1971)]

$$\sum_{l=1}^{m+n} \frac{\partial}{\partial t_l} \Gamma_{\mathcal{N}}^{(m,n)}(x_1, \ldots, x_{m+n}) = 0, \tag{4.144}$$

corresponding to the stationary condition (3.19) in the frequency domain. For $m = n = 1$, $\partial\Gamma_{\mathcal{N}}^{(1,1)}/\partial t_1 = -\partial\Gamma_{\mathcal{N}}^{(1,1)}/\partial t_2$ as it must be since $\Gamma_{\mathcal{N}}^{(1,1)}(t_1, t_2) = \Gamma_{\mathcal{N}}^{(1,1)}(t_2 - t_1)$ in this case.

4.7.2 Stationary conditions in phase space

First consider a one-mode field. If $\hat{H} = \hbar\omega\hat{a}^+\hat{a}$ and (2.31) is used in (4.140) we have

$$[\hat{H}, \hat{\varrho}] = \hbar\omega \sum_{n,m} \varrho(n, m)(n - m) | n\rangle \langle m | = \hat{0}, \tag{4.145}$$

that is

$$\langle k | [\hat{H}, \hat{\varrho}] | l\rangle = \hbar\omega\varrho(k, l)(k - l) = 0 \tag{4.146}$$

for all k and l. Consequently $\varrho(k, l)$ must have the form

$$\varrho(k, l) = \varrho(k, k)\,\delta_{kl}. \tag{4.147}$$

Thus we can see from (4.113) and (4.114a) that $\Phi_{\mathcal{N}}(\alpha)$ is independent of the phase φ of α. On the other hand, if $\Phi_{\mathcal{N}}(\alpha)$ is independent of the phase φ, then $\varrho(k, l)$ has the form (4.147) and (4.140) holds. Therefore for a one-mode field the stationary condition (4.140) is necessary and sufficient for the quasi-distribution $\Phi_{\mathcal{N}}(\alpha)$ to be independent of the phase φ of the complex amplitude α.

For a multimode field we can consider the renormalized hamiltonian (2.19a) and using (2.45) we obtain

$$[\hat{H}, \hat{\varrho}] = \sum_{\{n_\lambda\}} \sum_{\{m_\lambda\}} \varrho(\{n_\lambda\}, \{m_\lambda\}) [\sum_\lambda \hbar\omega_\lambda(n_\lambda - m_\lambda)] | \{n_\lambda\}\rangle \langle\{m_\lambda\} | = \hat{0}, \tag{4.148}$$

which leads to

$$\varrho(\{n_\lambda\}, \{m_\lambda\}) = \varrho_0(\{n_\lambda\}, \{m_\lambda\})\,\delta_{NM}, \tag{4.149}$$

where

$$N = \sum_\lambda \hbar\omega_\lambda n_\lambda, \qquad M = \sum_\lambda \hbar\omega_\lambda m_\lambda \tag{4.150}$$

and ϱ_0 is equal to ϱ where $N = M$. Substituting (4.149) into the multimode versions of (4.114a) and (4.113), we can obtain the corresponding $\Phi_{\mathcal{N}}(\{\alpha_\lambda\})$ for the stationary multimode field. This function will in general depend upon the phases $\{\varphi_\lambda\}$ of $\{\alpha_\lambda\}$.

If the additional condition [Kano (1964b, 1966)]

$$[\hat{\varrho}, \hat{a}_\lambda^+\hat{a}_\lambda] = \hat{0}, \qquad \text{(all } \lambda) \tag{4.151}$$

is satisfied, then

$$\varrho(\{n_\lambda\}, \{m_\lambda\}) = \varrho(\{n_\lambda\}, \{n_\lambda\})\,\delta_{\{n_\lambda\},\{m_\lambda\}} \tag{4.152}$$

and consequently $\Phi_{\mathcal{N}}(\{\alpha_\lambda\})$ is independent of the phases $\{\varphi_\lambda\}$. The Fock form of the density matrix (2.45) is diagonal in this case.

Analogous results can be obtained for the space dependence of the correlation functions from the condition $[\hat{\mathbf{P}}, \hat{\varrho}] = \hat{0}$ [Eberly and Kujawski (1967a)], where $\hat{\mathbf{P}}$ is given by (2.19b). This condition leads to correlation functions which are stationary in space (they are homogeneous and isotropic). If (4.151) holds, then $[\hat{H}, \hat{\varrho}] = \hat{0}$ and $[\hat{\mathbf{P}}, \hat{\varrho}] = \hat{0}$ and the field is stationary in time and space. This is evident also if we recall that $\Phi_{\mathcal{N}}(\{\alpha_\lambda\})$ is independent of the phases in this case [cf. for instance equation (5.26)]. In general, if a field is stationary in time, it need not be stationary in space, so that for example the function $\Gamma_{\mathcal{N}}^{(1,1)}(\mathbf{x}, \mathbf{x}, t_2 - t_1)$ will generally depend on $\mathbf{x}$.

Hence, the condition (4.151) is a necessary and sufficient condition for $\Phi_{\mathcal{N}}(\{\alpha_\lambda\})$ to be independent of the phases $\{\varphi_\lambda\}$, which are uniformly distributed in the interval $(0, 2\pi)$. The condition (4.140) is a necessary but not sufficient condition for $\Phi_{\mathcal{N}}(\{\alpha_\lambda\})$ to be independent of the phases $\{\varphi_\lambda\}$.

If $\Phi_{\mathcal{N}}(\{\alpha_\lambda\})$ is independent of the phases (the field is stationary in time and space), then $\mathrm{Tr}\,\{\hat{\varrho}\hat{a}_\lambda\} = \mathrm{Tr}\,\{\hat{\varrho}\hat{a}_\lambda^+\} = 0$ and

$$\mathrm{Tr}\,\{\hat{\varrho}\hat{A}(x)\} = 0, \tag{4.153}$$

and also

$$\Gamma_{\mathcal{N}}^{(m,n)}(x_1, \ldots, x_{m+n}) = 0, \qquad m \neq n, \tag{4.154}$$

since the integrals over phases, when the Glauber–Sudarshan representation is used, are equal to zero if $m \neq n$.

4.8 Ordering of field operators

In this and the preceding chapters we have dealt with particular orderings (normal, antinormal and symmetric). Now we outline a systematic approach to the correspondence of functions of operators and classical functions and develop an operator generalization of the classical Fourier analysis.

The problem of the ordering of field operators in quantum optics plays an important role in the context of the generalized phase-space descriptions of optical fields [for reviews, see Mandel and Wolf (1965), Klauder and Sudarshan (1968), Peřina (1972), Nussenzveig (1973)], laser theory [Lax (1967, 1968a, b), Lax and Louisell (1967), Lax and Yuen (1968), Haken (1970a, b), Louisell (1973)] and the photon statistics of nonlinear optical processes [Schubert and Wilhelmi (1980), Peřina (1980b)]. A systematic way of treating the problem of the ordering of field operators and the correspondence between functions of operators and classical functions has been developed by Agarwal and Wolf (1968a–c, 1970a–c), Wolf and Agarwal (1969), Lax (1968b) [see also Louisell (1970)], Agarwal (1969) and Cahill and Glauber (1969a, b). A multimode generalization has been also found [Peřina and Horák (1969a, b, 1970)].

4.8.1 Ω- and s-ordering and general decompositions

The completeness property of the displacement operator expressed by (4.37) enables us to decompose any operator $\hat{A}$ in terms of the displacement operator $\hat{D}(\alpha)$ in the form

$$\hat{A} = \frac{1}{\pi} \int \tilde{a}(\beta)\, \hat{D}^+(\beta)\, \mathrm{d}^2\beta, \tag{4.155}$$

where

$$\tilde{a}(\beta) = \mathrm{Tr}\,\{\hat{D}(\beta)\,\hat{A}\}. \tag{4.156}$$

Considering another operator $\hat{B}$, we can write

$$\mathrm{Tr}\,\{\hat{A}\hat{B}\} = \frac{1}{\pi}\int \tilde{a}(\beta)\,\tilde{b}(-\beta)\,\mathrm{d}^2\beta \tag{4.157}$$

and for $\hat{B} = \hat{A}^+$

$$\|\hat{A}\|^2 = \mathrm{Tr}\,\{\hat{A}^+\hat{A}\} = \frac{1}{\pi}\int |\tilde{a}(\beta)|^2\,\mathrm{d}^2\beta. \tag{4.158}$$

Thus, if $\tilde{a}(\beta)$ is a square integrable function, then the Hilbert – Schmidt norm $\|\hat{A}\| = [\mathrm{Tr}\,\{\hat{A}^+\hat{A}\}]^{1/2}$ is finite, and vice versa. In particular, for $\hat{A} = \hat{\varrho} = \hat{\varrho}^+$, $\mathrm{Tr}\,\hat{\varrho}^2 = \int |C(\beta)|^2\,\mathrm{d}^2\beta/\pi \leqq 1$, where $C(\beta)$ is the characteristic function (4.73); and consequently $0 \leqq |C(\beta)| \leqq 1$.

If $\hat{A} = \hat{\varrho}$, we have obtained a representation of the density matrix in terms of $\hat{D}(\beta)$, and $\tilde{a}(\beta)$ is obviously identical to $C(\beta)$ while $\varrho(\alpha^*, \alpha)/\pi$ is the Wigner function (4.79).

Following Agarwal and Wolf (1970a – c), we can introduce the operator $\hat{\Delta}(\alpha, \Omega)$ for Ω-ordering as

$$\hat{\Delta}(\alpha, \Omega) = \frac{1}{\pi}\int \Omega(\beta^*, \beta)\,\hat{D}(\beta)\exp(\alpha\beta^* - \alpha^*\beta)\,\mathrm{d}^2\beta = \frac{1}{\pi}\int \Omega(\beta^*, \beta)\exp[\beta(\hat{a}^+ - \alpha^*) - \beta^*(\hat{a} - \alpha)]\,\mathrm{d}^2\beta, \tag{4.159}$$

that is $\hat{\Delta}(\alpha, \Omega)/\pi$ is equal to the Fourier transform of the operator

$$\hat{D}(\beta, \Omega) = \hat{\Omega}\hat{D}(\beta) = \Omega(\beta^*, \beta)\,\hat{D}(\beta) \equiv \{\hat{D}(\beta)\}_\Omega, \tag{4.160}$$

where $\hat{\Omega}$ is an operator arranging the expression in an Ω-ordered form denoted by $\{\ \}_\Omega$. The function $\Omega(\beta^*, \beta)$ represents some filter function. Thus (4.159) can be written as

$$\hat{\Delta}(\alpha, \Omega) = \frac{1}{\pi}\int \hat{D}(\beta, \Omega)\exp(\alpha\beta^* - \alpha^*\beta)\,\mathrm{d}^2\beta = \pi\{\delta(\hat{a} - \alpha)\}_\Omega. \tag{4.161}$$

As a particular case we obtain the s-ordering,

$$\hat{D}(\beta, \Omega) \equiv \hat{D}(\beta, s) = \exp\left(\frac{s}{2}|\beta|^2\right)\hat{D}(\beta) = \exp\left(\frac{s-1}{2}|\beta|^2\right)\times \exp(\beta\hat{a}^+)\exp(-\beta^*\hat{a}) = \exp\left(\frac{s+1}{2}|\beta|^2\right)\exp(-\beta^*\hat{a})\exp(\beta\hat{a}^+), \tag{4.162}$$

where we have used (4.3). Hence, $s = 0$ for symmetric ordering, $s = +1$ for normal ordering and $s = -1$ for antinormal ordering.

Defining $\tilde{\Omega}$-ordering as the Ω-ordering with $\Omega(\beta^*, \beta) \to \Omega^{-1}(\beta^*, \beta)$ $(s \to -s)$, we can prove, using (4.37), that

$$\mathrm{Tr}\,\{\hat{D}(\beta, \Omega)\hat{D}(\gamma, \tilde{\Omega})\} = \mathrm{Tr}\,\{\hat{D}(\beta, s)\,\hat{D}(\gamma, -s)\} = \pi\delta(\beta + \gamma). \tag{4.163}$$

Using this completeness condition, we can make a similar decomposition to that given in (4.155), in terms of the Ω- or s-ordered displacement operator. From (4.161) and (4.163) we further have

$$\mathrm{Tr}\,\{\hat{\Delta}(\beta, \Omega)\,\hat{\Delta}(\gamma, \tilde{\Omega})\} = \pi\delta(\beta - \gamma). \tag{4.164}$$

Therefore we can decompose any operator $\hat{A}$ in the form

$$\hat{A} = \frac{1}{\pi}\int a(\beta, \tilde{\Omega})\,\hat{\Delta}(\beta, \Omega)\,\mathrm{d}^2\beta, \tag{4.165}$$

where

$$a(\beta, \tilde{\Omega}) = \mathrm{Tr}\,\{\hat{\Delta}(\beta, \tilde{\Omega})\,\hat{A}\}, \tag{4.166}$$

which provides, with respect to (4.161), the Ω-ordered form of the operator $\hat{A}$. Applying (4.164) we also have

$$\mathrm{Tr}\,\{\hat{\varrho}\hat{A}\} = \frac{1}{\pi}\int \Phi(\beta, \tilde{\Omega})\,a(\beta, \Omega)\,\mathrm{d}^2\beta, \tag{4.167}$$

where

$$\Phi(\beta, \tilde{\Omega}) = \mathrm{Tr}\,\{\hat{\varrho}\hat{\Delta}(\beta, \tilde{\Omega})\} = \pi\langle\{\delta(\hat{a} - \beta)\}_{\tilde{\Omega}}\rangle \tag{4.168}$$

and from (4.165)

$$\hat{\varrho} = \frac{1}{\pi}\int \Phi(\beta, \tilde{\Omega})\,\hat{\Delta}(\beta, \Omega)\,\mathrm{d}^2\beta. \tag{4.169}$$

Hence, the expectation value of an operator $\hat{A}$ can be calculated as a classical average of the function $a(\beta, \Omega)$ corresponding to the operator $\hat{A}$ via the $\tilde{\Omega}$-ordering, with the quasi-probability $\Phi(\beta, \tilde{\Omega})/\pi$ corresponding to the density matrix via the Ω-ordering. This generalizes the results previously discussed for the normal and antinormal orderings in Sec. 4.2.

Writing (4.168) in the form

$$\Phi(\alpha, \tilde{\Omega}) = \frac{1}{\pi}\int \mathrm{Tr}\,\{\hat{\varrho}\hat{D}(\beta, \tilde{\Omega})\}\exp(\alpha\beta^* - \alpha^*\beta)\,\mathrm{d}^2\beta, \tag{4.170}$$

where we have used (4.161), it is obvious that $\mathrm{Tr}\,\{\hat{\varrho}\hat{D}(\beta, \tilde{\Omega})\} = C(\beta, \tilde{\Omega})$ represents the characteristic function of the quasi-distribution $\Phi(\alpha, \tilde{\Omega})/\pi$ and

$$\Phi(\alpha, \tilde{\Omega}) = \frac{1}{\pi}\int C(\beta, \tilde{\Omega})\exp(\alpha\beta^* - \alpha^*\beta)\,\mathrm{d}^2\beta, \tag{4.171a}$$

$$C(\beta, \tilde{\Omega}) = \frac{1}{\pi}\int \Phi(\alpha, \tilde{\Omega})\exp(\alpha^*\beta - \alpha\beta^*)\,\mathrm{d}^2\alpha. \tag{4.171b}$$

The Ω-ordered moments $\langle \alpha^{*k}\alpha^l \rangle_\Omega$ can clearly be obtained as

$$\operatorname{Tr}\{\hat{\varrho}\{\hat{a}^{+k}\hat{a}^l\}_\Omega\} = \frac{\partial^{k+l}}{\partial\beta^k\,\partial(-\beta^*)^l} C(\beta, \Omega)\,|_{\beta=\beta^*=0} =$$

$$= \frac{1}{\pi}\int \Phi(\alpha, \Omega)\, \alpha^{*k}\alpha^l\, d^2\alpha = \langle \alpha^{*k}\alpha^l \rangle_\Omega. \qquad (4.172)$$

Adopting the s-ordering, it is evident that $\Phi(\alpha, s)/\pi$ and $C(\beta, s)$ are equal to the quasi-distributions and their characteristic functions, respectively, for normal ordering if $s = 1$, for antinormal ordering if $s = -1$ and for symmetric ordering if $s = 0$. For example $\Phi(\alpha, 1)/\pi = \Phi_{\mathcal{N}}(\alpha)$, $C(\beta, 1) = C_{\mathcal{N}}(\beta)$, etc.

In terms of the s-ordering an operator $\hat{A}$ can be decomposed, using (4.163), as

$$\hat{A} = \frac{1}{\pi}\int \operatorname{Tr}\{\hat{D}(\beta, -s)\,\hat{A}\}\,\hat{D}^{-1}(\beta, -s)\, d^2\beta = \sum_n\sum_m \{\hat{a}^{+n}\hat{a}^m\}_s\, A_{nm}, \qquad (4.173a)$$

where

$$A_{nm} = \frac{1}{\pi n!m!}\int \operatorname{Tr}\{\hat{D}(\beta)\,\hat{A}\}\exp\left(-\frac{s}{2}|\beta|^2\right)(-\beta)^n\beta^{*m}\, d^2\beta. \qquad (4.173b)$$

As a consequence of the presence of the factor $\exp(-s\,|\beta|^2/2)$ in the coefficients A_{nm}, it can be proved [Cahill and Glauber (1969a)] that all bounded operators $\hat{A}$ [(4.158) is finite] possess convergent s-ordered power series for $\operatorname{Re} s > 1/2$, i.e. when the ordering is closer to normal than to symmetric ordering.

A characteristic feature of the expansion (4.169) is that both the Ω- and $\tilde{\Omega}$-orderings [s-and ($-s$)-orderings] are involved. One can easily show that, adopting the s-ordering, $\hat{\Delta}(\beta, -1) = |\beta\rangle\langle\beta|$, while $\hat{\Delta}(\beta, 1)$ can be shown to be very singular. In the first case, when $\hat{\Delta}(\beta, -1)$ is regular, $\Phi_{\mathcal{N}}(\alpha)$ may have singularities, in the second case, when $\Phi_{\mathcal{A}}(\alpha)$ is regular, $\hat{\Delta}(\beta, 1)$ is singular. Thus extreme smoothness of one quantity leads to singular behaviour of the other. Only for the symmetric ordering ($s = 0$), are both the quantities $\hat{\Delta}(\beta, 0)$ and $\Phi_{\text{sym}}(\alpha) = \pi^{-1}\Phi(\alpha, 0)$ regular.

4.8.2 Connecting relations

Considering two orderings Ω_1 and Ω_2, we find from (4.159) and (4.168) that

$$\Phi(\alpha, \Omega_2) = \int \Phi(\beta, \Omega_1)\, K_{21}(\alpha - \beta)\, d^2\beta, \qquad (4.174)$$

where

$$K_{21}(\beta) = \frac{1}{\pi^2}\int \tilde{\Omega}_1(\gamma^*, \gamma)\,\Omega_2(\gamma^*, \gamma)\exp(\beta\gamma^* - \beta^*\gamma)\, d^2\gamma. \qquad (4.175)$$

The corresponding relation between characteristic functions reads

$$C(\beta, \Omega_2) = C(\beta, \Omega_1)\,\tilde{\Omega}_1(\beta^*, \beta)\,\Omega_2(\beta^*, \beta). \qquad (4.176)$$

For the s-ordering, $\Omega(\beta^*, \beta) = \exp(s\,|\beta|^2/2)$ and

$$C(\beta, s_2) = C(\beta, s_1)\exp\left[\frac{1}{2}(s_2 - s_1)|\beta|^2\right]. \qquad (4.177)$$

For K_{21} we have

$$K_{21}(\beta) = \frac{1}{\pi^2} \int \exp\left[\frac{1}{2}(s_2 - s_1)|\gamma|^2\right] \exp(\beta\gamma^* - \beta^*\gamma)\, \mathrm{d}^2\gamma =$$

$$= \frac{2}{\pi(s_1 - s_2)} \exp\left(-\frac{2|\beta|^2}{s_1 - s_2}\right), \qquad \mathrm{Re}\, s_1 > \mathrm{Re}\, s_2, \tag{4.178}$$

if we use the helpful integral

$$\int \exp(-s|\gamma|^2 + \alpha\gamma^* + \alpha'\gamma)\, \mathrm{d}^2\gamma = \frac{\pi}{s} \exp\left(\frac{\alpha\alpha'}{s}\right), \qquad \mathrm{Re}\, s > 0. \tag{4.179}$$

Hence

$$\Phi(\alpha, s_2) = \frac{2}{\pi(s_1 - s_2)} \int \Phi(\beta, s_1) \exp\left(-\frac{2|\alpha - \beta|^2}{s_1 - s_2}\right) \mathrm{d}^2\beta, \tag{4.180}$$

$$\mathrm{Re}\, s_1 > \mathrm{Re}\, s_2,$$

from which the relations (4.88a, b) and (4.71) follow with $s_1 = 0$, $s_2 = -1$, and $s_1 = 1$, $s_2 = 0$, and $s_1 = 1$, $s_2 = -1$ respectively. It can be seen quite generally from (4.180) that this Gaussian convolution tends to smooth out any unruly behaviour of the quasi-distribution $\Phi(\beta, s_1)/\pi$. For example for the coherent state and $s_1 = 1$, $\pi^{-1}\Phi(\beta, 1) \equiv \Phi_{\mathcal{N}}(\beta) = \delta(\beta - \gamma)$ and $\Phi(\alpha, s) = [2/(1 - s)] \exp[-2|\alpha - \gamma|^2/(1 - s)]$, $s_2 = s$, which is a regular function for $s \neq 1$. For $s = -1$, $\pi^{-1}\Phi(\alpha, -1) \equiv \Phi_{\mathcal{A}}(\alpha) = \pi^{-1} \exp(-|\alpha - \gamma|^2)$, for $s = 0$, $\pi^{-1}\Phi(\alpha, 0) \equiv \Phi_{\mathrm{sym}}(\alpha) = 2\pi^{-1} \exp(-2|\alpha - \gamma|^2)$.

The corresponding relations between moments are

$$\langle \hat{a}^{+k}\hat{a}^l \rangle_{s_2} = \frac{k!}{l!}\left(\frac{s_1 - s_2}{2}\right)^k \left\langle \hat{a}^{l-k} L_k^{l-k}\left(\frac{2\hat{a}^+\hat{a}}{s_2 - s_1}\right)\right\rangle_{s_1}, \qquad l \geqq k,$$

$$= \frac{l!}{k!}\left(\frac{s_1 - s_2}{2}\right)^l \left\langle \hat{a}^{+k-l} L_l^{k-l}\left(\frac{2\hat{a}^+\hat{a}}{s_2 - s_1}\right)\right\rangle_{s_1}, \qquad l \leqq k. \tag{4.181}$$

In particular cases, for $s_2 = 1$ and $s_1 = -1$,

$$\langle \hat{a}^{+k}\hat{a}^l \rangle_{\mathcal{N}} = \frac{k!}{l!}(-1)^k \langle \hat{a}^{l-k} L_k^{l-k}(\hat{a}^+\hat{a}) \rangle_{\mathcal{A}}, \qquad l \geq k, \tag{4.182a}$$

for $s_2 = -1$ and $s_1 = 1$

$$\langle \hat{a}^{+k}\hat{a}^l \rangle_{\mathcal{A}} = \frac{k!}{l!} \langle \hat{a}^{l-k} L_k^{l-k}(-\hat{a}^+\hat{a}) \rangle_{\mathcal{N}}, \qquad l \geq k, \tag{4.182b}$$

etc.

4.8.3 Multimode description

The general theory of operator ordering just developed may easily be extended to systems with any number of degrees of freedom. We prefer to formulate this theory with respect to the photocount statistics, employing the detection operator $\hat{\mathbf{A}}(x)$ and the number operator $\hat{n}_{Vt}$.

Applying equations (4.173a, b) to the operator $\exp(\mathrm{i}y\hat{a}^{+}\hat{a})$, we arrive at

$$\langle \exp(\mathrm{i}y\hat{n})\rangle = \langle \prod_{\lambda}^{M} \exp(\mathrm{i}y\hat{a}_{\lambda}^{+}\hat{a}_{\lambda})\rangle = \tag{4.183}$$

$$= \left[1 - \frac{1-s}{2}(1-\mathrm{e}^{\mathrm{i}y})\right]^{-M} \left\langle \exp\left[\frac{-\dfrac{2\hat{n}}{1-s}\dfrac{1-s}{2}(1-\mathrm{e}^{\mathrm{i}y})}{1-\dfrac{1-s}{2}(1-\mathrm{e}^{\mathrm{i}y})}\right]\right\rangle_{s},$$

where M represents the number of modes. This result is valid for an arbitrary volume V (whose linear dimensions are much larger than the wavelength) by the same argument which led to (2.51c–e).

Making use of the substitution $(\mathrm{e}^{\mathrm{i}y} - 1)/[1 - (1 - s)(1 - \mathrm{e}^{\mathrm{i}y})/2] \to \mathrm{i}y$ in (4.183), we obtain for the s-ordered characteristic function

$$\langle \exp(\mathrm{i}y\hat{n})\rangle_{s} = \left(1 - \frac{1-s}{2}\mathrm{i}y\right)^{-M} \left\langle \left[\frac{1+\dfrac{1+s}{2}\mathrm{i}y}{1-\dfrac{1-s}{2}\mathrm{i}y}\right]^{\hat{n}}\right\rangle. \tag{4.184}$$

Comparing (4.183) for two orderings $s = s_1$ and $s = s_2$ and using the above substitution again, we obtain

$$\langle \exp(\mathrm{i}y\hat{n})\rangle_{s_2} = \left(1 + \frac{s_2 - s_1}{2}\mathrm{i}y\right)^{-M} \left\langle \exp\left[\frac{\mathrm{i}y\hat{n}}{1+\dfrac{s_2-s_1}{2}\mathrm{i}y}\right]\right\rangle_{s_1}, \tag{4.185}$$

which is the generating function for the Laguerre polynomials L_r^{M-1} [Morse and Feshbach (1953), Chapter 6],

$$\langle \exp(\mathrm{i}y\hat{n})\rangle_{s_2} = \sum_{r=0}^{\infty} \frac{(\mathrm{i}y)^r}{\Gamma(r+M)}\left[\frac{s_1 - s_2}{2}\right]^r \left\langle L_r^{M-1}\left(\frac{2\hat{n}}{s_2 - s_1}\right)\right\rangle_{s_1}. \tag{4.186}$$

The relation between the distributions $P(W, s_2)$ and $P(W, s_1)$ is obtained directly from (4.185) using the Fourier transformation and the residue theorem

$$P(W, s_2) = \int \delta(\sum_{\lambda} |\alpha_\lambda|^2 - W) \prod_{\lambda} \Phi(\alpha_\lambda, s_2) \frac{\mathrm{d}^2\alpha_\lambda}{\pi} =$$

$$= \frac{2}{s_1 - s_2} \int_0^{\infty} \left(\frac{W}{W'}\right)^{(M-1)/2} \exp\left[-\frac{2(W+W')}{s_1 - s_2}\right] \times$$

$$\times I_{M-1}\left(4\frac{(WW')^{1/2}}{s_1 - s_2}\right) P(W', s_1)\,\mathrm{d}W', \qquad \mathrm{Re}\, s_1 > \mathrm{Re}\, s_2, \tag{4.187}$$

where $I_M(x)$ is the modified Bessel function.

From (4.184) we obtain

$$P(W, s) = \frac{1}{W} \exp\left(-\frac{2W}{1-s}\right)\left(\frac{2W}{1-s}\right)^M \sum_{n=0}^{\infty} \frac{n!p(n)}{[\Gamma(n+M)]^2}\left(\frac{s+1}{s-1}\right)^n \times$$
$$\times L_n^{M-1}\left(\frac{4W}{1-s^2}\right), \tag{4.188}$$

which can be inverted, using the orthogonality condition (3.118), in the form of the generalized photodetection equation [Peřina and Horák (1969b, 1970), Zardecki (1974)]:

$$p(n) = \frac{1}{\Gamma(n+M)}\left(\frac{2}{1+s}\right)^M\left(\frac{s-1}{s+1}\right)^n \int_0^{\infty} P(W, s)\, L_n^{M-1}\left(\frac{4W}{1-s^2}\right) \times$$
$$\times \exp\left(-\frac{2W}{1+s}\right) \mathrm{d}W. \tag{4.189}$$

For $s \to 1$, if the asymptotic formula $L_n^{M-1}(x) \underset{x\to\infty}{\simeq} \Gamma(n+M)(-x)^n/n!$ is employed, equation (4.189) reduces to the standard photodetection equation (3.85) involving normal ordering.

The relation between s_2- and s_1-ordered moments also follows from (4.185) or (4.186):

$$\langle \hat{n}^k \rangle_{s_2} = \frac{\mathrm{d}^k}{\mathrm{d}(\mathrm{i}y)^k} \langle \exp(\mathrm{i}y\hat{n}) \rangle_{s_2} \bigg|_{\mathrm{i}y=0} = \frac{k!}{\Gamma(k+M)}\left(\frac{s_1 - s_2}{2}\right)^k \times$$
$$\times \left\langle L_k^{M-1}\left(\frac{2\hat{n}}{s_2 - s_1}\right)\right\rangle_{s_1}. \tag{4.190}$$

A number of particular relations can be obtained for the normal, antinormal and symmetric orderings, such as

$$\langle \exp(\mathrm{i}y\hat{n}) \rangle_{\mathscr{A}} = (1 - \mathrm{i}y)^{-M} \left\langle \exp\left(\frac{\mathrm{i}y\hat{n}}{1 - \mathrm{i}y}\right)\right\rangle_{\mathscr{N}}, \tag{4.191a}$$

$$\langle \exp(\mathrm{i}y\hat{n}) \rangle_{\mathscr{N}} = (1 + \mathrm{i}y)^{-M} \left\langle \exp\left(\frac{\mathrm{i}y\hat{n}}{1 + \mathrm{i}y}\right)\right\rangle_{\mathscr{A}}, \tag{4.191b}$$

$$\langle \hat{n}^k \rangle_{\mathscr{A}} = \frac{k!}{\Gamma(k+M)} \langle L_k^{M-1}(-\hat{n}) \rangle_{\mathscr{N}}, \tag{4.192}$$

$$\langle \hat{n}^k \rangle_{\mathscr{A}} = \langle (\hat{n} + M)(\hat{n} + M + 1) \ldots (\hat{n} + M + k - 1) \rangle, \tag{4.193}$$

$$\langle \exp(\mathrm{i}y\hat{n}) \rangle_{\mathscr{A}} = \langle (1 - \mathrm{i}y)^{-\hat{n} - M} \rangle, \tag{4.194}$$

$$\langle \exp(\mathrm{i}y\hat{n}) \rangle = \langle \exp[-\mathrm{i}yM + \hat{n}(1 - \mathrm{e}^{-\mathrm{i}y})] \rangle_{\mathscr{A}}. \tag{4.195}$$

Equation (4.193) represents an M-mode generalization of (2.14b) and can be also obtained with the help of the commutation rules (2.51a, b) giving $\int_V \hat{\mathbf{A}}(\mathbf{x}, t) \,.\, \hat{\mathbf{A}}(\mathbf{x}, t)\, \mathrm{d}^3x = \hat{n}_{Vt} + M$, M being the number of modes in V. Note that successive factors in (2.51c) are decreased by unity, whereas those in (4.193) are increased by unity. The difference may be regarded as a reflection of the fact that normally ordered correlations correspond to photon absorption, whereas antinormally ordered correlations correspond to photon emission.

It is clear from (4.193) that measurements by quantum counters will be sensitive to the field only if the mean number of photons $\langle n \rangle$ is much larger than the number of modes, i.e. if the degeneracy parameter $\delta = \langle n \rangle / M$ is much larger than one. This is the case for laser fields, whereas for thermal fields, the quantum counter will not be a useful measuring device. However, for strong fields the difference between normally and antinormally ordered correlations vanishes and consequently both photoelectric detectors and quantum counters will give practically the same results.

Finally note that the convolution law (4.71) corresponds to the commutator $\hat{a}\hat{a}^+ - \hat{a}^+\hat{a} = \hat{1}$ as follows: $\hat{a}\hat{a}^+ \rightleftarrows \Phi_{\mathscr{A}}(\alpha)$, $\hat{a}^+\hat{a} \rightleftarrows \Phi_{\mathscr{N}}(\alpha)$ and the vacuum contribution $\hat{1} \rightleftarrows \exp(-|\alpha|^2) = \langle \alpha | 0 \rangle \langle 0 | \alpha \rangle = \langle \alpha | \mathscr{N}\{\exp(-\hat{a}^+\hat{a})\} | \alpha \rangle$ (i.e. $| 0 \rangle \langle 0 | = \mathscr{N}\{\exp(-\hat{a}^+\hat{a})\}$).

4.9 Interference of independent light beams

An interesting question is the interference of independent light beams. Although such interference effects are easy to observe in the region of radiowaves, it is more difficult to observe them in the optical region. However, it is clear that in principle no reasons exist for excluding this effect even in the optical region. In general, in the common part of two optical beams, interference fringes will be present (since both the fields are linear and superposition principle holds), but they will vary with time. If the resolving time of the detector is long the fringes will be washed out and no interference effect can be observed; if the resolving time is sufficiently short, fringes are visible. In this connection the question arises of whether this result contradicts the well-known remark of Dirac (1958) that "each photon interferes only with itself and interference between different photons never occurs".

The first demonstration of beats resulting from the superposition of incoherent light beams was given by Forrester, Gudmundsen and Johnson (1955), who used the two spectral components of a Zeeman doublet. It is interesting to note that this experiment was performed with non-degenerate light beams and the mean number of photons received on a coherence area, during the time in which a steady beat was observed, was much less than one. With the development of the laser, producing light beams with large values of the degeneracy parameter δ, such experiments became easier to perform [Javan et al. (1962), Lipsett and Mandel (1963, 1964)]. Interference fringes from the superposition of two independent laser beams have been observed by Magyar and Mandel (1963, 1964). The significance of a large value of the degeneracy parameter δ for these experiments can easily be seen. The number of photons defining the interference pattern in the receiving plane in a time less than the coherence time is limited by δ. When $\delta \ll 1$ it is difficult to think of interference fringes.

A number of papers have been addressed to the theory of interference from both the classical and quantum points of view [Paul et al. (1963), Paul (1964, 1966, 1967), Mandel (1964b), Richter et al. (1964), Jordan and Ghielmetti (1964), Korenman (1965), Mandel and Wolf (1965), Reynolds et al. (1969), Teich (1969), Walls (1977), Bertolotti and Sibilia (1980), Tatarskii (1983), Mandel (1983)].

Interesting experiments on the interference of independent photon beams were performed by Pfleegor and Mandel (1967a, b, 1968) and by Radloff (1968, 1971).

In the experiment of Pfleegor and Mandel interference fringes were measured under conditions where the light intensity was so low that the mean time interval between photons was large compared with their transit time through the measuring apparatus, i.e. there was a high probability that one photon was absorbed before the next one was emitted by one or other of the laser sources. Since the intensities were very low, a photon correlation technique was required to observe the interference fringes. The interference pattern was received on a stack of thin glass plates, each of which had a thickness corresponding to about a half fringe width. The plates were cut and arranged so that any light falling on the 1st, 3rd, 5th, etc. plate was fed to one photomultiplier, while light falling on the 2nd, 4th, 6th, etc. plate was fed to the other. When the half fringe spacing coincides with the plate thickness, and for example the fringe maxima fall on the odd-numbered plates, one photodetector will register nearly all the photons and the other almost none. The position of the fringe maxima is unpredictable and random, but if the number of photons registered by one photodetector increases, the number registered by the other must decrease, provided that the fringe spacing is right for the plates. Thus there must be a negative correlation between the numbers of counts from the two photodetectors and such a negative correlation was indeed observed. Since the experimental conditions were arranged so that one photon was absorbed before the next one was emitted, this result shows that the above mentioned Dirac statement is applicable to these experiments. In general experiments with independent light beams are in agreement with the Dirac statement since any „localization" of a photon in space-time automatically rules out the possibility of knowing its momentum, as a consequence of the uncertainty principle. Thus one cannot say to which beam a given photon belongs – each photon is to be considered as being partly in both the beams and interfering only with itself.

In the Radloff experiment too, the conditions were such that there was a high probability that only one photon was present in one exposure interval. After many such intervals interference fringes emerged.

Interference experiments with thermal sources have been performed by Haig and Sillitto (1968) and McMillan, R. M. Sillitto and W. Sillitto (1979), employing coincidence measurement for 144 hours.

In terms of the coherent states the interference experiments with independent light beams can be simply described. Let the field generated by the first source be described by $\Phi_{\mathcal{N}}^{(1)}(\{\alpha_{\lambda 1}\})$ and the field generated by the second source by $\Phi_{\mathcal{N}}^{(2)}(\{\alpha_{\lambda 2}\})$. The superposition of both these fields is then described by the convolution (4.128).

The expectation value of the intensity at a point $x \equiv (\mathbf{x}, t)$ of the interference pattern is

$$\begin{aligned} \langle \hat{A}^{(-)}(x) \hat{A}^{(+)}(x) \rangle &= I_1(x) + I_2(x) + \\ &+ 2 \operatorname{Re} \{ \int \Phi_{\mathcal{N}}^{(1)}(\{\alpha_{\lambda 1}\}) V^*(x, \{\alpha_{\lambda 1}\}) \, d^2\{\alpha_{\lambda 1}\} \times \\ &\times \int \Phi_{\mathcal{N}}^{(2)}(\{\alpha_{\lambda 2}\}) V(x, \{\alpha_{\lambda 2}\}) \, d^2\{\alpha_{\lambda 2}\} \}, \end{aligned} \tag{4.196}$$

where $I_j(x) = \langle \hat{A}^{(-)}(x, \{\alpha_{\lambda j}\}) \hat{A}^{(+)}(x, \{\alpha_{\lambda j}\}) \rangle$, $j = 1, 2$ are averaged intensities produced by each source if the other is absent. We restrict ourselves here and in the following to linearly polarized light for simplicity. The third term in (4.196) represents an interference term and if this term does not vanish one can observe the interference pattern. In general this term will not vanish if the fields produced by both the sources contain non-zero coherent components. When both the fields are in coherent states, $\Phi_{\mathcal{N}}^{(j)}(\{\alpha_{\lambda j}\}) = \prod_{\lambda} \delta(\alpha_{\lambda j} - \beta_{\lambda j})$, $j = 1, 2$, the interference term equals $2 \operatorname{Re} \{ V^*(x, \{\beta_{\lambda 1}\}) \times \times V(x, \{\beta_{\lambda 2}\}) \}$, and so it is non-vanishing. However, this assumes a knowledge of the phases of $\{\beta_\lambda\}$ which is not the case in the optical region where the phase information is almost always lost. Then the above quasi-distributions for the coherent states must be averaged over the phases giving distributions independent of the phases. The interference pattern then vanishes. However, for the correlation of intensities we obtain [Mandel (1964b)], as verified by Vajnshtejn et al. (1981),

$$\begin{aligned} \langle I(x_1) I(x_2) \rangle_{\mathcal{N}} &= \langle [| V(x_1, \{\alpha_{\lambda 1}\}) |^2 + | V(x_1, \{\alpha_{\lambda 2}\}) |^2]^2 \rangle_{\mathcal{N}} + \\ &+ 2 \langle | V(x_1, \{\alpha_{\lambda 1}\}) |^2 \rangle_{\mathcal{N}} \langle | V(x_1, \{\alpha_{\lambda 2}\}) |^2 \rangle_{\mathcal{N}} \times \\ &\times \cos [(\bar{\mathbf{k}}_1 - \bar{\mathbf{k}}_2) \cdot (\mathbf{x}_2 - \mathbf{x}_1) - c(\bar{k}_1 - \bar{k}_2)(t_2 - t_1)], \end{aligned} \tag{4.197}$$

exhibiting a periodic variation with $\mathbf{x}_2 - \mathbf{x}_1$ and $t_2 - t_1$, assuming quasi-monochromatic beams with a spread Δk which is much less than the mean wave number $\bar{k}$. Further we have assumed that $| \mathbf{x}_1 - \mathbf{x}_2 | \ll 1/\Delta k$ and $| t_1 - t_2 | \ll 1/c\Delta k$. The wave vectors and wave numbers $\bar{\mathbf{k}}_j$ and $\bar{k}_j$ refer to the jth beam ($j = 1, 2$). Thus, over a limited space-time region, the intensity correlation shows a cosinusoidal dependence on space and time, which can be interpreted both in terms of interference fringes and light beats [Mandel (1983), Teich, Saleh and Peřina (1983)]. The generating function for this case has been calculated by Klauder and Sudarshan (1968) (Section 10-2).

4.10 Two-photon coherent states, atomic coherent states and coherent states for general potentials

4.10.1 Two-photon coherent states

Two-photon coherent states $| \beta \rangle_g$ are introduced as eigenstates of the annihilation operator [Stoler (1970, 1971, 1972, 1975), Yuen (1975, 1976)]

$$\hat{b} = \mu \hat{a} + \nu \hat{a}^+ \qquad (\hat{a} = \mu^* \hat{b} - \nu \hat{b}^+), \tag{4.198}$$

$\hat{a}, \hat{a}^+$ being the photon annihilation and creation operators, respectively. As

$$[\hat{b}, \hat{b}^+] = \hat{1}, \tag{4.199}$$

it holds that

$$|\mu|^2 - |\nu|^2 = 1. \tag{4.200}$$

Hence

$$\hat{b} \mid \beta\rangle_g = \beta \mid \beta\rangle_g, \qquad {}_g\langle\beta \mid \hat{b}^+ = {}_g\langle\beta \mid \beta^*. \tag{4.201}$$

Consequently, a number of properties of the coherent states, such as the overcompleteness condition (4.18), the diagonal representation of the density matrix (4.63), etc. can be transferred to the two-photon coherent state formalism.

Further, one can prove that

$${}_g\langle\beta \mid \hat{a}^+\hat{a} \mid \beta\rangle_g = |\mu^*\beta - \nu\beta^*|^2 + |\nu|^2,$$

$$({}_g\langle 0 \mid \hat{a}^+\hat{a} \mid 0\rangle_g = |\nu|^2), \tag{4.202a}$$

$${}_g\langle\beta \mid (\Delta\hat{a}_1)^2 \mid \beta\rangle_g = \frac{1}{4}|\mu - \nu|^2, \tag{4.202b}$$

$${}_g\langle\beta \mid (\Delta\hat{a}_2)^2 \mid \beta\rangle_g = \frac{1}{4}|\mu + \nu|^2, \tag{4.202c}$$

where $\hat{a}_1 = (\hat{a} + \hat{a}^+)/2$ and $\hat{a}_2 = (\hat{a} - \hat{a}^+)/2\mathrm{i}$ are proportional to the generalized coordinate and momentum, respectively. Unlike the coherent state $|\alpha\rangle$, which satisfies

$$\langle\alpha \mid (\Delta\hat{a})^+(\Delta\hat{a}) \mid \alpha\rangle = 0, \tag{4.203a}$$

the two-photon coherent state $|\beta\rangle_g$ involves the quantum noise

$${}_g\langle\beta \mid (\Delta\hat{a})^+ (\Delta\hat{a}) \mid \beta\rangle_g = |\nu|^2, \tag{4.203b}$$

as follows from (4.202a). The coherent states are obtained for $\mu = 1$ and $\nu = 0$. We see from (4.202b) [(4.202c)] that ${}_g\langle\beta \mid (\Delta\hat{a}_1)^2 \mid \beta\rangle_g$ $({}_g\langle\beta \mid (\Delta\hat{a}_2)^2 \mid \beta\rangle_g)$ can be less than the corresponding quantity in the coherent state ($\mu = 1$, $\nu = 0$). However, the other of these quantities is then higher than for the coherent state. Such states are called the squeezed states. These states are intensively investigated [see e.g. Mandel (1982b), Becker et al. (1982), Kozierowski and Kielich (1983), Ficek et al. (1983), Wódkiewicz and Zubairy (1983), Milburn and Walls (1983), see also Abstracts of the Fifth Rochester Conference on Coherence and Quantum Optics (1983)]. If μ and ν are high and $\mu/\nu \to 1$ $(\mu/\nu \to -1)$, then ${}_g\langle\beta \mid (\Delta\hat{a}_1)^2 \mid \beta\rangle_g \times$ $\times ({}_g\langle\beta \mid (\Delta\hat{a}_2)^2 \mid \beta_g\rangle) \to 0$. This possibility of reducing quantum noise may be employed in optical communication [Yuen and Shapiro (1978a), Shapiro et al. (1979), Helstrom (1979), Shapiro (1980)[and in optical interferometers proposed to detect gravitational waves [Caves (1981), Walls and Zoller (1981), Walls and Milburn (1981), Loudon (1981)]. The coherent states as well as the two-photon coherent states minimize the uncertainty

$$\langle(\Delta\hat{a}_1)^2\rangle\,\langle(\Delta\hat{a}_2)^2\rangle \geqq \frac{1}{16}, \tag{4.204}$$

if $\mu = \delta\nu$, where δ is real.

The scalar products of the two-photon coherent states with the coherent states and the Fock states are [Yuen (1976)]

$$\langle\alpha\,|\,\beta\rangle_g = \mu^{-1/2}\exp\left(-\frac{1}{2}|\alpha|^2 - \frac{1}{2}|\beta|^2 - \frac{\nu}{2\mu}\alpha^{*2} + \frac{\nu^*}{2\mu}\beta^2 + \frac{1}{\mu}\alpha^*\beta\right) \tag{4.205a}$$

and

$$\langle n\,|\,\beta\rangle_g = (n!\mu)^{-1/2}\left(\frac{\nu}{2\mu}\right)^{n/2}\exp\left(-\frac{1}{2}|\beta|^2 + \frac{\nu^*}{2\mu}\beta^2\right)\times$$

$$\times H_n\left(\frac{\beta}{(2\mu\nu)^{1/2}}\right), \tag{4.205b}$$

H_n being the Hermite polynomial. For the photon-number distribution we have [Yuen (1976), Mišta et al. (1977)]

$$p(n) = |\langle n\,|\,\beta\rangle_g|^2 = \frac{1}{2^n n!}\frac{1}{|\mu|}\left(\frac{|\nu|}{|\mu|}\right)^n \times$$

$$\times\exp\left(-|\beta|^2 + \frac{\nu^*}{2\mu}\beta^2 + \frac{\nu}{2\mu^*}\beta^{*2}\right)H_n\left(\frac{\beta}{(2\mu\nu)^{1/2}}\right)H_n\left(\frac{\beta^*}{(2\mu^*\nu^*)^{1/2}}\right), \tag{4.206}$$

with the characteristic function

$$\langle\exp(isW)\rangle_{\mathcal{N}} = \tau^{-1/2}\exp\left\{\left(\frac{1+is}{\tau} - 1\right)|\beta|^2 + \left[1 - \frac{(1+is)^2}{\tau}\right]\times\right.$$

$$\left.\times\left(\frac{\nu^*}{2\mu}\beta^2 + \frac{\nu}{2\mu^*}\beta^{*2}\right)\right\}, \tag{4.207}$$

where $\tau = |\mu|^2 - (1+is)^2|\nu|^2$.

Yuen (1976) proved that the two-photon coherent states are generated by the interaction described by the hamiltonian

$$\hat{H} = \hbar\omega\hat{a}^+\hat{a} + g\hat{a}^2 + g^*\hat{a}^{+2} + f\hat{a} + f^*\hat{a}^+, \tag{4.208}$$

f and g being arbitrary classical functions, i.e. they are generated by two-photon stimulated emission, provided that the atomic variables are described classically [this assumption may not be always correct, see discussions by Golubev (1979) and Schubert and Vogel (1981), see Sec. 9.6]. Nevertheless, these states may be produced in the process of degenerate parametric amplification with a strong laser pumping field (Chapter 9). Such radiation exhibits naturally antibunching of photons and squeezing. The spatial behaviour of the two-photon coherent states has been investigated by Schubert and Vogel (1978).

We have finished a systematic treatment of the coherent-state technique connected with harmonic oscillators and free electromagnetic fields. Now we would like to show that the concept of coherent states is more general and that more general kinds of coherent states may be introduced, serving particularly for descriptions of the interaction of the electromagnetic field with atoms of matter.

4.10.2 Atomic coherent states

Radcliffe (1971) and Arecchi, Courtens, Gilmore and Thomas (1972, 1973) have shown that a free system of N two-level atoms can be described in terms of the so-called atomic coherent states, which have many properties in common with coherent states of the annihilation operator of the harmonic oscillator. They are eigenstates of the angular momentum operator and they correspond to a set of classical dipoles (classical currents), additionally involving physical vacuum fluctuations, and they are generated by a classical radiation field, rather as the coherent states are generated by classical currents (see Tab. 4.1).

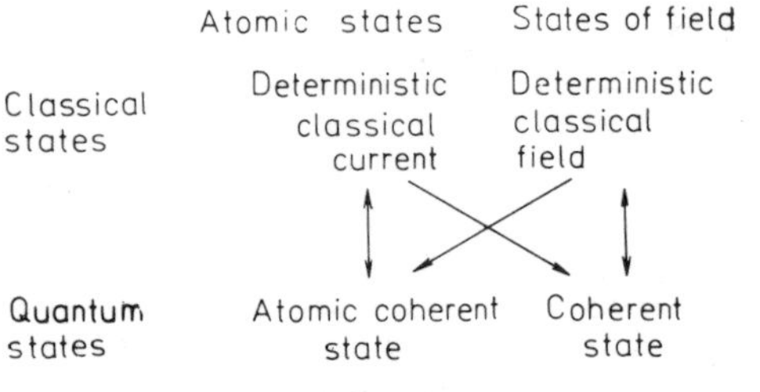

Tab. 4.1

Generation of the quantum states (one-fold arrows) and the correspondence principle (two-fold arrows).

We introduce the standard angular momentum operators $\mathscr{M}_1, \mathscr{M}_2, \mathscr{M}_3$ for N-atom system, which obey the commutation rules

$$[\mathscr{M}_1, \mathscr{M}_2] = i\mathscr{M}_3 \qquad \text{(and cyclic permutations)}, \tag{4.209a}$$

$$[\mathscr{M}_3, \mathscr{M}_\pm] = \pm\mathscr{M}_\pm \qquad (\mathscr{M}_\pm = \mathscr{M}_1 \pm i\mathscr{M}_2), \tag{4.209b}$$

$$[\mathscr{M}_+, \mathscr{M}_-] = 2\mathscr{M}_3; \tag{4.209c}$$

thus $\mathscr{M}_\pm$ are the raising and lowering operators, respectively. Now we can introduce the so-called Dicke states $| l, m\rangle$,

$$\mathscr{M}^2 | l, m\rangle = l(l+1) | l, m\rangle, \tag{4.210a}$$

$$\mathscr{M}_3 | l, m\rangle = m | l, m\rangle, \tag{4.210b}$$

where $\mathscr{M}^2 = \mathscr{M}_1^2 + \mathscr{M}_2^2 + \mathscr{M}_3^2$. From the theory of angular momentum it is known that l may be an integer or half-integer, and m, for a given l, varies from $-l$ to $+l$; the maximum value of l is $N/2$, that is

$$| m | \leqq l \leqq \frac{N}{2}. \tag{4.211}$$

The quantum number l is called the cooperation number and the ground state is defined as $\mathcal{M}_- \mid l, -l\rangle = 0$.

The corresponding hamiltonian of the radiation and the N-atom system in interaction is given by

$$\hat{H} = \hbar\omega\hat{a}^+\hat{a} + \hbar\omega_0\mathcal{M}_3 + \hbar G(\hat{a}\mathcal{M}_+ + \hat{a}^+\mathcal{M}_-), \tag{4.212}$$

ω_0 being the atom transition frequency and G a coupling constant. At resonance, $\omega = \omega_0$, and writing $g = G/\omega$, we have (omitting ω)

$$\hat{H} = \hbar\hat{a}^+\hat{a} + \hbar\mathcal{M}_3 + \hbar g(\hat{a}\mathcal{M}_+ + \hat{a}^+\mathcal{M}_-). \tag{4.213}$$

Following Schwinger, we can introduce boson operators $\hat{a}_j$, $[\hat{a}_j, \hat{a}_k^+] = \delta_{jk}\hat{1}$, $j, k = 1, 2$, so that

$$\mathcal{M}_+ = \hat{a}_2^+\hat{a}_1, \mathcal{M}_- = \hat{a}_1^+\hat{a}_2, \mathcal{M}_3 = \frac{1}{2}(\hat{a}_2^+\hat{a}_2 - \hat{a}_1^+\hat{a}_1). \tag{4.214}$$

These operators satisfy the commutation rules (4.209), and we have also

$$\mathcal{M}^2 = \hat{L}(\hat{L} + 1), \hat{L} = \frac{1}{2}(\hat{a}_1^+\hat{a}_1 + \hat{a}_2^+\hat{a}_2). \tag{4.215}$$

Substituting (4.214) in (4.213) we find [Walls and Barakat (1970), Nussenzveig (1973)]

$$\hat{H} = \hbar\omega\hat{a}^+\hat{a} + \hbar\omega_1\hat{a}_1^+\hat{a}_1 + \hbar\omega_2\hat{a}_2^+\hat{a}_2 + \hbar g(\hat{a}\hat{a}_1\hat{a}_2^+ + \hat{a}^+\hat{a}_1^+\hat{a}_2), \tag{4.216}$$

if $\omega = 1$, $\omega_1 = -1/2$ and $\omega_2 = 1/2$.

The atomic coherent states, or the Bloch states, can be defined from the ground state, in analogy to the displacement operator definition (4.1) of the coherent state $\mid\alpha\rangle$, as

$$\begin{aligned}\mid\vartheta, \varphi\rangle &= \exp(\xi\mathcal{M}_+ - \xi^*\mathcal{M}_-)\mid l, -l\rangle = \\ &= \left(\cos\frac{\vartheta}{2}\right)^{2l}\exp\left[e^{-i\varphi}\tan\left(\frac{\vartheta}{2}\right)\mathcal{M}_+\right]\mid l, -l\rangle,\end{aligned} \tag{4.217a}$$

where $\xi = \vartheta\exp(-i\varphi)/2$ [ϑ and φ being rotating angles in the angular momentum space (Bloch sphere)]. In terms of the Dicke states we have

$$\begin{aligned}\mid\vartheta, \varphi\rangle &= (1 + \mid\tau\mid^2)^{-l}\exp(\tau\mathcal{M}_+)\mid l, -l\rangle = \\ &= (1 + \mid\tau\mid^2)^{-l}\sum_{m=-l}^{+l}\binom{2l}{l+m}^{1/2}\tau^{l+m}\mid l, m\rangle,\end{aligned} \tag{4.217b}$$

where $\tau = \tan(\vartheta/2)\exp(-i\varphi)$; this is analogous to (4.4).

The atomic coherent states have the eigenvalue property

$$(\mathcal{M}\,.\,\mathbf{r})\mid\vartheta, \varphi\rangle = r\mid\vartheta, \varphi\rangle, \tag{4.218}$$

$\mathbf{r}$ being the unit vector in the direction (ϑ, φ). They are also minimum uncertainty states,

$$\langle(\Delta\mathcal{M}_1)^2\rangle\langle(\Delta\mathcal{M}_2)^2\rangle = \frac{1}{4}\langle\mathcal{M}_3^2\rangle. \tag{4.219}$$

The atomic coherent states satisfy the overcompleteness condition

$$\frac{2l+1}{4\pi}\int |\vartheta, \varphi\rangle \langle \vartheta, \varphi| \sin\vartheta \, d\vartheta \, d\varphi = \sum_{m=-l}^{m=+l} |l, m\rangle \langle l, m| = \hat{1}, \tag{4.220}$$

which can serve for representing vectors and operators in terms of the atomic coherent states.

The correspondence $|n_2, n_1\rangle \rightleftarrows |l, m\rangle$ holds, $\hat{a}_j^+ \hat{a}_j |n_2, n_1\rangle = n_j |n_2, n_1\rangle$, $j = 1, 2$, if

$$l = \frac{1}{2}(n_1 + n_2), \qquad m = \frac{1}{2}(n_2 - n_1). \tag{4.221}$$

It may be concluded that under the influence of a classical (non-fluctuating) current, an electromagnetic field initially in a coherent state (or in its ground state) will evolve into a coherent state, and under the influence of a classical (non-fluctuating) electromagnetic field, an atomic system initially in an atomic coherent state (or in its ground state) will evolve into an atomic coherent state (see Tab. 4.1).

It is obvious that the Dicke states correspond to the Fock states and the atomic coherent states (Bloch states) correspond to the coherent states (Glauber states), and $\hat{a} \rightleftarrows \mathcal{M}_-$, $\hat{a}^+ \rightleftarrows \mathcal{M}_+$.

A discussion of superradiance, based on these results, will be given in Chapter 6. The use of the atomic coherent states for mapping operator equations into classical differential forms has been discussed by Lugiato et al. (1977) [for classical-quantum correspondence for multilevel systems, see Gilmore (1975)]. Some general discussion of the origin of coherence in interacting systems has been given by Ernst (1969, 1976).

Finally we note that general coherent states for quantum systems described by general potentials can be introduced if, using integrals of motion, one can introduce generalized boson annihilation operators [Perelomov (1977), Malkin and Man'ko (1979), Nieto and Simmons (1978, 1979), Nieto et al. (1981)]. Such coherent states have proved to be useful also in integrated optics [Krivoshlykov and Sissakian (1979, 1980a, b, 1983)], because the coherent-state matrix elements can describe the propagation of the radiation field in optical waveguides and they can serve as generating functions for coupling mode coefficients. These general coherent states may be introduced in three different ways—as eigenstates of the generalized boson operators, or by using displacement operators, or as minimum-uncertainty states; in general, these three procedures need not be equivalent.

SPECIAL STATES OF THE ELECTROMAGNETIC FIELD

The general methods developed in the preceding chapters will be demonstrated now by applying them to the chaotic (Gaussian) field which is the most typical state in nature, to coherent laser light, and to the superposition of coherent and chaotic fields (i.e. of signal and noise).

5.1 Chaotic (Gaussian) light

5.1.1 Distributions and characteristic functions

The chaotic state can be defined as a state with maximal entropy

$$H = -\mathrm{Tr}\,\{\hat{\varrho}\log\hat{\varrho}\}, \tag{5.1}$$

assuming that $\mathrm{Tr}\,\hat{\varrho} = 1$ and $\mathrm{Tr}\,\{\hat{\varrho}\hat{a}^+\hat{a}\} = \langle n\rangle$. As we have seen, the density matrix for this state is a diagonal of the form (2.34) in terms of the Fock states with $\varrho(n)$ as members of the Bose–Einstein distribution (2.39), that is

$$\hat{\varrho} = \frac{\langle n\rangle^{\hat{a}^+\hat{a}}}{(1+\langle n\rangle)^{1+\hat{a}^+\hat{a}}}\,. \tag{5.2}$$

For thermal (blackbody) radiation (2.37) holds for $\langle n\rangle$ and we obtain (2.36) for $\hat{\varrho}$.

For a multimode field, in analogy to (2.36),

$$\hat{\varrho} = \frac{\prod_\lambda \exp(-\Theta_\lambda \hat{a}_\lambda^+\hat{a}_\lambda)}{\mathrm{Tr}\,\{\prod_\lambda \exp(-\Theta_\lambda \hat{a}_\lambda^+\hat{a}_\lambda)\}} = \frac{\exp\left(-\frac{1}{KT}\hat{H}\right)}{\mathrm{Tr}\left\{\exp\left(-\frac{1}{KT}\hat{H}\right)\right\}}, \tag{5.3}$$

where the renormalized hamiltonian $\hat{H}$ is given by (2.19a) with the vacuum energy $\sum_\lambda \hbar\omega_\lambda/2$ subtracted.

Because of the statistical independence of different modes of chaotic radiation, we can write

$$\hat{\varrho} = \prod_\lambda \frac{\langle n_\lambda\rangle^{\hat{a}_\lambda^+\hat{a}_\lambda}}{(1+\langle n_\lambda\rangle)^{1+\hat{a}_\lambda^+\hat{a}_\lambda}} = \sum_{\{n_\lambda\}}\prod_{\lambda'} \frac{\langle n_{\lambda'}\rangle^{n_{\lambda'}}}{(1+\langle n_{\lambda'}\rangle)^{1+n_{\lambda'}}}\,|\,\{n_\lambda\}\rangle\langle\{n_\lambda\}\,|. \tag{5.4}$$

In terms of the coherent states, making use of (4.50),

$$\hat{\varrho} = \int |\,\{\alpha_\lambda\}\rangle\langle\{\alpha_\lambda\}\,|\prod_\lambda \frac{\langle n_\lambda\rangle^{\hat{a}_\lambda^+\hat{a}_\lambda}}{(1+\langle n_\lambda\rangle)^{\hat{a}_\lambda^+\hat{a}_\lambda+1}}\,\frac{d^2\alpha_\lambda}{\pi}, \tag{5.5}$$

and if we substitute (4.49), we arrive at

$$\hat{\varrho} = \int \prod_{\lambda'} (\pi\langle n_{\lambda'}\rangle)^{-1} \exp\left(-\frac{|\alpha_{\lambda'}|^2}{\langle n_{\lambda'}\rangle}\right) |\{\alpha_\lambda\}\rangle \langle\{\alpha_\lambda\}| \, d^2\{\alpha_\lambda\}; \tag{5.6}$$

hence, we have arrived at the multimode analogue of (4.98) with $\beta_\lambda = 0$,

$$\Phi_{\mathcal{N}}(\{\alpha_\lambda\}) = \prod_\lambda (\pi\langle n_\lambda\rangle)^{-1} \exp\left(-\frac{|\alpha_\lambda|^2}{\langle n_\lambda\rangle}\right). \tag{5.7}$$

This is a multimode Gaussian distribution in the complex amplitudes α_λ and it is independent of the phases of α_λ. Thus the chaotic field is stationary in space and time.

The distribution (5.7) is consistent with the distribution (3.129); we see this by taking into account the fact that the global complex amplitude of the field is a linear superposition of the mode complex amplitudes α_λ given in (4.53) and remembering that the convolution of two Gaussian distributions is again a Gaussian distribution.

A calculation of the characteristic functional (3.69) for chaotic radiation will enable us to determine the correlation functions according to (3.70) and also the n-fold distribution (4.129). Setting

$$\varphi(x, \lambda) = \frac{(2\pi\hbar c)^{1/2}}{L^{3/2}} k^{-1/2} e_j^{(s)}(\mathbf{k}) \exp\left[i(\mathbf{k} \,.\, \mathbf{x} - ckt)\right], \tag{5.8}$$

the jth component of (2.28b) can be written in the form

$$\hat{A}_j^{(+)}(x) \equiv \hat{A}^{(+)}(x) = \sum_\lambda \varphi(x, \lambda)\, \hat{a}_\lambda. \tag{5.9}$$

Substituting (5.9) and (5.7) into (3.69) we obtain

$$\begin{aligned} C_{\mathcal{N}}\{y(x)\} &= \int \prod_\lambda (\pi\langle n_\lambda\rangle)^{-1} \exp\left(-\frac{|\alpha_\lambda|^2}{\langle n_\lambda\rangle}\right) \times \\ &\times \exp\left\{ \int \left[y(x) \sum_{\lambda'} \varphi(x, \lambda')\, \alpha_{\lambda'}^* - y^*(x) \sum_{\lambda'} \varphi(x, \lambda')\, \alpha_{\lambda'}\right] d^4x\right\} d^2\{\alpha_\lambda\} = \\ &= \int \prod_\lambda (\pi\langle n_\lambda\rangle)^{-1} \exp\left(-\frac{|\alpha_\lambda|^2}{\langle n_\lambda\rangle} + \alpha_\lambda^* y_\lambda - \alpha_\lambda y_\lambda^*\right) d^2\alpha_\lambda, \end{aligned} \tag{5.10}$$

where

$$y_\lambda = \int y(x)\, \varphi^*(x, \lambda)\, d^4x. \tag{5.11}$$

(If partial polarization is included, the integral over x includes a sum over the polarization indices j.) Using (4.179) we arrive at

$$\begin{aligned} C_{\mathcal{N}}\{y(x)\} &= \prod_\lambda \exp(-\langle n_\lambda\rangle\, |y_\lambda|^2) = \\ &= \exp\left[-\iint y^*(x') \sum_\lambda \langle n_\lambda\rangle\, \varphi(x', \lambda)\, \varphi^*(x, \lambda)\, y(x)\, d^4x\, d^4x'\right]. \end{aligned} \tag{5.12}$$

For the second-order correlation function we have

$$\Gamma_{\mathcal{N}}^{(1,1)}(x, x') = \sum_\lambda \langle n_\lambda\rangle\, \varphi(x', \lambda)\, \varphi^*(x, \lambda) \tag{5.13}$$

and we finally arrive at

$$C_{\mathcal{N}}\{y(x)\} = \exp\left[-\iint y(x)\, \Gamma_{\mathcal{N}}^{(1,1)}(x, x')\, y^*(x')\, \mathrm{d}^4x\, \mathrm{d}^4x'\right]. \tag{5.14}$$

The correlation function $\Gamma_{\mathcal{N}}^{(n,n)}$ is according to (3.70),

$$\Gamma_{\mathcal{N}}^{(n,n)}(x_1, \ldots, x_{2n}) = \sum_{\pi} \prod_{j=1}^{n} \Gamma_{\mathcal{N}}^{(1,1)}(x_j, x_{n+j}), \tag{5.15}$$

in agreement with (3.155).

Using the complex representation (4.38) of the δ-function, we can write (4.129) in the form

$$\begin{aligned} P_n(V_1, \ldots, V_n) &= \iint \Phi_{\mathcal{N}}(\{\alpha'_\lambda\}) \prod_{j=1}^{n} \exp(V_j'^* z_j - V_j' z_j^*) \times \\ &\times \exp(V_j z_j^* - V_j^* z_j) \frac{\mathrm{d}^2 z_j}{\pi^2}\, \mathrm{d}^2\{\alpha'_\lambda\} = \\ &= \int C_{\mathcal{N}}(\{z_j\}) \prod_{j=1}^{n} \exp(V_j z_j^* - V_j^* z_j) \frac{\mathrm{d}^2 z_j}{\pi^2}, \end{aligned} \tag{5.16}$$

where the characteristic function $C_{\mathcal{N}}(\{z_j\})$ is given by (5.14) with

$$y(x) = \sum_{j=1}^{n} z_j \delta(x - x_j). \tag{5.17}$$

Hence,

$$C_{\mathcal{N}}(\{z_j\}) = \exp\left[-\sum_{j,k=1}^{n} z_j \Gamma_{\mathcal{N}}^{(1,1)}(x_j, x_k)\, z_k^*\right], \tag{5.18}$$

which can be written, introducing the vectors

$$\hat{Z} = \begin{pmatrix} z_1 \\ \vdots \\ z_n \end{pmatrix}, \quad \hat{V} = \begin{pmatrix} V_1 \\ \vdots \\ V_n \end{pmatrix}, \quad \hat{V}' = \begin{pmatrix} V_1' \\ \vdots \\ V_n' \end{pmatrix} \tag{5.19}$$

and the covariance matrix

$$\mathscr{R} = \langle \hat{V}' \otimes \hat{V}'^{+} \rangle \tag{5.20}$$

($\otimes$ denotes the direct product), in the matrix form

$$C_{\mathcal{N}}(\hat{Z}) = \exp(-\hat{Z}^{+} \mathscr{R} \hat{Z}). \tag{5.21}$$

Thus we have from (5.16) the matrix form

$$P_n(V_1, \ldots, V_n) = \pi^{-2n} \int \exp(-\hat{Z}^{+} \mathscr{R} \hat{Z}) \exp(\hat{V}\hat{Z}^{+} - \hat{V}^{+}\hat{Z}) \prod_{j=1}^{n} \mathrm{d}^2 z_j, \tag{5.22}$$

which finally provides, using (4.179),

$$P_n(V_1, \ldots, V_n) = \frac{1}{\pi^n \operatorname{Det} \mathscr{R}} \exp(-\hat{V}^{+} \mathscr{R}^{-1} \hat{V}). \tag{5.23}$$

This is the multivariate Gaussian distribution, so that the quantized chaotic electromagnetic field is described as a Gaussian random process.

Quite similar results are obtainable for the s-ordering instead of the normal ordering. Applying the multimode analogue of (4.180) ($s_2 = s$, $s_1 = 1$) to (5.7), we obtain

$$\Phi(\{\alpha_\lambda\}, s) = \prod_\lambda \left(\langle n_\lambda \rangle + \frac{1-s}{2} \right)^{-1} \exp\left(- \frac{|\alpha_\lambda|^2}{\langle n_\lambda \rangle + \dfrac{1-s}{2}} \right), \tag{5.24}$$

which shows that all the results derived for chaotic light and normal ordering are also valid for s-ordering if $\langle n_\lambda \rangle$ is replaced by $\langle n_\lambda \rangle + (1-s)/2$. Hence, both the space-time quasi-probabilities and the phase space quasi-distributions are multivariate Gaussian distributions with positive-definite covariance matrices for $-1 \leqq s \leqq 1$, which includes the normal, symmetric and antinormal orderings.

5.1.2 The second-order correlation function for blackbody radiation

The second-order correlation tensors have been discussed by a number of authors, from classical as well as quantum points of view [Bourret (1960), Sarfatt (1963), Kano and Wolf (1962), Mehta (1963), Glauber (1963b, 1965), Kujawski (1968)]. Mehta and Wolf (1964, 1967) in particular developed a complete treatment of the electric, magnetic and mixed correlation tensors of the second order for blackbody radiation (including the spectral correlation tensors) and they showed by direct calculations that the classical and quantum correlation tensors are identical. The statistical properties of blackbody radiation in an unbounded domain were investigated by Holliday (1964), Holliday and Sage (1964) and Keller (1965) using the theory of functionals. Eberly and Kujawski (1967b) and Kujawski (1969) studied the properties of correlation tensors in uniformly moving coordinate systems, and the relativistic coherence theory of blackbody radiation was further discussed by Brevik and Suhonen (1968, 1970) and Eberly and Kujawski (1972). Here we discuss only some simple properties of the electric correlation tensor, particularly in connection with temporal coherence of blackbody radiation. A more detailed review can be found in the review article by Mandel and Wolf (1965).

The quantum electric correlation tensor $\mathscr{E}_{ij}(x_1, x_2)$ can be written in the form

$$\mathscr{E}_{ij}(x_1, x_2) = \mathrm{Tr}\, \{\hat{\varrho} \hat{E}_i^{(-)}(x_1)\, \hat{E}_j^{(+)}(x_2)\}. \tag{5.25}$$

The use of (5.13) and (2.2a) gives

$$\mathscr{E}_{ij}(x_1, x_2) = \frac{2\pi\hbar c}{L^3} \sum_{\mathbf{k}, s} \langle n_{\mathbf{k}s} \rangle k e_i^{(s)*}(\mathbf{k})\, e_j^{(s)}(\mathbf{k}) \times$$
$$\times \exp\left[i\mathbf{k} . (\mathbf{x}_2 - \mathbf{x}_1) - ick(t_2 - t_1) \right]. \tag{5.26}$$

As $\langle n_{\mathbf{k}s} \rangle \equiv \langle n_k \rangle$ [given by (2.37)], we can sum over s with the help of (2.11). Then replacing the sum $\sum_{\mathbf{k}}$ by the integral $(L/2\pi)^3 \int \ldots \mathrm{d}^3k$, we obtain

$$\begin{aligned}\mathscr{E}_{ij}(x_1, x_2) &= \mathscr{E}_{ij}(x_2 - x_1) = \mathscr{E}_{ij}(x) \\ &= \frac{\hbar c}{(2\pi)^2} \int \frac{k}{\exp(\Theta) - 1} \left(\delta_{ij} - \frac{k_i k_j}{k^2}\right) \exp(\mathrm{i}kx)\, \mathrm{d}^3k = \\ &= \frac{\hbar c}{(2\pi)^2} \left(-\delta_{ij}\Delta + \frac{\partial}{\partial x_i}\frac{\partial}{\partial x_j}\right) \int \frac{1}{k[\exp(\Theta) - 1]} \exp(\mathrm{i}kx)\, \mathrm{d}^3k, \end{aligned} \tag{5.27}$$

where $x \equiv (\mathbf{x}, \tau) = x_2 - x_1$ and x_i are cartesian components of $\mathbf{x}$. Introducing spherical coordinates for $\mathbf{k}$ with the polar axis along the direction $\mathbf{x}$, we obtain

$$\begin{aligned}\mathscr{E}_{ij}(x) = \frac{4\hbar c}{\pi} \sum_{n=1}^{\infty} \Bigg\{ &\frac{\delta_{ij}}{\left[\left(n\frac{\hbar c}{KT} + \mathrm{i}c\tau\right)^2 + r^2\right]^2} + \\ &+ 2\frac{x_i x_j - r^2\delta_{ij}}{\left[\left(n\frac{\hbar c}{KT} + \mathrm{i}c\tau\right)^2 + r^2\right]^3}\Bigg\},\end{aligned} \tag{5.28}$$

where $r = |\mathbf{x}|$. The corresponding degree of coherence is obtained in the form

$$\gamma_{ij}(x) = \frac{\mathscr{E}_{ij}(x)}{[\mathscr{E}_{ii}(0)\mathscr{E}_{jj}(0)]^{1/2}} = \frac{\mathscr{E}_{ij}(x)}{\mathscr{E}_{ii}(0)}, \tag{5.29}$$

where

$$\mathscr{E}_{ii}(0) = \frac{4\hbar c}{\pi}\left(\frac{KT}{\hbar c}\right)^4 \sum_{n=0}^{\infty} (n+1)^{-4} = \frac{4\hbar c}{\pi}\left(\frac{KT}{\hbar c}\right)^4 \frac{\pi^4}{90}; \tag{5.30}$$

here the sum is equal to the generalized Riemann ζ-function, $\zeta(4,1) = \pi^4/90$, defined as [Whittaker and Watson (1940), p. 266]

$$\zeta(s, a) = \sum_{n=0}^{\infty} (n + a)^{-s}. \tag{5.31}$$

As a special case we can consider temporal coherence by putting $\mathbf{x} = \hat{0}$ in (5.29), and we obtain

$$\gamma_{ij}(\hat{0}, \tau) = \delta_{ij}\frac{90}{\pi^4}\zeta\left(4, 1 + \mathrm{i}\frac{KT}{\hbar}\tau\right). \tag{5.32}$$

Hence any two orthogonal components $E_i(\mathbf{x}, t_1)$ and $E_j(\mathbf{x}, t_2)$ are completely uncorrelated. In this connection it should be noted that the function (5.32) has no zeros in the lower half of the complex τ-plane [Kano and Wolf (1962)], so the phase of this function can be reconstructed uniquely from its known modulus with the help of the dispersion relations [for a review see Peřina (1972)]. Thus in this case the spectrum of radiation can be determined from the visibility of the interference fringes using the Wiener–Khintchine theorem. Further discussions of coherence of blackbody radiation can be found [Steinle et al. (1975, 1976), Baltes, Steinle and

Pabst (1976), Baltes (1976, 1977), Baltes and Hilf (1976)], and much effort has been devoted to radiation from partially coherent sources and to the connection of partial coherence and radiometry with light of any state of coherence [Wolf (1978), Collet and Wolf (1979, 1980), Carter and Wolf (1981a, b), Baltes, Geist and Walther (1978), Friberg (1978a, b, 1979a, b, 1981b)].

Note that the higher-order correlation tensors are fully determined by the second-order correlation tensor in this case, by using (5.15).

5.1.3 Photocount statistics

Now we provide more general results for the photocount statistics of chaotic light, corresponding to those given in Chapter 3 [equations (3.138)–(3.142)].

Returning to the characteristic function (3.90), we obtain for Gaussian light

$$\langle \exp(\mathrm{i}sW_\eta)\rangle_{\mathcal{N}} = \int \exp\Big(-\sum_\lambda \frac{|\alpha_\lambda|^2}{\langle n_\lambda\rangle} + \mathrm{i}s\sum_{\lambda,\lambda'} \alpha_\lambda^* W_{\eta,\lambda\lambda'}\alpha_{\lambda'}\Big)\prod_\lambda \frac{\mathrm{d}^2\alpha_\lambda}{\pi\langle n_\lambda\rangle}, \tag{5.33}$$

where

$$W_{\eta,\lambda\lambda'} = \iint_0^T \iint_V \varphi^*(x',\lambda)\,\mathscr{S}(x'-x'')\,\varphi(x'',\lambda')\,\mathrm{d}^4x'\,\mathrm{d}^4x''. \tag{5.34}$$

Making use of the substitutions

$$\beta_\lambda = \frac{\alpha_\lambda}{\langle n_\lambda\rangle^{1/2}},\quad \langle n_\lambda\rangle^{1/2} W_{\eta,\lambda\lambda'}\langle n_{\lambda'}\rangle^{1/2} = U_{\lambda\lambda'}, \tag{5.35}$$

equation (5.33) becomes

$$\begin{aligned}\langle \exp(\mathrm{i}sW_\eta)\rangle_{\mathcal{N}} &= \int \exp\Big(-\sum_\lambda |\beta_\lambda|^2 + \mathrm{i}s\sum_{\lambda,\lambda'}\beta_\lambda^* U_{\lambda\lambda'}\beta_{\lambda'}\Big)\prod_\lambda \frac{\mathrm{d}^2\beta_\lambda}{\pi} = \\ &= \int \exp\Big(-\sum_\lambda |\beta_\lambda|^2 + \mathrm{i}s\sum_\lambda \mathscr{U}_\lambda|\beta_\lambda|^2\Big)\prod_\lambda\frac{\mathrm{d}^2\beta_\lambda}{\pi} = \prod_\lambda (1-\mathrm{i}s\mathscr{U}_\lambda)^{-1},\end{aligned} \tag{5.36}$$

where the $\mathscr{U}_\lambda$ are eigenvalues of the matrix $\hat{U} \equiv (U_{\lambda\lambda'})$. For a field in the normalization volume the quantity W ($W_\eta = \eta W$) given by (4.137) is suitable for the description of the field, and $\mathscr{U}_\lambda = \langle n_\lambda\rangle$. If moreover all the $\langle n_\lambda\rangle$ are assumed to be the same so that $\mathscr{U}_\lambda = \langle n\rangle/M$, where $\langle n\rangle$ is the complete mean number of photons for the whole field and M is the number of modes in the field, we arrive at (3.138).

The characteristic function (5.36) can be also written in the form

$$\begin{aligned}\langle \exp(\mathrm{i}sW_\eta)\rangle_{\mathcal{N}} &= \frac{1}{\mathrm{Det}(\hat{1}-\mathrm{i}s\hat{U})} = \exp\Big[-\sum_\lambda \log(1-\mathrm{i}s\mathscr{U}_\lambda)\Big] = \\ &= \exp\left[\sum_\lambda\sum_{n=1}^\infty \frac{(\mathrm{i}s\mathscr{U}_\lambda)^n}{n}\right] = \exp\left[\sum_{n=1}^\infty \frac{(\mathrm{i}s)^n}{n}\mathrm{Tr}\,\hat{U}^n\right] = \\ &= \exp\big[-\mathrm{Tr}\log(\hat{1}-\mathrm{i}s\hat{U})\big].\end{aligned} \tag{5.37}$$

Comparing this with the relation defining the cumulants $\varkappa_j^{(W_n)}$ [cf. (3.105)], we have for the cumulants

$$\varkappa_1^{(W_n)} = \operatorname{Tr} \hat{U} = \int\int_0^T \int\int_V \sum_\lambda \varphi^*(x', \lambda)\, \mathscr{S}(x' - x'')\, \varphi(x'', \lambda)\, \langle n_\lambda \rangle\, \mathrm{d}^4x'\, \mathrm{d}^4x'' =$$

$$= \int\int_0^T \int\int_V \mathscr{S}(x' - x'')\, \Gamma_{\mathscr{N}}^{(1,1)}(x', x'')\, \mathrm{d}^4x'\, \mathrm{d}^4x'', \tag{5.38a}$$

$$\varkappa_j^{(W_n)} = (j-1)!\, \operatorname{Tr} \hat{U}^j = (j-1)! \int\int \prod_{r=1}^{j} \mathscr{S}(x_r' - x_r'') \times$$

$$\times\, \Gamma_{\mathscr{N}}^{(1,1)}(x_r', x_{r+1}'')\, \mathrm{d}^4x_r'\, \mathrm{d}^4x_r'', \quad x_{j+1}' = x_1', j \geqq 2. \tag{5.38b}$$

Note that the first term in the series in (5.37) (is $\operatorname{Tr} \hat{U}$) corresponds to the fully coherent field. The presence of the other terms abolishes the full coherence of the field. If $\mathscr{S}(x' - x'') = \eta\delta(x'' - x')\, \delta(\mathbf{x}' - \mathbf{x}_0)$, $\mathbf{x}_0$ being a fixed point, and if a stationary field is assumed, then

$$\varkappa_1^{(W_n)} = \eta T \langle I \rangle_{\mathscr{N}}, \tag{5.39a}$$

$$\varkappa_j^{(W_n)} = (j-1)!\, (\eta\langle I \rangle_{\mathscr{N}})^j \int_0^T \ldots \int_0^T \gamma_{\mathscr{N}}(t_1 - t_2)\, \gamma_{\mathscr{N}}(t_2 - t_3) \ldots$$

$$\times\, \gamma_{\mathscr{N}}(t_j - t_1)\, \mathrm{d}t_1\, \mathrm{d}t_2 \ldots \mathrm{d}t_j, \qquad j \geqq 2. \tag{5.39b}$$

These expressions for the cumulants will be generalized to the superposition of chaotic and coherent light in Sec. 5.3.

In order to calculate the characteristic function (5.36) we must find the eigenvalues $\mathscr{U}_\lambda$ of the matrix $\hat{U}$ with elements given in (5.35), where $W_{\eta, \lambda\lambda'}$ is given by (5.34). This problem was considered by Bédard (1966a), Jakeman and Pike (1968), Dialetis (1969a) and Rousseau (1969) for Gaussian Lorentzian light, by Jaiswal and Mehta (1969) [see also Mehta (1970)] for partially polarized light, and by Jakeman and Pike (1969a) and Peřina and Peřinová (1971) for the superposition of coherent and Gaussian Lorentzian light. Considering a complete system of orthonormal functions $\{\varphi_\lambda(\mathbf{x}, t)\}$ over the interval $(0, T)$ and the volume V, we can write

$$\hat{A}^{(+)}(x) = \sum_\lambda \hat{b}_\lambda \varphi_\lambda(x), \tag{5.40a}$$

where

$$\hat{b}_\lambda = \int_0^T \int_V \hat{A}^{(+)}(x)\, \varphi_\lambda^*(x)\, \mathrm{d}^4x. \tag{5.40b}$$

Thus for a field stationary in space and time

$$\operatorname{Tr} \{\hat{\varrho} \hat{b}_\lambda^+ \hat{b}_{\lambda'}\} = \mathscr{U}_\lambda \delta_{\lambda\lambda'} =$$

$$= \int\int_0^T \int\int_V \operatorname{Tr} \{\hat{\varrho} \hat{A}^{(-)}(x)\, \hat{A}^{(+)}(x')\}\, \varphi_\lambda(x)\, \varphi_{\lambda'}^*(x')\, \mathrm{d}^4x\, \mathrm{d}^4x'. \tag{5.41}$$

Multiplying this equation by $\varphi_{\lambda'}(x'')$ and summing over λ' we obtain the integral equation determining $\mathscr{U}_\lambda$ and $\varphi_\lambda(x)$,

$$U_\lambda \varphi_\lambda(x'') = \int_0^T \int_V \Gamma_{\mathcal{N}}^{(1,1)*}(x'', x)\, \varphi_\lambda(x)\, \mathrm{d}^4 x, \tag{5.42}$$

since the completeness condition $\sum_\lambda \varphi_\lambda^*(x')\, \varphi_\lambda(x'') = \delta(x' - x'')$ holds. The eigenvalues $\mathscr{U}_\lambda$ will depend on the spectral and spatial properties of the radiation since (5.42) has the correlation function $\Gamma_{\mathcal{N}}^{(1,1)*}(x'', x) = \Gamma_{\mathcal{N}}^{(1,1)*}(x'' - x)$ as its kernel. The homogeneous integral equation in the time domain was solved for a spectral line with a Lorentzian profile in the previously quoted papers.

It is obvious that exact calculation of the characteristic function is a very difficult problem although all the mathematics is relatively simple if $T \ll \tau_c \approx 1/\Delta\nu$ or $T \gg \tau_c$. In the first case the correlation function is practically constant over the time interval $(0, T)$ so that $\varkappa_j^{(W_\eta)} = (j-1)!\, (\eta \langle I \rangle_{\mathcal{N}} T)^j$ [$\mathscr{S}(x' - x'') = \eta\delta(x'' - x')\, \delta(\mathbf{x}' - \mathbf{x})$ is assumed] and

$$\langle \exp(\mathrm{i}sW_\eta) \rangle_{\mathcal{N}} = \exp\left[\sum_{n=1}^{\infty} \frac{(\mathrm{i}s)^n}{n} (\eta \langle I \rangle_{\mathcal{N}} T)^n \right] =$$

$$= \exp\left[-\log(1 - \mathrm{i}s\eta \langle I \rangle_{\mathcal{N}} T)\right] = \frac{1}{1 - \mathrm{i}s\eta \langle I \rangle_{\mathcal{N}} T} . \tag{5.43}$$

This corresponds to (3.135) for the Bose–Einstein distribution (3.133). In the second case we obtain (3.138) with $p(n, T)$ given by (3.140) or (3.169) ($M = T/\tau_c$). Hence formula (3.169) holds for $T \gg \tau_c$ and also for $T \ll \tau_c$ ($M = 1$ in this case). This suggests that perhaps it represents a good approximation for all T. Indeed it was shown by Bédard et al. (1967) that it represents a good approximation to $p(n, T)$ for all T if

$$M = \frac{T}{\xi(T)} = \frac{T^2}{2 \int_0^T (T - t')\, |\gamma_{\mathcal{N}}(t')|^2\, \mathrm{d}t'} , \tag{5.44}$$

where (3.165) has been used. The spectrum of the radiation enters in the Mandel–Rice formula (3.140) through M only. Similar results have been obtained for the superposition of coherent and chaotic fields, which are discussed in Sec. 5.3.

Another expression for $p(n, T)$ for a Lorentzian line shape, when

$$\langle n_\lambda \rangle \hbar\omega_\lambda = \frac{\text{constant}}{(\omega_k - \bar{\omega})^2 + (\Delta\nu)^2} \tag{5.45}$$

and $T \gg \tau_c$, was derived by Glauber (1963c, 1965, 1966a) (see also Klauder and Sudarshan (1968), p. 225):

$$p(n, T) = \frac{1}{n!} \left(\frac{2\Omega T}{\pi} \right)^{1/2} \left(\frac{\Delta\nu\eta \langle I \rangle_{\mathcal{N}} T}{\Omega} \right)^n K_{n-1/2}(\Omega T) \exp(\Delta\nu T), \tag{5.46}$$

where $\Omega = [(\Delta\nu)^2 + 2\eta \langle I \rangle_{\mathcal{N}} \Delta\nu]^{1/2}$, and $K_{n-1/2}$ is the modified Hankel function of half-integral order. The asymptotic form of (5.46) is

$$p(n, T) = \frac{\langle n \rangle}{(2\pi\mu n^3)^{1/2}} \exp\left[-\frac{1}{2\mu}\left(n^{1/2} - \frac{\langle n \rangle}{n^{1/2}}\right)^2\right], \tag{5.47}$$

where $\mu = \eta\langle I \rangle_{\mathcal{N}}/\Delta\nu$ and $\langle n \rangle = \eta\langle I \rangle_{\mathcal{N}}T$. The formula (5.46) was also compared with the exact solution [Bédard et al. (1967)] and was found to be in a good agreement for $T/\tau_c \gtrsim 10$. Yet another asymptotic expression for $p(n, T)$ has been suggested by McLean and Pike (1965).

Experimental verification of the theory of photocount statistics for chaotic light was performed by a number of authors [Arecchi (1965), Arecchi, Berné and Burlamacchi (1966), Arecchi, Berné, Sona and Burlamacchi (1966), Freed and Haus (1965), Martienssen and Spiller (1966a), Arecchi (1969), Pike (1969, 1970), Bendjaballah (1971)]. A measurement of the time evolution of a stationary Gaussian field by means of joint photocount distributions was carried out by Arecchi, Berné and Sona (1966). Further investigations of the photocount statistics have been performed by Bures et al. (1971, 1972a–c) and Cantrell and Fields (1973).

Finally we give the s-ordered forms of some of the present expressions. It is clear from (5.24) that the equations related to a particular s-ordering can be obtained by substituting $\langle n_\lambda \rangle \to \langle n_\lambda \rangle + (1 - s)/2$ in the corresponding expressions. Thus we obtain from (4.185), considered for $s_1 = 1$ and $s_2 = s$, and for (3.138)

$$\langle \exp(\mathrm{i}yW) \rangle_s = \left[1 - \mathrm{i}y\left(\frac{\langle n \rangle}{\eta M} + \frac{1-s}{2}\right)\right]^{-M}, \tag{5.48}$$

where $\langle n \rangle$ is the mean number of counts. This leads to

$$P(W, s) = \frac{1}{\left(\frac{\langle n \rangle}{\eta M} + \frac{1-s}{2}\right)^M} \frac{W^{M-1}}{\Gamma(M)} \exp\left(-\frac{W}{\frac{\langle n \rangle}{\eta M} + \frac{1-s}{2}}\right) \tag{5.49}$$

and

$$\langle W^k \rangle_s = \left(\frac{\langle n \rangle}{\eta M} + \frac{1-s}{2}\right)^k \frac{\Gamma(k + M)}{\Gamma(M)}. \tag{5.50}$$

Corresponding to the bunching effect for chaotic photons, occurring as a consequence of the Bose–Einstein statistics with the requirement of maximum entropy, an antibunching effect occurs for chaotic fermions [Agarwal (1975)], even though no coherent states exist for fermions. This was discussed by Glauber (1970) and Bénard (1969, 1970a, b, 1975), using the theory of Goldberger and Watson (1965), valid for bosons as well as fermions. The coherence properties of fermions have been considered by Ledinegg (1967), and number fluctuations of interacting particles by Pusey (1979). Recently Ledinegg and Schachinger (1983) have considerered coherence of neutron fields.

5.2 Laser radiation

The statistical properties of laser radiation above the threshold of oscillation are qualitatively different from those of Gaussian radiation. This fact was first recognized by Golay (1961). A complete investigation of the statistical and coherence properties of laser radiation requires the solution of equations of motion for a system consisting of radiation in interaction with the atoms of lasing active matter and reservoirs, including pumping light, cavity losses, lattice vibrations, etc. The main difference between laser radiation and thermal radiation lies in the fact that laser radiation is produced mainly by stimulated emission rather than by spontaneous emission and that very strong coupling of modes exists, reflected in the nonlinearity of the equations of motion. Only the nonlinear theory of the laser is able to explain successfully all its properties. The nonlinearity also plays a decisive role in the stability of the laser. In the literature the statistical properties of laser radiation are treated in three ways, using

i) the Langevin equations for the complex amplitudes (these are the usual equations of motion with stochastic forces describing a Markoff process added),

ii) the classical Fokker–Planck equation for the probability distribution, and

iii) the master equation for the density matrix.

The last method is a completely quantum one, and using the coherent-state technique, the generalized Fokker–Planck equation and the Heisenberg–Langevin equations can be derived from it, including all the quantum properties of the system. These methods will be discussed in greater detail in Chapter 7, although they will be applied to nonlinear optical processes rather than to the laser, which has been treated in many review papers [Haken (1967, 1970a, b, 1972, 1978), Lax (1967, 1968a, b), Risken (1968, 1970), Scully and Lamb (1967, 1968), Willis (1966), Lax and Louisell (1967), Lax and Yuen (1968), Fleck (1966a, b), Korenman (1966, 1967), Paul (1966, 1969), Lax and Zwanziger (1973), Nussenzveig (1973), Loudon (1973), Louisell (1973), Sargent, Scully and Lamb (1974), Yariv (1967, 1975)]. Experimental verifications of the laser theory were reviewed by Armstrong and Smith (1967), Pike (1970), Pike and Jakeman (1974) and Arecchi and Degiorgio (1972). Concerning laser radiation we provide only basic results and some comparisons with experiment.

5.2.1 Ideal laser model

As we have mentioned, a field in the coherent state is produced by classical non-fluctuating currents. The radiation field is connected with the dipole transitions of atoms in the active medium of the laser. The polarization of these atoms oscillates with the external field and they radiate energy into the field. The whole active medium has an oscillating polarization density and, since the time derivative of the polarization density gives the current distribution, the radiation field can be regarded as a product of oscillating classical non-fluctuating currents if the laser is stabilized. Therefore the radiation field of the ideal laser is in a coherent state $|\beta\rangle$. All this is demonstrated in Tab. 4.1 on p. 106 (two-fold arrows show the correspondence principle and one-fold arrows indicate the generation of the corresponding quantum state).

The density matrix for such an ideal laser field is

$$\hat{\varrho} = | \beta \rangle \langle \beta | \tag{5.51}$$

and

$$\Phi_{\mathcal{N}}(\alpha) = \delta(\alpha - \beta) = \frac{\delta(|\alpha| - |\beta|)}{|\alpha|} \delta(\varphi - \psi), \tag{5.52}$$

where $\alpha = |\alpha| \exp(i\varphi)$, $\beta = |\beta| \exp(i\psi)$ and the factor $1/|\alpha|$ appears here since $d(\operatorname{Re}\alpha)\, d(\operatorname{Im}\alpha) = |\alpha|\, d|\alpha|\, d\varphi$. The distribution (5.52) expresses the fact that both the intensity $|\alpha|^2$ and the phase φ of the complex amplitude α are perfectly stabilized.

In the optical domain we usually have no information about the phase φ of oscillations and we have to assume that the phase is uniformly distributed in the interval $(0, 2\pi)$. Thus

$$\Phi_{\mathcal{N}}(\alpha) = \frac{1}{2\pi} \int_0^{2\pi} \frac{\delta(|\alpha| - |\beta|)}{|\alpha|} \delta(\varphi - \psi)\, d\varphi = \\ = \frac{\delta(|\alpha| - |\beta|)}{2\pi |\alpha|} = \frac{1}{\pi} \delta(|\alpha|^2 - |\beta|^2). \tag{5.53}$$

For the multimode function $\Phi_{\mathcal{N}}(\{\alpha_\lambda\})$ we have

$$\Phi_{\mathcal{N}}(\{\alpha_\lambda\}) = \frac{\delta(|\alpha_\lambda| - |\beta_\lambda|)}{2\pi |\alpha_\lambda|} \prod_{\lambda' \neq \lambda} \frac{\delta(|\alpha_{\lambda'}|)}{2\pi |\alpha_{\lambda'}|}, \tag{5.54}$$

i.e. the λth mode represents the stabilized one-mode laser field and the other modes are in the vacuum state. As we have shown, superpositions of coherent states with various phases are coherent in the sense of the full coherence conditions (3.53) for $m = n$, and this is also true for (5.54), which is independent of phases and describes a stationary field. This distribution corresponds to the intensity of a perfectly stabilized laser where the phase is uniformly distributed [cf. equation (3.151) and Fig. 3.5b]. The corresponding photocount distribution is Poissonian and is given by (3.152).

The corresponding density matrix elements $\varrho(n, m)$ for (5.52) are, from (4.66)

$$\varrho(n, m) = \frac{\beta^n \beta^{*m}}{(n!m!)^{1/2}} \exp(-|\beta|^2) \tag{5.55}$$

and the normally ordered moments are given by

$$\langle \hat{a}^{+k} \hat{a}^l \rangle = \alpha^{*k} \alpha^l. \tag{5.56}$$

Putting $m = n$ in (5.55) we obtain the Poissonian probability $p(n) = \varrho(n, n)$ that n photons are present in the field. The quasi-distribution $\Phi(\alpha, s)$ and the corresponding moments follow from (4.180) and (4.181) with $s_1 = 1$ and $s_2 = s$,

$$\Phi(\alpha, s) = \frac{2}{1 - s} \exp\left(- \frac{2|\alpha - \beta|^2}{1 - s} \right) \tag{5.57}$$

and

$$\langle \hat{a}^{+k}\hat{a}^{l}\rangle_s = \frac{l!}{k!}\left(\frac{1-s}{2}\right)^l \beta^{*k-l} L_l^{k-l}\left(\frac{2|\beta|^2}{s-1}\right), \qquad k \geqq l. \tag{5.58}$$

The corresponding relations for the symmetric and antinormal orderings follow with $s = 0$ and $s = -1$, respectively.

To obtain the quantities $p(n)$, $\langle \hat{a}^{+k}\hat{a}^{l}\rangle = \delta_{kl}\langle \hat{a}^{+k}\hat{a}^{k}\rangle$, $\Phi(\alpha, s)$ and $\langle \hat{a}^{+k}\hat{a}^{l}\rangle_s = \delta_{kl}\langle \hat{a}^{+k}\hat{a}^{k}\rangle_s$ corresponding to the phase averaged distribution (5.53) or (5.54), we may consider the more general distribution

$$\Phi_{\mathcal{N}}(\{\alpha_\lambda\}) = \prod_{\lambda=1}^{M} \frac{\delta(|\alpha_\lambda|^2 - |\beta_\lambda|^2)}{\pi} \prod_{\lambda=M+1}^{\infty} \frac{\delta(|\alpha_\lambda|)}{2\pi|\alpha_\lambda|}. \tag{5.59}$$

The statistical properties of this model for $M = 2$ and for arbitrary M were investigated both theoretically and experimentally by Bertolotti et al. (1966, 1967).

The distribution $P(W, s)$ in the normalization volume can be calculated as follows

$$\begin{aligned} P(W, s) &= \int \Phi(\{\alpha'_\lambda\}, s)\, \delta\Big(W - \sum_{\lambda=1}^{M} |\alpha'_\lambda|^2\Big) \prod_{\lambda=1}^{M} \frac{d^2\alpha'_\lambda}{\pi} = \\ &= \frac{1}{2\pi} \int_{-\infty}^{+\infty} \exp(-iyW) \Big\langle \exp\Big(iy \sum_{\lambda=1}^{M} |\alpha'_\lambda|^2\Big)\Big\rangle_s dy, \end{aligned} \tag{5.60}$$

where the characteristic function equals, considering M modes only in (5.59),

$$\begin{aligned} &\Big\langle \exp\Big(iy \sum_{\lambda=1}^{M} |\alpha'_\lambda|^2\Big)\Big\rangle_s = \\ &= \left(\frac{2}{1-s}\right)^M \int \prod_{\lambda=1}^{M} \exp\left(-\frac{2|\alpha'_\lambda - \beta_\lambda|^2}{1-s} + iy|\alpha'_\lambda|^2\right) \frac{d^2\alpha'_\lambda}{\pi} \end{aligned} \tag{5.61}$$

(the vacuum terms $\delta(|\alpha_\lambda|)/2\pi|\alpha_\lambda|$ in (5.59) can be obtained by putting $\beta_\lambda = 0$ in the corresponding modes). Writing $|\alpha'_\lambda - \beta_\lambda|^2 = |\alpha'_\lambda|^2 + |\beta_\lambda|^2 - 2|\alpha'_\lambda||\beta_\lambda| \times \cos(\varphi'_\lambda - \psi_\lambda)$ we have

$$\begin{aligned} \Big\langle \exp\Big(iy \sum_{\lambda=1}^{M} |\alpha'_\lambda|^2\Big)\Big\rangle_s &= \int_0^\infty \prod_{\lambda=1}^{M} \frac{2}{1-s} \exp\left(-\frac{2|\beta_\lambda|^2}{1-s}\right) \times \\ &\times \exp\left(-\frac{2|\alpha'_\lambda|^2}{1-s} + iy|\alpha'_\lambda|^2\right) I_0\left(4\frac{|\alpha'_\lambda||\beta_\lambda|}{1-s}\right) d|\alpha'_\lambda|^2, \end{aligned} \tag{5.62}$$

where I_0 is the modified Bessel function of zero order. Expressing I_0 in a power series and integrating over $|\alpha'_\lambda|^2$ we arrive at

$$\langle \exp(iyW)\rangle_s = \left(1 - iy\frac{1-s}{2}\right)^{-M} \exp\left(\frac{iy\langle n_c\rangle}{1 - iy\dfrac{1-s}{2}}\right), \tag{5.63}$$

where $\langle n_c\rangle = \sum_\lambda |\beta_\lambda|^2$ is the mean number of photons in the coherent field. This expression is the generating function for the Laguerre polynomials $L_n^{M-1}(x)$ since for any A and B

$$(1-\mathrm{i}yB)^{-M}\exp\left(\frac{\mathrm{i}yA}{1-\mathrm{i}yB}\right)=\sum_{n=0}^{\infty}\frac{(\mathrm{i}yB)^n}{\Gamma(n+M)}L_n^{M-1}\left(-\frac{A}{B}\right) \tag{5.64a}$$

$$=(1+B)^{-M}\exp\left(-\frac{A}{1+B}\right)\sum_{n=0}^{\infty}\frac{1}{\Gamma(n+M)}\times$$

$$\times\left(1+\frac{1}{B}\right)^{-n}(1+\mathrm{i}y)^nL_n^{M-1}\left(-\frac{A}{B(B+1)}\right). \tag{5.64b}$$

Equation (5.63) follows directly from (4.185) ($s_2 = s$, $s_1 = 1$), taking into account that

$$P_{\mathcal{N}}(W)=\delta(W-\langle n_c\rangle), \tag{5.65}$$

and this corresponds to

$$p(n)=\frac{(\sum_\lambda|\beta_\lambda|^2)^n}{n!}\exp(-\sum_\lambda|\beta_\lambda|^2). \tag{5.66}$$

Substituting (5.63) into (5.60) [or (5.65) into (4.187)], we obtain

$$P(W,s)=\frac{2}{1-s}\left(\frac{W}{\langle n_c\rangle}\right)^{(M-1)/2}\times$$

$$\times\exp\left[-\frac{2(W+\langle n_c\rangle)}{1-s}\right]I_{M-1}\left(4\frac{(W\langle n_c\rangle)^{1/2}}{1-s}\right), \tag{5.67}$$

and from (5.63) and (5.64a)

$$\langle W^k\rangle_s=\frac{d^k}{d(\mathrm{i}y)^k}\langle\exp(\mathrm{i}yW)\rangle_s\Big|_{\mathrm{i}y=0}=$$

$$=\frac{k!}{\Gamma(k+M)}\left(\frac{1-s}{2}\right)^kL_k^{M-1}\left(\frac{2\langle n_c\rangle}{s-1}\right). \tag{5.68}$$

For normal ordering $s = 1$ and

$$\langle\exp(\mathrm{i}yW)\rangle_{\mathcal{N}}=\exp(\mathrm{i}y\langle n_c\rangle), \tag{5.69}$$

i.e. the photon statistics are Poissonian given by (5.66),

$$p(n)=\frac{\langle n_c\rangle^n}{n!}\exp(-\langle n_c\rangle) \tag{5.70}$$

independently of M and the normal moments are

$$\langle W^k\rangle_{\mathcal{N}}=\langle n_c\rangle^k. \tag{5.71}$$

Equations (5.67) and (5.68) tend to (5.65) and (5.71) for $s\to 1$.

5.2.2 Real laser model

A model more realistic than that of the ideal laser was investigated by Glauber (1965) and Klauder and Sudarshan (1968) (Sec. 9-2). This model leads to a Lorentzian spectrum of laser radiation and is based on diffusion of the phase.

Assuming the phase of the radiation field to diffuse and to be represented by $\varphi(t)$, we have

$$\langle \hat{a}^+(t)\,\hat{a}(t+\tau)\rangle = \langle |\alpha|^2\rangle_{\mathcal{N}} \exp(-\mathrm{i}\bar{\omega}\tau)\,\langle \exp[\mathrm{i}\varphi(t) - \mathrm{i}\varphi(t+\tau)]\rangle, \tag{5.72}$$

where $\varphi(t) - \varphi(t+\tau) = \int_0^\tau \Delta\omega(t')\,\mathrm{d}t'$. If $\Delta\omega(t)$ is a Markoffian process, $\langle \Delta\omega(t')\,\Delta\omega(t'')\rangle = 2D\delta(t'-t'')$, D being the diffusion constant, and if it is Gaussian with $\Delta\omega(t) = \sum_n c_n\varphi_n(t)$, $\varphi_n(t)$ being an orthonormal system and c_n real Gaussian random variables with the standard deviation σ_n, then

$$\begin{aligned}\langle \hat{a}^+(t)\,\hat{a}(t+\tau)\rangle &= \langle |\alpha|^2\rangle_{\mathcal{N}} \exp(-\mathrm{i}\bar{\omega}\tau)\exp\left(-\frac{1}{2}\sum_n \lambda_n^2\sigma_n^2\right) = \\ &= \langle |\alpha|^2\rangle_{\mathcal{N}} \exp(-\mathrm{i}\bar{\omega}\tau)\exp\left[-\frac{1}{2}\iint_0^\tau \langle \Delta\omega(t')\,\Delta\omega(t'')\rangle\,\mathrm{d}t'\,\mathrm{d}t''\right] = \\ &= \langle |\alpha|^2\rangle_{\mathcal{N}} \exp(-\mathrm{i}\bar{\omega}\tau - D|\tau|),\end{aligned} \tag{5.73}$$

where $\lambda_n = \int_0^\tau \varphi_n(t')\,\mathrm{d}t'$, $\sigma_n^2 = \langle c_n^2\rangle$. The Wiener–Khintchine theorem (3.20b) leads to the Lorentzian spectral shape (3.20d) (Sec. 3.2.5) with $\Delta\nu = D$.

Now we can review briefly the main results of the nonlinear laser theory based on the Van der Pol oscillator, as developed by Risken (1965, 1966), Hempstead and Lax (1967), Fleck (1966a, b), Scully and Lamb (1966), Weidlich, Risken and Haken (1967), Lax (1967) and Carmichael and Walls (1974).

When equations of motion are derived [e.g. Louisell (1973)] for the system of the radiation field, atoms of the lasing medium, and the reservoir variables describing the pumping and losses, one finds, under physically reasonable assumptions, that the complex amplitude of the one-mode radiation field satisfies the Langevin equation [the dependence on $\exp(-\mathrm{i}\omega t)$ is eliminated]

$$\frac{\mathrm{d}\alpha}{\mathrm{d}t} = (w - |\alpha|^2)\,\alpha + L, \tag{5.74}$$

where w is the pumping parameter (the gain coefficient minus the damping constant) and L is a Markoffian Langevin force, $\langle L^*(t)\,L(t')\rangle = 4\delta(t-t')$, describing all noises, such as pumping, spontaneous emission, etc. The nonlinear term $(-|\alpha|^2)\,\alpha$ is responsible for the saturation of the laser field. The corresponding Fokker–Planck equation reads (cf. Chapter 7)

$$\frac{\partial \Phi_{\mathcal{N}}}{\partial t} = -\left\{\frac{\partial}{\partial\alpha}[(w - |\alpha|^2)\,\alpha\Phi_{\mathcal{N}}] + \text{c.c.}\right\} + 4\frac{\partial^2\Phi_{\mathcal{N}}}{\partial\alpha\,\partial\alpha^*}, \tag{5.75}$$

where c.c. denotes the complex conjugate term, and introducing $\alpha = r \exp(\mathrm{i}\varphi)$ we arrive at

$$\frac{\partial \Phi_{\mathcal{N}}}{\partial t} + \frac{1}{r}\frac{\partial}{\partial r}[(w - r^2) r^2 \Phi_{\mathcal{N}}] =$$

$$= \frac{1}{r}\frac{\partial}{\partial r}\left(r \frac{\partial \Phi_{\mathcal{N}}}{\partial r}\right) + \frac{1}{r^2}\frac{\partial^2 \Phi_{\mathcal{N}}}{\partial \varphi^2}. \tag{5.76}$$

The pumping parameter $w < 0$ below threshold, $w = 0$ at threshold and $w > 0$ above the oscillation threshold. The steady-state solution is

$$\Phi_{\mathcal{N}}(r) = \frac{N}{2\pi} \exp\left(\frac{w^2}{4}\right) \exp\left(-\frac{1}{4}(r^2 - w)^2\right), \tag{5.77}$$

where the normalization constant N is given by

$$\frac{1}{N} = \int_0^\infty \exp\left(-\frac{r^4}{4} + w\frac{r^2}{2}\right) r \, \mathrm{d}r. \tag{5.78}$$

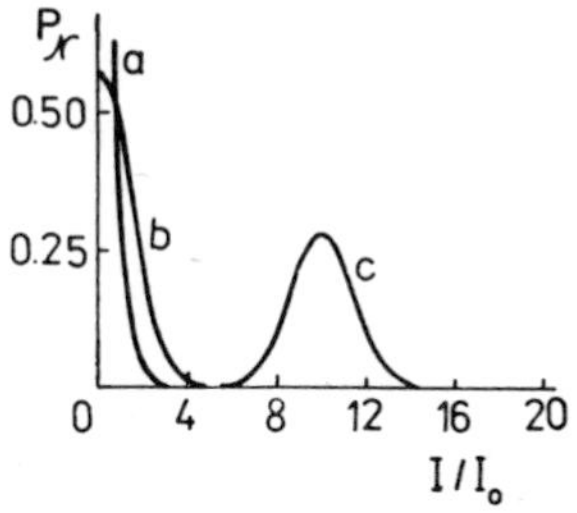

Fig. 5.1 — The intensity probability distribution $P_{\mathcal{N}}$ as a function of the normalized intensity I/I_0 for (a) $w = -3$ (below threshold), (b) $w = 0$ (at threshold) and (c) $w = 10$ (above threshold).

Substituting $r^2 = I/(\pi^{1/2} I_0)$, I being the intensity of the field and I_0 the average intensity at threshold, we can rewrite (5.77) in the form

$$P_{\mathcal{N}}(I) = \frac{2}{\pi I_0} \frac{1}{1 + \operatorname{erf}(w)} \exp\left[-\frac{1}{\pi I_0^2}(I - \pi^{1/2} w I_0)^2\right], \quad I \geqq 0, \tag{5.79}$$

where $\operatorname{erf}(w)$ is the error function, $\operatorname{erf}(w) = (2/\pi^{1/2}) \int_0^w \exp(-x^2)\,\mathrm{d}x$. The dependence of the distribution $P_{\mathcal{N}}(I)$ on the pumping parameter w is shown in Fig. 5.1. Note that an interesting description of such families of distributions may be based on urn models [Blažek (1979)]. The mean intensity can be expressed as

$$\langle I \rangle_{\mathcal{N}} = I_0 \left[\pi^{1/2} w + \frac{\exp(-w^2)}{1 + \operatorname{erf}(w)}\right]. \tag{5.80}$$

The photocount distribution has been calculated by Smith and Armstrong (1966a) [see also Armstrong and Smith (1967)] and by Bédard (1967b) and it can be expressed in the form

$$p(n, T) = \frac{N\mu^n}{(2\pi)^{1/2}} \exp\left[\frac{1}{4}(\mu^2 - 2J - 2w^2)\right] D_{-n-1}(\mu - 2^{1/2}w), \quad (5.81)$$

where

$$\mu = \left(\frac{\pi}{2}\right)^{1/2} \eta I_0 T, \qquad J = \pi^{1/2}\eta I_0 T w, \qquad N = \frac{2}{1 + \mathrm{erf}(w)}, \quad (5.82)$$

and $D_n(z)$ is the parabolic cylindric function [Gradshteyn and Ryzhik (1965), p. 1064]. The factorial moments are [Bédard (1967b), Arecchi, Rodari and Sona (1967)]

$$\langle W_\eta^k \rangle_{\mathcal{N}} = \frac{N\mu^k k!}{(2\pi)^{1/2}} \exp\left(-\frac{w^2}{2}\right) D_{-k-1}(-2^{1/2}w). \quad (5.83)$$

There are three typical regions of operation for the laser. Below threshold $w < 0$, and if $| w |$ is sufficiently large, then

$$P_{\mathcal{N}}(I) \approx \text{constant} \exp\left(-\frac{2 | w | I}{\pi^{1/2} I_0}\right), \quad (5.84)$$

which is a negative exponential distribution; thus the laser radiation below threshold is Gaussian. Near and at threshold the distribution (5.79), corresponding to the non-linear oscillator, is appropriate. Well above threshold, where the linearized theory is appropriate and $w > 0$, one can write

$$P_{\mathcal{N}}(I) = \text{constant} \exp\left[-\frac{4w}{\pi^{1/2} I_0}(I^{1/2} - \pi^{1/4} w^{1/2} I_0^{1/2})^2\right], \quad (5.85)$$

so the amplitude distribution $P_{\mathcal{N}}(I^{1/2})$ is a Gaussian function centered at $I^{1/2} = (\pi^{1/2} w I_0)^{1/2}$ and it tends to the distribution $\delta(I - \langle I \rangle_{\mathcal{N}})$ in the limit $w \to \infty$, valid for an ideal stabilized laser. The distribution (5.79) can be regarded as the smoothed δ-distribution $\delta(I - \pi^{1/2} w I_0)$ in the form of a Gaussian distribution (cf. Fig. 3.5b). Such a model of laser statistics was proposed [Bédard (1966b)] and other models were discussed [Sillitto (1968), Mandel (1965), Troup (1965b)]. The evolution of the quantum statistics of stimulated and spontaneous emission has been dealt with by Rockower, Abraham and Smith (1978) and questions of saturation have been discussed by Chung, Huang and Abraham (1980).

Now we can mention some experimental studies of laser statistics which verify the theory.

i) *Below threshold*

In this region photons obey Bose–Einstein statistics. The first experimental research on laser light below threshold was done by Freed and Haus (1965) and Smith and Armstrong (1966b), who verified the validity of the Bose–Einstein distribution (3.133) for $T \ll \tau_c$. Freed and Haus (1965) also measured sub-threshold photocount statistics for $T \gg \tau_c$, which leads to verification of (5.47) and (3.140) with M given by (5.44) [for a review, see Armstrong and Smith (1967)]. Similar results were obtained by Arecchi (1965), who started with a stabilized laser field above threshold scattered from a moving scattering plate.

ii) *Near and at threshold*

Near and at threshold the nonlinear theory must be used to describe the laser photon statistics, and equations (5.79), (5.81) and (5.83) are appropriate. The photocount statistics in this region have been measured by Smith and Armstrong (1966a) (Fig. 5.2), who also measured the reduced factorial moment $H_2 = \langle n(n-1)\rangle/\langle n\rangle^2 - 1 = \langle W^2\rangle_{\mathcal{N}}/\langle W\rangle_{\mathcal{N}}^2 - 1 = \varkappa_2^{(W)}/\varkappa_1^{(W)2}$, which goes from 1 (for Gaussian light well below threshold) to 0 (for an ideal amplitude stabilized laser well above threshold).

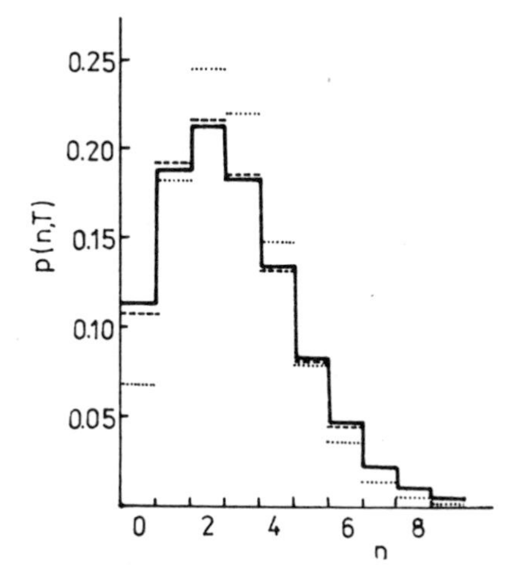

Fig. 5.2 — Photocount distribution observed just above the threshold (solid line); the Poisson distribution (dotted lines) and the nonlinear oscillator distribution (dashed lines), giving the best fit, are also shown; for $n = 7$, 8 and 9 the nonlinear oscillator and observed distributions coincide (after Smith and Armstrong, 1966a, Phys. Rev. Lett. **16**, 1169).

Similar results were obtained by Arecchi, Rodari and Sona (1967) and Pike (1969, 1970). Also, measurements of the third reduced factorial moment $H_3 = \langle n(n-1)(n-2)\rangle/\langle n\rangle^3 - 1 = \langle W^3\rangle_{\mathcal{N}}/\langle W\rangle_{\mathcal{N}}^3 - 1$ (which goes from 5 for Gaussian light to 0 for laser light well above threshold) were in very good agreement with the theory [Arecchi (1969)]. Further verifications of the validity of the nonlinear theory for the laser were performed by Chang et al. (1967, 1968, 1969), who determined the second, third and fourth factorial cumulants from the measured photocount distribution. The normalized factorial cumulants $\varkappa_j^{(W)}/\varkappa_1^{(W)j}$ ($\equiv Q_j$) used by Chang et al. vary from $(j-1)!$ for Gaussian light to 0 for laser light well above threshold ($j > 1$, $Q_1 = 1$). These measurements verify the theory in the threshold region from 1/10 to 10 times the threshold intensity and their results are shown in Fig. 5.3. Davidson and Mandel (1967, 1968) and Davidson (1969) measured the sixth-order correlation function at the threshold region using the correlation technique. These correlation measurements have been continued by many authors [Arecchi, Giglio and Sona (1967), Meltzer and Mandel (1970, 1971), Meltzer, Davis and Mandel (1970), Jakeman, Oliver, Pike, Lax and Zwanziger (1970), Arecchi and Degiorgio

(1971), Corti and Degiorgio (1974, 1976a, b), Corti, Degiorgio and Arecchi (1973)], and all their experimental results also are in very good agreement with the nonlinear laser theory [Cantrell and Smith (1971), Lax and Zwanziger (1970, 1973)].

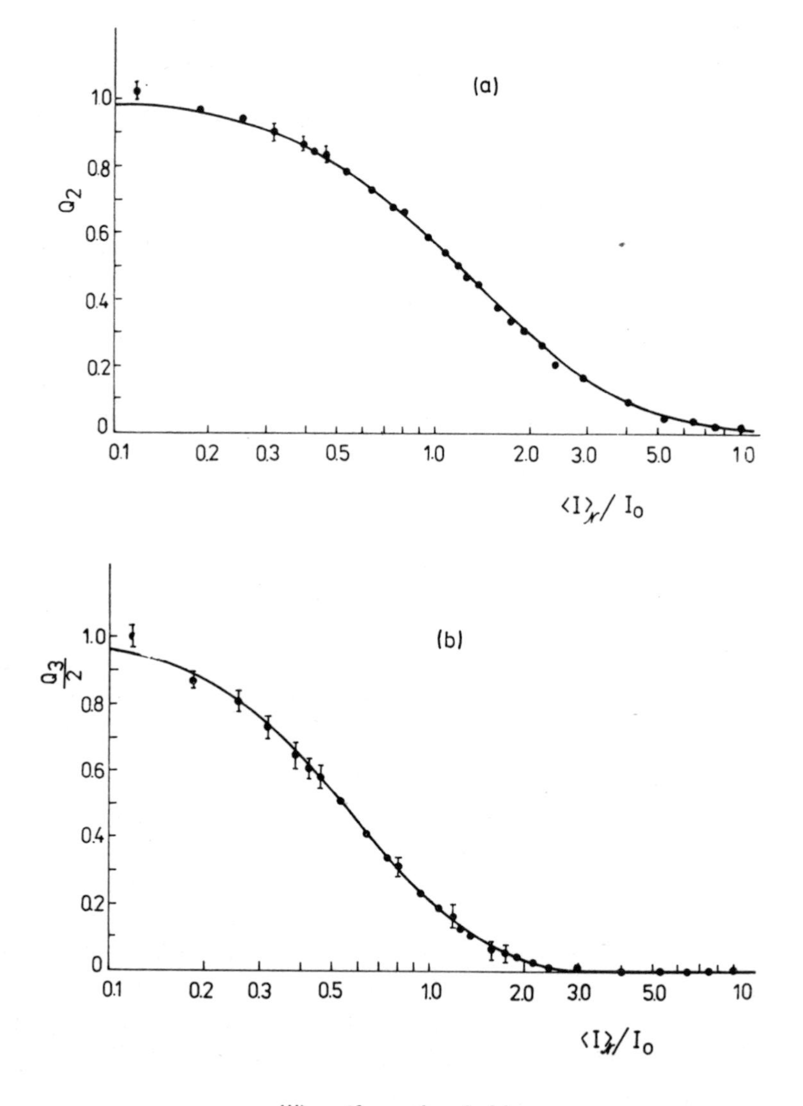

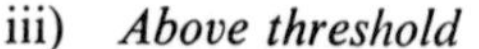

iii) *Above threshold*

In the region above threshold, where $I/I_0 > 5$, the model of the superposition of fully coherent radiation (signal) and Gaussian radiation (noise), as described by the distribution (5.85) (which is discussed in greater detail in the next section), is appropriate for the description of the laser statistics. Important experimental work has been done in this region by Arecchi, Berné and Burlamacchi (1966), Arecchi, Berné, Sona and Burlamacchi (1966), Smith and Armstrong (1966a), Freed and Haus (1966), Martiens-

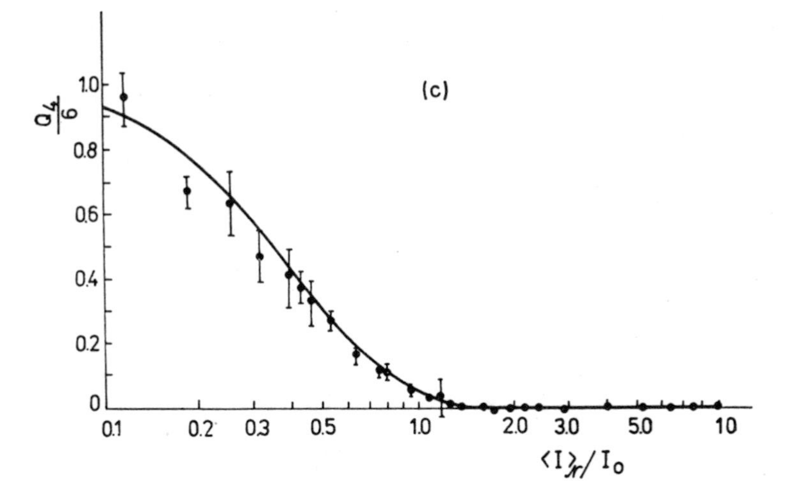

Fig. 5.3 — Normalized second (a), third (b) and fourth (c) factorial cumulants of laser light plotted as functions of the normalized intensity $\langle I \rangle_{\mathcal{N}}/I_0$ in the threshold region. The curves are the theoretical predictions and the dots are the experimental data (after Chang, Korenman, Alley and Detenbeck, 1969, Phys. Rev. **178**, 612).

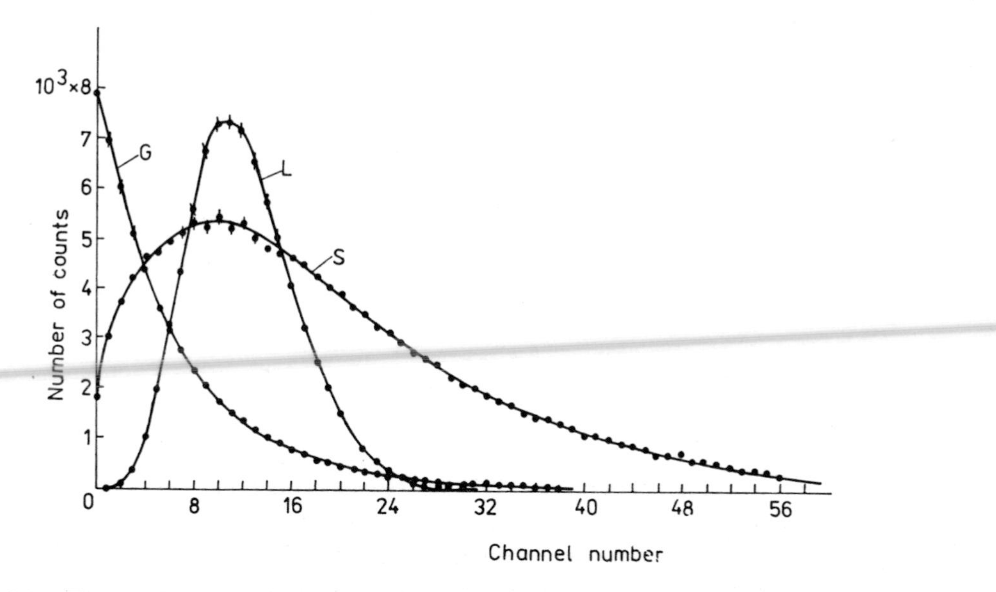

Fig. 5.4 — The experimental and theoretical photocount distributions for laser light (L), Gaussian light (G) and their superposition (S) (after Arecchi, Berné, Sona and Burlamacchi, 1966, IEEE J. Quant. Electr. **QE-2**, 341; Copyright © 1966 IEEE).

sen and Spiller (1966b), Magill and Soni (1966), and Ruggieri, Cummings and Lachs (1972). Photocount distributions observed by Arecchi, Berné, Sona and Burlamacchi (1966) are shown in Fig. 5.4. These results confirm the validity of the superposition model of coherent and chaotic light for laser light above threshold where the linear approximation holds ($I/I_0 > 5$). The curve L in Fig. 5.4 confirms that the photocount distribution is Poissonian well above threshold, so that in this region the laser field is in a phase-averaged coherent state. The ideal laser model is then appropriate.

The photocount distribution of modulated laser beams above threshold has been examined by Pearl and Troup (1968), Teich and Vannucci (1978) and Prucnal and Teich (1979), and it will be discussed in the next section.

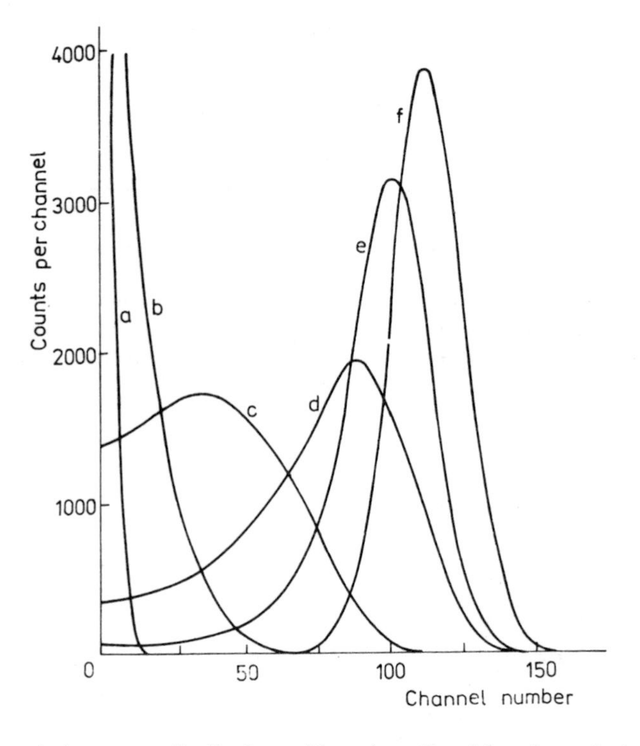

Fig. 5.5 — Observed photocount distributions with various time delays for a single-mode gas laser in transient operating conditions for 2.6 μs (curve *a*), 3.7 μs (curve *b*), 4.3 μs (curve *c*), 5 μs (curve *d*), 5.6 μs (curve *e*) and 8.8 μs (curve *f*) (after Arecchi, Degiorgio and Querzola, 1967, Phys. Rev. Lett. **19**, 1168).

The transient solution of the laser Fokker – Planck equation (5.76) was investigated by Risken and Vollmer (1967), Gnutzmann (1969, 1970) and Lax and Zwanziger (1970, 1973) and the transient photocount statistics [Zardecki, Bures and Delisle (1972)] were measured by Arecchi, Degiorgio and Querzola (1967), Arecchi and Degiorgio (1971), and Jakeman, Oliver, Pike, Lax and Zwanziger (1970). The transient photocount distributions obtained by Arecchi, Degiorgio and Querzola (1967) are shown in Fig. 5.5; they vary from the Bose – Einstein distributions *a*, *b*, corresponding to operation below threshold, to the stabilized Poisson distribution *f* well above threshold.

5.3 Superposition of coherent and chaotic fields

The photocount statistics of a superposition of coherent and chaotic fields have been studied by a number of authors because of their importance in describing the statistics of laser light above threshold and of scattered light. This model is useful also in other branches of physics as a model of the superposition of signal and noise. The first papers dealing with this subject in a quantum way were published by Lachs (1965), Troup (1966) and Glauber (1966a). In these papers the photocount distribution and its factorial moments for the superposition of one-mode coherent and narrow-band chaotic fields with the same frequencies were derived. Another approach, based on calculation of the normal correlation functions, was developed by Morawitz (1965, 1966). These results were generalized to multimode fields by Peřina (1967b, 1968a, b), Lachs (1967, 1971), Peřina and Mišta (1968b) and Fillmore (1969). An extension with respect to arbitrary orderings of the field operators was given by Peřina and Horák (1969a, b, 1970). Jakeman and Pike (1969a) pointed out that additional spectral information can be obtained using heterodyne detection, the chaotic field being superimposed, before detection, on a known coherent component. In this way the central frequency and halfwidth of the spectrum can be determined. However, the general heterodyne detection problem for chaotic light includes the case when the central frequency of the chaotic field and the frequency of the coherent field do not coincide. One-mode results of these authors were extended and generalized to multimode fields by Peřina and Horák (1969a). Experimental results have been obtained by Jakeman, Oliver and Pike (1971b). These investigations have been continued by a number of authors [Korenman (1967), Aldridge (1969), Fillmore and Lachs (1969), Jaiswal and Mehta (1970), Mehta and Jaiswal (1970), Solimeno, Corti and Nicoletti (1970), Cantrell (1971), Mišta, Peřina and Peřinová (1971), Horák, Mišta and Peřina (1971a, b), Diament and Teich (1971), Laxpati and Lachs (1972), Mišta (1971), Ruggieri, Cummings and Lachs (1972), Saleh (1975a, b, 1978)], including partial polarization [Peřina, Peřinová and Mišta (1971, 1972)], multiphoton processes [Mišta and Peřina (1971)] and propagation of radiation through random media [Diament and Teich (1971), Peřina and Peřinová (1972)].

The quasi-distribution $\Phi_{\mathcal{N}}(\{\alpha_\lambda\})$ for the superposition of coherent and chaotic fields can be derived as the convolution of the Gaussian distribution (5.7) and the δ-distribution $\prod_\lambda \delta(\alpha_\lambda - \beta_\lambda)$ corresponding to the coherent state $|\{\beta_\lambda\}\rangle$, according to the quantum superposition law (4.128), which leads to

$$\Phi_{\mathcal{N}}(\{\alpha_\lambda\}) = \prod_\lambda (\pi \langle n_{ch\lambda}\rangle)^{-1} \exp\left(-\frac{|\alpha_\lambda - \beta_\lambda|^2}{\langle n_{ch\lambda}\rangle}\right), \tag{5.86}$$

which is the multimode form of (4.98). We write here and in the following $\langle n_{ch\lambda}\rangle$ instead of $\langle n_\lambda\rangle$ for the mean number of photons in the mode λ of the chaotic field to distinguish it from the mean number $\langle n_{c\lambda}\rangle = |\beta_\lambda|^2$ of photons in the mode λ of the coherent field. The multimode form of (4.180) ($s_2 = s$ and $s_1 = 1$) gives

$$\Phi(\{\alpha_\lambda\}, s) = \prod_\lambda \left(\langle n_{ch\lambda}\rangle + \frac{1-s}{2}\right)^{-1} \exp\left(-\frac{|\alpha_\lambda - \beta_\lambda|^2}{\langle n_{ch\lambda}\rangle + \frac{1-s}{2}}\right). \quad (5.87)$$

This equation reduces to (5.24) for a chaotic field if $\{\beta_\lambda\} = 0$. For $s = -1$ we obtain the quasi-distribution $\Phi_{\mathscr{A}}(\{\alpha_\lambda\})$ for the superposition of coherent and chaotic fields [cf. (4.101)].

5.3.1 One-mode field

First we assume a one-mode field, omitting the mode index λ for simplicity. Substituting $\Phi_{\mathscr{N}}(\alpha)$ into (4.66), we obtain, on performing the integration, $\varrho(n, m)$ [in the form (5.90)], which leads to the correct $\Phi_{\mathscr{N}}(\alpha)$ in the form (4.98) again. Or $\varrho(n, m)$ can be calculated from the generating function

$$\begin{aligned} C(z^*, z, \lambda) &= \int (\pi\langle n_{ch}\rangle)^{-1} \exp\left(-\frac{|\alpha - \beta|^2}{\langle n_{ch}\rangle}\right) \times \\ &\times \exp(-\lambda|\alpha|^2 + z^*\alpha + z\alpha^*)\, d^2\alpha = \\ &= (\lambda\langle n_{ch}\rangle + 1)^{-1} \exp\left(-\frac{|\beta|^2\lambda}{\lambda\langle n_{ch}\rangle + 1}\right) \times \\ &\times \exp\left(\frac{\langle n_{ch}\rangle}{\lambda\langle n_{ch}\rangle + 1}|z|^2 + \frac{\beta^* z + \beta z^*}{\lambda\langle n_{ch}\rangle + 1}\right), \end{aligned} \quad (5.88)$$

where (4.179) has been used and $\lambda \geqq 0$ is a real number. Making use of the identity [Morse and Feshbach (1953)]

$$(1 + t)^l \exp(-xt) = \sum_{n=0}^{\infty} \frac{t^n}{\Gamma(l+1)} L_n^{l-n}(x), \quad (5.89)$$

we obtain $\varrho(n, m)$ in the form [Mollow and Glauber (1967a)]

$$\begin{aligned} \varrho(n, m) &= (n!\,m!)^{-1/2} \frac{\partial^n}{\partial z^{*n}} \frac{\partial^m}{\partial z^m} C(z^*, z, 1)\Big|_{z=z^*=0} = \\ &= \frac{1}{m!}\left(\frac{n!}{m!}\right)^{1/2} \frac{1}{\langle n_{ch}\rangle + 1} \exp\left(-\frac{|\beta|^2}{\langle n_{ch}\rangle + 1}\right) \frac{\langle n_{ch}\rangle^n}{(\langle n_{ch}\rangle + 1)^m} \beta^{*m-n} \times \\ &\times L_n^{m-n}\left(-\frac{|\beta|^2}{\langle n_{ch}\rangle(\langle n_{ch}\rangle + 1)}\right), \qquad m \geqq n; \end{aligned} \quad (5.90)$$

if $n > m$, the condition $\varrho(n, m) = \varrho^*(m, n)$ can be used. The normal moments can be calculated as [Mollow and Glauber (1967a), Peřina and Mišta (1968b)]

$$\begin{aligned} \langle \hat{a}^{+k}\hat{a}^l\rangle &= \frac{\partial^k}{\partial z^k} \frac{\partial^l}{\partial z^{*l}} C(z^*, z, 0)\Big|_{z=z^*=0} = \\ &= \frac{l!}{k!} \langle n_{ch}\rangle^l \beta^{*k-l} L_l^{k-l}\left(-\frac{|\beta|^2}{\langle n_{ch}\rangle}\right), \qquad k \geqq l; \end{aligned} \quad (5.91)$$

if $k < l$, then $\langle \hat{a}^{+k}\hat{a}^l\rangle = \langle \hat{a}^{+l}\hat{a}^k\rangle^*$. This is in agreement with (5.58) with $(1 - s)/2$ replaced by $\langle n_{ch}\rangle$.

5.3.2 Multimode field—characteristic generating function

The characteristic function $\langle \exp(\mathrm{i}sW)\rangle_{\mathcal{N}}$ (here and in the following we put $\eta = 1$ for simplicity), which is the generating function for the photocount distribution $p(n) = p(n, T)$ and its factorial moments $\langle W^k\rangle_{\mathcal{N}}$, as expressed by (3.102) and (3.103), can be calculated, employing (5.87), (5.61)–(5.63), in the form

$$\langle \exp(\mathrm{i}sW)\rangle_{\mathcal{N}} = \prod_{\lambda} (1 - \mathrm{i}s\langle n_{ch\lambda}\rangle)^{-1} \exp\left(\frac{\mathrm{i}s\langle n_{c\lambda}\rangle}{1 - \mathrm{i}s\langle n_{ch\lambda}\rangle}\right), \tag{5.92}$$

where the integrated intensity is given by (4.137) and we first consider fully polarized radiation. This is the product of the generating functions for the Laguerre polynomials $L_n^M(x)$, as expressed in equations (5.64a, b).

If the detection time T is much less than the coherence time τ_c and the detection area S is much smaller than the coherence area S_c, $\Gamma_{\mathcal{N}}^{(1,1)}(t'' - t) \approx \Gamma_{\mathcal{N}}^{(1,1)}(0) \times \exp[-\mathrm{i}\bar{\omega}(t - t'')]$ in (5.42), assuming point detectors and quasi-monochromatic light. Hence, only one eigenvalue $\langle n_{ch1}\rangle = \Gamma_{\mathcal{N},ch}^{(1,1)}(0)\, T = \langle n_{ch}\rangle$ is non-zero, with the corresponding eigenfunction $\varphi_1(t) = T^{-1/2} \exp(-\mathrm{i}\bar{\omega}t)$. Thus from (5.41) $\langle n_{c1}\rangle = (\langle n_c\rangle/T^2) \iint_0^T \exp(\mathrm{i}\omega_c(t - t')) \exp(-\mathrm{i}\bar{\omega}t) \exp(\mathrm{i}\bar{\omega}t')\, \mathrm{d}t\, \mathrm{d}t' = \langle n_c\rangle \varkappa^2$, since $\Gamma_{\mathcal{N},c}^{(1,1)}(t, t') = \langle n_c\rangle \exp[\mathrm{i}\omega_c(t - t')]$, $\bar{\omega}$ being the mean frequency of the chaotic light and ω_c the frequency of the coherent light; further $\langle n_{c2}\rangle = \langle n_c\rangle - \langle n_{c1}\rangle = \langle n_c\rangle \times (1 - \varkappa^2)$, $\varkappa = \sin(\Omega/2)/(\Omega/2)$ and $\Omega = (\bar{\omega} - \omega_c)\, T$, provided that the chaotic light has a Lorentzian spectrum [the degree of coherence is given by (3.20e) (Sec. 3.2.5)]. Consequently (5.92) can be written in the form

$$\langle \exp(\mathrm{i}sW)\rangle_{\mathcal{N}} = (1 - \mathrm{i}s\langle n_{ch}\rangle)^{-1} \exp\left[\frac{\mathrm{i}s\langle n_c\rangle \varkappa^2}{1 - \mathrm{i}s\langle n_{ch}\rangle} + \mathrm{i}s\langle n_c\rangle (1 - \varkappa^2)\right], \tag{5.93a}$$

where $\langle n_c\rangle = |\beta|^2 T$ and $\langle n_{ch}\rangle$ are the mean numbers of detected coherent and chaotic photons respectively. The quantity $\varkappa$ clearly characterizes the separation of the frequencies $\bar{\omega}$ and ω_c.

For arbitrary T and S we may introduce the multimode characteristic function [Peřina and Horák (1969a)]

$$\langle \exp(\mathrm{i}sW)\rangle_{\mathcal{N}} = \prod_{\lambda}^{M} (1 - \mathrm{i}s\langle n_{ch\lambda}\rangle)^{-1} \exp\left[\frac{\mathrm{i}s\langle n_{c\lambda}\rangle \varkappa_\lambda^2}{1 - \mathrm{i}s\langle n_{ch\lambda}\rangle} + \mathrm{i}s\langle n_{c\lambda}\rangle (1 - \varkappa_\lambda^2)\right], \tag{5.93b}$$

where M represents the number of degrees of freedom. This characteristic function can formally be obtained from the $2M$-mode function (5.92) if one makes the substitutions $\langle n_{c\lambda}\rangle \to \langle n_{c\lambda}\rangle \varkappa_\lambda^2$, $\langle n_{c\lambda+M}\rangle \to \langle n_{c\lambda}\rangle(1 - \varkappa_\lambda^2)$, $\lambda = 1, \dots, M$ and $\langle n_{ch\lambda}\rangle = 0$ for $\lambda = M + 1, \dots, 2M$.

It is generally a difficult problem to determine $\langle n_{ch\lambda}\rangle$ for a given spectrum, as discussed in Sec. 5.1 for chaotic radiation. This problem was considered for the particular case of the superposition of coherent and chaotic fields and for point detectors and a Lorentzian spectrum by Jakeman and Pike (1969a), Jaiswal and Mehta (1970) and Mehta and Jaiswal (1970). Exact recursion formulae for the photocount distribution and for its factorial moments have been derived by Laxpati and Lachs (1972). Here we propose approximate, simple and closed formulae involving the assumption of a uniform spectrum of the chaotic component, $\langle n_{ch\lambda}\rangle = \langle n_{ch}\rangle/M$, $\langle n_{ch}\rangle$ being the total mean number of chaotic photons. They are obtained in the spirit of the Mandel–Rice approximate formula for chaotic radiation [Bédard, Chang and Mandel (1967)], as discussed in Sec. 5.1. When the signal-to-noise ratio $\langle n_c\rangle/\langle n_{ch}\rangle \geqq 4$, their accuracy is better than 1 % [Peřina, Peřinová, Lachs and Braunerová (1973)]. The spectral properties of the radiation are again included through the number of degrees of freedom $M \geqq 1$. In this way we arrive at the multimode characteristic function

$$\langle \exp(\mathrm{i}sW)\rangle_{\mathcal{N}} = \left(1 - \mathrm{i}s\frac{\langle n_{ch}\rangle}{M}\right)^{-M} \exp\left[\frac{\mathrm{i}s\langle n_c\rangle \varkappa^2}{1 - \mathrm{i}s\dfrac{\langle n_{ch}\rangle}{M}} + \mathrm{i}s\langle n_c\rangle (1 - \varkappa^2)\right], \tag{5.94}$$

where $\langle n_c\rangle = \sum_{\lambda}^{M} \langle n_{c\lambda}\rangle$, $\varkappa^2 = \sum_{\lambda}^{M} \varkappa_\lambda^2 \langle n_{c\lambda}\rangle/\langle n_c\rangle$.

Assuming that the partially polarized chaotic component has the degree of polarization P and is superimposed on the coherent component whose polarization direction is φ with respect to the positive x-direction of the main polarization system (x, y), we obtain, in the main polarization system (x, y) [Peřina, Peřinová and Mišta (1971), Mišta, Peřina and Braunerová (1973)]

$$\langle \exp(\mathrm{i}sW)\rangle_{\mathcal{N}} = \prod_{\lambda}^{M} (1 - \mathrm{i}s\langle n_{ch\lambda 1}\rangle)^{-1} (1 - \mathrm{i}s\langle n_{ch\lambda 2}\rangle)^{-1} \times \\ \times \exp\left[\frac{\mathrm{i}s\langle n_{c\lambda 1}\rangle \varkappa_\lambda^2}{1 - \mathrm{i}s\langle n_{ch\lambda 1}\rangle} + \frac{\mathrm{i}s\langle n_{c\lambda 2}\rangle \varkappa_\lambda^2}{1 - \mathrm{i}s\langle n_{ch\lambda 2}\rangle} + \mathrm{i}s\frac{B}{M}\right], \tag{5.95a}$$

where $\langle n_{ch\lambda 1,2}\rangle = \langle n_{ch\lambda}\rangle (1 \pm P)/2$, $\langle n_{c\lambda 1}\rangle = \langle n_{c\lambda}\rangle \cos^2 \varphi$, $\langle n_{c\lambda 2}\rangle = \langle n_{c\lambda}\rangle \sin^2 \varphi$, $B = \langle n_c\rangle (1 - \varkappa^2)$ and the polarization cross-spectral purity of chaotic light is assumed, $\gamma^{(ch)}_{\mathcal{N},ij}(\tau) = \gamma^{(ch)}_{\mathcal{N},ij}(0)\, \gamma^{(ch)}_{\mathcal{N}}(\tau)$ [Mandel and Wolf (1961)]. In the case of uniform noise we have

$$\langle \exp(\mathrm{i}sW)\rangle_{\mathcal{N}} = \left(1 - \mathrm{i}s\frac{\langle n_{ch1}\rangle}{M}\right)^{-M} \left(1 - \mathrm{i}s\frac{\langle n_{ch2}\rangle}{M}\right)^{-M} \times \\ \times \exp\left[\frac{\mathrm{i}s\langle n_{c1}\rangle \varkappa^2}{1 - \mathrm{i}s\dfrac{\langle n_{ch1}\rangle}{M}} + \frac{\mathrm{i}s\langle n_{c2}\rangle \varkappa^2}{1 - \mathrm{i}s\dfrac{\langle n_{ch2}\rangle}{M}} + \mathrm{i}sB\right], \tag{5.95b}$$

where $\langle n_{ch1,2}\rangle = \langle n_{ch}\rangle (1 \pm P)/2$, $\langle n_{c1}\rangle = \langle n_c\rangle \cos^2 \varphi$, $\langle n_{c2}\rangle = \langle n_c\rangle \sin^2 \varphi$.

These characteristic functions together with the integrated probability distribution, the photocount distribution, its factorial moments and its cumulants can characterize systems where a coherent signal is embedded in noise. This includes laser radiation above threshold and scattered laser radiation. The formulae containing the frequency detuning parameter can be applied to heterodyne detection of chaotic radiation [Jakeman and Pike (1969a), Teich (1977)], where a coherent component is superimposed on the chaotic radiation before detection. This makes it possible to determine the spectral parameters $\Delta\nu$ and $\bar{\omega}$, provided that ω_c is known.

If $\langle n_{c\lambda}\rangle = 0$ we can describe chaotic radiation, whereas if $\langle n_{ch\lambda}\rangle = 0$ we can describe coherent radiation.

5.3.3 Integrated intensity probability distribution

The integrated intensity probability distribution $P_{\mathcal{N}}(W)$ can be determined, as in (5.60), by the use of a Fourier transformation and the residue theorem,

$$P_{\mathcal{N}}(W) = \frac{1}{2\pi} \int_{-\infty}^{+\infty} \langle \exp(\mathrm{i}sW')\rangle_{\mathcal{N}} \exp(-\mathrm{i}sW)\,\mathrm{d}s. \tag{5.96}$$

In cases of different $\langle n_{ch\lambda}\rangle$, the corresponding expressions are rather complex and can be found in the literature [Peřina, Peřinová and Mišta (1972), Peřina and Peřinová (1971)]. For the case of uniform noise we obtain from (5.94) [Peřina and Horák (1969a)]

$$\begin{aligned} P_{\mathcal{N}}(W) &= \frac{M}{\langle n_{ch}\rangle}\left(\frac{W-B}{\langle n_c\rangle \varkappa^2}\right)^{(M-1)/2} \exp\left(-\frac{W+\langle n_c\rangle \varkappa^2 - B}{\langle n_{ch}\rangle} M\right) \times \\ &\times I_{M-1}\left(2|\varkappa| M \frac{[\langle n_c\rangle (W-B)]^{1/2}}{\langle n_{ch}\rangle}\right), \qquad W \geqq B, \\ &= 0, \qquad W < B, \end{aligned} \tag{5.97a}$$

where I_M is the modified Bessel function. If the chaotic component is partially polarized [Peřina, Peřinová and Mišta (1971)]

$$\begin{aligned} P_{\mathcal{N}}(W) &= M \frac{\langle n_{ch2}\rangle^{M-1}}{\langle n_{ch1}\rangle^{M}} \left(\frac{W-B}{\langle n_{c2}\rangle \varkappa^2}\right)^{M-1/2} \times \\ &\times \exp\left[-\frac{W-B}{\langle n_{ch2}\rangle} M - \frac{M\varkappa^2 \langle n_{c1}\rangle}{\langle n_{ch1}\rangle} - \frac{M\varkappa^2\langle n_{c2}\rangle}{\langle n_{ch2}\rangle}\right] \sum_{n=0}^{\infty} \frac{1}{\Gamma(n+M)} \times \\ &\times \left[\left(1 - \frac{\langle n_{ch2}\rangle}{\langle n_{ch1}\rangle}\right)^2 \frac{W-B}{\langle n_{c2}\rangle \varkappa^2}\right]^{n/2} L_n^{M-1}\left(\frac{M\varkappa^2 \langle n_{c1}\rangle \langle n_{c2}\rangle}{\langle n_{ch1}\rangle (\langle n_{ch2}\rangle - \langle n_{ch1}\rangle)}\right) \times \\ &\times I_{n+2M-1}\left(\frac{2M|\varkappa|}{\langle n_{ch2}\rangle} [\langle n_{c2}\rangle (W-B)]^{1/2}\right), \qquad W \geqq B, \\ &= 0, \qquad W < B. \end{aligned} \tag{5.97b}$$

The behaviour of $P_{\mathcal{N}}(W)$ for fully polarized light is shown in Fig. 5.6, on the basis of (5.97a). The curves are narrower and higher with increasing M (also with increasing $\langle n_c\rangle$ and Ω). Since $\langle \exp(\mathrm{i}sW)\rangle_{\mathcal{N}}$ tends to $\exp(\mathrm{i}s\langle n\rangle)$ ($\langle n\rangle = \langle n_c\rangle + \langle n_{ch}\rangle$) as

$M \to \infty$, $P_{\mathcal{N}}(W) \to \delta(W - \langle n \rangle)$. The curves are also narrower and higher if P decreases [the number of polarization degrees of freedom is $M = 2/(1 + P^2)$, see (3.149)]. The dotted curves in Fig. 5.6 show for comparison the corresponding integrated intensity probability distribution for the case when the detection is performed by means of quantum counters (Sec. 3.4), which operate by stimulated emission, there being an additional contribution from the physical vacuum.

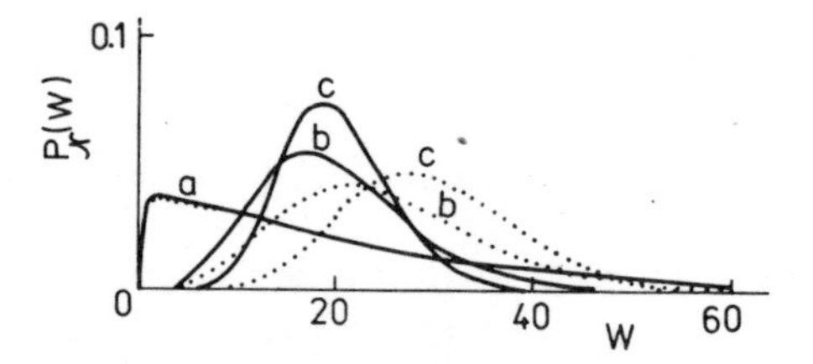

Fig. 5.6 — The normal integrated intensity probability distribution for $\langle n_c \rangle / \langle n_{ch} \rangle = 10/10$, $\Omega = 0$ and for $M = 1, 5$ and 10 (curves a, b and c respectively). The probability distribution for a photon counter is shown by dotted lines (after Horák, Mišta and Peřina, 1971b, Czech. J. Phys. **B21**, 614).

5.3.4 The photocount distribution

Applying (5.64b) and (3.102) to (5.93b) and (5.95a) we arrive at the photocount distributions for polarized and partially polarized fields [Peřina and Horák (1969a)]

$$p(n) = p(0) \sum_{j=0}^{n} \frac{B^{n-j}}{(n-j)!} \sum_{\sum_{\lambda}^{M} n_\lambda = j} \prod_{\lambda}^{M} \frac{1}{n_\lambda !} \left(1 + \frac{1}{\langle n_{ch\lambda} \rangle}\right)^{-n_\lambda} \times$$

$$\times L_{n_\lambda}^{0} \left(- \frac{\langle n_{c\lambda} \rangle \varkappa_\lambda^2}{\langle n_{ch\lambda} \rangle (\langle n_{ch\lambda} \rangle + 1)} \right), \tag{5.98a}$$

$$p(0) = \langle \exp(-W) \rangle_{\mathcal{N}} =$$

$$= \exp(-B) \prod_{\lambda}^{M} (1 + \langle n_{ch\lambda} \rangle)^{-1} \exp\left(- \frac{\langle n_{c\lambda} \rangle \varkappa_\lambda^2}{1 + \langle n_{ch\lambda} \rangle} \right),$$

and

$$p(n) = p(0) \sum_{j=0}^{n} \frac{B^{n-j}}{(n-j)!} \sum_{\sum_{\lambda}^{M} (n_\lambda + n'_\lambda) = j} \prod_{\lambda}^{M} \frac{1}{n_\lambda ! n'_\lambda !} \left(1 + \frac{1}{\langle n_{ch\lambda 1} \rangle}\right)^{-n_\lambda} \times$$

$$\times \left(1 + \frac{1}{\langle n_{ch\lambda 2} \rangle}\right)^{-n'_\lambda} L_{n_\lambda}^{0} \left(- \frac{\langle n_{c\lambda 1} \rangle \varkappa_\lambda^2}{\langle n_{ch\lambda 1} \rangle (\langle n_{ch\lambda 1} \rangle + 1)} \right) \times$$

$$\times L_{n'_\lambda}^{0} \left(- \frac{\langle n_{c\lambda 2} \rangle \varkappa_\lambda^2}{\langle n_{ch\lambda 2} \rangle (\langle n_{ch\lambda 2} \rangle + 1)} \right), \tag{5.98b}$$

$$p(0) = \exp(-B) \prod_{\lambda}^{M} (1 + \langle n_{ch\lambda 1} \rangle)^{-1} (1 + \langle n_{ch\lambda 2} \rangle)^{-1} \times$$

$$\times \exp\left(- \frac{\langle n_{c\lambda 1} \rangle \varkappa_\lambda^2}{1 + \langle n_{ch\lambda 1} \rangle} - \frac{\langle n_{c\lambda 2} \rangle \varkappa_\lambda^2}{1 + \langle n_{ch\lambda 2} \rangle} \right).$$

These expressions are substantially simplified if the noise component is uniform. Then from (5.94) and (5.95b) in the same way as above, or by applying the identity

$$\sum_{\sum_\lambda^M m_\lambda = j} \prod_\lambda^M \frac{1}{\Gamma(m_\lambda + l_\lambda + 1)} L_{m_\lambda}^{l_\lambda}(x_\lambda) =$$

$$= \frac{1}{\Gamma(j + \sum_\lambda^M l_\lambda + M)} L_j^{\sum_\lambda^M l_\lambda + M - 1} (\sum_\lambda^M x_\lambda) \tag{5.99}$$

and writing

$$\sum_{\sum_\lambda^M (n_\lambda + n'_\lambda) = j} = \sum_{i=0}^{j} \sum_{\sum_\lambda^M n_\lambda = i} \sum_{\sum_\lambda^M n'_\lambda = j-i}, \tag{5.100}$$

we obtain the following practical expressions [Peřina and Horák (1969a)]

$$p(n) = p(0) \sum_{j=0}^{n} \frac{1}{(n-j)!\,\Gamma(j+M)} B^{n-j} \left(1 + \frac{M}{\langle n_{ch} \rangle}\right)^{-j} \times$$
$$\times L_j^{M-1}\left(-\frac{\langle n_c \rangle \varkappa^2 M^2}{\langle n_{ch} \rangle (\langle n_{ch} \rangle + M)}\right), \tag{5.101a}$$
$$p(0) = \left(1 + \frac{\langle n_{ch} \rangle}{M}\right)^{-M} \exp\left(-\frac{\langle n_c \rangle M + \langle n_{ch} \rangle B}{\langle n_{ch} \rangle + M}\right),$$

or for $\varkappa = 1$ ($\Omega = 0$) [Peřina (1967b)]

$$p(n) = \frac{1}{\Gamma(n+M)} \left(1 + \frac{\langle n_{ch} \rangle}{M}\right)^{-M} \left(1 + \frac{M}{\langle n_{ch} \rangle}\right)^{-n} \times$$
$$\times \exp\left(-\frac{\langle n_c \rangle M}{\langle n_{ch} \rangle + M}\right) L_n^{M-1}\left(-\frac{\langle n_c \rangle M^2}{\langle n_{ch} \rangle (\langle n_{ch} \rangle + M)}\right); \tag{5.101b}$$

if the chaotic component is partially polarized [Peřina, Peřinová and Mišta (1971)]

$$p(n) = p(0) \sum_{j=0}^{n} \frac{B^{n-j}}{(n-j)!} \left(1 + \frac{M}{\langle n_{ch2} \rangle}\right)^{-j} \sum_{i=0}^{j} \frac{1}{\Gamma(i+M)\Gamma(j+M-i)} \times$$
$$\times \left[\frac{\langle n_{ch1} \rangle (\langle n_{ch2} \rangle + M)}{\langle n_{ch2} \rangle (\langle n_{ch1} \rangle + M)}\right]^i L_i^{M-1}\left(-\frac{\langle n_{c1} \rangle \varkappa^2 M^2}{\langle n_{ch1} \rangle (\langle n_{ch1} \rangle + M)}\right) \times$$
$$\times L_{j-i}^{M-1}\left(-\frac{\langle n_{c2} \rangle \varkappa^2 M^2}{\langle n_{ch2} \rangle (\langle n_{ch2} \rangle + M)}\right), \tag{5.101c}$$
$$p(0) = \left(1 + \frac{\langle n_{ch1} \rangle}{M}\right)^{-M} \left(1 + \frac{\langle n_{ch2} \rangle}{M}\right)^{-M} \times$$
$$\times \exp\left(-\frac{\langle n_{c1} \rangle \varkappa^2}{1 + \frac{\langle n_{ch1} \rangle}{M}} - \frac{\langle n_{c2} \rangle \varkappa^2}{1 + \frac{\langle n_{ch2} \rangle}{M}} - B\right).$$

Finally we present the formula for purely chaotic partially polarized light, which follows from (5.101c),

$$p(n) = [\Gamma(M)]^{-2}\left(1 + \frac{\langle n_{ch1}\rangle}{M}\right)^{-M}\left(1 + \frac{\langle n_{ch2}\rangle}{M}\right)^{-M}\left(1 + \frac{M}{\langle n_{ch2}\rangle}\right)^{-n} \times$$

$$\times \sum_{j=0}^{n} \frac{\Gamma(j + M)\,\Gamma(n + M - j)}{j!(n - j)!}\left(\frac{1 + M/\langle n_{ch2}\rangle}{1 + M/\langle n_{ch1}\rangle}\right)^{j}. \qquad (5.101d)$$

The photocount distribution $p(n)$ $(= p(n, T),\ \eta = 1)$ for fully polarized light is shown in Fig. 5.7, based on (5.101a). We observe that the curves are narrower and

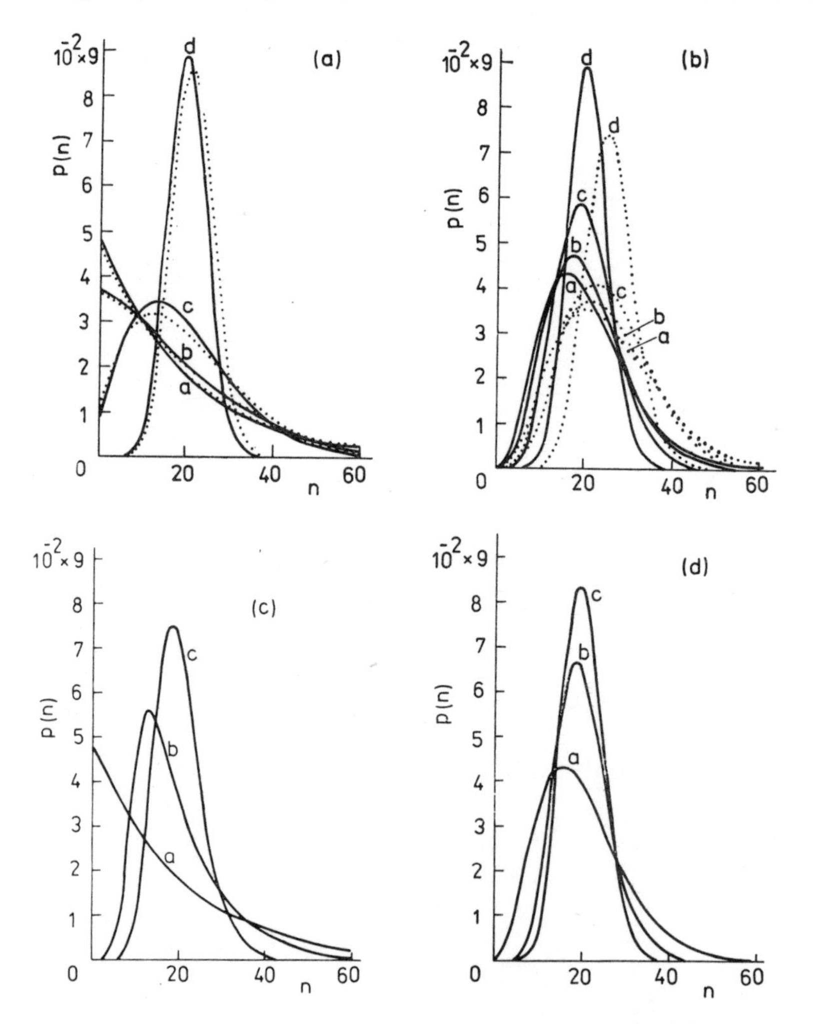

Fig. 5.7 — The photocount distribution for (a) $M = 1$, $\Omega = 0$, (b) $M = 5$, $\Omega = 0$, (c) $M = 1$, $\Omega = 100$ and (d) $M = 5$, $\Omega = 100$. The curves *a*, *b*, *c* and *d* are shown for $\langle n_c\rangle/\langle n_{ch}\rangle = 0/20$, 10/10, 16/4 and 20/0 respectively ($\langle n_c\rangle + \langle n_{ch}\rangle = 20$). In figures (c) and (d) the Poisson distribution is not shown. The dotted curves in figures (a) and (b) correspond to measurements by photon counters (after Horák, Mišta and Peřina, 1971b, Czech. J. Phys. **B21**, 614).

higher with increasing M, Ω and signal-to-noise ratio $\langle n_c \rangle / \langle n_{ch} \rangle$. In the limit as $M \to \infty$ the photocount distribution tends to the Poisson distribution $p(n) = \langle n \rangle^n \times \times \exp(-\langle n \rangle)/n!$, $\langle n \rangle = \langle n_c \rangle + \langle n_{ch} \rangle$. As discussed in Sec. 5.2, since 1966 a number of measurements have been performed to verify the validity of this model for laser light above threshold [Armstrong and Smith (1967), Pike (1970), Arecchi and Degiorgio (1972), Ruggieri, Cummings and Lachs (1972), Pike and Jakeman (1974)], for cases where the ratio of the intensity to the threshold intensity is greater than about 5. The dependence of $p(n)$ on the degree of polarization P has also been investigated [Peřina,

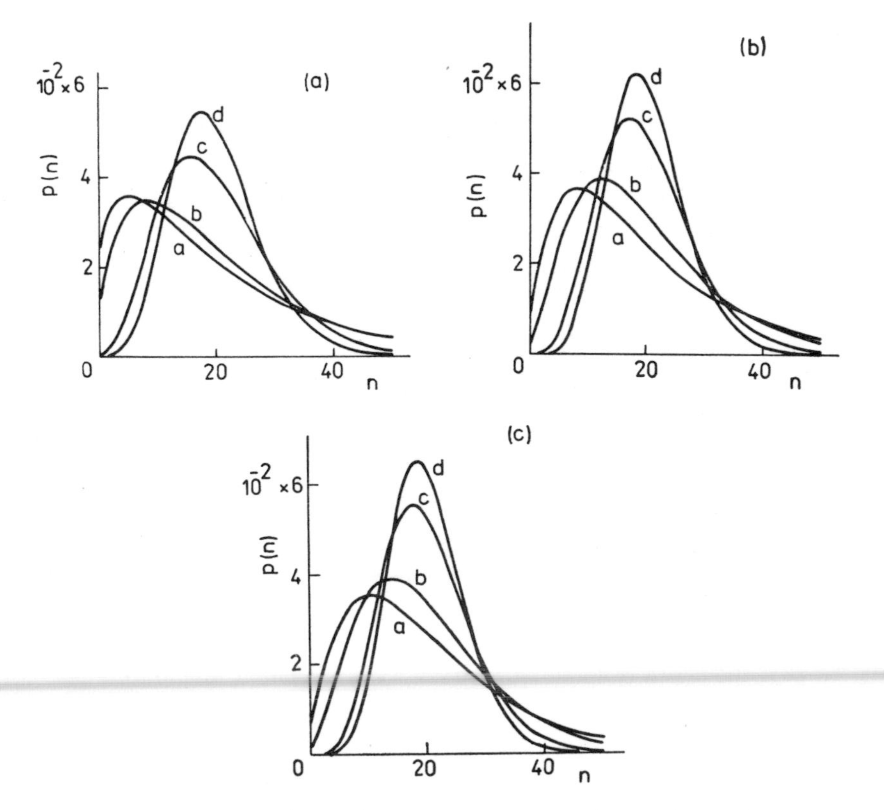

Fig. 5.8 — The photocount distribution for chaotic Lorentzian light and $\gamma = 0.1$, 1, 5 and 10 (curves a, b, c and d respectively) for (a) $P = 1$, (b) $P = 0.5$ and (c) $P = 0$ (after Peřina, Peřinová and Mišta, 1972, Opt. Acta **19**, 579).

Peřinová and Mišta (1971, 1972), Mišta, Peřina and Braunerová (1973), Aoki and Sakurai (1979)]. In Fig. 5.8 we demonstrate the dependence of the photocount distribution on P and $\gamma = \Delta\nu T$ for chaotic light with a Lorentzian spectrum. It is seen that the curves are narrower and higher as P decreases. Nevertheless, it has been shown [Mišta et al. (1973)] that for $\varphi = 90°$ the dependence on P may be just the opposite. Generally the peak of $p(n)$ increases with increasing φ ($0 \leq \varphi \leq 90°$). Also the superposition of one-mode coherent light and of chaotic light consisting of two

Lorentzian spectral lines has been investigated (Tornau and Echtermeyer (1973), Mehta and Gupta (1975), Mišta and Peřina (1977a)). The photocount distribution for this case is shown in Fig. 5.9, where $\bar{\omega}_j$ are the mean frequencies of the chaotic components having halfwidths $\Delta\nu_j$ $(j = 1, 2)$, $\gamma_j = \Delta\nu_j T$, $\Omega_j = (\bar{\omega}_j - \omega_c)\,T$ and $\langle n_{chj}\rangle$ are the mean photon numbers corresponding to the spectral lines.

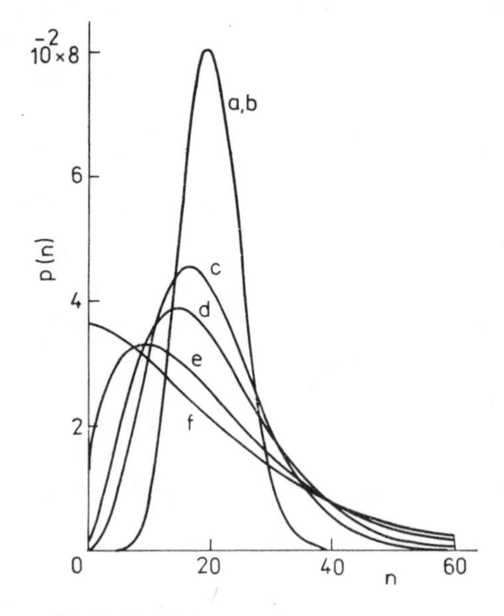

Fig. 5.9 — The photocount distribution for $\langle n_c\rangle = 10$, $\langle n_{ch1}\rangle = \langle n_{ch2}\rangle = 5$, $\gamma_1 = \gamma_2 = \gamma$, $\Omega_1 = \Omega_2 = \Omega$ and the curves are given for (a) $\gamma = 100$, $\Omega = 10$, (b) $\gamma = 100$, $\Omega = 0$, (c) $\gamma = 1$, $\Omega = 10$, (d) $\gamma = 0.01$, $\Omega = 10$, (e) $\gamma = 1$, $\Omega = 0$ and (f) $\gamma = 0.01$, $\Omega = 0$ (after Mišta and Peřina, 1977a, Czech. J. Phys. **B27**, 373).

An interesting behaviour of the photocount distribution occurs for modulated fields. If m is the depth of modulation, then the photocount distribution $p(n, m)$ for square-wave modulation is

$$p(n, m) = \frac{1}{2}\left[p(n, \langle n\rangle \to \langle n\rangle(1 - m)) + p(n, \langle n\rangle \to \langle n\rangle(1 + m))\right], \tag{5.102a}$$

for triangular modulation

$$p(n, m) = \int_{\langle n\rangle(1-m)}^{\langle n\rangle(1+m)} p(n, \langle n'\rangle)\frac{d\langle n'\rangle}{2m\langle n\rangle}, \tag{5.102b}$$

and for cosinusoidal modulation

$$p(n, m) = \int_0^{\pi/2} p(n, \langle n\rangle \to \langle n\rangle(1 + m\cos(2\vartheta)))\frac{2\,d\vartheta}{\pi}. \tag{5.102c}$$

Further details of the photocount statistics of modulated fields have been reviewed by Saleh (1978). The photocount distribution for modulated light beams has been examined by Fray et al. (1967), Pearl and Troup (1968), Diament and Teich (1970a), Bendjaballah and Perrot (1971, 1973), Picinbono (1971), Mišta (1973), Kitazima (1974), Teich and Vannucci (1978) and Prucnal and Teich (1979). Fig. 5.10 indicates

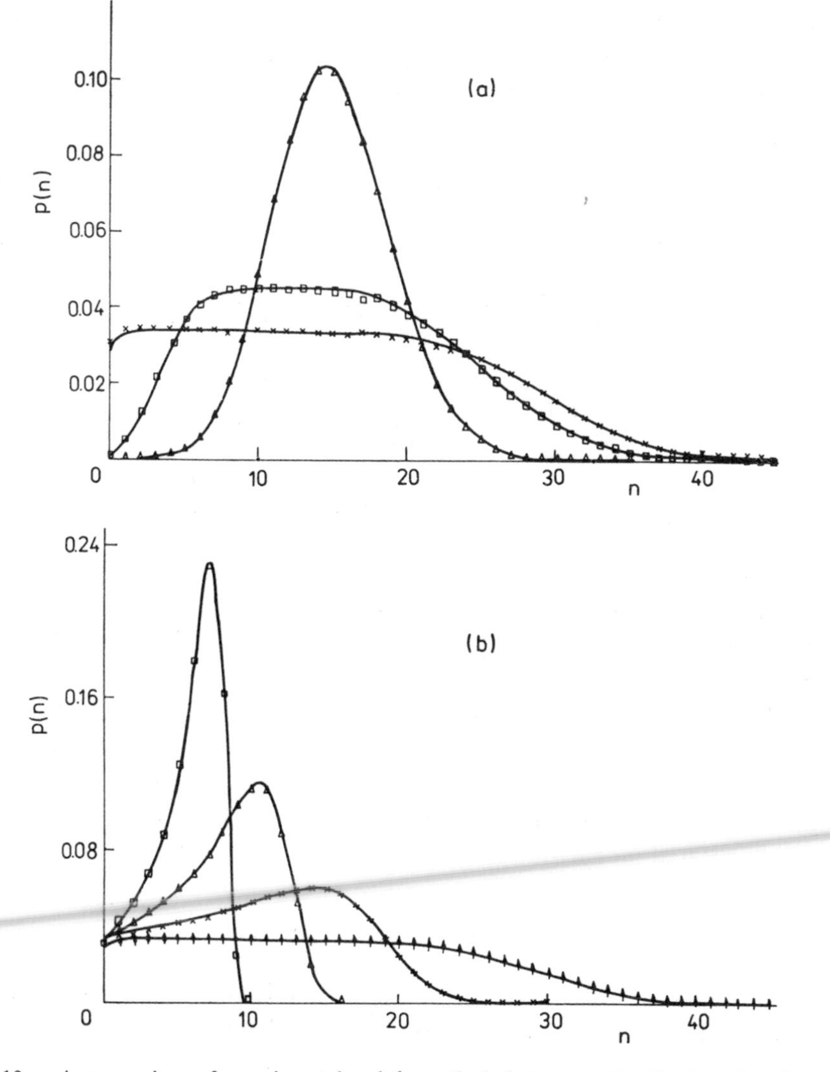

Fig. 5.10 — A comparison of experimental and theoretical photocount distributions for triangularly modulated coherent light with $\langle n_c \rangle \approx 15$ for (a) $m = 0$ ($\triangle$), $m = 0.74$ ($\square$) and $m = 0.99$ ($\times$) in the absence of the dead-time effect, (b) with the dead-time effect if $m \approx 1$ and $\tau_D/T = 0$ ($\uparrow$), 0.02 ($\times$), 0.05 ($\triangle$), 0.1 ($\square$) (after Teich and Vannucci, 1978, J. Opt. Soc. Am. **68**, 1338).

the comparison of the theoretical photocount distribution (solid curves) and experimental data for triangularly modulated coherent radiation, in the absence of dead-time effect and with the dead-time τ_D [Teich and Vannucci (1978)]. We observe that the photocount distribution can be extremely flat, and that the modulation leads to broadening of the curves and the bunching phenomenon is accentuated. With increasing M the photocount distribution is narrower and higher as above. On the other hand, the dead-time decreases both the mean and variance, leading to a reduction in the number of counts and a kind of "antibunching". The photocount distributions of extremely weak exponentially modulated luminescence radiation, having a typical two-peak bistable behaviour up to a threshold modulation frequency,

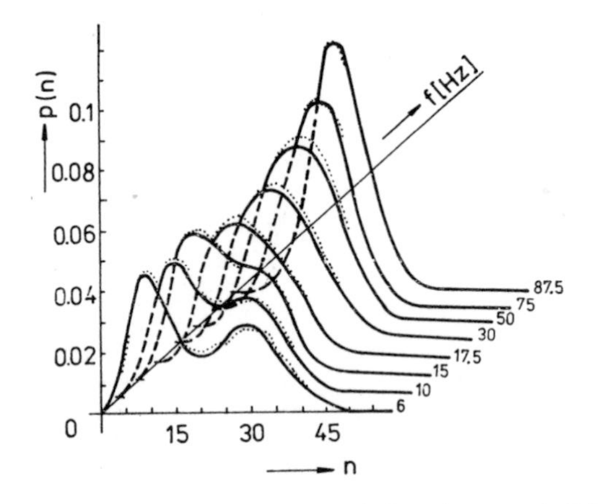

Fig. 5.11 — The modulation frequency dependence of the photocount distribution with an exponential modulation of light; the theoretical distributions are shown by full curves and the experimental distributions by dotted curves (after Koňák et al., 1982, Opt. Acta **29**, 1105).

have been employed by Koňák et al. (1982) to determine decay times. A comparison of the theoretical (full curves) and experimental (dotted curves) photocount distributions for various modulation frequencies for a slow luminophore is given in Fig. 5.11. The theoretical photocount distribution is described by the Mandel–Rice formula with an exponential modulation reflecting transient states of the luminophore.

5.3.5 Factorial moments

The factorial moments can be obtained from the characteristic functions by applying (5.64a) and (3.103). Thus from (5.93b) [Peřina and Horák (1969a)]

$$\langle W^k \rangle_{\mathcal{N}} = k! \sum_{j=0}^{k} \frac{B^{k-j}}{(k-j)!} \sum_{\sum_{\lambda}^{M} m_\lambda = j} \prod_{\lambda}^{M} \frac{\langle n_{ch\lambda} \rangle^{m_\lambda}}{m_\lambda!} L_{m_\lambda}^{0} \left(-\frac{\langle n_{c\lambda} \rangle \varkappa_\lambda^2}{\langle n_{ch\lambda} \rangle} \right) \quad (5.103a)$$

and from (5.95a) [Peřina, Peřinová and Mišta (1971)]

$$\langle W^k \rangle_{\mathcal{N}} = k! \sum_{j=0}^{k} \frac{B^{k-j}}{(k-j)!} \sum_{\sum_{\lambda}^{M} (m_\lambda + m'_\lambda) = j} \prod_{\lambda}^{M} \frac{1}{m_\lambda! m'_\lambda!} \langle n_{ch\lambda 1} \rangle^{m_\lambda} \langle n_{ch\lambda 2} \rangle^{m'_\lambda} \times$$

$$\times L^0_{m\lambda}\left(-\frac{\langle n_{c\lambda 1}\rangle \varkappa_\lambda^2}{\langle n_{ch\lambda 1}\rangle}\right) L^0_{m'\lambda}\left(-\frac{\langle n_{c\lambda 2}\rangle \varkappa_\lambda^2}{\langle n_{ch\lambda 2}\rangle}\right). \tag{5.103b}$$

In the case when the chaotic component is uniform we obtain in the same way from (5.94) and (5.95b), or applying the identities (5.99) and (5.100) [Peřina and Horák (1969a)],

$$\langle W^k\rangle_{\mathcal{N}} = k! \sum_{j=0}^{k} \frac{B^{k-j}}{(k-j)!\,\Gamma(j+M)} \left(\frac{\langle n_{ch}\rangle}{M}\right)^j L_j^{M-1}\left(-\frac{\langle n_c\rangle M\varkappa^2}{\langle n_{ch}\rangle}\right); \tag{5.104a}$$

and in particular for $\bar{\omega} = \omega_c$ $(B = 0)$ [Peřina (1967b)]

$$\langle W^k\rangle_{\mathcal{N}} = \frac{k!}{\Gamma(k+M)} \left(\frac{\langle n_{ch}\rangle}{M}\right)^k L_k^{M-1}\left(-\frac{\langle n_c\rangle M}{\langle n_{ch}\rangle}\right). \tag{5.104b}$$

For a partially polarized chaotic component [Peřina, Peřinová and Mišta (1971)]

$$\langle W^k\rangle_{\mathcal{N}} = k! \sum_{j=0}^{k} \frac{B^{k-j}}{(k-j)!} \left(\frac{\langle n_{ch2}\rangle}{M}\right)^j \sum_{i=0}^{j} \frac{1}{\Gamma(i+M)\,\Gamma(j+M-i)} \times$$

$$\times \left(\frac{\langle n_{ch1}\rangle}{\langle n_{ch2}\rangle}\right)^i L_i^{M-1}\left(-\frac{\langle n_{c1}\rangle \varkappa^2 M}{\langle n_{ch1}\rangle}\right) L_{j-i}^{M-1}\left(-\frac{\langle n_{c2}\rangle \varkappa^2 M}{\langle n_{ch2}\rangle}\right), \tag{5.104c}$$

and if the coherent component is zero,

$$\langle W^k\rangle_{\mathcal{N}} = [\Gamma(M)]^{-2} \left(\frac{\langle n_{ch2}\rangle}{M}\right)^k \times$$

$$\times \sum_{j=0}^{k} \binom{k}{j} \Gamma(j+M)\,\Gamma(k+M-j) \left(\frac{\langle n_{ch1}\rangle}{\langle n_{ch2}\rangle}\right)^j. \tag{5.104d}$$

As we noted earlier equations such as (5.103a, b) require the solution of a Fredholm integral equation, which determines $\langle n_{ch\lambda}\rangle$ if the temporal and spatial spectra of the noise component are given. For this reason, assuming the chaotic component to have a uniform spectrum, the use of the M-mode formulae is more convenient. However, these formulae are generally only approximate. They become accurate for narrow-band fields and $T \ll \tau_c$, $S \ll S_c$ (S_c being the coherence area), while for broad-band fields and $T \gg \tau_c$, $S \gg S_c$ one has $M = TS/\tau_c S_c$. When exact formulae are determined, taking the spectral properties into account accurately, one can examine the accuracy of the approximate formulae.

We restrict ourselves to the temporal analysis, because the spatial analysis may be performed in a similar way [Bures et al. (1972a), Zardecki et al. (1972), Peřina and Mišta (1974), Peřina (1977), Saleh (1978)]. We assume point detectors ($S \ll S_c$) so that the factorial moments are expressed in the form [cf. (3.11b)]

$$\langle W^k\rangle_{\mathcal{N}} = \int_0^T \cdots \int_0^T \sum_{j_1} \cdots \sum_{j_k} \Gamma^{(k,k)}_{\mathcal{N}, j_1 \ldots j_k}(\mathbf{x}, \ldots, \mathbf{x}, t_1, \ldots, t_k)\, dt_1 \ldots dt_k, \tag{5.105}$$

where $\mathbf{x}$ is a fixed point specifying the position of the photodetectors, and for simplicity

$\Gamma^{(k,k)}_{\mathcal{N},j_1\ldots j_k j_k\ldots j_1}(x_1,\ldots,x_k,x_k,\ldots,x_1) \equiv \Gamma^{(k,k)}_{\mathcal{N},j_1\ldots j_k}(x_1,\ldots,x_k)$. If we calculate these correlation functions and make use of the quasi-distribution (5.86), we obtain

$$\Gamma^{(k,k)}_{\mathcal{N},j_1\ldots j_k}(x_1,\ldots,x_k) = \int \prod_{\lambda}^{M} (\pi\langle n_{ch\lambda}\rangle)^{-1} \exp\left(-\frac{|\gamma_\lambda|^2}{\langle n_{ch\lambda}\rangle}\right)\times$$
$$\times [V^*_{j_1}(x_1) + B^*_{j_1}(x_1)] \ldots [V^*_{j_k}(x_k) + B^*_{j_k}(x_k)][V_{j_k}(x_k) + B_{j_k}(x_k)] \ldots$$
$$\ldots [V_{j_1}(x_1) + B_{j_1}(x_1)]\, \mathrm{d}^2\{\gamma_\lambda\}, \tag{5.106}$$

where $V_j(x) \equiv V_j(x, \{\gamma_\lambda\})$ is a Gaussian process and $B_j(x) \equiv B_j(x, \{\beta_\lambda\})$ is the coherent field, $\gamma_\lambda = \alpha_\lambda - \beta_\lambda$. Applying the factorization theorem (5.15) for chaotic fields, we successively obtain

$$\Gamma^{(1,1)}_{\mathcal{N},j_1}(x_1) = \Gamma^{(ch)}_{\mathcal{N},j_1}(x_1) + |B_{j_1}(x_1)|^2,$$
$$\Gamma^{(2,2)}_{\mathcal{N},j_1 j_2}(x_1,x_2) = (\Gamma^{(ch)}_{\mathcal{N},j_1}(x_1) + |B_{j_1}(x_1)|^2)(\Gamma^{(ch)}_{\mathcal{N},j_2}(x_2) + |B_{j_2}(x_2)|^2) +$$
$$+ |\Gamma^{(ch)}_{\mathcal{N},j_1 j_2}(x_1,x_2)|^2 + 2\,\mathrm{Re}\,\{\Gamma^{(ch)}_{\mathcal{N},j_1 j_2}(x_1,x_2)\, B_{j_1}(x_1)\, B^*_{j_2}(x_2)\}, \tag{5.107}$$

etc., where $\Gamma^{(ch)}_{\mathcal{N}}$ is the second-order normal correlation function $\Gamma^{(1,1)}_{\mathcal{N}}$ for chaotic radiation. Thus from (5.105)

$$\langle W\rangle_{\mathcal{N}} = \langle n_{ch}\rangle + \langle n_c\rangle,$$
$$\langle W^2\rangle_{\mathcal{N}} = \langle n\rangle^2 + \langle n_{ch}\rangle^2 \frac{1+P^2}{2}\mathcal{I}_1 + 2\langle n_{ch}\rangle\langle n_c\rangle \frac{1+P\cos(2\varphi)}{2}\overline{\mathcal{I}}_1,$$
$$\langle W^3\rangle_{\mathcal{N}} = \langle n\rangle^3 + 3\langle n\rangle\langle n_{ch}\rangle^2 \frac{1+P^2}{2}\mathcal{I}_1 + 2\langle n_{ch}\rangle^3 \frac{1+3P^2}{4}\mathcal{I}_2 +$$
$$+ 6\langle n\rangle\langle n_{ch}\rangle\langle n_c\rangle \frac{1+P\cos(2\varphi)}{2}\overline{\mathcal{I}}_1 + 6\langle n_{ch}\rangle^2\langle n_c\rangle\times$$
$$\times \frac{1+P^2+2P\cos(2\varphi)}{4}\overline{\mathcal{I}}_2, \tag{5.108}$$

where $\langle n\rangle = \langle W\rangle_{\mathcal{N}} = \langle n_{ch}\rangle + \langle n_c\rangle$, $\langle n_{ch}\rangle = \langle I_{ch}\rangle T$, $\langle n_c\rangle = I_c T$, $\langle I_{ch}\rangle$ and I_c being the mean intensities of the chaotic and the coherent components respectively, $\langle I_{ch}\rangle = \Gamma^{(ch)}_{\mathcal{N},xx} + \Gamma^{(ch)}_{\mathcal{N},yy} = \langle I_{ch,x}\rangle + \langle I_{ch,y}\rangle$. (The photoefficiency η can be taken into account if $\langle n_{ch}\rangle$ and $\langle n_c\rangle$ are multiplied by η.) We have also assumed the stationary condition for the field and the polarization cross-spectral purity [Mandel and Wolf (1961)] (i.e. the independence of the polarization and coherence properties of the field), expressed here as

$$\Gamma_{\mathcal{N},j_1 j_2}(\tau) = \Gamma_{\mathcal{N},j_1 j_2}\,\gamma_{\mathcal{N}}(\tau); \tag{5.109}$$

here $\Gamma_{\mathcal{N},j_1 j_2} = \Gamma_{\mathcal{N},j_1,j_2}(0)$, $\gamma_{\mathcal{N}}(\tau)$ being the degree of temporal coherence. We note that with the main polarization system perpendicular to the direction of propagation, $\langle I_{ch,x}\rangle = \langle I_{ch}\rangle(1+P)/2$, $\langle I_{ch,y}\rangle = \langle I_{ch}\rangle(1-P)/2$, $\Gamma^{(ch)}_{\mathcal{N},xy} = 0$, $I_{c,x} = I_c\cos^2\varphi$, $I_{c,y} = I_c\sin^2\varphi$; we have also applied the following identities involving the polarization properties

$$\sum_{j_1} \dots \sum_{j_k} \Gamma^{(ch)}_{\mathcal{N}, j_1 j_2} \Gamma^{(ch)}_{\mathcal{N}, j_2 j_3} \dots \Gamma^{(ch)}_{\mathcal{N}, j_k j_1} = \langle I_{ch} \rangle^k \left[\left(\frac{1+P}{2} \right)^k + \left(\frac{1-P}{2} \right)^k \right],$$

$$\sum_{j_1} \dots \sum_{j_k} \Gamma^{(ch)}_{\mathcal{N}, j_1 j_2} \dots \Gamma^{(ch)}_{\mathcal{N}, j_{k-1} j_k} \Gamma^{(c)}_{\mathcal{N}, j_k j_1} =$$

$$= \langle I_{ch} \rangle^{k-1} I_c \left[\left(\frac{1+P}{2} \right)^{k-1} \cos^2 \varphi + \left(\frac{1-P}{2} \right)^{k-1} \sin^2 \varphi \right], \qquad (5.110)$$

and

$$\mathcal{I}_1 = T^{-2} \iint_0^T | \gamma^{(ch)}_{\mathcal{N}}(t_1 - t_2) |^2 \, dt_1 \, dt_2 = 2T^{-2} \int_0^T (T - \tau) | \gamma^{(ch)}_{\mathcal{N}}(\tau) |^2 \, d\tau,$$

$$\overline{\mathcal{I}}_1 = T^{-2} \operatorname{Re} \iint_0^T \gamma^{(ch)}_{\mathcal{N}}(t_1 - t_2) \gamma^{(c)}_{\mathcal{N}}(t_2 - t_1) \, dt_1 \, dt_2 =$$

$$= 2T^{-2} \operatorname{Re} \int_0^T (T - \tau) \gamma^{(ch)}_{\mathcal{N}}(\tau) \gamma^{(c)*}_{\mathcal{N}}(\tau) \, d\tau, \qquad (5.111)$$

$$\mathcal{I}_2 = T^{-3} \operatorname{Re} \iiint_0^T \gamma^{(ch)}_{\mathcal{N}}(t_1 - t_2) \gamma^{(ch)}_{\mathcal{N}}(t_2 - t_3) \gamma^{(ch)}_{\mathcal{N}}(t_3 - t_1) \, dt_1 \, dt_2 \, dt_3,$$

$$\overline{\mathcal{I}}_2 = T^{-3} \operatorname{Re} \iiint_0^T \gamma^{(ch)}_{\mathcal{N}}(t_1 - t_2) \gamma^{(ch)}_{\mathcal{N}}(t_2 - t_3) \gamma^{(c)}_{\mathcal{N}}(t_3 - t_1) \, dt_1 \, dt_2 \, dt_3;$$

here $\gamma^{(ch)}_{\mathcal{N}}$ and $\gamma^{(c)}_{\mathcal{N}}$ are the degrees of temporal coherence for chaotic and coherent fields respectively, corresponding to the correlation functions $\Gamma^{(ch)}_{\mathcal{N}}$ and $\Gamma^{(c)}_{\mathcal{N}}$. Note that for $k = 2$ in the first identity in (5.110), $M = 2/(1 + P^2)$ follows for the number of polarization degrees of freedom. An experiment for measuring the third moment $\langle W^3 \rangle_{\mathcal{N}}$ with the help of three photodetectors was also discussed [Peřina and Mišta (1968b)].

Provided that the spectrum of the chaotic light is Lorentzian, the integrals (5.111) can be expressed in the form [Jaiswal and Mehta (1970), Horák, Mišta and Peřina (1971a)]

$$\mathcal{I}_1 = \frac{1}{\gamma} + \frac{1}{2\gamma^2} (\exp(-2\gamma) - 1),$$

$$\overline{\mathcal{I}}_1 = 2 \left\{ \frac{\gamma}{\gamma^2 + \Omega^2} + \frac{\exp(-\gamma) [(\gamma^2 - \Omega^2) \cos \Omega - 2\gamma\Omega \sin \Omega] + \Omega^2 - \gamma^2}{(\gamma^2 + \Omega^2)^2} \right\},$$

$$\mathcal{I}_2 = \frac{3}{2\gamma^3} [\gamma - 1 + (\gamma + 1) \exp(-2\gamma)],$$

$$\overline{\mathcal{I}}_2 = 2 \left\{ \frac{2\gamma^2}{(\gamma^2 + \Omega^2)^2} - \frac{\gamma^2 - \Omega^2}{2\gamma(\gamma^2 + \Omega^2)^2} - \frac{\gamma(3\gamma^2 - 5\Omega^2)}{(\gamma^2 + \Omega^2)^3} + \right.$$

$$+ \exp(-\gamma) \cos(\Omega) \left[\frac{4\gamma + \gamma^2 - \Omega^2}{(\gamma^2 + \Omega^2)^2} - \frac{8\gamma\Omega^2}{(\gamma^2 + \Omega^2)^3} \right] - \exp(-\gamma) \times$$

$$\left. \times \sin(\Omega) \left[\frac{2\gamma}{(\gamma^2 + \Omega^2)^2} + \frac{8\gamma^2}{(\gamma^2 + \Omega^2)^3} \right] - \frac{\exp(-2\gamma)}{2\gamma(\gamma^2 + \Omega^2)} \right\}, \qquad (5.112)$$

where $\gamma = \Delta\nu T$ and $\Omega = (\bar{\omega} - \omega_c) T$ again.

The increase of the right-hand sides in (5.108) over $\langle n \rangle^2$ and $\langle n \rangle^3$ in $\langle W^2 \rangle_{\mathcal{N}}$ and $\langle W^3 \rangle_{\mathcal{N}}$ implies that the Hanbury Brown–Twiss and the bunching effects are non-zero; the term $\langle n_{ch} \rangle^2 \mathscr{I}_1(1 + P^2)/2$ in $\langle W^2 \rangle_{\mathcal{N}}$ and the terms $3\langle n_{ch} \rangle^3 \mathscr{I}_1(1 + P^2)/2$ and $2\langle n_{ch} \rangle^3 \mathscr{I}_2(1 + 3P^2)/4$ in $\langle W^3 \rangle_{\mathcal{N}}$ represent the Hanbury Brown–Twiss terms for chaotic light, the remaining terms being interference terms between the coherent

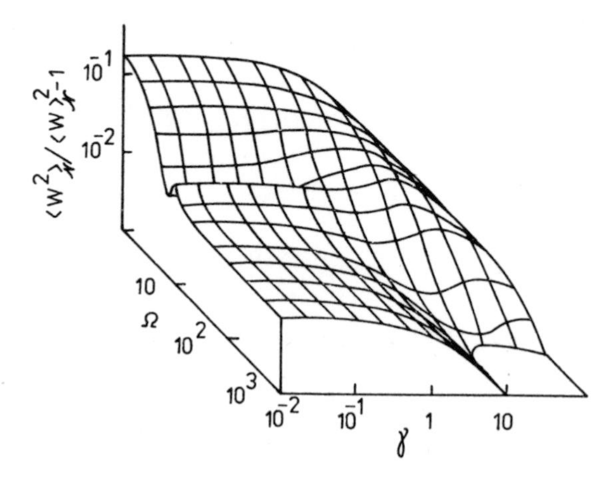

Fig. 5.12 — The dependence of the second reduced factorial moment on γ and Ω for $\langle n_c \rangle = 18$, $\langle n_{ch} \rangle = 2$ (after Horák, Mišta and Peřina, 1971a, J. Phys. **A4**, 231).

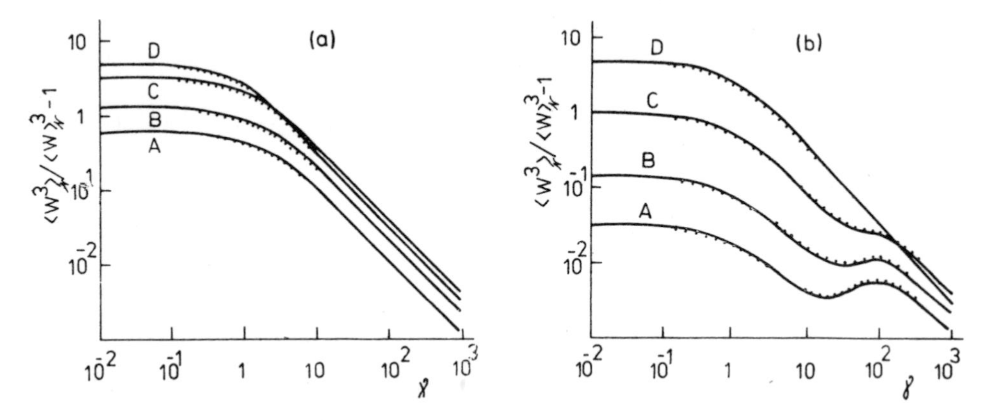

Fig. 5.13 — Dependence of the third reduced factorial moment on γ for $\langle n_c \rangle / \langle n_{ch} \rangle = 18/2$ (A), 16/4 (B), 10/10 (C) and 0/20 (D) ($\langle n_c \rangle + \langle n_{ch} \rangle = 20$) and for (a) $\Omega = 0$, (b) $\Omega = 100$; the exact values are represented by solid curves and the approximate ones are represented by dotted curves (after Horák, Mišta and Peřina, 1971a, J. Phys. **A4**, 231).

and chaotic fields. The bunching effect for the superposition of coherent and chaotic fields has been explicitly demonstrated by Lachs and Voltmer (1976).

The behaviour of the second reduced factorial moment $\langle W^2 \rangle_{\mathcal{N}} / \langle W \rangle^2_{\mathcal{N}} - 1$ as a function of γ and Ω for fully polarized light is shown in Fig. 5.12. Its values lie between $2! - 1 = 1$ for chaotic light and 0 for coherent light. The behaviour of the third reduced factorial moment as a function of γ is given in Fig. 5.13. The dotted curves represent approximate values which are discussed in Sec. 5.3.7. These quantities behave similarly for a two-line chaotic component, and the accuracy of the approximate formulae is roughly the same [Mišta and Peřina (1977a)]. The dependence of the third reduced factorial moment on the degree of polarization P and on γ for the superposition of coherent and chaotic fields can be seen in Fig. 5.14.

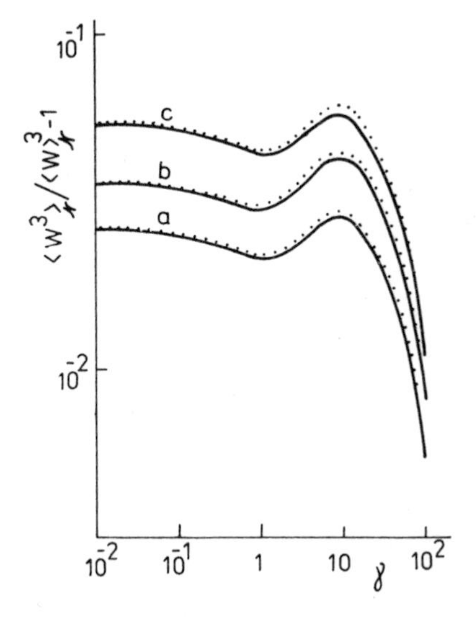

Fig. 5.14 — The third reduced factorial moment as a function of γ for $\langle n_c \rangle = 18$, $\langle n_{ch} \rangle = 2$, $\Omega = 10$ and (a) $P = 0$, (b) $P = 0.5$ and (c) $P = 1$; the exact values are shown by solid curves and the approximate ones by dotted curves (after Peřina, Peřinová and Mišta, 1972, Opt. Acta **19**, 579).

The photon statistics of multiphoton processes have been discussed in a number of papers [Teich and Diament (1969), Mišta and Peřina (1971), Mišta (1971), Peřina, Peřinová and Mišta (1972)]. Multiphoton absorption makes it possible to increase the order of the correlation function which is to be determined, and correlation functions of higher orders, measured with the help of multiphoton absorption, are a source of information about the form of picosecond pulses [Greenhow and Schmidt (1974)]. The exact second factorial moment for the superposition of coherent and chaotic fields of arbitrary spectrum, arbitrary detection time intervals and for N-photon absorption process has been calculated [Mišta (1971)], and approximate formulae for two-photon absorption have been also proposed [Mišta and Peřina (1971), Peřina, Peřinová and Mišta (1972)]. The photocount distribution and its factorial moments are, in the last case and for fully polarized radiation, expressed in terms of the parabolic cylindric functions $D_n(z)$ as follows

$$p(n) = \frac{1}{n!}\left(\frac{a}{2^{1/2}}\right)^M \exp\Bigl[-a\langle n_c\rangle \varkappa^2 - \langle n_c\rangle^2 \; 1 - \varkappa^2)^2 +$$

$$+ \frac{1}{8}[a + 2\langle n_c\rangle(1 - \varkappa^2)]^2\Bigg]\sum_{k=0}^{2n}\binom{2n}{k}\frac{[\langle n_c\rangle(1-\varkappa^2)]^{2n-k}}{2^{k/2}}\times \tag{5.113a}$$

$$\times\sum_{l=0}^{\infty}\frac{(a^2\varkappa^2\langle n_c\rangle)^l\Gamma(k+l+M)}{l!\,\Gamma(l+M)\,2^{l/2}}D_{-k-l-M}\left(\frac{1}{2^{1/2}}[a+2\langle n_c\rangle(1-\varkappa^2)]\right),$$

$$\langle W^k\rangle_{\mathcal{N}} = \tag{5.113b}$$

$$= \sum_{l=0}^{2k}\frac{(2k)!}{(2k-l)!\,\Gamma(l+M)}[\langle n_c\rangle(1-\varkappa^2)]^{2k-l}a^{-l}L_l^{M-1}(-a\varkappa^2\langle n_c\rangle),$$

where $a = [M(M+1)/\langle W\rangle_{\mathcal{N}}]^{1/2}$. Such an approach provides a relatively simple description of the process, compared to rigorous descriptions based on the study of the interaction of radiation with matter [Tornau and Bach (1974), Loudon (1973), McNeil and Walls (1974), Simaan and Loudon (1975, 1978), Schubert and Wilhelmi (1976, 1978, 1980)]. The second and third reduced factorial moments for chaotic light are shown in Fig. 5.15 for two-photon absorption (solid curves) and for one-photon absorption for comparison (broken curves). The accuracy of the approximate formulae is also evident from the figure (the approximate values of the third reduced factorial moment are shown by the dotted curve).

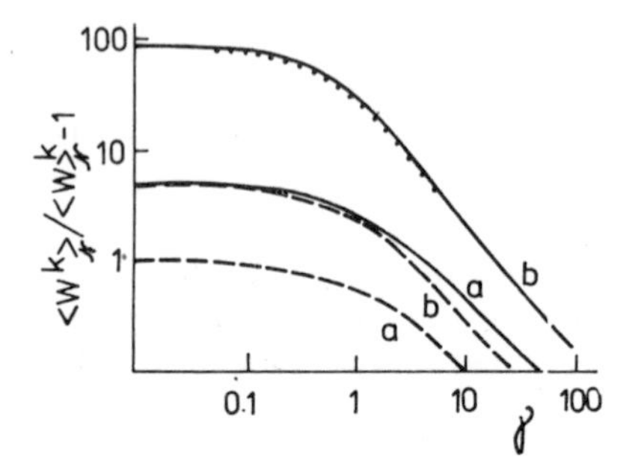

Fig. 5.15 — The reduced (a) second ($k = 2$), (b) third ($k = 3$) factorial moments for one- and two-photon absorption of chaotic light; two-photon values are shown by solid curves, one-photon values are shown by dashed curves. The approximate values of the third reduced factorial moment are represented by the dotted curve (Mišta and Peřina, 1971, Opt. Comm. **2**, 441).

Expressions (5.108) have been verified experimentally by Jakeman, Oliver and Pike (1971b) and Aoki, Endo, Takayanagi and Sakurai (1976).

5.3.6 Factorial cumulants

The factorial cumulants $\varkappa_l^{(W)}$ describe the pure bunching effect of the lth order, without lower-order contributions. They are related to the factorial moments as follows [see equations (3.108a–c)]

$$\langle W\rangle_{\mathcal{N}} = \varkappa_1^{(W)},$$

$$\langle W^2\rangle_{\mathcal{N}} = \varkappa_1^{(W)2} + \varkappa_2^{(W)},$$

$$\langle W^3 \rangle_{\mathcal{N}} = \varkappa_1^{(W)3} + 3\varkappa_1^{(W)}\varkappa_2^{(W)} + \varkappa_3^{(W)}, \tag{5.114}$$

etc. The factorial cumulants are obtained from the definition [Jaiswal and Mehta (1970), Cantrell (1971), Peřina, Peřinová and Mišta (1972), Saleh (1978)]

$$\begin{aligned}
\varkappa_l^{(W)} &= \frac{\mathrm{d}^l}{\mathrm{d}(\mathrm{i}s)^l} \log \langle \exp(\mathrm{i}sW) \rangle_{\mathcal{N}} \bigg|_{\mathrm{i}s=0} = \\
&= (l-1)! \left[\left(\frac{1+P}{2} \right)^l + \left(\frac{1-P}{2} \right)^l \right] \sum_\lambda \langle n_{ch\lambda} \rangle^l + \\
&+ l! \left[\left(\frac{1+P}{2} \right)^{l-1} \cos^2 \varphi + \left(\frac{1-P}{2} \right)^{l-1} \sin^2 \varphi \right] \sum_\lambda \langle n_{c\lambda} \rangle \langle n_{ch\lambda} \rangle^{l-1} = \\
&= (l-1)! \frac{\langle n_{ch} \rangle^l}{T^l} \int_0^T \ldots \int_0^T \sum_{j_1} \ldots \sum_{j_l} \gamma_{\mathcal{N}, j_1 j_2}^{(ch)}(t_1 - t_2) \ldots \\
&\ldots \gamma_{\mathcal{N}, j_l j_1}^{(ch)}(t_l - t_1)\, \mathrm{d}t_1 \ldots \mathrm{d}t_l + l! \frac{\langle n_{ch} \rangle^{l-1} \langle n_c \rangle}{T^l} \times \\
&\times \int_0^T \ldots \int_0^T \sum_{j_1} \ldots \sum_{j_l} \gamma_{\mathcal{N}, j_1 j_2}^{(ch)}(t_1 - t_2) \ldots \\
&\ldots \gamma_{\mathcal{N}, j_{l-1} j_1}^{(ch)}(t_{l-1} - t_l)\, \gamma_{\mathcal{N}, j_l j_1}^{(c)}(t_l - t_1)\, \mathrm{d}t_1 \ldots \mathrm{d}t_l,
\end{aligned} \tag{5.115}$$

for $l = 1, 2, \ldots$ ($\varkappa_0^{(W)} = 0$). Here the Mercer theorem has been used,

$$\begin{aligned}
\Gamma_{\mathcal{N}}^{(ch)}(x_1, x_2) &= \sum_\lambda \langle n_{ch\lambda} \rangle \varphi_\lambda^*(x_1) \varphi_\lambda(x_2), \\
\Gamma_{\mathcal{N}}^{(c)}(x_1, x_2) &= \sum_\lambda \langle n_{c\lambda} \rangle \varphi_\lambda^*(x_1) \varphi_\lambda(x_2),
\end{aligned} \tag{5.116}$$

as follows from (5.40a, b) and from the definition of the correlation functions. The first term in (5.115) corresponds to the chaotic field [see equations (5.39a, b)], the second one is the interference term between the chaotic and coherent fields. For a fully coherent field $\varkappa_1^{(W)} = \langle n_c \rangle$ and $\varkappa_l^{(W)} = 0$, $l \geqq 2$.

In the case of a uniform chaotic component [Peřina, Peřinová and Mišta (1971)]

$$\begin{aligned}
\varkappa_l^{(W)} &= (l-1)! \left(\frac{\langle n_{ch} \rangle}{M} \right)^{l-1} \left\{ \left[\left(\frac{1+P}{2} \right)^l + \left(\frac{1-P}{2} \right)^l \right] \langle n_{ch} \rangle + \right. \\
&\left. + l\varkappa^2 \left[\left(\frac{1+P}{2} \right)^{l-1} \cos^2 \varphi + \left(\frac{1-P}{2} \right)^{l-1} \sin^2 \varphi \right] \langle n_c \rangle \right\} + \delta_{l1} B,
\end{aligned} \tag{5.117}$$

where δ_{l1} is the Kronecker symbol. For a Lorentzian spectrum integrals (5.112) may be employed.

5.3.7 Accuracy of approximate M-mode formulae

The Mandel–Rice formula (3.140) for the photocount distribution is useful in the approximate description of fully polarized chaotic light with arbitrary spectral composition and for arbitrary detection time intervals (and detection areas) if the degrees of freedom parameter M is adjusted so that the exact and approximate second factorial moments coincide, i.e. if (5.44) holds [Bédard, Chang and Mandel (1967), Mehta (1970), Lachs (1971)]. By analogy, the above given M-mode formulae are playing a similar role to approximations for the statistical descriptions of the superposition of coherent and chaotic fields [Horák, Mišta and Peřina (1971a, b), Peřina, Peřinová and Mišta (1971, 1972), Mišta, Peřina and Braunerová (1973), Peřina,

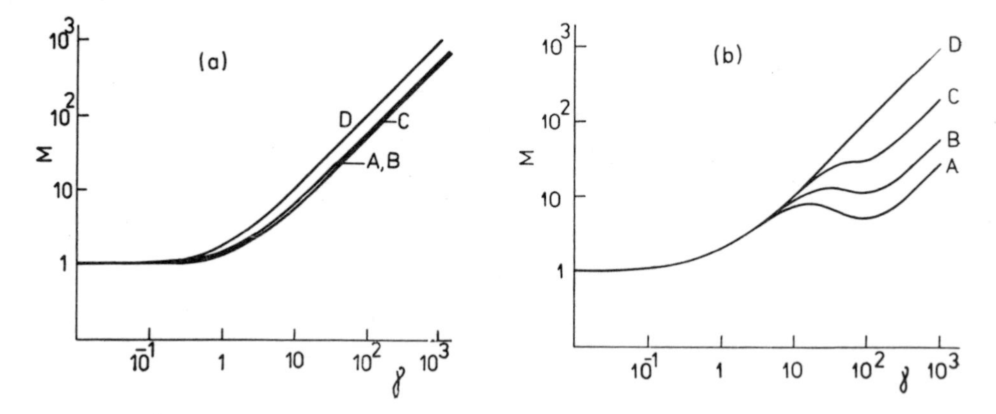

Fig. 5.16 — The dependence of the number of degrees of freedom on γ for (a) $\Omega = 0$, (b) $\Omega = 100$ and for $\langle n_c \rangle / \langle n_{ch} \rangle = 18/2$ (A), 16/4 (B), 10/10 (C) and 0/20 (D) ($\langle n_c \rangle + \langle n_{ch} \rangle = 20$) (after Horák, Mišta and Peřina, 1971a, J. Phys. **A4**, 231).

Peřinová, Lachs and Braunerová (1973), Saleh (1975a, b, 1978)]. Comparing the exact moment $\langle W^2 \rangle_{\mathcal{N}}$ from (5.108) with the approximate one obtained from (5.104c) for $k = 2$,

$$\langle W^2 \rangle_{\mathcal{N}} = \langle W \rangle_{\mathcal{N}}^2 + \\ + \frac{1}{M} \left[\langle n_{ch} \rangle^2 \frac{1 + P^2}{2} + 2 \langle n_{ch} \rangle \langle n_c \rangle \varkappa^2 \frac{1 + P \cos(2\varphi)}{2} \right], \quad (5.118)$$

we arrive at the value of the parameter M [Peřina, Peřinová and Mišta (1971, 1972)]

$$M = \frac{1 + 2\varkappa^2 \dfrac{\langle n_c \rangle}{\langle n_{ch} \rangle} \dfrac{1 + P \cos(2\varphi)}{1 + P^2}}{\mathcal{J}_1 + 2 \dfrac{\langle n_c \rangle}{\langle n_{ch} \rangle} \dfrac{1 + P \cos(2\varphi)}{1 + P^2} \bar{\mathcal{J}}_1}. \quad (5.119)$$

Thus M includes the temporal (as well as spatial) spectral and polarization properties, and substituting this value of M into the above formulae we can determine approximately all the quantities of the photocount statistics. The dependence of M on γ can be seen in Fig. 5.16 for fully polarized light. We observe that $M \to 1$ for $\gamma \to 0$ and that M is proportional to γ for $\gamma \gg 1$. In the case of purely chaotic light $\langle n_c \rangle = 0$ and $M = 1/\mathcal{J}_1$, i.e. we have (5.44), and (5.101b) reduces to the Mandel–Rice formula.

In Figs. 5.13–5.15, which represent the third reduced factorial moment, the approximate values (dotted curves) are also given and one can see good agreement between the exact and the approximate values. In addition, the exact and the approximate photocount distributions have been compared by using the Laxpati–Lachs recursion formulae [Peřina, Peřinová, Lachs and Braunerová (1973)], as is illustrated in Fig. 5.17. The relative errors for these cases are shown in Fig. 5.18. The values are given for a Lorentzian chaotic component, for which the agreement is the poorest

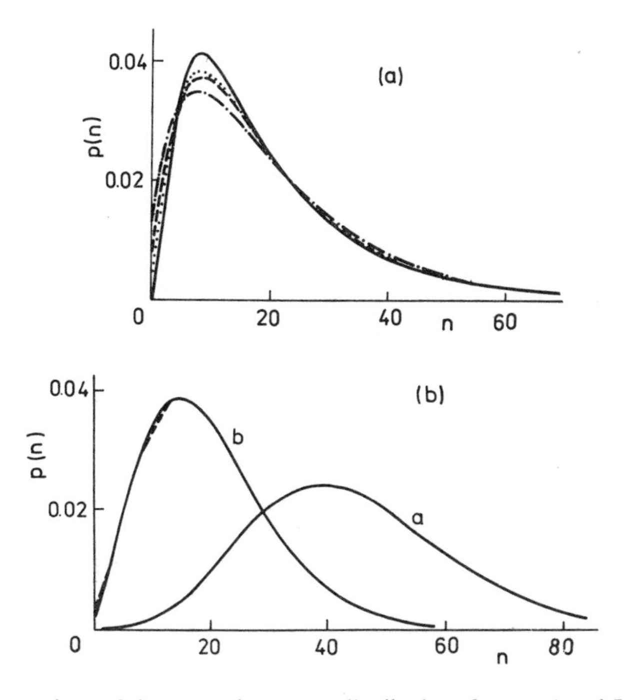

Fig. 5.17 — A comparison of the exact photocount distributions for $\gamma = 1$ and Lorentzian (solid curves), rectangular (dashed curve) and Gaussian (dotted curve) spectra with the corresponding approximate values (dot-dashed curve) and for (a) $\langle n_c \rangle = 2$, $\langle n_{ch} \rangle = 18$, (b) $\langle n_c \rangle = 40$, $\langle n_{ch} \rangle = 4$ (curve *a*, all curves coincide), $\langle n_c \rangle = 16$, $\langle n_{ch} \rangle = 4$ (curve *b*, for Lorentzian spectrum only, solid curve is exact, dashed curve approximate) (after Peřina, Peřinová, Lachs and Braunerová, 1973, Czech. J. Phys. **B23**, 1008).

(Fig. 5.17). In general the approximate formulae give better agreement for rectangular and Gaussian spectra than for the Lorentzian. Their accuracy is better than 1 %, provided that the signal-to-noise ratio $\langle n_c \rangle / \langle n_{ch} \rangle$ is greater than about 4. Taking into account that most communication and radar systems operate with values of this ratio in excess of 10, this approximation is adequate. This is particularly true for systems employing laser radiation above the threshold. A decrease of P [increase of $M = 2/(1 + P^2)$] and detuning of ϖ and ω_c lead also to an increase of the accuracy [Horák, Mišta and Peřina (1971a, b), Peřina, Peřinová and Mišta (1972), Mišta, Peřina and Braunerová (1973)].

In Fig. 5.19 a dependence on P of the statistics of the superposition of coherent and chaotic fields is presented in terms of the entropy

$$H = -\sum_{n=0}^{\infty} p(n) \log p(n). \tag{5.120}$$

The entropy decreases (the photocount distribution is more peaked and the uncertainty in the number of photons is decreasing) with increasing γ and Ω and decreasing P.

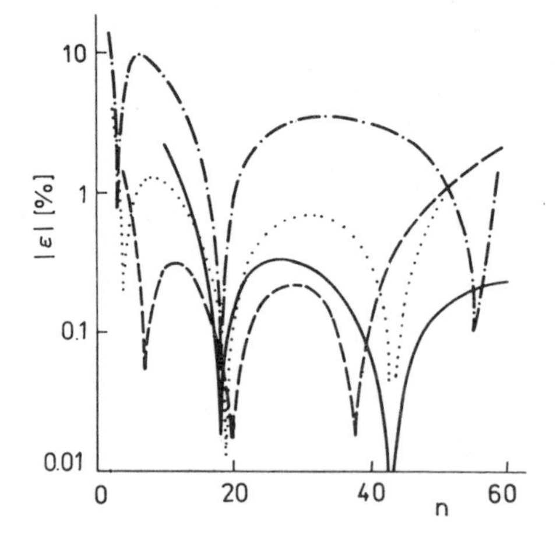

Fig. 5.18 — The absolute value of the relative error of the approximate formula (5.101b) for $\gamma = 1$ and $\langle n_c \rangle = 40$, $\langle n_{ch} \rangle = 4$ (solid curve), $\langle n_c \rangle = 18$, $\langle n_{ch} \rangle = 2$ (dashed curve), $\langle n_c \rangle = 16$, $\langle n_{ch} \rangle = 4$ (dotted curve) and $\langle n_c \rangle = 10$, $\langle n_{ch} \rangle = 10$ (dot-dashed curve) (after Peřina, Peřinová, Lachs and Braunerová, 1973, Czech. J. Phys. **B23**, 1008).

Equation (5.101b), with M given by (5.119), has been used to propose more general definitions of coherence time, coherence area and coherence volume [Peřina and Mišta (1974), cf. also Mandel (1981a) and the discussion following equation (3.179)]. Such definitions make it possible to include higher-order coherence effects, which is not possible with definitions based on interferometric and traditional spectroscopic measurements. Employing the Van Cittert–Zernike theorem of coherence theory [Born and Wolf (1965), Beran and Parrent (1964), Peřina (1972)], explicit corrections have been obtained [Peřina and Mišta (1974)], taking into account intensity interference between coherent and chaotic radiation. This approach was originally suggested by Bures, Delisle and Zardecki (1972a) and Zardecki, Delisle and Bures (1972, 1973) for chaotic radiation. One found that the factorization property of the coherence volume as the product of coherence area and coherence length is not strictly valid even if the radiation is cross-spectrally pure, $\gamma_{\mathcal{N}}(\mathbf{x}_1, \mathbf{x}_2, \tau) = \gamma_{\mathcal{N}}(\mathbf{x}_1, \mathbf{x}_2, 0)\, \gamma_{\mathcal{N}}(\mathbf{x}_1, \mathbf{x}_1, \tau)$ expressing the independence of spatial and temporal coherence. If r is the distance of a plane, uniform source of area σ from the parallel plane in which the state of coherence is examined, and if $\bar{\lambda}$ is the mean wavelength of radiation, then the coherence volume is determined by

$$V_c = \frac{ScT}{M} =$$

$$= c\,\frac{S(\bar{\lambda}r)^{-2}\int_{-\infty}^{+\infty} |\gamma_{\mathcal{N}}^{(ch)}(\tau)|^2\,d\tau + 2(\langle n_c\rangle/\langle n_{ch}\rangle)\,g_{\mathcal{N}}^{(ch)}(\omega_c)}{S(\bar{\lambda}r)^{-2} + 2\langle n_c\rangle/\langle n_{ch}\rangle}\,\frac{(\bar{\lambda}r)^2}{\sigma}, \qquad (5.121)$$

where $g_{\mathcal{N}}^{(ch)}$ is the normalized spectrum of chaotic light [the Fourier transform of $\gamma_{\mathcal{N}}^{(ch)}(\tau)$]. For purely chaotic light ($\langle n_c\rangle = 0$) we have for the coherence volume $V_c = c\int_{-\infty}^{+\infty} |\gamma_{\mathcal{N}}^{(ch)}(\tau)|^2\,d\tau(\bar{\lambda}r)^2/\sigma$, i.e. it is the product of the coherence length $l_c = c\int_{-\infty}^{+\infty} |\gamma_{\mathcal{N}}^{(ch)}(\tau)|^2\,d\tau$ and the coherence area $S_c = (\bar{\lambda}r)^2/\sigma$.

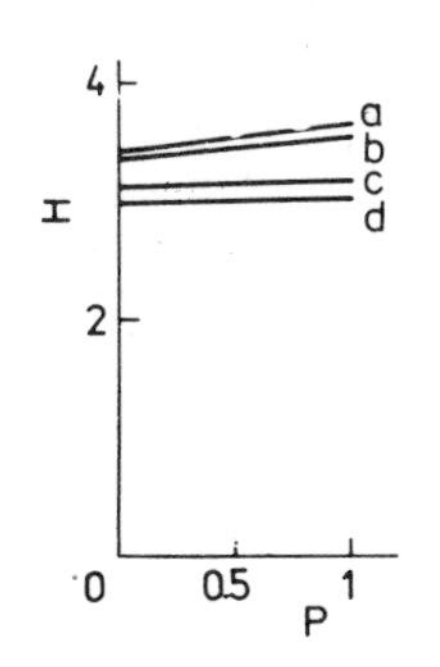

Fig. 5.19 — The dependence of the entropy H for the superposition of coherent and chaotic light on the degree of polarization P, $\langle n_c\rangle = 18$, $\langle n_{ch}\rangle = 2$ and (a) $\gamma = 0.01$, $\Omega = 0$, (b) $\gamma = 1$, $\Omega = 0$, (c) $\gamma = 10$, $\Omega = 0$, (d) $\gamma = 0.01$, $\Omega = 10$, also $\gamma = 10$, $\Omega = 10$ (after Peřina, Peřinová and Mišta, 1972, Opt. Acta **19**, 579).

Finally note that further results concerning the photocount statistics in random media will be discussed in Chapter 8. Some theoretical aspects of the photon statistics of the superposition of coherent and chaotic fields have been considered by Funke (1974) and Funke and Hoffmann (1976). A canonical distribution for the superposition of signal and noise has been discussed by Yarunin (1978).

CHAPTER 6

REVIEW OF NONLINEAR OPTICAL PHENOMENA

In this chapter we deal with fundamental nonlinear optical phenomena, such as the generation of harmonics, parametric oscillation and amplification, frequency conversion, self-focusing, Raman, Brillouin and hyper-Raman scattering and multiphoton absorption and emission; we also provide a brief review of resonant coherent phenomena, such as superradiance, self-induced transparency and photon echoes. A more detailed quantum statistical treatment is then given in Chapter 9. The treatment of this chapter is rather brief and introductory and further details concerning the traditional description of nonlinear optical phenomena can be found in the monographs and reviews mentioned in the Introduction.

We consider nonlinear interactions of various orders of nonlinearity and we distinguish between those nonlinear optical interactions where the medium plays only a parametric role (optical parametric processes) and nonlinear interactions where the variables of the active medium occur explicitly (Raman scattering, multiphoton absorption and emission, etc.).

6.1 General classical description

The classical phenomenological description of nonlinear optical phenomena is based on the Maxwell equations for the electric and magnetic field vectors **E** and **H** in the nonlinear medium. By the standard method we can obtain from them the equivalent wave equation, usually for **E**, in the nonlinear medium

$$\Delta \mathbf{E}(\mathbf{x}, t) - \mu_0 \varepsilon_0 \frac{\partial^2 \mathbf{E}(\mathbf{x}, t)}{\partial t^2} = \mu_0 \frac{\partial^2 \mathbf{P}(\mathbf{x}, t)}{\partial t^2}, \tag{6.1a}$$

$$\nabla \cdot \mathbf{E}(\mathbf{x}, t) = -\frac{1}{\varepsilon_0} \nabla \cdot \mathbf{P}(\mathbf{x}, t), \tag{6.1b}$$

where ε_0 and μ_0 are the permittivity and permeability constants of the vacuum; a non-magnetic and non-conducting medium is assumed, without external charges. If the intensity of the radiation is sufficiently high, the medium has a nonlinear response to the radiation, and the generalized electric polarization vector **P** can be written in the frequency domain as

$$\mathbf{P}(\omega_i) = \chi^{(1)}(\omega_i) \cdot \mathbf{E}(\omega_i) + \sum_{j,k} \chi^{(2)}(\omega_i = \omega_j + \omega_k) : \mathbf{E}(\omega_j) \mathbf{E}(\omega_k) + \\ + \sum_{j,k,l} \chi^{(3)}(\omega_i = \omega_j + \omega_k + \omega_l) \vdots \mathbf{E}(\omega_j) \mathbf{E}(\omega_k) \mathbf{E}(\omega_l) + \dots, \tag{6.2}$$

where $\chi^{(n)}$ are the susceptibility tensors of order $(n + 1)$, and multiple scalar products are expressed. Note that in the time domain, (6.2) is represented by a multifold convolution. Now we can treat the nonlinear processes of various orders, classified on the basis of (6.2). Since $\mathbf{E}(t)$ and $\mathbf{P}(t)$ are real vectors, $\chi^{(n)}$ must fulfil the cross-symmetry conditions, such as $\chi^{(1)*}(\omega_j) = \chi^{(1)}(-\omega_j)$, etc.

6.2 The second-order phenomena

The second-order nonlinear optical phenomena are characterized by the second term in (6.2). Assuming propagation along the z-axis, and monochromatic waves, we obtain from (6.1a) for the parametric interaction of three waves, provided that $k\,\mathrm{d}E_i/\mathrm{d}z \gg \mathrm{d}^2E_i/\mathrm{d}z^2$ $(k = |\mathbf{k}|)$,

$$\frac{\mathrm{d}E_{1i}}{\mathrm{d}z} = -\frac{\mathrm{i}\omega_1}{2}\left(\frac{\mu_0}{\varepsilon_1}\right)^{1/2} \chi^{(2)}_{ijk} E_{3j} E^*_{2k} \exp(\mathrm{i}\Delta k z), \tag{6.3a}$$

$$\frac{\mathrm{d}E_{2k}}{\mathrm{d}z} = -\frac{\mathrm{i}\omega_2}{2}\left(\frac{\mu_0}{\varepsilon_2}\right)^{1/2} \chi^{(2)}_{kij} E^*_{1i} E_{3j} \exp(\mathrm{i}\Delta k z), \tag{6.3b}$$

$$\frac{\mathrm{d}E_{3j}}{\mathrm{d}z} = -\frac{\mathrm{i}\omega_3}{2}\left(\frac{\mu_0}{\varepsilon_3}\right)^{1/2} \chi^{(2)}_{jik} E_{1i} E_{2k} \exp(-\mathrm{i}\Delta k z), \tag{6.3c}$$

where i, j, k represent the cartesian components, $\omega_j = k_j/(\mu_0\varepsilon_j)^{1/2}$ and $\Delta k = k_3^{(j)} - k_2^{(k)} - k_1^{(i)}$ represents the phase mismatch, $k_j^{(i)}$ being the ith polarization component of the wave vector of the jth wave. Further, the frequency resonance condition $\omega_3 = \omega_1 + \omega_2$ holds.

If one cannot distinguish between waves 1 and 2, this three-wave interaction reduces to the degenerate case, described by the following set of coupled equations

$$\frac{\mathrm{d}E_{1i}}{\mathrm{d}z} = -\mathrm{i}\omega_1\left(\frac{\mu_0}{\varepsilon_1}\right)^{1/2} \chi^{(2)}_{ijk} E^*_{1k} E_{2j} \exp(\mathrm{i}\Delta k z), \tag{6.4a}$$

$$\frac{\mathrm{d}E_{2j}}{\mathrm{d}z} = -\mathrm{i}\frac{\omega_1}{2}\left(\frac{\mu_0}{\varepsilon_2}\right)^{1/2} \chi^{(2)}_{jik} E_{1i} E_{1k} \exp(-\mathrm{i}\Delta k z), \tag{6.4b}$$

where $\Delta k = k_2^{(j)} - 2k_1^{(i)}$ and $\omega_2 = 2\omega_1$. Compared to (6.3), where the subfrequency modes are identified as 1 and 2 and the sum-frequency mode as 3, we have here denoted the subfrequency mode as 1 and the sum-frequency mode as 2.

Thus equations (6.3) describe, in various channels, the process of sum-frequency generation, if radiation of frequency $\omega_3 = \omega_1 + \omega_2$ is generated from the sub-frequency radiations; the splitting of the radiation of the frequency ω_3 into two radiations with subfrequencies ω_1 and ω_2 (if both the modes 1 and 2 start from the vacuum fluctuations, we speak of the parametric generation process; if the signal mode 1 is amplified, whereas the idler mode 2 starts from the vacuum fluctuations we speak of the parametric amplification process); frequency down-conversion, $\omega_2 = \omega_3 - \omega_1$, radiations of the frequencies ω_3 and ω_1 being introduced; or frequency up-conversion, $\omega_3 = \omega_1 + \omega_2$, if radiations of frequencies ω_1 and ω_2 are introduced.

In a simplified description, the idler mode 2 may be interpreted as a phonon mode and we can describe Brillouin and Raman scattering. A fully quantum description of all of these processes starts with the hamiltonian (4.216) and will be developed in Chapter 9. Similarly in the degenerate case, equations (6.4) describe sum-frequency generation $\omega_2 = 2\omega_1$ (pairs of red photons produce blue photons) or subharmonic generation $\omega_1 = \omega_2/2$.

Consider second harmonic generation, described by equations (6.4a, b). If the complex amplitude $E_2 = 0$ for $z = 0$ and if the length of the nonlinear crystal is L, then, considering the pumping wave E_1 as a constant, we see from (6.4b) that the second harmonic intensity is

$$| E_{2j}(z) |^2 \sim | E_{1i} |^2 | E_{1k} |^2 \frac{\sin^2(\Delta kL/2)}{(\Delta kL/2)^2}, \tag{6.5}$$

that is the second harmonic intensity is proportional to the product of the pumping subharmonic intensities modulated by the $(\sin(x)/x)^2$-function, dependent on the phase mismatch. We can introduce the coherence length $L_c = \pi/\Delta k$ giving the distance over which a systematic exchange of energy between the pumping radiation and the signal occurs.

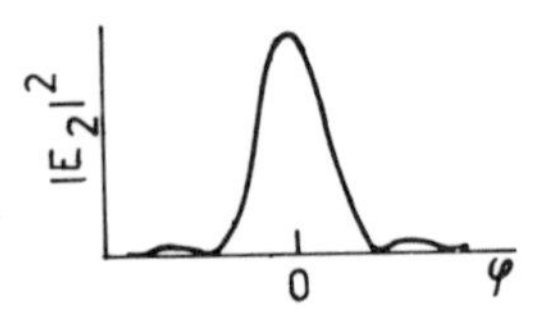

Fig. 6.1 — The dependence of the second harmonic intensity on the angle φ of deviation from the synchronization direction.

From equations (6.4a, b) the following conservation law follows simply, the polarization indices being omitted:

$$\frac{1}{g_1}\frac{\mathrm{d}| E_1 |^2}{\mathrm{d}z} + \frac{4}{g_2}\frac{\mathrm{d}| E_2 |^2}{\mathrm{d}z} = 0, \tag{6.6a}$$

that is

$$\frac{1}{g_1}| E_1(z) |^2 + \frac{4}{g_2}| E_2(z) |^2 =$$

$$= \frac{1}{g_1}| E_1(0) |^2 + \frac{4}{g_2}| E_2(0) |^2, \tag{6.6b}$$

or using $\omega_2 = 2\omega_1$ and assuming $\varepsilon_1 = \varepsilon_2$ for permittivities, then $g_2 = 2g_1$ and

$$| E_1(z) |^2 + 2 | E_2(z) |^2 = | E_1(0) |^2 + 2 | E_2(0) |^2; \tag{6.6c}$$

here $g_j = \omega_j \mu_0^{1/2} \chi_{ijk}^{(2)} / 2\varepsilon_j^{1/2}$. The equations of motion (6.4a, b) can be solved in terms of hyperbolic or elliptic functions [Bloembergen (1965), Zernike and Midwinter (1973)].

In Fig. 6.1 we see the dependence of the second harmonic intensity $|E_2|^2$ on the angular deviation from the synchronization direction, determined by $\Delta\mathbf{k} = 0$; the halfwidth of the curve varies from several minutes to several degrees for various nonlinear crystals. In Fig. 6.2a we see the dependence of the modulus of the complex amplitude on z for the fundamental frequency ω_1 and the second harmonic frequency

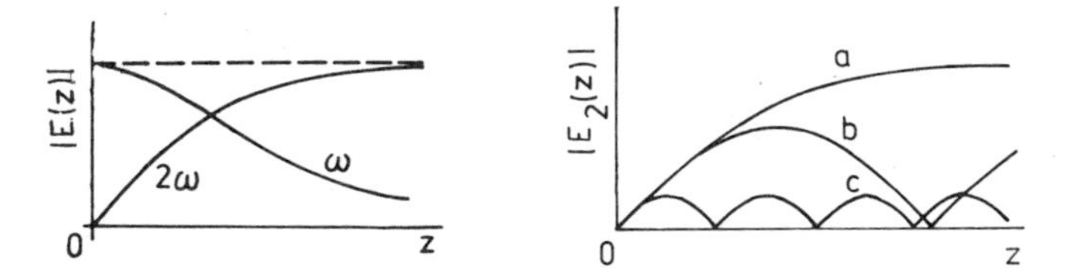

Fig. 6.2 — The real amplitude as a function of distance travelled in the nonlinear medium for (a) the first and second harmonics with $\Delta k = 0$, (b) the second harmonic wave with phase mismatch $\Delta k = 0$ (curve *a*), $\Delta k = 2$ (curve *b*) and $\Delta k = 10$ (curve *c*).

$\omega_2 = 2\omega_1$ for $\Delta\mathbf{k} = 0$, and in Fig. 6.2b we see this dependence for the second harmonic amplitude if $\Delta\mathbf{k} \neq 0$. The synchronization condition $\Delta\mathbf{k} = 0$ can be fulfilled in anisotropic crystals, if the fundamental wave corresponds to the ordinary beam and the second harmonic wave to the extraordinary beam, as shown in Fig. 6.3. In the synchronization direction $\Delta\mathbf{k} = 0$ we have for the refractive indices $n(2\omega) = n(\omega)$ and $k(2\omega) = 2k(\omega)$. The coherence length is $L_c = \lambda_1/4[n(2\omega) - n(\omega)]$. Systematic studies of the spatial synchronization of optical beams in optical parametric processes have been performed [Chmela (1971) and references therein] and the aperture effects and effects of the intensity distribution in the pumping beams have been investigated as well [Chmela (1974)]. Questions of the dependence of the efficiencies of the second harmonic generation on the initial statistics of the pumping field have been discussed also [Crosignani, Di Porto and Solimeno (1972), Chmela (1973)]

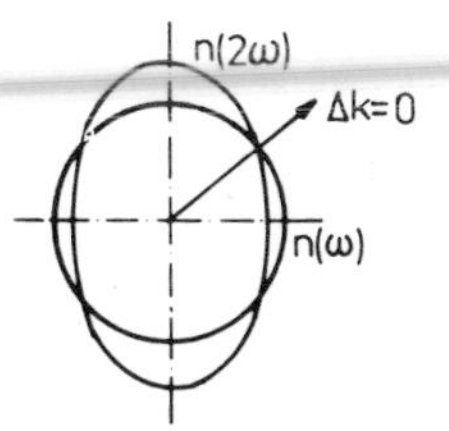

Fig. 6.3 — The synchronization direction in an anisotropic crystal.

Similar results can be derived for non-degenerate parametric processes governed by the equations of motion (6.3a–c). From them we obtain the conservation laws

$$\frac{1}{g_1}\frac{\mathrm{d}|E_1|^2}{\mathrm{d}z} = \frac{1}{g_2}\frac{\mathrm{d}|E_2|^2}{\mathrm{d}z} = -\frac{1}{g_3}\frac{\mathrm{d}|E_3|^2}{\mathrm{d}z}. \tag{6.7}$$

Again the equations of motion (6.3a–c) can be solved in terms of hyperbolic or elliptic functions [Bloembergen (1965), Zernike and Midwinter (1973)].

Consider now the case of the parametric amplifier, provided that the pumping radiation of frequency ω_3 is so strong that it remains practically unchanged during the nonlinear interaction. Then the solution for the subfrequency radiations has the form (E_3 is included in g, $\Delta \mathbf{k} = 0$)

$$E_1(z) = E_1(0)\,\mathrm{ch}(gz) - \mathrm{i}E_2^*(0)\,\mathrm{sh}(gz),$$
$$E_2(z) = E_2(0)\,\mathrm{ch}(gz) - \mathrm{i}E_1^*(0)\,\mathrm{sh}(gz), \tag{6.8}$$

where $g_1 = g_2 = g$. If the idler mode 2 starts with zero amplitude, both the signal mode 1 and the idler mode 2 are amplified, since

$$n_1(z) = |E_1(z)|^2 = |E_1(0)|^2\,\mathrm{ch}^2(gz) \underset{gz\to\infty}{\simeq} \frac{|E_1(0)|^2}{4}\exp(2gz),$$
$$n_2(z) = |E_2(z)|^2 = |E_1(0)|^2\,\mathrm{sh}^2(gz) \underset{gz\to\infty}{\simeq} \frac{|E_1(0)|^2}{4}\exp(2gz). \tag{6.9}$$

On the other hand, in the case of frequency conversion with a strong pumping radiation of frequency ω_2, we obtain the periodic solution (E_2 is included in g, $\Delta \mathbf{k} = 0$)

$$E_1(z) = E_1(0)\cos(gz) - \mathrm{i}E_3(0)\sin(gz),$$
$$E_3(z) = E_3(0)\cos(gz) - \mathrm{i}E_1(0)\sin(gz). \tag{6.10}$$

For the up-conversion process, $\omega_1 + \omega_2 \to \omega_3$ ($E_3(0) = 0$), for down-conversion $\omega_3 \to \omega_1 + \omega_2$ ($E_1(0) = 0$), and

$$n_1(z) = |E_3(0)|^2\sin^2(gz),$$
$$n_3(z) = |E_3(0)|^2\cos^2(gz) \tag{6.11}$$

and $n_1(z) + n_3(z) = n_3(0)$ [cf. (6.7)].

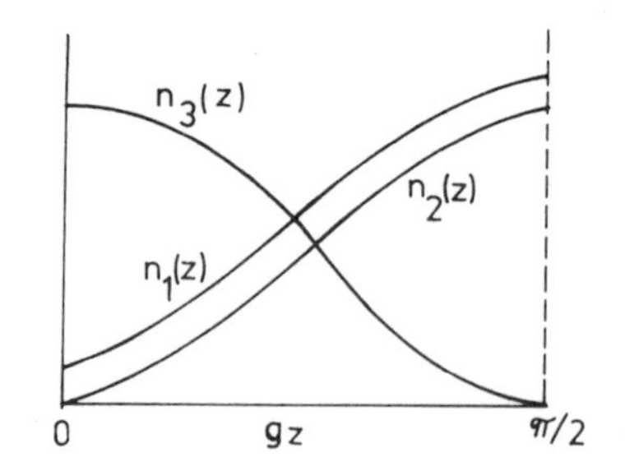

Fig. 6.4 — Parametric amplication of light with frequency ω_1, ω_2 is the frequency of the idler mode and ω_3 is the frequency of pumping.

A more realistic treatment taking into account the depletion of the pumping radiation provides results shown in Fig. 6.4 for the parametric amplification process, and in Fig. 6.5 for the frequency up- and down-conversion. Note that in the later stage of the process the inverse process starts.

The above equations describe also Brillouin or Raman scattering, in which laser radiation of frequency ω_L is scattered by acoustical vibrations or molecular vibrations (phonons) of the medium, whose typical frequency is ω_V. In this process scattered modes arise having Stokes and anti-Stokes frequencies

$$\omega_S = \omega_L - \omega_V \tag{6.12a}$$

and

$$\omega_A = \omega_L + \omega_V \tag{6.12b}$$

respectively, or, if the laser field is extremely strong, higher-order frequencies of the scattered modes can occur, $\omega_S = \omega_L - k\omega_V$, $\omega_A = \omega_L + k\omega_V$, $k \geqq 1$ being an integer. If the laser field is assumed to be strong, then the Stokes interaction determined by (6.12a) corresponds to the parametric amplification process, whereas the anti-Stokes interaction related to (6.12b) corresponds to the frequency conversion process.

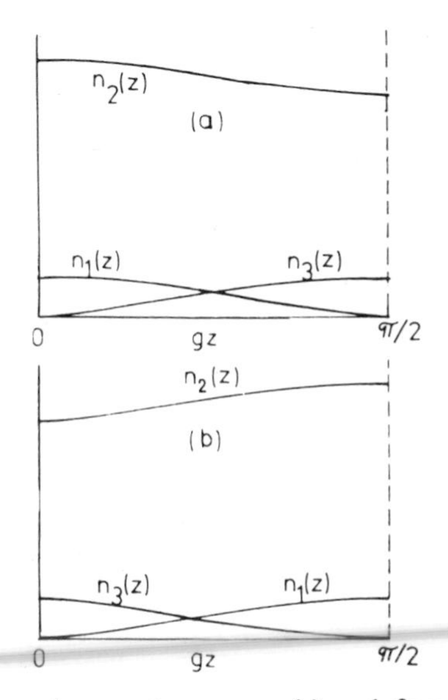

Fig. 6.5 — Frequency up-conversion $\omega_1 + \omega_2 \to \omega_3$ (a) and frequency down-conversion $\omega_3 \to \omega_1 + \omega_2$ (b), ω_2 is the frequency of pumping.

6.3 The third- and higher-order phenomena

Third and higher harmonic generation can be considered in a similar way, with $\omega_2 = k\omega_1$, $k \geqq 3$ being integral. In the course of the inversion process, the kth subharmonic frequency is generated, $\omega_1 = \omega_2/k$. Also higher-order parametric processes can be considered, including the generation of sum and difference frequency fields, with more than two fields.

The important class of third-order nonlinear optical processes is represented by stimulated Raman scattering. Compared to the optical parametric processes, where the medium is involved parametricly, stimulated Raman scattering is a direct resonant interaction of the incident laser field with the active medium. In general, in a scattering process, a photon of a frequency $\omega_1(\mathbf{k}_1)$ is absorbed, while a photon of frequency $\omega_2(\mathbf{k}_2)$ is emitted by the medium. The radiation state changes from its initial state $|i\rangle$ to its final state $|f\rangle$, the active medium makes a transition as well. The excitation of the medium can include an entropy wave without changing the frequency of the incident light (Rayleigh scattering), pressure acoustic waves (Brillouin scattering), a phonon, magnon, plasmon or electric excitation (Raman scattering), concentration variations (concentration scattering), etc. The transition probability is given by (3.68) which defines spontaneous scattering if $\langle n_2 \rangle = 0$ and stimulated scattering if $\langle n_2 \rangle \gg \gg 0$. For stimulated scattering, the Stokes and anti-Stokes radiations are coherent and coupled, and taking into account an effective internal vibration frequency ω_V of the atoms and molecules, the frequency conditions (6.12a, b) hold and consequently

$$2\omega_L = \omega_S + \omega_A, \tag{6.13}$$

which expresses the coupling of the laser, Stokes and anti-Stokes modes. Thus the corresponding equations of motion for the Stokes and anti-Stokes amplitudes E_S and E_A can be written in the form

$$\frac{\mathrm{d}E_S}{\mathrm{d}z} = g_S |E_L|^2 E_S + g_{SA} E_L^2 E_A^* \exp(\mathrm{i}\Delta\mathbf{k}\,.\,\mathbf{x}),$$

$$\frac{\mathrm{d}E_A}{\mathrm{d}z} = -g_A |E_L|^2 E_A - g_{SA} E_L^2 E_S^* \exp(\mathrm{i}\Delta\mathbf{k}\,.\,\mathbf{x}), \tag{6.14}$$

where g_S, g_A and g_{SA} are the Stokes, anti-Stokes and mutual coupling constants, E_L is the complex amplitude of the laser field, and $\Delta\mathbf{k} = 2\mathbf{k}_L - \mathbf{k}_S - \mathbf{k}_A$ characterizes the phase mismatch.

Qualitatively we can make the following conclusions concerning the number of photons in the Stokes and anti-Stokes modes in the course of the Raman scattering interaction. With respect to the results of Sec. 3.1 and Sec. 3.4, the probability $W_e^{(S)}$ of emission of a Stokes photon is proportional to $\langle \hat{a}_S \hat{a}_V \hat{a}_L^+ \hat{a}_L \hat{a}_V^+ \hat{a}_S^+ \rangle \sim (n_V + 1) \times \times n_L(n_S + 1)$ and the probability $W_a^{(S)}$ of absorption of a Stokes photon is proportional to $\langle \hat{a}_S^+ \hat{a}_V^+ \hat{a}_L \hat{a}_L^+ \hat{a}_V \hat{a}_S \rangle \sim n_V n_S (n_L + 1)$; thus we have for the change in the number of Stokes photons

$$\frac{\mathrm{d}n_S}{\mathrm{d}t} = g_S[(n_V + 1)\, n_L(n_S + 1) - n_V n_S(n_L + 1)], \tag{6.15a}$$

n_V being the number of phonons. In the same way we obtain for anti-Stokes photons $W_e^{(A)} \sim n_V n_L(n_A + 1)$ and $W_a^{(A)} \sim (n_V + 1)(n_L + 1)\, n_A$, and for the change in the number of anti-Stokes photons

$$\frac{\mathrm{d}n_A}{\mathrm{d}t} = g_A[n_V n_L(n_A + 1) - (n_V + 1)(n_L + 1)\, n_A], \tag{6.15b}$$

g_S and g_A being constants. Further, it holds that

$$\frac{\mathrm{d}n_S}{\mathrm{d}t} = -\frac{\mathrm{d}n_L}{\mathrm{d}t}, \qquad \frac{\mathrm{d}n_A}{\mathrm{d}t} = -\frac{\mathrm{d}n_L}{\mathrm{d}t} \tag{6.15c}$$

and $n_V = (\exp(h\nu_V/KT) - 1)^{-1} \approx \exp(-h\nu_V/KT)$. In spontaneous scattering $n_S = n_A = 0$ and $\mathrm{d}/\mathrm{d}t = (\mathrm{d}/\mathrm{d}z)(\mathrm{d}z/\mathrm{d}t) = c\,\mathrm{d}/\mathrm{d}z$ and we have from (6.15a, b)

$$\frac{\mathrm{d}n_L}{\mathrm{d}z} = -\frac{g_S}{c}(n_V + 1)\,n_L, \tag{6.16a}$$

$$\frac{\mathrm{d}n_L}{\mathrm{d}z} = -\frac{g_A}{c}\,n_V n_L. \tag{6.16b}$$

Hence

$$n_L(z) = n_L(0)\exp\left(-\frac{g_S}{c}(n_V + 1)\,z\right),$$

$$n_L(z) = n_L(0)\exp\left(-\frac{g_A}{c}\,n_V z\right), \tag{6.17a}$$

and using the conservation laws $n_L(z) + n_S(z) = n_L(0)$ and $n_L(z) + n_A(z) = n_L(0)$ following from (6.15c) we arrive at

$$n_S(z) = n_L(0)\left[1 - \exp\left(-\frac{g_S}{c}(n_V + 1)\,z\right)\right],$$

$$n_A(z) = n_L(0)\left[1 - \exp\left(-\frac{g_A}{c}\,n_V z\right)\right]; \tag{6.17b}$$

that is, the numbers of Stokes and anti-Stokes photons are saturated by the number of laser photons. On the other hand, for stimulated scattering $n_L, n_S, n_A \gg 1$ and

$$\frac{\mathrm{d}n_S}{\mathrm{d}z} = \frac{g_S}{c}\,n_S n_L, \tag{6.18a}$$

$$\frac{\mathrm{d}n_A}{\mathrm{d}z} = -\frac{g_A}{c}\,n_A n_L. \tag{6.18b}$$

Thus in the stimulated process the Stokes photons are amplified since

$$n_S(z) = n_S(0)\exp\left(\frac{g_S}{c}\,n_L z\right), \tag{6.19a}$$

whereas the anti-Stokes photons are attenuated,

$$n_A(z) = n_A(0)\exp\left(-\frac{g_A}{c}\,n_L z\right) \tag{6.19b}$$

[cf. the first terms on the right-hand sides of (6.14)].

A more detailed discussion, based on the exact quantum theory, will be presented in Chapter 9.

With very powerful laser beams hyper-Raman scattering may be realized, which is a higher-order nonlinear process. Using two laser beams with frequencies ω_1 hand

ω_2, the Stokes and anti-Stokes modes have frequencies $\omega_{S,A} = \omega_1 + \omega_2 \mp \omega_V$, whereas in the degenerate case $\omega_{S,A} = 2\omega_L \mp \omega_V$. A systematic treatment of this phenomenon is given in Chapter 9.

Other interesting nonlinear optical phenomena are multiphoton absorption and emission. Consider first n-photon absorption. In the course of the n-photon absorption, the atomic system makes a transition from the ground state to an exciting state, simultaneously absorbing n photons, in general of frequencies $\omega_1, \ldots, \omega_n$ (of course it may be that $\omega_1 = \ldots = \omega_n = \omega$). The transition probability is proportional (see Sec. 3.1) to the normal correlation function $\Gamma_{\mathcal{N}}^{(n,n)}(x, \ldots, x) = \langle \hat{A}^{(-)n}(x) \hat{A}^{(+)n}(x) \rangle$.

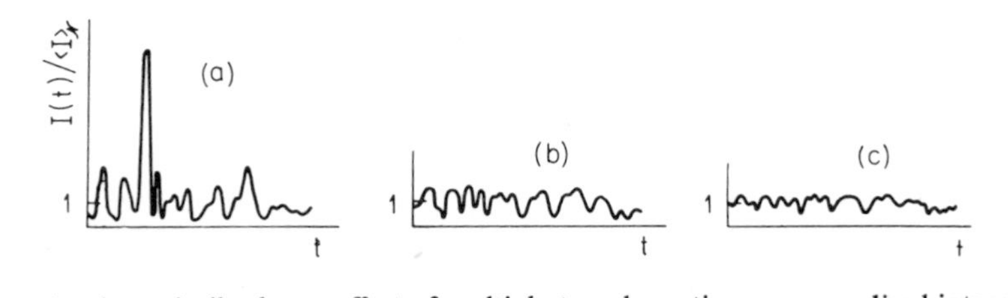

Fig. 6.6 — A schematically shown effect of multiphoton absorption on normalized intensity fluctuations, (a), caused by two-photon absorption, (b), and three-photon absorption, (c).

It is equal to $\langle \hat{A}^{(-)}(x) \hat{A}^{(+)}(x) \rangle^n = \langle I(x) \rangle^n$ for coherent radiation, and to $n! \langle I(x) \rangle^n$ for chaotic radiation; thus multiphoton absorption is more effective (for n-photon absorption $n!$ times) for chaotic radiation than for coherent radiation, which is a consequence of the bunching of chaotic photons. The presence in the transition probability of n-photon absorption of the factor $n!$ for chaotic radiation, or for laser radiation with a very high number of modes, has been experimentally tested and verified by Jakeman, Oliver and Pike (1968b), Kovarskii (1974), Krasinski, Chudzyński and Majewski (1974, 1976) and Glódź (1978) [cf. also Schubert and Wilhelmi (1980) for a review], and as quoted above by Shiga and Inamura (1967) and Teich, Abrams and Gandrud (1970). Further, it has been shown experimentally by LeCompte, Mainfray and Manus (1974) and LeCompte, Mainfray, Manus and Sanchez (1975) that the eleventh-order ionization of Xe-atoms with laser radiation composed of 100 modes is $10^{6.9 \pm 0.3}$ times more probable than for one-mode laser radiation, which corresponds approximately to 11! for chaotic radiation. The sensitivity of multiphoton absorption to fluctuations of the incident light and to higher powers is demonstrated in Figs. 6.6 and 6.7. In Fig. 6.6 we see the smoothing effect of fluctuations of radiation by two- and three-photon absorbers [Weber (1971), Schubert and Wilhelmi (1976)]. Fig. 6.7 shows the probability distribution of the normalized intensity for Gaussian light and for Gaussian light passed through two-photon and three-photon absorbers [Schubert and Wilhelmi (1976)]. Mutliphoton ionization of atoms in strong stochastic fields has been discussed by Kraynov and Todirashku (1980).

Phenomenologically, n-photon absorption can be described, neglecting one-photon process, by the differential equation

$$\frac{dI(z)}{dz} = -\beta I^n(z), \tag{6.20}$$

β being a positive proportionality constant. With some simplifications ($\beta z \gg 1$),

$$I(z) = [(n-1)\beta z]^{-1/(n-1)}. \tag{6.21}$$

Multiphoton absorption processes are of great practical importance. They make possible the construction of filters which are transparent to low-power radiation and which absorb high-power radiation. Further, in combination with tunable lasers, two-photon absorption is a powerful spectroscopic technique with high resolution.

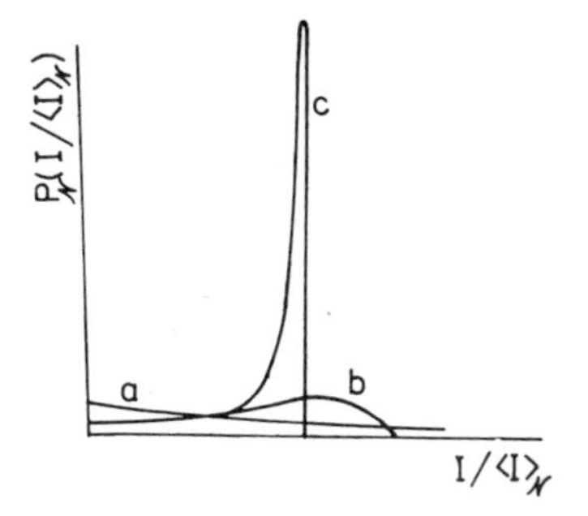

Fig. 6.7 — A schematically shown reduction of the normalized intensity probability distribution of Gaussian light, (a), by two-photon absorption, (b), and three-photon absorption, (c).

Multiphoton absorption is the inverse process to multiphoton emission; a transition scheme for two-photon stimulated emission is shown in Fig. 6.8. This process is discussed in greater detail in Chapter 9, where the quantum statistical properties are included.

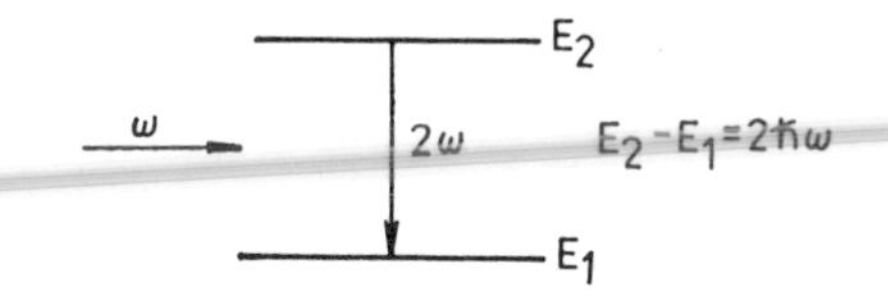

Fig. 6.8 — Two-photon stimulated emission.

Another third-order nonlinear effect is self-focusing of laser beams. This effect is caused by the dependence of the refractive index or the permittivity on the intensity of the radiation, $\varepsilon = \bar{\varepsilon} + \varepsilon_2 |\mathbf{E}|^2$, $\bar{\varepsilon}$ and $\varepsilon_2 > 0$ being constants. If a laser beam has a Gaussian intensity profile, it propagates slower in the central part than at the borders (Fig. 6.9) so the beam is focused at a distance z_f. However, if the beam propagates through finite apertures, diffraction of the beam occurs also, and both effects are superimposed. If they just compensate one another, the beam propagates without changing its cross-section over long distances, and it is self-trapped. Since $\Delta\varepsilon = \varepsilon - \bar{\varepsilon} = \varepsilon_2 |\mathbf{E}|^2$ has to be high enough in order to have an observable effect, pulsed lasers are usually used. One may distinguish two cases: (i) The laser pulse is larger than a typical response time for $\Delta\varepsilon$, so that the response of $\Delta\varepsilon$ to the laser intensity variations can be considered as instantaneous and we have quasi-steady state self-focusing; (ii) the laser pulse length is comparable with, or shorter than, the response time of $\Delta\varepsilon$, and then a transient self-focusing occurs.

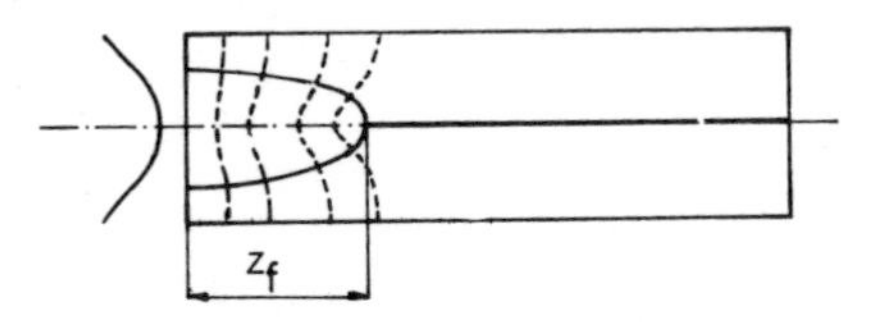

Fig. 6.9 — Self-focusing of a laser beam.

The self-focusing of the beam is described by the wave equation

$$\Delta \mathbf{E} - \frac{1}{v^2}\frac{\partial^2 \mathbf{E}}{\partial t^2} = \frac{\varepsilon_2}{\bar{\varepsilon} v^2}\frac{\partial^2(|\mathbf{E}|^2\mathbf{E})}{\partial t^2}, \tag{6.22}$$

where $v^2 = (\bar{\varepsilon}\mu)^{-1} = (\bar{\varepsilon}\mu_0)^{-1}$ is the speed of light in the nonlinear medium. Considering only one polarization component, we obtain after some simplifications [Yariv (1967)]

$$E(z) = E(0)\exp\left(-\mathrm{i}\frac{k\varepsilon_2}{4\bar{\varepsilon}}|E(0)|^2 z\right), \tag{6.23}$$

which corresponds to the excess $(k\varepsilon_2/4\bar{\varepsilon})\,|E(0)|^2$ in the wave number k and leads to self-focusing. For z_f one can obtain

$$z_f = R^2\bar{\varepsilon}\left(\frac{\pi v}{\varepsilon_2 P}\right)^{1/2}, \tag{6.24}$$

and for the threshold power P_c of the phenomenon

$$P_c = \frac{\bar{\varepsilon}^2 v \lambda^2}{2\pi\varepsilon_2}, \tag{6.25}$$

P being the laser power and R the beam radius.

In the transient case the front of the pulse remains unchanged and only the back part of the pulse is made to be narrower, so that the pulse takes a horn-shaped form, propagating thereafter without any change.

The self-focusing phenomenon is important in the design of high-power lasers, where it may be dangerous and optical breakdown of the active medium may ensue. On the other hand, it may be very significant for optical communication, leading to the transfer of energy in narrow beams without losses.

Further details concerning self-focusing can be found in reviews by Akhmanov, Khokhlov and Sukhorukov (1972), Svelto (1974) and Shen (1976).

Recently, much effort has been devoted to the development of the so-called dynamical holography or adaptive optics, which represents a combination of the holographic principle of imagery with a reference beam, and nonlinear optics [Yariv

(1975), Bespalov (1979), Adaptive Optics (1977)—a special issue of the Journal of American Optical Society]. This phenomenon can be explained as shown in Fig. 6.10. Two opposite pumping waves E_1 and E_2 of frequency ω are in nonlinear interaction with weak waves E_3 and E_4 also of frequency ω. The variation of the signal waves E_3 and E_4 along the nonlinear medium is shown in Fig. 6.11. We see that the signal

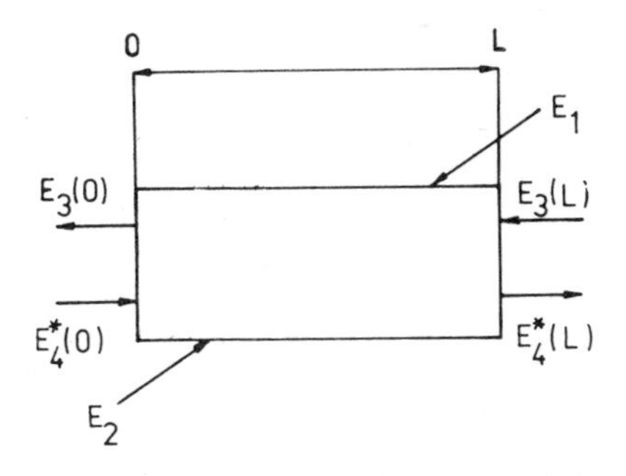

Fig. 6.10 — An outline of four-wave mixing.

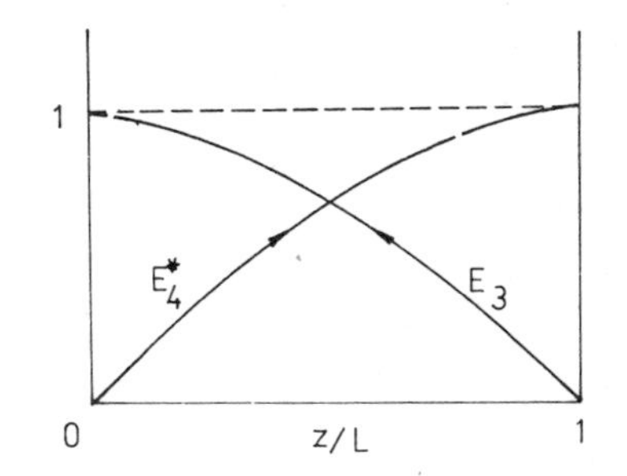

Fig. 6.11 — A schematic behaviour of the amplified signal waves.

waves E_3 and E_4 are amplified and at the planes 0 and L they have opposite phases (the amplifier reversing the wave front). Compared to the three-field interaction, in this four-wave interaction the synchronization condition is not needed and any form of the wave front can be reversed.

6.4 Transient coherent optical effects

It has been recognized that there exists a close analogy between magnetic resonance and optical resonance phenomena. In general, any two-level system can be treated like a (1/2)-spin system. Magnetic transient coherent phenomena are now used routinely for relaxation studies in the radio- and micro-wave range. Similar transient resonant phenomena exist also in the optical region and have been the subject of intensive research in recent years [Courtens (1972), Nussenzveig (1973), Sargent, Scully and Lamb (1974), Slusher (1974), Allen and Eberly (1975), Butylkin, Kaplan, Khronopulo and Yakubovitch (1977), Kujawski and Eberly (1978), Schubert and Wilhelmi (1978)]. The role of fluctuations in nonlinear pulse propagation has been discussed by Crosignani, Papas and Di Porto (1980). The Maxwell equations (or the equivalent wave equation) for the radiation field, coupled to the Bloch equations for the two-level atomic active system, provide the usual basis for the description of the phenomena associated with the propagation of very short optical pulses through nonlinear media.

6.4.1 Self-induced transparency

It has been shown by McCall and Hahn (1967, 1969) theoretically as well as experimentally that, for sufficiently intense light pulses of a suitable shape, the medium, due to nonlinear effects, can behave as completely transparent. This phenomenon is called self-induced transparency and arises when the front edge of the pulse can excite the medium and the back edge then stimulates emission. Then the pulse propagates without any change, it is only delayed. A short pulse is necessary to avoid relaxation processes.

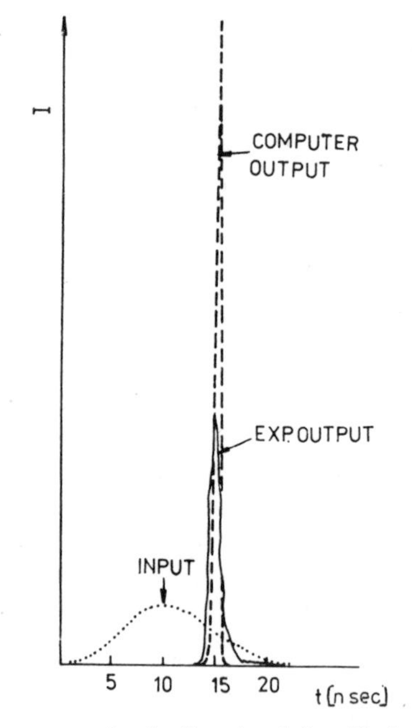

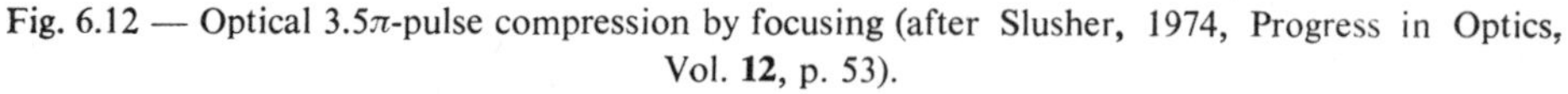

Fig. 6.12 — Optical 3.5π-pulse compression by focusing (after Slusher, 1974, Progress in Optics, Vol. **12**, p. 53).

We define the pulse area

$$\vartheta(z) = \frac{\mathscr{P}}{\hbar} \int_{-\infty}^{+\infty} \mathscr{E}(z, t)\, \mathrm{d}t, \tag{6.26}$$

where $\mathscr{E}$ represents the complex envelope function of the pulse and $\mathscr{P}$ is the matrix element of the dipole moment. Then it follows from the Maxwell and Bloch equations [McCall and Hahn (1967, 1969)] that

$$\frac{d\vartheta(z)}{dz} = \frac{a}{2} \sin \vartheta(z); \tag{6.27}$$

for absorbing media ($a < 0$) and weak pulses ($\vartheta \ll 1$), the well-known Beer's law is obtained

$$|\vartheta(z)|^2 = |\vartheta(0)|^2 \exp(-|a| z). \tag{6.28}$$

The solution of (6.27) has the shape of hyperbolic secant, which remains unchanged during propagation through the nonlinear medium. Pulses with $\vartheta < \pi$ are attenuated to zero, whereas pulses with $2\pi > \vartheta > \pi$ are amplified to the 2π-pulse. Arbitrary-shaped pulses are split into a number of 2π-pulses, which then propagate without any change. In Fig. 6.12 the compression of the 3.5π-pulse to 2π-pulse is demonstrated [Gibbs and Slusher (1971), Slusher (1974)].

6.4.2 Photon echo

The optical photon echo is analogous to the spin echo in magnetic resonance. The principle of the phenomenon is sketched in Fig. 6.13. Application of the first resonant coherent $\pi/2$-pulse causes the rotation of the dipoles $\vec{P}$ from the direction z by 90° about $\vec{\mathscr{E}}$. The dipoles then precess around z with a frequency ω. As a result of inhomogeneous broadening, different dipoles precess with slightly different frequencies. At a time τ after the first pulse we apply the second resonant π-pulse. Each dipole is now rotated by 180° about $\vec{\mathscr{E}}$. After the pulse is over, the dipoles again

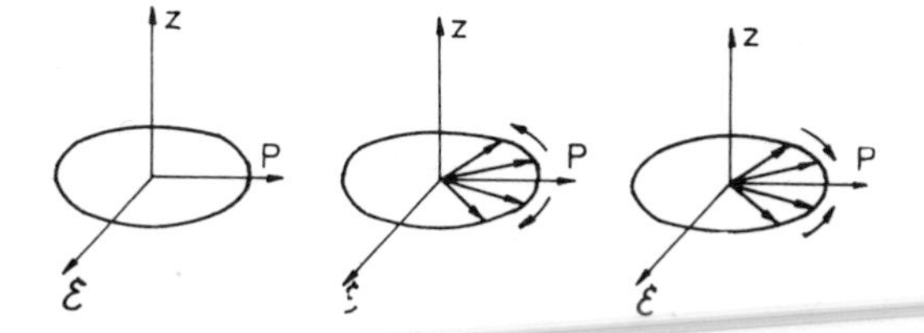

Fig. 6.13 — A sketch of the photon echo phenomenon.

precess around $\vec{z}$, but in the opposite direction. Thus, at time τ after the second pulse, all the dipoles create one giant dipole which radiates a giant pulse, representing the photon echo.

6.4.3 Superradiance

As shown by Dicke (1954, 1964), in a system composed of N atoms, of which $N/2$ are excited, correlation of the atoms can occur via their interaction with the electromagnetic field. The system then represents a big dipole in an atomic coherent state which spontaneously emits radiation in a cooperative way, the intensity being proportional to N^2, whereas the resulting intensity of ordinary spontaneous emission is proportional to N.

This phenomenon is described by the hamiltonian (4.212), (4.213) or (4.216). For the matrix element of $\mathcal{M}_-$, as defined by (4.209b), in the Dicke states we have

$$|\langle l, m-1 | \mathcal{M}_- | l, m\rangle|^2 = (l+m)(l-m+1). \tag{6.29}$$

Hence, if the state is fully excited (all N atoms are excited), then $l = m = N/2$ (from (4.221) $n_1 = 0$, $n_2 = N$) and from (6.29) we have for the resulting intensity

$$I = NI_0, \tag{6.30}$$

I_0 being the spontaneous emission rate for a single excited atom. However, if $l = N/2$, $m = 0$ ($n_1 = n_2 = N/2$, i.e. only half of the atoms are excited), then from (6.29)

$$I = \frac{N}{2}\left(\frac{N}{2} + 1\right) I_0 \approx \frac{N^2}{4} I_0, \tag{6.31}$$

provided that N is large. This means that the atoms radiate in phase and the rate is therefore proportional to N^2. The state $|N/2, 0\rangle$ represents the superradiant state (the state of cooperative spontaneous emission) and it requires a large value of the cooperation number l and small values of $|m|$.

For the fully excited state $|N/2, N/2\rangle$ Bose–Einstein statistics are appropriate and $\langle(\Delta n)^2\rangle \approx \langle n\rangle(\langle n\rangle + 1)$, whereas for the superradiant state $|N/2, 0\rangle$ photons obey the Poissonian statistics as a coherent field, with $\langle(\Delta n)^2\rangle \approx \langle n\rangle$. The statistical properties of superradiance have been discussed by Eberly and Rehler (1970) and Karczewski (1976). Subharmonic superradiance has been also considered [Sczaniecki and Buchert (1978)]. A number of papers have been devoted to the investigation of the interaction of N atoms with the radiation field [Orszag (1979), Kumar and Mehta (1980, 1981)], including phase transitions in cooperative atomic systems [Walls, Drummond, Hassan and Carmichael (1978)]. A source of nonclassical radiation exhibiting photon antibunching composed of two-level atoms in the Dicke state has been investigated by Kumar, Mehta and Agarwal (1981).

CHAPTER 7

HEISENBERG—LANGEVIN AND MASTER EQUATIONS APPROACHES TO THE STATISTICAL PROPERTIES OF RADIATION INTERACTING WITH MATTER

In this chapter we introduce the Heisenberg–Langevin approach, the master equation approach, and the related generalized Fokker–Planck equation approach, to the statistical properties of radiation in interaction with matter. First we consider a model of the interaction of a one-mode radiation field with an infinite reservoir boson system (the quantum theory of damping). Then we consider the interaction of the electromagnetic field with a system of two-level atoms. The methods presented in this chapter will be applied to propagation of radiation in random and nonlinear media in Chapters 8 and 9.

7.1 The Heisenberg—Langevin approach

Consider a one-mode radiation field of a frequency ω described in the Heisenberg picture by the annihilation and creation operators $\hat{a}$ and $\hat{a}^+$ respectively and an infinite boson system with the annihilation and creation operators $\hat{b}_j$ and $\hat{b}_j^+$ respectively and with the corresponding frequencies ψ_j. If the coupling constant of the interaction is $\varkappa_j$, then the hamiltonian of the compound system can be written in the form

$$\hat{H} = \hat{H}_0 + \hat{H}_{\mathrm{int}}, \tag{7.1a}$$

where

$$\hat{H}_0 = \hbar\omega\left(\hat{a}^+\hat{a} + \frac{1}{2}\right) + \sum_j \hbar\psi_j\left(\hat{b}_j^+\hat{b}_j + \frac{1}{2}\right) \tag{7.1b}$$

is the free hamiltonian and

$$\hat{H}_{\mathrm{int}} = \sum_j (\hbar\varkappa_j\hat{b}_j\hat{a}^+ + \hbar\varkappa_j^*\hat{b}_j^+\hat{a}) \tag{7.1c}$$

is the hamiltonian of the interaction between the one-mode radiation field and the infinite reservoir system $\{\hat{b}_j\}$.

The Heisenberg equations (2.54a) for this interaction read

$$\frac{\mathrm{d}\hat{M}}{\mathrm{d}t} = \frac{1}{\mathrm{i}\hbar}[\hat{M}, \hat{H}] = -\mathrm{i}\omega\frac{\partial\hat{M}}{\partial\hat{a}}\hat{a} + \mathrm{i}\omega\hat{a}^+\frac{\partial\hat{M}}{\partial\hat{a}^+} - \mathrm{i}\frac{\partial\hat{M}}{\partial\hat{a}}\sum_j \varkappa_j\hat{b}_j + \\ + \mathrm{i}\frac{\partial\hat{M}}{\partial\hat{a}^+}\sum_j \varkappa_j^*\hat{b}_j^+, \tag{7.2a}$$

$$\frac{\mathrm{d}\hat{b}_j}{\mathrm{d}t} = -\mathrm{i}\psi_j\hat{b}_j - \mathrm{i}\varkappa_j^*\hat{a}, \tag{7.2b}$$

where $\hat{M}(\hat{a}^+, \hat{a})$ is an arbitrary operator and we have used the identities

$$[\hat{M}, \hat{a}^+\hat{a}] = [\hat{M}, \hat{a}^+]\,\hat{a} + \hat{a}^+[\hat{M}, \hat{a}] \tag{7.3}$$

and (4.32).

If $\hat{M} = \hat{a}$, (7.2a) reduces to

$$\frac{d\hat{a}}{dt} = -i\omega\hat{a} - i\sum_l \varkappa_l \hat{b}_l. \tag{7.4}$$

Writing the solution of (7.2b) in the form

$$\hat{b}_l(t) = \hat{b}_l(0)\exp(-i\psi_l t) - i\varkappa_l^* \int_0^t \hat{a}(t')\exp[i\psi_l(t'-t)]\,dt' \tag{7.5}$$

and substituting it into (7.4), we obtain, in the Wigner–Weisskopf approximation,

$$\frac{d\hat{a}}{dt} = -i\left(\omega' - i\frac{\gamma}{2}\right)\hat{a} + \hat{L}, \tag{7.6}$$

where we have assumed the light quasi-monochromatic so that $\hat{a}(t') \approx \hat{a}(t) \times \times \exp[i\omega(t-t')]$, replaced the integration over the interval $(0, t)$ by $(0, \infty)$ and used

$$\int_0^\infty \exp[i(\omega - \psi_l)\,t]\,dt = \lim_{\substack{\varepsilon>0\\ \varepsilon\to 0}} \frac{1}{-i(\omega-\psi_l)+\varepsilon} =$$

$$= \frac{iP}{\omega - \psi_l} + \pi\delta(\omega - \psi_l). \tag{7.7}$$

Here P denotes the Cauchy principal value of the integral, $\omega' = \omega + \Delta\omega$ and the frequency shift (which may usually be neglected) and the damping constant are determined by

$$\Delta\omega = P\int_{-\infty}^{+\infty} \frac{|\varkappa(\psi)|^2\varrho(\psi)}{\omega - \psi}\,d\psi, \tag{7.8a}$$

$$\gamma = 2\pi\,|\varkappa(\omega)|^2\varrho(\omega), \tag{7.8b}$$

$\varrho(\omega)$ being the density function of the reservoir oscillators, the sum over l has been replaced by the integration over ψ (if the density of reservoir oscillators is sufficiently high) and the Langevin force $\hat{L}(t)$ has the form

$$\hat{L}(t) = -i\sum_l \varkappa_l \hat{b}_l(0)\exp(-i\psi_l t). \tag{7.9}$$

$\hat{L}(t)$ has the following properties:

$$\begin{aligned}
\langle\hat{L}(t)\rangle &= \langle\hat{L}^+(t)\rangle = 0,\\
\langle\hat{L}^+(t)\,\hat{L}(t')\rangle &= \gamma\langle n_d\rangle\,\delta(t-t'),\\
\langle\hat{L}(t)\,\hat{L}^+(t')\rangle &= \gamma(\langle n_d\rangle + 1)\,\delta(t-t'),\\
\langle\hat{L}(t)\,\hat{L}(t')\rangle &= \langle\hat{L}^+(t)\,\hat{L}^+(t')\rangle = 0,
\end{aligned} \tag{7.10}$$

provided that the reservoir spectrum is flat so that the mean number of reservoir oscillators (phonons) in the mode l is $\langle n_d \rangle = \langle \hat{b}_l^+(0)\, \hat{b}_l(0) \rangle$ independently of l, and the reservoir oscillators form a chaotic system.

The Heisenberg–Langevin equation (7.6) may be solved, for instance, by using the Laplace transform method, in the form

$$\hat{a}(t) = u(t)\, \hat{a}(0) + \sum_l w_l(t)\, \hat{b}_l(0), \tag{7.11}$$

where

$$u(t) = \exp\left(-\mathrm{i}\omega' t - \gamma t/2\right),$$
$$w_l(t) = \varkappa_l \exp\left(-\mathrm{i}\psi_l t\right) \frac{1 - \exp\left[\mathrm{i}(\psi_l - \omega')\, t - \gamma t/2\right]}{\psi_l - \omega' + \mathrm{i}\gamma/2}. \tag{7.12}$$

We can verify (again assuming a very high density of reservoir modes, and using the residue theorem) that

$$|u(t)|^2 + \sum_l |w_l(t)|^2 = 1, \tag{7.13a}$$

that is

$$\sum_l |w_l(t)|^2 = 1 - \exp(-\gamma t). \tag{7.13b}$$

Consequently the commutation rule is preserved for all times,

$$\left[\hat{a}(t), \hat{a}^+(t)\right] = \hat{1} \tag{7.14}$$

(similarly with the other rules). Therefore the character of the particles is preserved, although their statistics can change with time since the radiation mode represents an open non-equilibrium system. Neglecting the Langevin force in (7.6), $\hat{a}(t) = \hat{a}(0) \times \exp(-\mathrm{i}\omega' t - \gamma t/2)$ and we have the incorrect commutation rule $[\hat{a}(t), \hat{a}^+(t)] = \hat{1} \exp(-\gamma t)$, which violates the boson character of photons (it may be considered to be approximately valid if $t \ll 1/\gamma$). Since $\Delta\omega \ll \omega$ we neglect $\Delta\omega$ in the following.

From (7.11) and (7.13a, b) it follows that

$$\langle \hat{n}(t) \rangle = \langle \hat{a}^+(t)\, \hat{a}(t) \rangle = \exp(-\gamma t) \langle \hat{n}(0) \rangle + \langle n_d \rangle \left[1 - \exp(-\gamma t)\right], \tag{7.15}$$

the first term being the radiation term, the second one represents the reservoir contribution. Writing $\langle\ \rangle_R$ for the average over the reservoir variables only, we can show that

$$\frac{\mathrm{d}\langle \hat{n}(t) \rangle_R}{\mathrm{d}t} = -\gamma \langle \hat{n}(t) \rangle_R + \gamma \langle n_d \rangle, \tag{7.16}$$

which is satisfied by (7.15), and that the following important identity holds

$$\langle \hat{L}^+(t)\, \hat{a}(t) + \hat{a}^+(t)\, \hat{L}(t) \rangle_R = \gamma \langle n_d \rangle. \tag{7.17}$$

Having found the solution (7.11), we can calculate the normal quantum characteristic function (4.75), which can be written, in view of (2.57), in the forms

$$C_{\mathcal{N}}(\beta, t) = \mathrm{Tr}\left\{\hat{\varrho}(t) \exp\left[\beta \hat{a}^+(0)\right] \exp\left[-\beta^* \hat{a}(0)\right]\right\} =$$

$$= \mathrm{Tr}\,\{\varrho(0)\exp[\beta\hat{a}^+(t)]\exp[-\beta^*\hat{a}(t)]\} =$$
$$= \int \Phi_{\mathcal{N}}(\xi, 0)\exp(\beta u^*\xi^* - \beta^* u\xi)\,\mathrm{d}^2\xi \times$$
$$\times \prod_l \int \exp\left(-\frac{|\eta_l|^2}{\langle n_{dl}\rangle} + \beta w_l^*\eta_l^* - \beta^* w_l\eta_l\right)\frac{\mathrm{d}^2\eta_l}{\pi\langle n_{dl}\rangle} =$$
$$= \langle \exp[-\langle n_d\rangle[1 - \exp(-\gamma t)]\,|\beta|^2 + \beta\xi^*(t) - \beta^*\xi(t)]\rangle; \qquad (7.18)$$

we have used the Glauber–Sudarshan representation (4.123) of the density matrix $\hat{\varrho}(0) = \hat{\varrho}_{\mathrm{rad}}(0)\,\hat{\varrho}_{\mathrm{reserv}}(0)$ in terms of the coherent states $|\xi\rangle$ and $|\{\eta_j\}\rangle$ as eigenstates of $\hat{a}$ and $\hat{b}_j$. Since the reservoir represents the chaotic system,

$$\Phi_{\mathcal{N},\,\mathrm{reserv}}(\{\eta_l\}, 0) = \prod_l (\pi\langle n_{dl}\rangle)^{-1}\exp\left(-\frac{|\eta_l|^2}{\langle n_{dl}\rangle}\right); \qquad (7.19)$$

here $\langle n_{dl}\rangle \equiv \langle n_d\rangle$ is independent of l, the brackets in (7.18) denote the average over the initial complex field amplitude $\xi = \xi(0)$ for which the probability distribution is $\Phi_{\mathcal{N}}(\xi, 0)$ and

$$\xi(t) = u(t)\,\xi = \xi\exp(-\mathrm{i}\omega t - \gamma t/2). \qquad (7.20)$$

The quasi-distribution $\Phi_{\mathcal{N}}(\alpha, t)$ is derived using the Fourier transformation (4.77) together with the integral (4.179),

$$\Phi_{\mathcal{N}}(\alpha, t) =$$
$$= [\pi\langle n_d\rangle[1 - \exp(-\gamma t)]]^{-1}\left\langle \exp\left[-\frac{|\alpha - \xi(t)|^2}{\langle n_d\rangle[1 - \exp(-\gamma t)]}\right]\right\rangle. \qquad (7.21a)$$

The quasi-distribution $\Phi_{\mathcal{A}}(\alpha, t)$ related to antinormal ordering [cf. (4.74) and (4.71)] is

$$\Phi_{\mathcal{A}}(\alpha, t) = \frac{1}{\pi[\langle n_d\rangle[1 - \exp(-\gamma t)] + 1]} \times$$
$$\times \left\langle \exp\left[-\frac{|\alpha - \xi(t)|^2}{\langle n_d\rangle[1 - \exp(-\gamma t)] + 1}\right]\right\rangle. \qquad (7.21b)$$

Note that in the Schrödinger picture the density matrix is expressed in terms of the coherent states as

$$\hat{\varrho}(t) = \int \Phi_{\mathcal{N}}(\alpha, t)\,|\alpha\rangle\langle\alpha|\,\mathrm{d}^2\alpha. \qquad (7.22)$$

If the initial field is in the coherent state $|\xi\rangle$, the angle brackets in (7.21a, b) [and in (7.18)] can be omitted. The time dependence of $\Phi_{\mathcal{N}}(\alpha, t)$ is shown in Fig. 7.1 for this case. The amplitude of the initial coherent state $|\xi\rangle$ is continuously attenuated and the quantum noise $\langle n_d\rangle[1 - \exp(-\gamma t)]$ increases with time to the saturation value $\langle n_d\rangle$ for $t \to \infty$, when the coherent energy of the radiation mode is transferred to the reservoir. The photocount generating function, the photocount distribution and its factorial moments are given by (5.93a) with $\varkappa = 1$, (5.101b) and (5.104b) with $M = 1$, $\langle n_c\rangle = |\xi(t)|^2$ and $\langle n_{ch}\rangle = \langle n_d\rangle[1 - \exp(-\gamma t)]$.

As another example of the application of the Heisenberg approach we consider the interaction described by the hamiltonian (4.56). The Heisenberg equations are

$$i\hbar \frac{d\hat{a}_j(t)}{dt} = \sum_k f_{jk}(t)\,\hat{a}_k(t) + g_j(t), \tag{7.23}$$

which do not contain any creation operators. Their solution is expressed, using the perturbation technique [equation (2.62)], as

$$\hat{a}_j(t) = \sum_m U_{jm}(t)\,\hat{a}_m(0) - q_j(t), \tag{7.24}$$

where the time-dependent operator is

$$\hat{U}(t) = \hat{1} + \sum_{n=1}^{\infty}\left(-\frac{i}{\hbar}\right)^n \int_0^t dt'_1 \dots \int_0^{t'_{n-1}} dt'_n \hat{f}(t'_1) \dots \hat{f}(t'_n), \tag{7.25}$$

$\hat{f}$ being the matrix (f_{jk}) and

$$q_j(t) = \frac{i}{\hbar} \sum_{m,n} U_{jm}(t) \int_0^t U^{+}_{mn}(t')\, g_n(t')\, dt'. \tag{7.26}$$

Since the annihilation operators $\hat{a}_j(t)$ in (7.24) are dependent only on the initial annihilation operators $\hat{a}_m(0)$, the normal characteristic function is also in the normal form in the initial operators and the Fourier transform $\Phi_{\mathcal{N}}(\{\alpha_j\}, t)$ has the form of

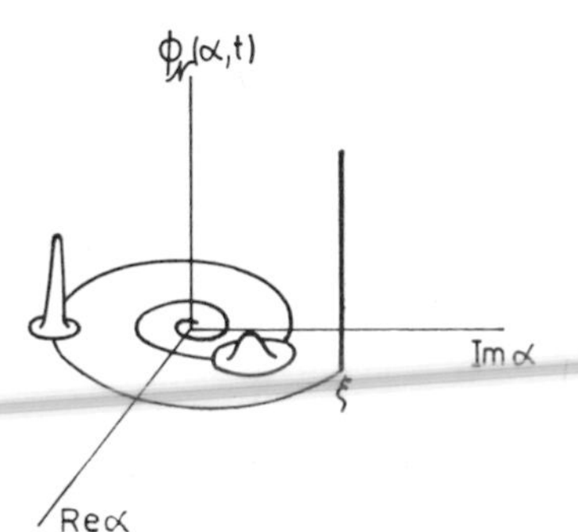

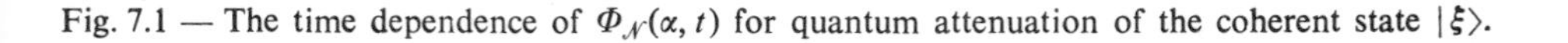

Fig. 7.1 — The time dependence of $\Phi_{\mathcal{N}}(\alpha, t)$ for quantum attenuation of the coherent state $|\xi\rangle$.

a δ-function, provided that the initial field is in a coherent state (including the vacuum state). Thus the hamiltonian (4.56) preserves the coherent state and generates the coherent state from the vacuum state. More generally, the initial statistics of the field remain unchanged by this interaction [Webber (1968, 1969), Robl (1967, 1968)], as is seen by substituting (7.24) in the multimode version of the normal characteristic function (4.75) and performing the Fourier transformation (4.77), giving

$$\Phi_{\mathcal{N}}(\{\alpha_j\}, t) = \Phi_{\mathcal{N}}(\{\sum_m U^{+}_{jm}(t)\,\alpha_m + Q_j(t)\}, 0), \tag{7.27}$$

where

$$Q_j(t) = \frac{i}{\hbar} \sum_m \int_0^t U_{jm}^+(t)\, g_m(t')\, dt'. \tag{7.28}$$

Thus, if the field is in the coherent state $| \{\beta_j\}\rangle$ for $t = 0$, $\Phi_{\mathcal{N}}(\{\alpha_j\}, 0) = \prod_j \delta(\alpha_j - \beta_j)$ and $\Phi_{\mathcal{N}}(\{\alpha_j\}, t) = \prod_j \delta(\alpha_j - \alpha_j(t))$, where $\alpha_j(t) = \sum_m U_{jm}(t)\, \beta_m - q_j(t)$; if $\beta_m = 0$, we see that classical currents indeed produce a field in a coherent state. Also, Gaussian distributions remain Gaussian for all times. Making use of (4.138), we arrive at

$$\begin{aligned} P_{\mathcal{N}}(W, t) &= \frac{1}{2\pi} \int_{-\infty}^{+\infty} \exp(-isW)\, ds \int \Phi_{\mathcal{N}}(\{\alpha_j\}, t) \exp(is \sum_j |\alpha_j|^2)\, d^2\{\alpha_j\} = \\ &= \frac{1}{2\pi} \int_{-\infty}^{+\infty} \exp(-isW)\, ds \int \Phi_{\mathcal{N}}(\{\gamma_j + Q_j(t)\}, 0) \times \\ &\times \exp(is \sum_j |\gamma_j|^2)\, d^2\{\gamma_j\}, \end{aligned} \tag{7.29}$$

where we have substituted $\gamma_j = \sum_m U_{jm}^+ \alpha_m$ (Det $\hat{U} = 1$, $\sum_j |\alpha_j|^2 = \sum_j |\gamma_j|^2$, $\hat{U}$ being unitary). If all currents are zero, $g_m(t) = 0$ and $P_{\mathcal{N}}(W, t) = P_{\mathcal{N}}(W, 0)$ and consequently the photocount distribution and its factorial moments are time independent. Applying (7.29) to the distribution (5.86), which is valid for the superposition of coherent and chaotic fields, we arrive again at the superposition of a coherent field with the complex amplitudes $\beta_j - Q_j(t)$ and a chaotic field with the mean number of photons $\langle n_{chj}\rangle$. Consequently, this interaction changes only the coherent component of the field and modulates it [Horák, Mišta and Peřina (1971c)].

Further investigations of the quantum statistics of the single-photon interaction of light with matter have been performed by Chandra and Prakash (1969) and Carusotto (1975).

7.2 The master equation and generalized Fokker—Planck equation approaches

An alternative equivalent approach developed in the framework of the Schrödinger picture can be based on the equation of motion for the density matrix,

$$i\hbar \frac{\partial \hat{\varrho}}{\partial t} = [\hat{H}_0 + \hat{H}_{\text{int}}, \hat{\varrho}]. \tag{7.30}$$

The Gaussian property of the reservoir, leading to the Markoffian $\hat{L}(t)$, as expressed in (7.10), makes it possible to perform two iterations when solving (7.30) for the reduced density matrix $\hat{\varrho}$ (with traced reservoir operators) and to calculate easily the related expressions $\langle[\hat{H}_{\text{int}}, \hat{\varrho}]\rangle_R = 0$ ($\langle \hat{b}_j\rangle = \langle \hat{b}_j^+\rangle = 0$) and $\langle[\hat{H}_{\text{int}}, [\hat{H}_{\text{int}}, \hat{\varrho}]]\rangle_R$; in this way we arrive at the master equation for the reduced density matrix

$$\frac{\partial \hat{\varrho}}{\partial t} = -i\omega([\hat{a}, \hat{\varrho}\hat{a}^+] + [\hat{a}^+, \hat{a}\hat{\varrho}]) + \frac{\gamma}{2}(\langle n_d\rangle + 1)([\hat{a}\hat{\varrho}, \hat{a}^+] +$$

$$+ [\hat{a}, \hat{\varrho}\hat{a}^+]) + \frac{\gamma}{2}\langle n_d\rangle ([\hat{a}^+\hat{\varrho}, \hat{a}] + [\hat{a}^+, \hat{\varrho}\hat{a}]) =$$

$$= \left(\frac{\gamma}{2} - i\omega\right)[\hat{a}, \hat{\varrho}\hat{a}^+] - \left(\frac{\gamma}{2} + i\omega\right)[\hat{a}^+, \hat{a}\hat{\varrho}] -$$

$$- \frac{\gamma\langle n_d\rangle}{2}\{[\hat{a}^+, [\hat{a}, \hat{\varrho}]] + [\hat{a}, [\hat{a}^+, \hat{\varrho}]]\}. \tag{7.31}$$

Performing antinormal ordering in this equation with the help of (4.32) and using (4.92), we obtain the generalized Fokker–Planck equation

$$\frac{\partial \Phi_{\mathcal{N}}}{\partial t} = -\frac{\partial}{\partial \alpha}(A_\alpha \Phi_{\mathcal{N}}) - \frac{\partial}{\partial \alpha^*}(A_\alpha^* \Phi_{\mathcal{N}}) + 2\frac{\partial^2}{\partial \alpha\, \partial \alpha^*}(D_{\alpha\alpha^*}\Phi_{\mathcal{N}}) =$$

$$= \left(\frac{\gamma}{2} + i\omega\right)\frac{\partial}{\partial \alpha}(\alpha \Phi_{\mathcal{N}}) + \left(\frac{\gamma}{2} - i\omega\right)\frac{\partial}{\partial \alpha^*}(\alpha^* \Phi_{\mathcal{N}}) +$$

$$+ \gamma\langle n_d\rangle \frac{\partial^2 \Phi_{\mathcal{N}}}{\partial \alpha\, \partial \alpha^*}. \tag{7.32a}$$

Using the Fourier transformation (4.75) we have the equation for the normal characteristic function

$$\frac{\partial C_{\mathcal{N}}}{\partial t} = -\left(\frac{\gamma}{2} - i\omega\right)\beta \frac{\partial C_{\mathcal{N}}}{\partial \beta} - \left(\frac{\gamma}{2} + i\omega\right)\beta^* \frac{\partial C_{\mathcal{N}}}{\partial \beta^*} -$$

$$- \gamma\langle n_d\rangle\, |\beta|^2\, C_{\mathcal{N}}. \tag{7.32b}$$

This equation can be solved by the standard method of characteristics in the form (7.18) and $\Phi_{\mathcal{N}}$, given by (7.21a), satisfies the Fokker–Planck equation (7.32a). Thus both approaches are fully equivalent.

The Fokker–Planck equation (7.32a) directly corresponds to the Heisenberg–Langevin equation (7.6), because the drift vector A_α and the diffusion constant $D_{\alpha\alpha^*}$ are determined by

$$A_\alpha = \left\langle \frac{d\alpha}{dt} \right\rangle_R = -\left(\frac{\gamma}{2} + i\omega\right)\alpha,$$

$$2D_{\alpha\alpha^*} = \left\langle \frac{d\alpha^*\alpha}{dt} \right\rangle_R - \left\langle \alpha^* \frac{d\alpha}{dt} \right\rangle_R - \left\langle \frac{d\alpha^*}{dt}\alpha \right\rangle_R = \gamma\langle n_d\rangle, \tag{7.33}$$

whereas the diffusion constants such as $D_{\alpha\alpha}$ and $D_{\alpha^*\alpha^*}$ are zero. If the antinormal order is adopted, $2D_{\alpha^*\alpha} = \gamma(\langle n_d\rangle + 1)$.

Further details concerning the interaction of light with reservoirs can be found in papers by Senitzky (1967a, b, 1968, 1969, 1973), Mollow (1968b), Peřina, Peřinová, Mišta and Horák (1974), among others. An alternative quantum theory of damping, involving quadratic terms in $\hat{a}$ and $\hat{a}^+$ in the hamiltonian, has been considered by Colegrave and Abdalla (1981, 1983). Also, the quantum statistical properties of a randomly modulated harmonic oscillator have been investigated [Crosignani, DiPorto and Solimeno (1969), Mollow (1970)].

7.3 The interaction of radiation with the atomic system of a nonlinear medium

The interaction of radiation with the atomic system of a nonlinear medium can be described by the interaction hamiltonian [McNeil and Walls (1974)]

$$\hat{H}_{\text{int}} = \sum_j \hbar\mu^{(n)}(\mathbf{x}_j)\,\hat{c}_{2j}^{+}\hat{c}_{1j}\,\hat{O}^{(n)} + \text{h.c.},$$

$$\hat{O}^{(n)} = \prod_{l=1}^{m} \hat{a}_l^{+} \prod_{k=m+1}^{n} \hat{a}_k, \tag{7.34}$$

where h.c. means the Hermitian conjugate terms, $\hat{a}_l$ and $\hat{a}_l^{+}$ are again the annihilation and creation operators of a photon in the lth radiation mode, $\hat{c}_{\lambda j}$ and $\hat{c}_{\lambda j}^{+}$ are the annihilation and creation operators of the λth level of the jth atom and $\mu^{(n)}(\mathbf{x}_j)$ is the coupling constant proportional to the n-photon transition matrix element. This hamiltonian describes m emission and $n-m$ absorption events during one atomic transition. In quantum optics we are mostly interested in the properties of radiation, and the atomic variables may be eliminated in the same way as the above reservoir variables; thus the Heisenberg–Langevin equations or the generalized Fokker–Planck equation can be derived. If virtual electronic transitions are taken into account and real transitions may be neglected [Graham (1970)], an effective hamiltonian $\hat{H}_{\text{int,eff}}$ may be derived with the atomic variables eliminated and the nonlinear optical process is described by the Heisenberg–Langevin equations

$$\mathrm{i}\hbar \frac{\mathrm{d}\hat{a}_j}{\mathrm{d}t} = [\hat{a}_j, \hat{H}_0 + \hat{H}_{\text{int,eff}}] + \mathrm{i}\hbar\hat{L}_j, \tag{7.35}$$

where $\hat{L}_j$ are the Langevin forces arising in the elimination procedure, usually represented by Markoffian processes. If also real transitions are involved, the effective hamiltonian cannot be obtained; however, one can always perform the elimination of atomic and reservoir variables in order to derive directly the Heisenberg–Langevin equations or the master quation, including the radiation variables only. In the Markoff approximation, the master equation for the reduced density matrix $\hat{\varrho}$ can be derived in the same way as (7.31) in the form [Shen (1967), Haken (1970a, b), Agarwal (1973)]

$$\frac{\partial\hat{\varrho}}{\partial t} = K\{N_1([\hat{O}^{(n)}\hat{\varrho}, \hat{O}^{(n)+}] + [\hat{O}^{(n)}, \hat{\varrho}\hat{O}^{(n)+}]) - $$
$$- N_2([\hat{O}^{(n)}, \hat{O}^{(n)+}\hat{\varrho}] + [\hat{\varrho}\hat{O}^{(n)}, \hat{O}^{(n)+}])\}, \tag{7.36}$$

where K is a constant related to $\mu^{(n)}$ and N_1 and N_2 are the occupation numbers of the atomic levels 1 and 2 respectively, under the condition of thermal equilibrium. If $\hat{O}^{(n)} = \hat{a}$ ($m = 0$, $n = 1$) and $K = \gamma/2$, $N_1 = \langle n_d \rangle + 1$, $N_2 = \langle n_d \rangle$, we just arrive at the interaction part of the Fokker–Planck equation (7.31) (with the exp $(-\mathrm{i}\omega t)$-dependence eliminated) for the damped harmonic oscillator. The same procedure as above leads to the generalized Fokker–Planck equation

$$\frac{\partial \Phi}{\partial t} = \sum_j \left[-\frac{\partial}{\partial \alpha_j}(A_j \Phi) \right] + \sum_{i,j} \frac{\partial^2}{\partial \alpha_i \partial \alpha_j}(D_{ij}\Phi), \tag{7.37}$$

where for instance $\alpha_1 = \alpha$, $\alpha_2 = \alpha^*$, etc., and the drift vectors $A_j(A_\alpha, A_{\alpha^*}$, etc.) and the diffusion constants $D_{ij}(D_{\alpha\alpha^*}, D_{\alpha\alpha}, D_{\alpha^*\alpha^*}$, etc.) are determined as

$$A_j = \left\langle \frac{\mathrm{d}\hat{a}_j}{\mathrm{d}t} \right\rangle_{R,\,\mathrm{atoms}},$$

$$2D_{ij} = \left\langle \frac{\mathrm{d}\hat{a}_i \hat{a}_j}{\mathrm{d}t} \right\rangle_{R,\,\mathrm{atoms}} - \left\langle \hat{a}_i \frac{\mathrm{d}\hat{a}_j}{\mathrm{d}t} \right\rangle_{R,\,\mathrm{atoms}} - \left\langle \frac{\mathrm{d}\hat{a}_i}{\mathrm{d}t} \hat{a}_j \right\rangle_{R,\,\mathrm{atoms}}. \tag{7.38}$$

This follows after taking into account that for any operator $\hat{M}$

$$\left\langle \frac{\mathrm{d}\hat{M}}{\mathrm{d}t} \right\rangle \doteq \left\langle \sum_j \frac{\partial \hat{M}}{\partial \hat{a}_j} \frac{\Delta \hat{a}_j}{\Delta t} + \sum_{i>j} \frac{\partial^2 \hat{M}}{\partial \hat{a}_i \partial \hat{a}_j} \frac{\Delta \hat{a}_i \Delta \hat{a}_j}{\Delta t} + \sum_i \frac{1}{2} \frac{\partial^2 \hat{M}}{\partial \hat{a}_i^2} \frac{(\Delta \hat{a}_i)^2}{\Delta t} \right\rangle \tag{7.39}$$

and integrating by parts, then we have (7.37) with

$$A_j = \left\langle \frac{\Delta \hat{a}_j}{\Delta t} \right\rangle, 2D_{ij} = \left\langle \frac{\Delta \hat{a}_i \Delta \hat{a}_j}{\Delta t} \right\rangle, \tag{7.40}$$

giving (7.38), as $\Delta(\hat{a}_i \hat{a}_j) = \hat{a}_i(t + \Delta t)\, \hat{a}_j(t + \Delta t) - \hat{a}_i(t)\, \hat{a}_j(t) = \Delta \hat{a}_i \Delta \hat{a}_j + (\Delta \hat{a}_i)\, \hat{a}_j + \hat{a}_i \Delta \hat{a}_j$. For instance, the most general one-mode generalized Fokker–Planck equation has the form

$$\frac{\partial \Phi}{\partial t} = -\frac{\partial}{\partial \alpha}(A_\alpha \Phi) - \frac{\partial}{\partial \alpha^*}(A_{\alpha^*}\Phi) + \frac{\partial^2}{\partial \alpha^2}(D_{\alpha\alpha}\Phi) + \frac{\partial^2}{\partial \alpha^{*2}}(D_{\alpha^*\alpha^*}\Phi) + 2\frac{\partial^2}{\partial \alpha\, \partial \alpha^*}(D_{\alpha^*\alpha}\Phi), \tag{7.41}$$

where $A_\alpha = \langle \Delta \hat{a}/\Delta t \rangle$, $A_{\alpha^*} = \langle \Delta \hat{a}^+/\Delta t \rangle$, $2D_{\alpha\alpha} = \langle (\Delta \hat{a})^2/\Delta t \rangle$, $2D_{\alpha^*\alpha^*} = \langle (\Delta \hat{a}^+)^2/\Delta t \rangle$, $2D_{\alpha^*\alpha} = \langle \Delta \hat{a}^+ \Delta \hat{a}/\Delta t \rangle \neq 2D_{\alpha\alpha^*} = \langle \Delta \hat{a} \Delta \hat{a}^+/\Delta t \rangle$.

If antinormal ordering is performed in (7.36) and the multimode version of (4.92), $\Phi_{\mathscr{N}}(\{\alpha_j\}) = \varrho^{(\mathscr{A})}(\{\hat{a}_j^+ \to \alpha_j^*\}, \{\hat{a}_j \to \alpha_j\})/\pi^M$, M being the number of modes, is used, the generalized Fokker–Planck equation for $\Phi_{\mathscr{N}}$ is obtained; if the normal ordering is carried out and the multimode version of (4.94b), $\Phi_{\mathscr{A}}(\{\alpha_j\}) = \varrho^{(\mathscr{N})}(\{\hat{a}_j^+ \to \alpha_j^*\}, \{\hat{a}_j \to \alpha_j\})/\pi^M$ is used, we derive the generalized Fokker–Planck equation for $\Phi_{\mathscr{A}}$. These procedures of deriving the generalized Fokker–Planck equations will be demonstrated explicitly in the following chapters. In general we prefer to use the quasi-distribution $\Phi_{\mathscr{A}}$, which is always well behaved, whereas the Glauber–Sudarshan quasi-distribution $\Phi_{\mathscr{N}}$ does not usually exist in interaction problems. General procedures for obtaining the equations of motion for $\Phi_{\mathscr{N}}$ for general classes of hamiltonians have been developed by Crosignani, Ganiel, Solimeno and Di Porto (1971a, b) and Graham (1973). An interesting use of the Fokker–Planck equation has been suggested by Arecchi and Politi (1980) to measure transient fluctuations, by fitting the experimental and theoretical moments (see Figs. 7.2 and 7.3).

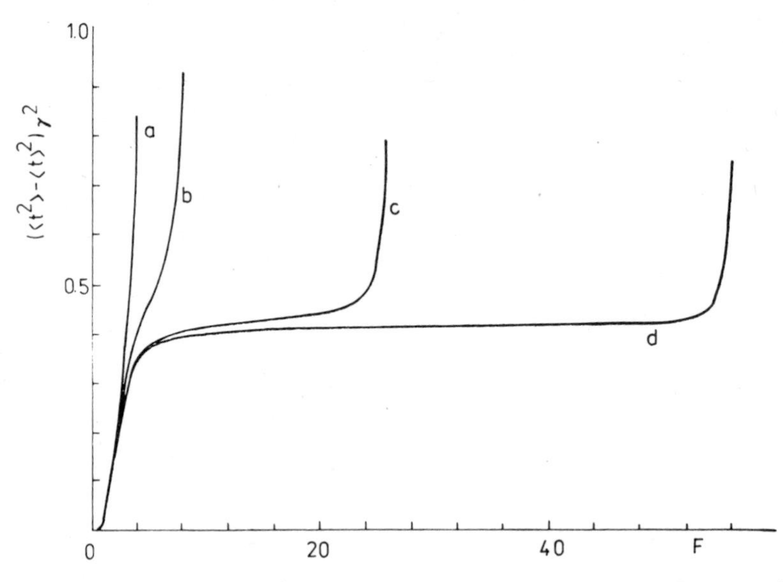

Fig. 7.2 — Model of a laser transient, $\dot{\alpha} = \gamma\alpha - \beta \mid \alpha \mid^2 \alpha + L$, $\langle L^*(0)\, L(t)\rangle = D\delta(t)$, $A = \gamma/(\beta D)^{1/2}$. The variance in stochastic crossing time is plotted against the threshold. Curves $c - d$ correspond to $A = 4$, 8, 25.8 and 55.9 respectively; F represents the threshold amplitude normalized to $(D/\gamma)^{1/2}$ (after Arecchi, Politi and Ulivi, 1982, Phys. Lett. **87A**, 333).

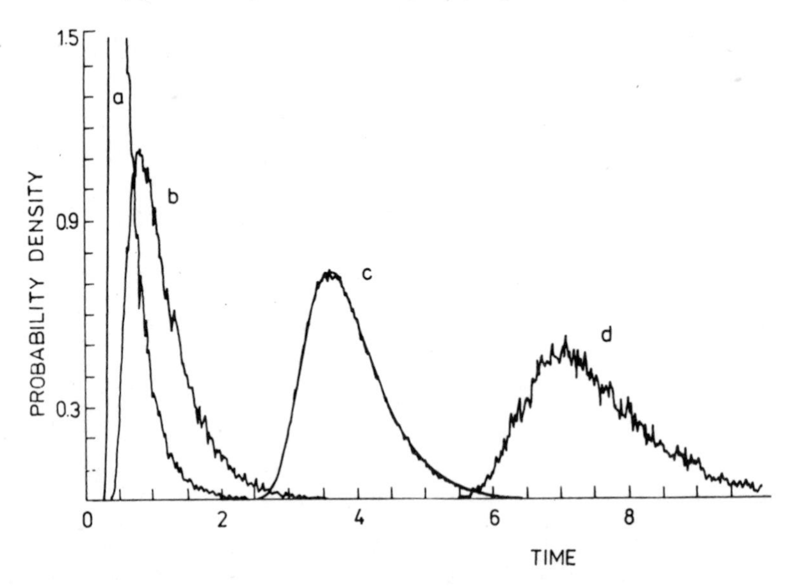

Fig. 7.3 — Experimental time probability distributions versus time normalized to $1/\gamma$ ($\gamma = 807\ \mathrm{s}^{-1}$, $\beta = 337\ \mathrm{s}^{-1}/\mathrm{V}^2$, D is so that $A = 152$). Curves $a - d$ correspond to the threshold values 0.016, 0.028, 0.436 and 1.550 V respectively. The wavy lines are experimental records, the smooth lines a and c represent shifted theoretical values $2 \exp\left[-2(t - t_0) - e^{-2(t - t_0)}\right]$ from the asymptotic approximation (Haake, Haus and Glauber (1981)). The initial and final regions strongly diverge from that approximation and only in the region which is flat in Fig. 7.2 is good agreement obtained (curve c) (after Arecchi, Politi and Ulivi, 1982, Phys. Lett. **87A**, 333).

Finally note that an alternative method for the investigation of the interaction of light and matter can be developed in terms of the Fock states. The master equation (7.36) may then be solved recursively by applying the Laplace transformation. Such a method has been developed by Scully and Lamb for the laser [see Sargent and Scully (1972)] and has been employed in two- and multiphoton absorption and Raman scattering processes [see e.g. Loudon (1973), Simaan and Loudon (1975, 1978), Simaan (1975, 1978), Gupta and Mohanty (1980), Mohanty and Gupta (1981a – c)]; in second- and third-harmonic generation [Nayak and Mohanty (1977), Nayak (1980, 1982)]; and in two-photon emission [Nayak and Mohanty (1979), Gupta and Mohanty (1981)]. However, the phase information about the field is definitely lost in such formulations, unless all the off-diagonal Fock matrix elements of the density matrix are determined. Therefore we prefer to use the coherent-state technique and the generalized Fokker – Planck equation approach.

A quantum theory of propagation of the electromagnetic field, particularly for strong fields, has been developed by Gordov and Tvorogov (1978, 1980).

The presented brief review of the Heisenberg – Langevin, master equation and the Fokker – Planck equation methods, is sufficient for our purpose. More extended reviews are available in the literature [Lax (1968a), Haken (1970b), Scully and Whitney (1972), Agarwal (1973), Haake (1973), Louisell (1973), Sargent, Scully and Lamb (1974), Davies (1976)].

CHAPTER 8

QUANTUM STATISTICS OF RADIATION IN RANDOM MEDIA

We can now apply the general methods developed above to the investigation of the quantum statistical properties of radiation interacting with random media, particularly with Gaussian and log-normal media, such as the turbulent atmosphere in certain regimes of fluctuation. We will not go into the nature and the structure of turbulence, which has been intesively investigated recently [e.g. Haken (1978), Sinaya and Shilnikov (1981), Jansen and Oberman (1981)], however we present a theory treating radiation in a fully quantal way, including self-radiation of media.

8.1 Phenomenological description of propagation of radiation through turbulent atmosphere and Gaussian media

Before reviewing the quantum dynamical description of the quantum statistics of radiation propagating through random media, we discuss some results obtained for the turbulent atmosphere [Diament and Teich (1970b, 1971), Solimeno, Corti and Nicoletti (1970), Rosenberg and Teich (1972), Peřina and Peřinová (1972), Lachs and Laxpati (1973)] and Gaussian media [Bertolotti, Crosignani and Di Porto (1970), Peřina and Peřinová (1972)], based on a modification of the Mandel photodetection equation (3.85). Namely, if the effect of a random medium is considered as a modulation of the intensity I_0 of the incident radiation, then the resulting intensity is $I = \mathscr{K} I_0$, where $\mathscr{K}$ is a fluctuating quantity characteristic of the random medium. This means that the fluctuations of $\mathscr{K}$ must be much slower than those of I_0 and we can then write for the photocount distribution

$$p(n) = \iint_0^\infty \frac{(\mathscr{K} W_0)^n}{n!} \exp(-\mathscr{K} W_0) P_{\mathscr{N}}(W_0) \bar{P}(\mathscr{K}) \,\mathrm{d}W_0 \,\mathrm{d}\mathscr{K} =$$
$$= \int_0^\infty p_0(n, \mathscr{K}) \bar{P}(\mathscr{K}) \,\mathrm{d}\mathscr{K}, \tag{8.1}$$

where $P_{\mathscr{N}}(W_0)$ is the probability distribution of the integrated intensity W_0 of the incident radiation, $p_0(n, \mathscr{K})$ is the photocount distribution of the incident radiation, its mean photon number being considered as a fluctuating quantity $(\langle n_c \rangle + \langle n_{ch} \rangle) \mathscr{K}$, where $\langle n_c \rangle$ and $\langle n_{ch} \rangle$ are the mean numbers of incident coherent and chaotic photons, respectively, and $\bar{P}(\mathscr{K})$ is the probability distribution characterizing the random medium. If the incident radiation is coherent, then the outgoing radiation just obeys the statistics of the random medium. If $T \ll \tau_c$, then $P_{\mathscr{N}}(W_0)$ is simply determined by the probability distribution of the intensity I_0.

In the turbulent atmosphere $\bar{P}(\mathscr{K})$ is usually assumed to be in the form of the log-normal distribution [e.g. Diament and Teich (1970b)]

$$\bar{P}(\mathscr{K}) = \frac{1}{(2\pi)^{1/2}\,\sigma\mathscr{K}} \exp\left[-\frac{(\log \mathscr{K} + \sigma^2/2)^2}{2\sigma^2} \right] \tag{8.2}$$

and for the moments one has

$$\langle \mathscr{K}^n \rangle = \langle \exp(n\varphi) \rangle = \exp\left[\sigma^2 n(n-1)/2\right], \qquad n = 0, \pm 1, \ldots, \tag{8.3}$$

where σ is the standard deviation of $\varphi = \log \mathscr{K}$, whose saturation value in the turbulent atmosphere is about $\sigma = 3/2$ [Tatarskii (1967), Strohbehn (1971)] and $\langle \mathscr{K} \rangle = \langle \exp \varphi \rangle = 1$, which expresses the energy conservation law. We assume this probability distribution even if only some regimes of fluctuations exist in which this distribution holds [De Wolf (1973a–c), Furutsu (1972), Tatarskii (1967)]. However here we will follow the quantum description, which is microscopial as regards radiation, whereas the medium is described macroscopically by matrix elements of the fluctuating permittivity irrespective of the structure of the turbulence. Classical treatments of the structure of turbulence are rather complex, and the literature is extensive [Tatarskii (1967, 1970), Beran and Ho (1969), Ho (1969), Strohbehn (1971, 1978), Furutsu and Furuhama (1973), Klyackin and Tatarskii (1973), Klyackin (1975), Rytov, Kravcov and Tatarskii (1978), Semenov and Arsenyan (1978), Furutsu (1976), Tatarskii and Zavorotnyi (1980), Leader (1981), Hill and Clifford (1981), Clifford and Hill (1981), Gurvich et al. (1979)].

The photocount distribution (8.1) has been calculated using the log-normal distribution (8.2) for the initial photocount distribution p_0 as given for the superposition of coherent and chaotic fields in (5.101a, b) with M determined by (5.119) ($P = 1$, $\varphi = 0$), employing the saddle-point method [Diament and Teich (1970b), Peřina and Peřinová (1972), Saleh (1978)]. In this manner one can obtain

$$p(n) = p_0(n, S)\left[1 - \sigma^2 q_2(n, S)\right]^{-1/2} \exp\left[-\frac{1}{2}\sigma^2 q_1^2(n, S) \right], \tag{8.4}$$

where S is the position of the stationary point satisfying the equation

$$\log S = \log \langle n \rangle - \frac{1}{2}\sigma^2 + \sigma^2 q_1(n, S), \tag{8.5}$$

where $\langle n \rangle = \langle n_c \rangle + \langle n_{ch} \rangle$ and

$$q_m(n, S) = \left. \frac{\partial^m \log p_0(n, K)}{\partial(\log K)^m} \right|_{K=S}, \tag{8.6}$$

$K = \mathscr{K}\langle n \rangle$. We write (5.101a) in the form

$$p_0(n, K) = \\ = \left[M(1 + y)\,H\right]^M K^n \exp\left[-\frac{yK}{1 + y}\; 1 - \varkappa^2 KH) \right] \hat{F}\{L_j^{M-1}(x)\}, \tag{8.7}$$

where the operator $\hat{F}\{...\}$ is defined as

$$\hat{F}\{...\} = \sum_{j=0}^{n} \frac{1}{(n-j)!\,\Gamma(j+M)} \left[\frac{y(1-\varkappa^2)}{1+y}\right]^{n-j} H^j\{...\} \tag{8.8}$$

and

$$y = \frac{\langle n_c \rangle}{\langle n_{ch} \rangle}, \qquad K = \mathscr{K}(\langle n_c \rangle + \langle n_{ch} \rangle), \qquad x = -y(1+y)\varkappa^2 HM^2,$$
$$H = [K + (1+y)M]^{-1}. \tag{8.9}$$

(The photoefficiency η is assumed to be unity again for simplicity.) From (8.6) it follows that

$$q_1 = n - SHM - \frac{yS}{1+y}(1 - \varkappa^2 SH) + y\varkappa^2 S^2 H^2 M + S\bar{A},$$
$$q_2 = -(1+y)SH^2M^2 + \frac{yS}{1+y}(\varkappa^2 - 1) -$$
$$- xSH^2[S - (1+y)M] + S\bar{A} - S^2\bar{A}^2 + S^2\bar{B}, \tag{8.10}$$

where

$$\bar{A} = \frac{\hat{F}\{H[-jL_j^{M-1}(x) + xL_{j-1}^{M}(x)]\}}{\hat{F}\{L_j^{M-1}(x)\}}, \tag{8.11}$$
$$\bar{B} = \frac{\hat{F}\{H^2[j(j+1)L_j^{M-1}(x) - 2(j+1)xL_{j-1}^{M}(x) + x^2L_{j-2}^{M+1}(x)]\}}{\hat{F}\{L_j^{M-1}(x)\}}.$$

In (8.10), H is given as in (8.9) with $K \to S$. The above equations can be simplified if $\varkappa = 1$ (the frequencies ω_c and $\bar{\omega}$ coincide). In this case

$$q_1 = [n(1+y) - S]HM + xSHA,$$
$$q_2 = -(n+M)(1+y)H^2SM - xSH^2[S - (1+y)M]A + x^2S^2H^2B, \tag{8.12}$$

where

$$A = 1 + \frac{L_{n-1}^{M}(x)}{L_n^{M-1}(x)}, \qquad B = \frac{L_{n-1}^{M+1}(x)}{L_n^{M-1}(x)} - \left[\frac{L_{n-1}^{M}(x)}{L_n^{M-1}(x)}\right]^2. \tag{8.13}$$

For $n = 0$ and 1 the quantities A and B are determined by (8.13) where zero stands for undefined polynomials (with negative lower suffix). For $M = 1$ (a narrow-band chaotic field and $T \ll \tau_c$) we have the expressions derived by Diament and Teich (1971); x here is obtained from that given in (8.9) for $\varkappa = 1$. In (8.13) also the identity $(d/dx)\,L_n^M(x) = -L_{n-1}^{M+1}(x)$ has been applied.

For purely chaotic light ($y = 0$)

$$p_0(n, K) = \frac{\Gamma(n+M)}{n!\,\Gamma(M)} \frac{M^M K^n}{(K+M)^{n+M}}, \tag{8.14}$$

and

$$q_1 = \frac{n-S}{S+M}M, \qquad q_2 = -\frac{n+M}{(S+M)^2}SM; \tag{8.15}$$

here $M = 1/\mathscr{J}_1$. For $M = 1$ we obtain the expressions derived by Diament and Teich (1970b) for chaotic narrow-band light ($T \ll \tau_c$). For $y \to \infty$ ($\langle n_{ch} \rangle \to 0$),

$$p_0(n, K) = \frac{K^n}{n!} \exp(-K), \tag{8.16}$$

$q_1(n, S) = n - S, q_2(n, S) = -S$, and we have the case of initially coherent radiation.

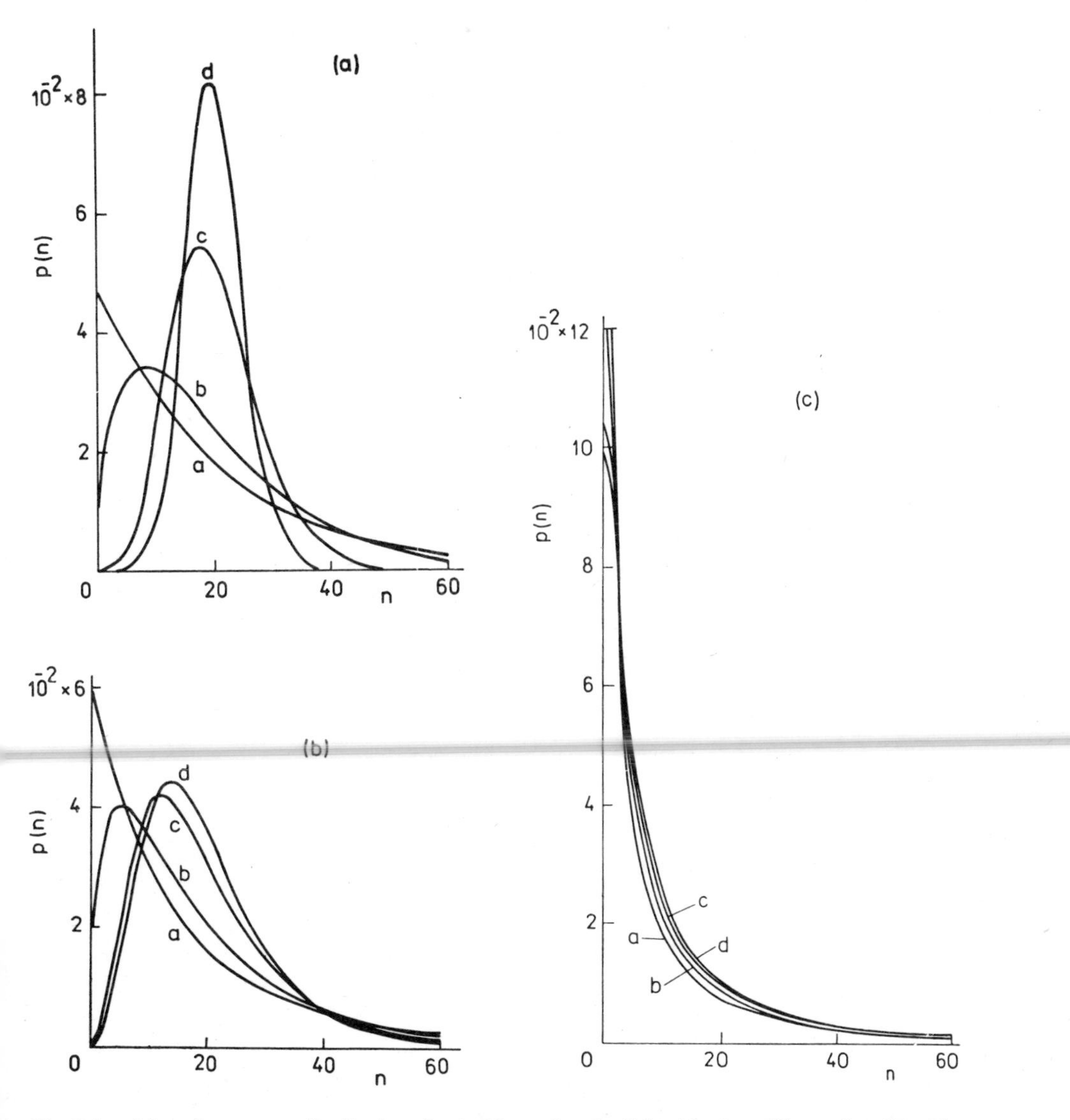

Fig. 8.1 — The photocount distribution for incident chaotic light ($\langle n_{ch} \rangle = 20$) passing through a turbulent atmosphere for $\gamma = 0.01$, 1, 10 and 100 (curves *a*—*d* respectively) and (a) $\sigma = 0$ (no fluctuations), (b) $\sigma = 1/2$ (intermediate level of fluctuations) and (c) $\sigma = 3/2$ (saturated fluctuations) (after Peřina and Peřinová, 1972, Czech. J. Phys. **B22**, 1085).

The factorial moments can be derived in closed form using (8.1) and (8.3):

$$\langle W^k\rangle_{\mathcal{N}} = \exp\left[\frac{1}{2}\sigma^2 k(k-1)\right]\frac{\langle n\rangle^k}{(1+y)^k}\sum_{j=0}^{k}\frac{k!}{(k-j)!\,\Gamma(j+M)}\times$$
$$\times[y(1-\varkappa^2)]^{k-j}M^{-j}L_j^{M-1}(-y\varkappa^2 M), \tag{8.17a}$$

and for $\varkappa = 1$ ($\omega_c = \bar{\omega}$)

$$\langle W^k\rangle_{\mathcal{N}} = \exp\left[\frac{1}{2}\sigma^2 k(k-1)\right]\frac{k!}{M^k\Gamma(k+M)}\frac{\langle n\rangle^k}{(1+y)^k}L_k^{M-1}(-yM); \tag{8.17b}$$

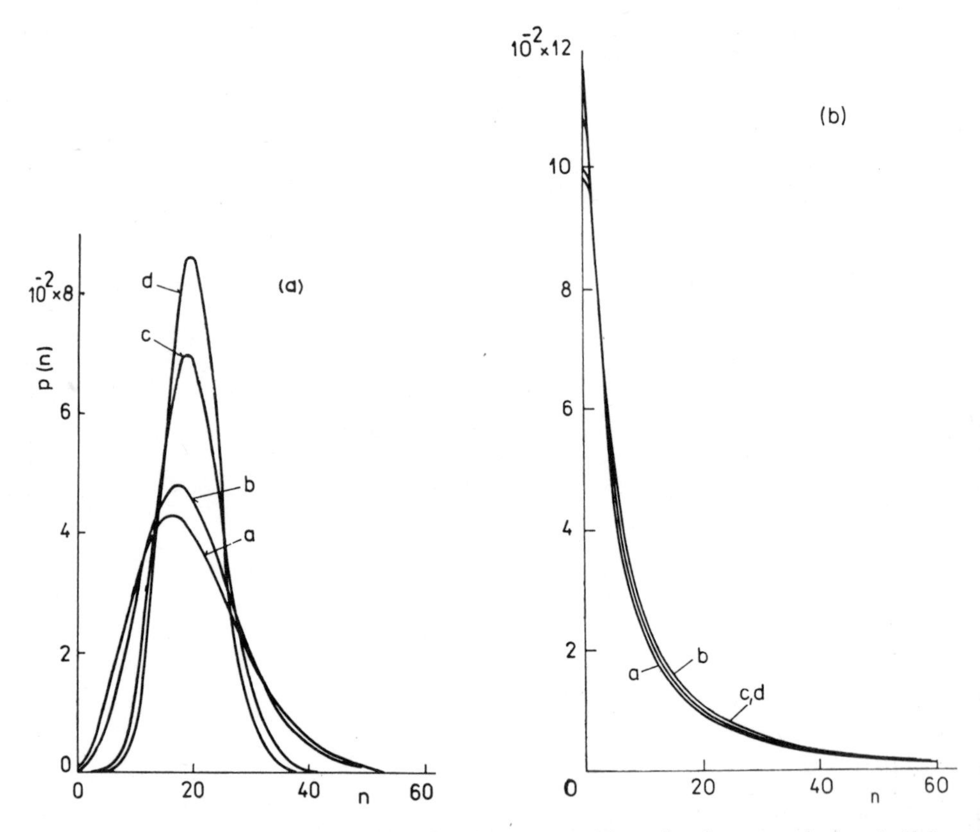

Fig. 8.2 — The photocount distribution for the superposition of coherent and chaotic light ($\langle n_c\rangle + \langle n_{ch}\rangle = 20$, $\langle n_c\rangle/\langle n_{ch}\rangle = 9$, γ as in Fig. 8.1) for (a) $\sigma = 0$, (b) $\sigma = 3/2$ (after Peřina and Peřinová, 1972, Czech. J. Phys. **B22**, 1085).

for $k = 1$, $\langle W\rangle_{\mathcal{N}} = \langle n\rangle$. The photocount distribution obtained from (8.4) is shown in Fig. 8.1 for chaotic light ($y = 0$, $\langle n\rangle = 20$) and in Fig. 8.2 for the superposition of coherent and chaotic fields ($y = 9$, $\langle n\rangle = 20$), for various $\gamma = \Delta\nu T$. In fact Fig. 8.1a represents the Mandel–Rice formula (3.140). We generally find that an increase in the level of turbulence broadens the photocount distribution (the photons are more bunched) and shifts the peak of the photocount distribution to lower n (the same behaviour of the photocount distribution of radiation in a random medium is demonstrated in Figs. 8.4 and 8.5b, for incident radiation in the Fock state and in the coherent state). For the saturation value of σ the number $n = 0$ is the most probable one and the spectral information is practically lost in the photocount distribution. For the factorial moments we have, using (8.3),

$$\langle W^k \rangle_{\mathcal{N}} = \exp\left[\sigma^2 k(k-1)/2\right] \langle W_0^k \rangle_{\mathcal{N}}, \tag{8.18}$$

where $\langle W_0^k \rangle_{\mathcal{N}}$ is the factorial moment of the incident radiation. The normalized factorial moments $\langle W^k \rangle_{\mathcal{N}} / \langle W \rangle_{\mathcal{N}}^k$ are shown in Fig. 8.3 as functions of γ for chaotic

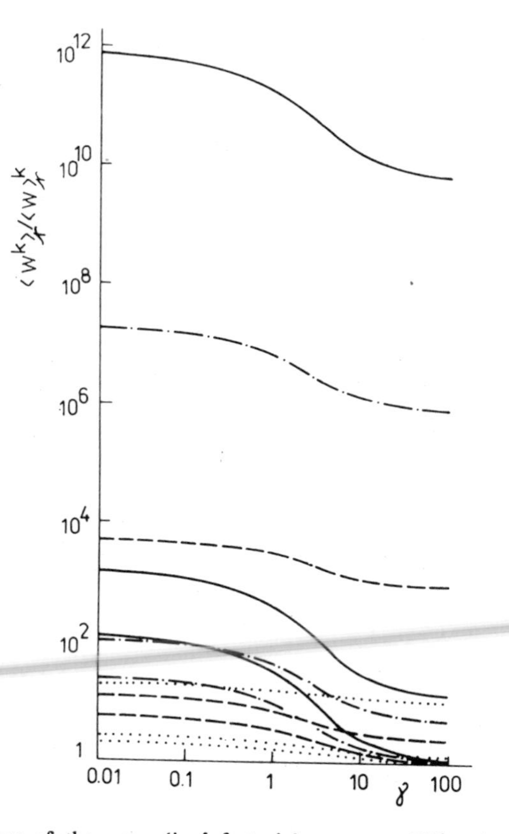

Fig. 8.3 — The dependence of the normalized factorial moments $\langle W^k \rangle_{\mathcal{N}} / \langle W \rangle_{\mathcal{N}}^k$ on γ for $k = 2$ (dotted curves), $k = 3$ (dashed curves), $k = 4$ (dot-dashed curves) and $k = 5$ (solid curves), chaotic light ($\langle n_{ch} \rangle = 20$) and $\sigma = 0$, 1/2 and 3/2 (after Peřina and Peřinová, 1972, Czech. J. Phys. **B22**, 1085).

light, based on (8.17b). The curves corresponding to the moments of the same order are shown successively (from the bottom of the figure) with increasing level of turbulence σ. The curves for the superposition of coherent and chaotic light lie between the curves shown and the asymptotic lines $\langle W^k \rangle_{\mathcal{N}} / \langle W \rangle_{\mathcal{N}}^k = \exp\left[\sigma^2 k(k-1)/2\right]$, corresponding to coherent incident light.

In the same way we can treat the scattering of light by a Gaussian medium where [Bertolotti, Crosignani and Di Porto (1970)]

$$p(n) = \int_0^\infty p_0(n, n'_c, n'_{ch})\, \bar{P}(\mathscr{K})\, d\mathscr{K}, \tag{8.19}$$

with $n'_c = \langle n_c \rangle \mathscr{K}$, $n'_{ch} = \langle n_{ch} \rangle \mathscr{K}$ and [Mandel (1969)]

$$\bar{P}(\mathscr{K}) = \frac{1}{\langle \mathscr{K} \rangle} \exp\left(-\frac{\mathscr{K}}{\langle \mathscr{K} \rangle}\right). \tag{8.20}$$

The photocount distribution $p(n)$ is expressed in terms of the Whittaker functions $W_{k,l}$ [Whittaker and Watson (1940)] in the rather complicated form [Bertolotti, Crosignani and Di Porto (1970), Peřina and Peřinová (1972)]

$$p(n) = \frac{1}{A}\left(\frac{MA}{\langle n'_{ch} \rangle}\right)^{M/2} \exp\left[\frac{M}{\langle n'_{ch} \rangle}\left(\frac{1}{2}A - \varkappa^2 \langle n'_c \rangle\right)\right] \times$$
$$\times \sum_{i=0}^{n} \binom{n}{i} \left[\frac{M \langle n'_c \rangle^2 (1 - \varkappa^2)}{\langle n'_{ch} \rangle A}\right]^{(n-i)/2} \sum_{k=0}^{\infty} \frac{\Gamma(k + i + M)}{k!\, \Gamma(k + M)} \times$$
$$\times \left[\frac{\varkappa^2 M \langle n'_c \rangle}{\langle n'_{ch} \rangle} \left(\frac{MA}{\langle n'_{ch} \rangle}\right)^{1/2}\right]^k W_{-\frac{k+i+M+n}{2},\, \frac{k+i+M-n-1}{2}} \left(\frac{MA}{\langle n'_{ch} \rangle}\right), \tag{8.21}$$

where $\langle n'_{ch} \rangle = \langle n_{ch} \rangle \langle \mathscr{K} \rangle$, $\langle n'_c \rangle = \langle n_c \rangle \langle \mathscr{K} \rangle$, $A = \langle n_c \rangle (1 - \varkappa^2) + 1$. The factorial moments have the simple form

$$\langle W^k \rangle_{\mathscr{N}} = k!\, \langle \mathscr{K}^k \rangle \langle W_0^k \rangle_{\mathscr{N}}, \tag{8.22}$$

where $\langle W_0^k \rangle_{\mathscr{N}}$ is given, for instance, by (5.104a).

For incident coherent radiation, $\langle n_{ch} \rangle = 0$ and

$$p(n) = \frac{\langle n'_c \rangle^n}{(1 + \langle n'_c \rangle)^{1+n}}, \tag{8.23a}$$

$$\langle W^k \rangle_{\mathscr{N}} = k!\, \langle n'_c \rangle^k, \tag{8.23b}$$

so the initially coherent radiation is transformed by the Gaussian medium into chaotic radiation.

For chaotic incident light $\langle n_c \rangle = 0$ and

$$p(n) = \frac{n!}{\langle n'_{ch} \rangle^{1/2}} \exp\left[(2\langle n'_{ch} \rangle)^{-1}\right] W_{-n-1/2,0}(\langle n'_{ch} \rangle^{-1}), \tag{8.24a}$$

$$\langle W^k \rangle_{\mathscr{N}} = (k!)^2 \langle n'_{ch} \rangle^k, \tag{8.24b}$$

so the resulting radiation has enhanced fluctuations (enhanced bunching) compared with the original chaotic radiation ($\langle W^k \rangle_{\mathscr{N}} = k!\, \langle n_{ch} \rangle^k$ for chaotic radiation, while $\langle W^k \rangle_{\mathscr{N}} = \langle n_c \rangle^k$ for coherent radiation).

The photon statistics of radiation in the turbulent atmosphere have been further investigated by Teich and Rosenberg (1971), Rosenberg and Teich (1972), Lachs and Laxpati (1973) and Prucnal (1980).

8.2 The hamiltonian for radiation interacting with a random medium

We characterize the random medium by its fluctuating dielectric permittivity $\varepsilon(\nu, \mathbf{x}, t)$, which depends on the frequency ν, on the position of a point $\mathbf{x}$ and on time t, and we assume for simplicity that the mean value $\langle \varepsilon \rangle = 1$. We set $\varepsilon = 1 + \varepsilon'$, so that ε' represents the fluctuating part of the permittivity and $\langle \varepsilon' \rangle = 0$. We will neglect the dispersion of the medium. We can then write the hamiltonian in a volume V in the form [Louisell, Yariv and Siegman (1961), Yariv (1967), Crosignani, Di Porto and Solimeno (1971), Peřina and Peřinová (1976a)]

$$\hat{H}_{rad} = \frac{1}{8\pi} \int_V [\hat{\mathbf{E}}^2(\mathbf{x}, t) + \hat{\mathbf{H}}^2(\mathbf{x}, t)] \, d^3x + \frac{1}{8\pi} \int_V \varepsilon'(\mathbf{x}, t) \hat{\mathbf{E}}^2(\mathbf{x}, t) \, d^3x =$$
$$= \sum_l \hbar\omega_l(\hat{a}_l^+ \hat{a}_l + 1/2) - \sum_{l,m} \hbar K'_{lm}(t) (\hat{a}_l^+ - \hat{a}_l)(\hat{a}_m^+ - \hat{a}_m), \tag{8.25}$$

where (2.2a, b) and (2.4) have been used. The matrix elements of the fluctuating part of the permittivity, which are characteristic of the medium, are given by

$$K'_{lm}(t) = \frac{(\omega_l \omega_m)^{1/2}}{16\pi c^2} \int_V \varepsilon'(\mathbf{x}, t) \, \mathbf{u}_l(\mathbf{x}) \, . \, \mathbf{u}_m(\mathbf{x}) \, d^3x. \tag{8.26}$$

The first term in (8.25) represents the free-field hamiltonian of the radiation, while the second is the interaction hamiltonian. Further we introduce the spectral function

$$K'_{lm}(\vartheta) = \frac{1}{2\pi} \int_{-\infty}^{+\infty} K'_{lm}(t) \exp(i\vartheta t) \, dt, \tag{8.27}$$

which satisfies the cross-symmetry condition $K'_{lm}(-\vartheta) = K'^*_{lm}(\vartheta)$, because $K'^*_{lm}(t)$ is real and symmetric in l, m. In addition we assume, for simplicity, that $K'_{lm}(-t) = K'_{lm}(t)$, so that $K'_{lm}(-\vartheta) = K'_{lm}(\vartheta)$ is real. For fast fluctuations we take into account the resonant terms only, so that

$$\hat{H}_{rad} = \sum_l \hbar\omega_l(\hat{a}_l^+ \hat{a}_l + 1/2) + \sum_{l,m} \hbar K'_{lm} \times$$
$$\times [\hat{a}_l^+ \hat{a}_m \exp[-i(\vartheta_l - \vartheta_m) t] - \hat{a}_l \hat{a}_m \exp[i(\vartheta_l + \vartheta_m) t] + \text{h.c.}], \tag{8.28}$$

where we have assumed that $\hat{a}_l(t)$ behaves approximately as $\hat{a}_l \exp(-i\vartheta_l t)$, $\vartheta_l = \omega_l$, and also, for simplicity, that $K'_{lm}(\vartheta_l - \vartheta_m) \approx K'_{lm}(\vartheta_l + \vartheta_m) \approx K'_{lm}$ (i.e. the spectrum of fluctuations is assumed to be flat), which is always the case for slow fluctuations ($\vartheta_l = 0$).

We consider various regimes of the fluctuation of ε. If $|\varepsilon'| \ll 1$, the fluctuations are weak, while for $|\varepsilon'| \approx 1$, they are strong; in addition, for fluctuations slow compared to optical vibrations, ε depends only parametrically on t through $\mathbf{x}$ [Tatarskii (1971)] and consequently the K'_{lm} are independent of t. Finally we consider fast fluctuations. Also, the adiabatic behaviour of $K'_{lm}(t)$ ($|dK'_{lm}(t)/dt| \ll |K'_{lm}(t)|$) was taken into account [Peřina, Peřinová and Horák (1973a, b), Peřina, Peřinová and Mišta (1974), Peřina, Peřinová, Mišta and Horák (1974)]. Then we distinguish weakly and strongly inhomogeneous media. In the first case the inhomogeneities are much larger than the wavelength of the radiation and consequently ε depends slowly on $\mathbf{x}$, off-diagonal elements (8.26) may be neglected (i.e. fluctuations of the direction of propagation may be neglected) and the radiation modes may be considered as practically independent.

In order to describe the loss mechanism, we introduce, as in Chapter 7, the additional hamiltonian

$$\hat{H}_{\text{add}} = \sum_{j,l} \hbar\psi_l^{(j)}(\hat{c}_l^{(j)+}\hat{c}_l^{(j)} + 1/2) + \\ + \sum_{j,l} [\hbar\varkappa_{jl}^*\hat{c}_l^{(j)+}(\hat{a}_j + \Theta\hat{a}_j^+ \exp(-\mathrm{i}2\vartheta_j t)) + \text{h.c.}], \tag{8.29}$$

where the first summation represents the free hamiltonian of the reservoir, composed of damping chaotic oscillators having the boson annihilation operators $\hat{c}_l^{(j)}$ and frequencies $\psi_l^{(j)}$ (there is an independent set of damping oscillators for every radiation mode j), and the second summation represents the interaction hamiltonian, $\varkappa_{jl}$ being the coupling constants between the radiation and the reservoir system. In (8.29) we have included the rotating terms $\hat{a}_j\hat{c}_l^{(j)}$ and $\hat{a}_j^+\hat{c}_l^{(j)+}$ [see also Agarwal (1973)] that are usually neglected ($\Theta = 0$ in this case; if $\Theta = 1$ they are included). The reservoir system includes atoms of the medium causing the damping, scattered modes of radiation, etc.

The coupling of modes is characterized by off-diagonal elements K'_{lm}, $l \neq m$. In order to solve the Heisenberg equations and to determine the quantum statistics including the coupling of modes, we need to diagonalize the radiation hamiltonian (8.28) [Peřina, Peřinová and Horák (1973a), Peřina and Peřinová (1976a)]. Introducing new variables

$$\hat{A}_j(t) = \hat{a}_j(t)\exp(\mathrm{i}\vartheta_j t), \\ \hat{C}_l^{(j)}(t) = \hat{c}_l^{(j)}(t)\exp(\mathrm{i}\vartheta_j t) \tag{8.30a}$$

and a real unitary transformation matrix $\hat{U}$, we can introduce the following new operators $\hat{B}_j$ and $\hat{D}_l^{(j)}$, by

$$\hat{A}_j = \sum_l U_{jl}\hat{B}_l, \\ \hat{C}_l^{(j)} = \sum_m U_{jm}\hat{D}_l^{(m)}, \tag{8.30b}$$

in terms of which the renormalized hamiltonians have the forms

$$\hat{H}_{\text{rad}} = \hbar(\omega - \vartheta)\sum_j \hat{B}_j^+\hat{B}_j - \hbar\sum_j \lambda_j(\hat{B}_j^+ - \hat{B}_j)^2, \tag{8.31a}$$

$$\hat{H}_{\text{add}} = \hbar\sum_{j,l}(\psi_l - \vartheta)\hat{D}_l^{(j)+}\hat{D}_l^{(j)} + \hbar\sum_{j,l}[(\varkappa_l^*\hat{D}_l^{(j)+} + \Theta\varkappa_l\hat{D}_l^{(j)})\hat{B}_j + \text{h.c.}]. \tag{8.31b}$$

$\hat{H}_{\text{rad}}$ is diagonal in the new variables, λ_j are eigenvalues of the matrix $\hat{K}' = (K'_{lm})$, i.e. they are elements of the diagonal matrix $\Lambda = \hat{U}^T\hat{K}'\hat{U}$ ($\hat{U}^T$ is the transpose to $\hat{U}$, $\hat{U}^T\hat{U} = \hat{1}$), and we have assumed for simplicity that $\omega_1 \approx \omega_2 \approx \ldots \approx \omega_j \approx \ldots \approx \omega$

$(\vartheta_1 \approx \vartheta_2 \approx \ldots \approx \vartheta_j \approx \ldots \approx \vartheta)$, $\varkappa_{jl} \equiv \varkappa_l$ and $\psi_l^{(j)} \equiv \psi_l$ independently of j. For the case of two modes and with $K'_{11} = K'_{22} = K'$ we find that

$$\hat{U} = 2^{-1/2}\begin{pmatrix} 1 & 1 \\ 1 & -1 \end{pmatrix} \tag{8.32}$$

and $\lambda_{1,2} = K' \pm K'_{12}$.

8.3 Heisenberg—Langevin equations and the generalized Fokker—Planck equation

The Heisenberg equations

$$\begin{aligned} i\hbar \frac{d\hat{a}_j}{dt} &= [\hat{a}_j, \hat{H}_{rad} + \hat{H}_{add}], \\ i\hbar \frac{d\hat{c}_l^{(j)}}{dt} &= [\hat{c}_l^{(j)}, \hat{H}_{add}] \end{aligned} \tag{8.33}$$

can be written, making use of the above substitutions, as

$$\begin{aligned} i\frac{d\hat{B}_j}{dt} &= (\omega - \vartheta + 2\lambda_j)\,\hat{B}_j - 2\lambda_j \hat{B}_j^+ + \sum_l (\varkappa_l \hat{D}_l^{(j)} + \Theta \varkappa_l^* \hat{D}_l^{(j)+}), \\ i\frac{d\hat{D}_l^{(j)}}{dt} &= (\psi_l - \vartheta)\,\hat{D}_l^{(j)} + \varkappa_l^*(\hat{B}_j + \Theta \hat{B}_j^+), \end{aligned} \tag{8.34}$$

corresponding to the hamiltonians (8.31a, b). As in Chapter 7 we obtain the following Heisenberg – Langevin equations for the radiation variables:

$$i\frac{d\hat{B}_j}{dt} = (\omega - \vartheta + 2\lambda_j - i\gamma_j/2)\,\hat{B}_j - (2\lambda_j - i\Theta\gamma_j/2)\,\hat{B}_j^+ + i\hat{L}_j, \tag{8.35}$$

together with the Hermitian conjugates, where we neglect the frequency shifts and

$$\gamma_j = 2\pi\,|\,\varkappa_j(\omega_j \mathscr{K}_j^{1/2})\,|^2 \varrho_j(\omega_j \mathscr{K}_j^{1/2}) \tag{8.36}$$

are the damping constants [cf. (7.8b)] and

$$\mathscr{K}_j = 1 + \frac{4\lambda_j}{\omega}; \tag{8.37}$$

assuming that $\varkappa_{jl} \approx \varkappa_l$, $\mathscr{K}_j \approx \mathscr{K}$, $\omega_j \approx \omega$, $\varrho_j \approx \varrho$, all the γ_j are practically the same. The Langevin forces are expressed as

$$\hat{L}_j(t) = -i\sum_l (\varkappa_l \hat{D}_l^{(j)} \exp[-i(\psi_l - \vartheta)\,t] + \Theta \varkappa_l^* \hat{D}_l^{(j)+} \exp[i(\psi_l - \vartheta)\,t]), \tag{8.38}$$

where $\hat{D}_l^{(j)}$ are the initial operators [for $\Theta = \vartheta = 0$ we have a multimode version of (7.9)]. The Langevin forces obey the following relations, which are consequence of the Gaussian property of $\{\hat{D}_l^{(j)}\}$:

$$\langle \hat{L}_j(t) \rangle = \langle \hat{L}_j^+(t) \rangle = 0, \tag{8.39a}$$

$$\langle \hat{L}_j(t)\,\hat{L}_k^+(t')\rangle = \langle \hat{L}_k^+(t')\,\hat{L}_j(t)\rangle = -\langle \hat{L}_j(t)\,\hat{L}_k(t')\rangle =$$
$$= -\langle \hat{L}_j^+(t)\,\hat{L}_k^+(t')\rangle = \delta_{jk}\gamma_j(2\langle n_{dj}\rangle + 1)\,\delta(t - t') \tag{8.39b}$$

if $\Theta = 1$; while for $\Theta = 0$ we have, instead of (8.39b),

$$\langle \hat{L}_j(t)\,\hat{L}_k^+(t')\rangle = \delta_{jk}\gamma_j(\langle n_{dj}\rangle + 1)\,\delta(t - t'),$$
$$\langle \hat{L}_j(t)\,\hat{L}_k(t')\rangle = \langle \hat{L}_j^+(t)\,\hat{L}_k^+(t')\rangle = 0, \tag{8.39c}$$

corresponding to (7.10); here $\langle n_{dj}\rangle = \langle \hat{D}_l^{(j)+}\hat{D}_l^{(j)}\rangle$ (independently of l, assuming a flat reservoir spectrum) is the mean number of damping oscillators coupled to the jth radiative mode. The Langevin forces ensure the validity of the commutation rules for all times, as discussed in Chapter 7.

Since the quasi-distribution $\Phi_{\mathcal{N}}(\{\alpha_j\}, t)$ (in the Schrödinger picture) appropriate to normal ordering of field operators does not exist in general in interaction problems for some times, we prefer to adopt the quasi-distribution $\Phi_{\mathcal{A}}(\{\alpha_j\}, t)$ appropriate to antinormal ordering of the field operators, which is always well behaved and non-negative, as discussed in Chapter 4. These multimode quasi-distributions are related to their characteristic functions by the multimode Fourier transformations, by analogy with (4.77) and (4.78),

$$\Phi_{\mathcal{N},\mathcal{A}}(\{\alpha_j\}, t) = \int C_{\mathcal{N},\mathcal{A}}(\{\beta_j\}, t)\prod_j \exp(\alpha_j\beta_j^* - \alpha_j^*\beta_j)\,\frac{\mathrm{d}^2\beta_j}{\pi^2}\,, \tag{8.40}$$

where the multimode normal and antinormal characteristic functions are defined, by analogy with (4.75) and (4.76), as

$$C_{\mathcal{N},\mathcal{A}}(\{\beta_j\}, t) = \mathrm{Tr}\,\{\hat{\varrho}(t)\,\mathcal{N},\mathcal{A}\prod_j \exp[-\beta_j^*\hat{a}_j(0)]\exp[\beta_j\hat{a}_j^+(0)]\} =$$
$$= \mathrm{Tr}\,\{\hat{\varrho}(0)\,\mathcal{N},\mathcal{A}\prod_j \exp[-\beta_j^*\hat{a}_j(t)]\exp[\beta_j\hat{a}_j^+(t)]\} =$$
$$= \int \Phi_{\mathcal{N},\mathcal{A}}(\{\alpha_j\}, t)\prod_j \exp(-\alpha_j\beta_j^* + \alpha_j^*\beta_j)\,\mathrm{d}^2\alpha_j, \tag{8.41}$$

$\mathcal{N}$ and $\mathcal{A}$ being the normal and antinormal ordering operators again, and $\hat{\varrho}(t), \hat{a}_j(0)$ and $\hat{\varrho}(0)$, $\hat{a}_j(t)$ are the density matrices and the annihilation operators in the Schrödinger and Heisenberg pictures respectively.

Making use of these tools in the same way as in Chapter 7, we arrive at the generalized Fokker–Planck equation appropriate for the propagation of radiation of arbitrary quantum statistics through a random medium [Peřina and Peřinová (1975, 1976a)]:

$$\frac{\partial \Phi_{\mathcal{A}}}{\partial t} = \Bigg\{\sum_j \Bigg[\frac{\partial}{\partial B_j}\Bigg\{\Bigg[\frac{\gamma_j}{2} + \mathrm{i}(\omega - \vartheta + 2\lambda_j)\Bigg]B_j - \Bigg(\Theta\,\frac{\gamma_j}{2} + \mathrm{i}2\lambda_j\Bigg)B_j^*\Bigg\} +$$
$$+ \mathrm{c.c.} + \gamma_j(\langle n_{dj}\rangle + 1 + \Theta\langle n_{dj}\rangle)\,\frac{\partial^2}{\partial B_j\,\partial B_j^*} -$$
$$- \Bigg\{[\mathrm{i}\lambda_j + \Theta\gamma_j(\langle n_{dj}\rangle + 1/2)]\frac{\partial^2}{\partial B_j^2} + \mathrm{c.c.}\Bigg\}\Bigg]\Bigg\}\,\Phi_{\mathcal{A}}; \tag{8.42a}$$

for the corresponding antinormal characteristic function

$$\frac{\partial C_{\mathscr{A}}(\{\beta_j\}, t)}{\partial t} =$$
$$= \left\{ \sum_j \left[\left\{ \left[-\frac{\gamma_j}{2} + i(\omega - \vartheta + 2\lambda_j) \right] \beta_j - \left(\Theta \frac{\gamma_j}{2} + i2\lambda_j \right) \beta_j^* \right\} \frac{\partial}{\partial \beta_j} + \right.\right.$$
$$+ \text{c.c.} - \gamma_j(\langle n_{dj} \rangle + 1 + \Theta \langle n_{dj} \rangle) |\beta_j|^2 +$$
$$\left.\left. + \{[i\lambda_j - \Theta \gamma_j(\langle n_{dj} \rangle + 1/2)] \beta_j^2 + \text{c.c.}\} \right] \right\} C_{\mathscr{A}}(\{\beta_j\}, t). \tag{8.42b}$$

Here c.c. means the complex conjugate, B_j is the eigenvalue of $\hat{B}_j$ in the coherent state and B_j^* is its complex conjugate, $\langle n_{dj} \rangle = \langle n_d(\omega \mathscr{K}_j^{1/2}) \rangle$. The drift vectors in (8.42a) are determined, as discussed in Chapter 7, by $\langle \dot{B}_j \rangle_R$, the coefficient of $\partial^2/\partial B_j \, \partial B_j^*$ is equal to $2D_j$, where the diffusion constant $D_j = \gamma_j(\langle n_{dj} \rangle + 1/2)$ from (8.39b), or $D_j = \gamma_j (\langle n_{dj} \rangle + 1)/2$ from (8.39c), and this term represents the reservoir contribution. The coefficient of $\partial^2/\partial B_j^2$ is composed of two terms, the reservoir term which equals the diffusion constant $- \gamma_j(\langle n_{dj} \rangle + 1/2)$ and is non-zero only when $\Theta = 1$, and the term which arises from the nonlinear term $-\hbar\lambda_j(\hat{B}_j^2 + \hat{B}_j^{+2})$ in the hamiltonian (8.31a). The latter term is derived from the expression $i\lambda_j[\hat{B}_j^2 + \hat{B}_j^{+2}, \hat{\varrho}]$ in the equation of motion for the density matrix, if normal ordering is performed by means of the identities (4.32), $[\hat{B}_j^2, \hat{\varrho}] = \hat{B}_j[\hat{B}_j, \hat{\varrho}] + [\hat{B}_j, \hat{\varrho}] \hat{B}_j = \hat{B}_j \, \partial\hat{\varrho}/\partial\hat{B}_j^+ + (\partial\hat{\varrho}/\partial\hat{B}_j^+) \hat{B}_j$, $[\hat{B}_j^{+2}, \hat{\varrho}] = -(\partial\hat{\varrho}/\partial\hat{B}_j) \hat{B}_j^+ - \hat{B}_j^+ \partial\hat{\varrho}/\partial\hat{B}_j$, if *c*-numbers B_j and B_j^* are substituted for the operators $\hat{B}_j$ and $\hat{B}_j^+$ and if the standard procedure is applied.

Since the reservoir spectrum is flat, and if we assume a strongly inhomogeneous medium (otherwise $K_{12}' \approx 0$) and two modes for simplicity, then γ_1 and γ_2 ($\langle n_{d1} \rangle$ and $\langle n_{d2} \rangle$) will differ from each other only if $4 |K_{12}'| \approx \omega$ [cf. equations (8.26) and (2.77)], i.e. only if the fluctuations are strong ($|\varepsilon'| \approx 1$). In this case, applying the transformation (8.32) to (8.42a), strong coupling of modes occurs [Carmichael and Walls (1973), Peřina and Peřinová (1976a)]. For weak fluctuations $|\varepsilon'| \ll 1$ and consequently $4 |K_{12}'| \ll \omega$, regardless of the fact that the medium is weakly or strongly inhomogeneous; then $\gamma_1 \approx \gamma_2$ ($\langle n_{d1} \rangle \approx \langle n_{d2} \rangle$) and the mode coupling terms are missing so the modes can be treated independently.

8.4 Solutions of the generalized Fokker—Planck equation and the Heisenberg—Langevin equations

The solution of (8.42b) can be written, adopting for instance the standard method of characteristics, in the following form [Peřina and Peřinová (1975, 1976a)]

$$C_{\mathscr{A}}(\{\beta_j\}, t) = \langle \prod_j \exp[-B_j(t) |\beta_j|^2 + C_j'^*(t) \beta_j^2/2 + C_j'(t) \beta_j^{*2}/2 +$$
$$+ \xi_j'^*(t) \beta_j - \xi_j'(t) \beta_j^*] \rangle, \tag{8.43}$$

in full agreement with the Heisenberg–Langevin approach [Peřina, Peřinová and Horák (1973a, b), Peřina, Peřinová and Mišta (1974), Peřina, Peřinová, Mišta and Horák (1974)]. Here $\xi_j'(t)$, $B_j(t)$ and $C_j'(t)$ are functions of time [the differential equations for them can be derived by substituting (8.43) into (8.42b)], which have been determined for the various fluctuation regimes discussed above [Peřina and Peřinová (1975, 1976a)]. In general, they are rather complicated and we prefer therefore to give here only those for slow, small fluctuations. The Fourier transformation (8.40) provides the averaged shifted multimode Gaussian distribution

$$\Phi_{\mathscr{A}}(\{b_j\}, t) = \left\langle \prod_j (\pi K_j^{1/2}(t))^{-1} \exp\left[-\frac{B_j(t)}{K_j(t)} | b_j - \xi_j(t) |^2 + \right.\right.$$
$$\left.\left. + \frac{C_j^*(t)(b_j - \xi_j(t))^2 + \text{c.c.}}{2K_j(t)} \right] \right\rangle, \tag{8.44}$$

where we have returned to the original variables $b_j = B_j \exp(-i\vartheta t)$, $C_j(t) = C_j'(t) \exp(-i2\vartheta t)$, $\xi_j(t) = \xi_j'(t) \exp(-i\vartheta t)$, $K_j(t) = B_j^2(t) - | C_j(t) |^2$, and the angle brackets in (8.43) and (8.44) denote the average over the initial complex amplitudes $\xi_j(0) \equiv \xi_j$ with the initial probability distribution $\Phi_{\mathscr{N}}(\{\xi_j\}, 0)$.

If we solve the Heisenberg–Langevin equations (8.35), we obtain for the operators $\hat{b}_j(t) = \hat{B}_j(t) \exp(-i\vartheta t)$

$$\hat{b}_j(t) = u_j'(t)\hat{b}_j + v_j'(t)\hat{b}_j^+ + \sum_l (w_{jl}'(t)\hat{d}_l^{(j)} + z_{jl}'(t)\hat{d}_l^{(j)+}), \tag{8.45}$$

where the functions $u_j'(t)$, $v_j'(t)$, $w_{jl}'(t)$ and $z_{jl}'(t)$ have been determined explicitly and are related to the above functions $\xi_j'(t)$, $B_j(t)$ and $C_j'(t)$ [Peřina and Peřinová (1975, 1976a)]; $\hat{b}_j = \hat{B}_j$, $\hat{d}_l^{(j)} = \hat{D}_l^{(j)}$ are the initial operators. Substituting the solution (8.45) in the antinormal characteristic function (8.41), performing the normal ordering in the initial operators with the help of the Baker–Hausdorff identity (4.3), applying the Glauber–Sudarshan representation for $\hat{\varrho}(0)$, viz.,

$$\hat{\varrho}(0) = \int \Phi_{\mathscr{N}}(\{\xi_j\}, 0)\, \Phi_{\mathscr{N}}(\{\eta_{jl}\}, 0) \,|\{\xi_j\}\rangle\, |\{\eta_{jl}\}\rangle \times$$
$$\times \langle\{\eta_{jl}\}| \,\langle\{\xi_j\}| \,\mathrm{d}^2\{\xi_j\}\, \mathrm{d}^2\{\eta_{jl}\} \tag{8.46}$$

(ξ_j and η_{jl} are eigenvalues of $\hat{b}_j$ and $\hat{d}_l^{(j)}$ in the initial coherent state $|\{\xi_j\}\rangle\, |\{\eta_{jl}\}\rangle$) and averaging over the Gaussian $\{\eta_{jl}\}$, in the same way as in Chapter 7, we just arrive at (8.43) and (8.44).

Although all these solutions have been obtained explicitly [Peřina and Peřinová (1975, 1976a)], they are rather elaborate and consequently we concentrate on the simplest case of slow fluctuations ($\vartheta = 0$), neglecting also the small rotating terms between the radiation and the reservoir ($\Theta = 0$), and restricting ourselves to small fluctuations; further, we neglect the off-diagonal elements K_{lm}', $l \neq m$ (i.e. we neglect the fluctuations of the direction of propagation in weakly inhomogeneous media). Thus $\lambda_j = K_j'$ $(= K_{jj}')$, from (8.37) and (8.26) we have $\mathscr{K}_j = 1 + 4K_j'/\omega_j = (1/4\pi c^2) \int_V \varepsilon(\mathbf{x}, t)\, \mathbf{u}_j^2(\mathbf{x})\, \mathrm{d}^3x$, $\hat{B}_j = \hat{b}_j = \hat{A}_j = \hat{a}_j$ and $\hat{D}_l^{(j)} = \hat{d}_l^{(j)} = \hat{C}_l^{(j)} = \hat{c}_l^{(j)}$.

Equation (8.35) gives in this case

$$\frac{d\hat{a}_j}{dt} = -(i\omega_j\mathscr{K}_j^{1/2}\,\mathrm{ch}\,(\varphi_j/2) + \gamma_j/2)\,\hat{a}_j + \\ + i\omega_j\mathscr{K}_j^{1/2}\,\mathrm{sh}\,(\varphi_j/2)\,\hat{a}_j^+ + \hat{L}_j, \tag{8.47}$$

where $\omega_j + 2K_j' = \omega_j\mathscr{K}_j^{1/2}\,\mathrm{ch}\,(\varphi_j/2)$, $2K_j' = \omega_j\mathscr{K}_j^{1/2}\,\mathrm{sh}\,(\varphi_j/2)$ and $\varphi_j = \log\mathscr{K}_j$.

We note that if the $\hat{a}_l$ for scattered radiation modes are interpreted as reservoir operators $\hat{b}_l^{(j)}$, then the K_{jl}' corresponding to these modes play the role of coupling constants $\varkappa_{jl}$ [cf. (8.28) and (8.29)], and it follows from (8.36) that $\gamma_j = 2\pi\langle K_j'^2\rangle\,\varrho_j = 2\pi\sigma^2\varrho_j\,(\langle K_{jl}'\rangle = 0$ since $\langle\varepsilon'\rangle = 0)$, if we assume a Gaussian probability distribution for K_j' with standard deviation σ; this means that the damping constants are proportional to the square of the standard deviation, a result obtained earlier by means of more complicated classical considerations, based on diffraction of light [Tatarskii (1967)].

Assuming that the fluctuations of ε' are Gaussian, then the fluctuations of ε are log-normal ($\varepsilon' \approx \log\varepsilon$); and similarly if the fluctuations of K_{jl}' are Gaussian, fluctuations of $\mathscr{K}_j$ are log-normal [for a discussion of sums of independent log-normally distributed random variables, see Barakat (1976)]. The latter quantity corresponds to the fluctuating quantity used in Sec. 8.1. Hence, the probability distribution $\bar{P}(\mathscr{K}_j)$ and the moments $\langle\mathscr{K}_j^n\rangle$ are expressed in equations (8.2) and (8.3).

Equation (8.47) can be solved, together with its Hermitian conjugate, in the form (8.45):

$$\hat{a}_j(t) = u_j(t)\,\hat{a}_j + v_j(t)\,\hat{a}_j^+ + \sum_l (w_{jl}(t)\,\hat{c}_l^{(j)} + z_{jl}(t)\,\hat{c}_l^{(j)+}), \tag{8.48}$$

where

$$u_j(t) = \exp(-\gamma_j t/2)\left[\cos(\omega_j\mathscr{K}_j^{1/2}t) - i\,\mathrm{ch}\,(\varphi_j/2)\sin(\omega_j\mathscr{K}_j^{1/2}t)\right], \\ v_j(t) = i\exp(-\gamma_j t/2)\,\mathrm{sh}\,(\varphi_j/2)\sin(\omega_j\mathscr{K}_j^{1/2}t); \tag{8.49}$$

we shall not need the explicit expressions for $w_{jl}(t)$ and $z_{jl}(t)$, which satisfy the identities

$$i\sum_l [\varkappa_{jl}^* w_{jl}\exp(i\psi_l^{(j)}t) - \varkappa_{jl}w_{jl}^*\exp(-i\psi_l^{(j)}t)] = \gamma_j, \tag{8.50a}$$

$$\sum_l \varkappa_{jl}z_{jl}\exp(-i\psi_l^{(j)}t) = 0; \tag{8.50b}$$

these quantities are rather complicated [Peřina, Peřinová, Mišta and Horák (1974), Peřina and Peřinová (1976a)] and we only need to use

$$\sum_l |w_{jl}(t)|^2 \approx 1 - \exp(-\gamma_j t), \tag{8.51}$$

since $z_{jl} \approx 0$ in the optical region. Further, it follows from the commutation rule $[\hat{a}_j(t), \hat{a}_j^+(t)] = \hat{1}$ that $|u_j|^2 - |v_j|^2 + \sum_l (|w_{jl}|^2 - |z_{jl}|^2) = 1$ (and further identities follow from the other commutation rules, such as $[\hat{a}_j(t), \hat{a}_k^+(t)] = \hat{0}$, $j \neq k$, $[\hat{a}_j(t), \hat{a}_k(t)] = \hat{0}$, etc.). Note that the second term in (8.48) is related to self-radiation in the medium, since if we neglect the reservoir terms, we have for the vacuum expectation value of the number operator $\langle 0 | \hat{a}_j^+(t) \hat{a}_j(t) | 0 \rangle = | v_j(t) |^2$, which is the response of the medium in the absence of any incident radiation. This enables us to consider such a process as a prototype of the amplification process [Yuen (1976)]. As described above, we may now substitute the solution (8.48) into the antinormal characteristic function (8.41), perform the normal ordering in the initial operators, use the diagonal representation (8.46) of the density matrix, and average over the reservoir variables with the Gaussian probability distribution (7.19),

$$\Phi_{\mathcal{N}}(\{\eta_{jl}\}, 0) = \prod_{j,l} (\pi\langle n_{dj}\rangle)^{-1} \exp\left(-\frac{|\eta_{jl}|^2}{\langle n_{dj}\rangle}\right), \tag{8.52}$$

where $\langle n_{dj}\rangle = (\exp(\hbar\psi^{(j)}/KT) - 1)^{-1}$, T being the absolute temperature of the reservoir. We then arrive at the expressions (8.43) and (8.44) where [$C_j'(t) = C_j(t)$, $\xi_j'(t) = \xi_j(t)$ since $\vartheta = 0$ in this case]

$$\begin{aligned} B_j(t) &= | u_j(t) |^2 + (1 + \langle n_{dj}\rangle)(1 - \exp(-\gamma_j t)), \\ C_j(t) &= u_j(t)\, v_j(t), \\ \xi_j(t) &= u_j(t)\, \xi_j + v_j(t)\, \xi_j^*. \end{aligned} \tag{8.53}$$

If the incident radiation is in the coherent state $|\{\xi_j'\}\rangle$, then $\Phi_{\mathcal{N}}(\{\xi_j\},0) = \prod_j \delta(\xi_j - \xi_j')$ and we may omit the angle brackets in (8.43) and (8.44), substituting $\xi_j \to \xi_j'$. More generally, the form of these functions is preserved if the incident radiation can be represented by the superposition of coherent and chaotic fields [Peřina, Peřinová and Mišta (1974)]. Thus in these cases the statistics of radiation propagating through a random medium is described by a multimode superposition of coherent and chaotic fields with the real and imaginary parts of the complex amplitudes a_j correlated. The correlation vanishes if $C_j(t) \approx 0$ ($v_j(t) \approx 0$), so that $K_j(t) \approx B_j^2(t)$, $B_j(t) = 1 + \langle n_{dj}\rangle (1 - \exp(-\gamma_j t))$ and

$$\Phi_{\mathcal{A}}(\{a_j\}, t) = \prod_j (\pi B_j(t))^{-1} \exp\left[-\frac{|a_j - \xi_j(t)|^2}{B_j(t)}\right], \tag{8.54}$$

corresponding to (7.21b). In this case $\Phi_{\mathcal{N}}(\{a_j\}, t)$ exists for all times and is equal to (8.54) with $B_j(t) = \langle n_{dj}\rangle (1 - \exp(-\gamma_j t))$ and $\xi_j(t) = \xi_j \exp(-i\omega_j t - \gamma_j t/2)$, in agreement with (7.21a).

If the off-diagonal elements of K_{jl}' are taken into account, $b_j = \sum_j U_{jl} a_l$ is to be substituted into $\Phi_{\mathcal{A}}(\{b_j\}, t)$. Then the intermode coupling is obtained and the properties of such quasi-distributions including their time evolution can be discussed [Peřina, Peřinová and Horák (1973a), Peřina, Peřinová and Mišta (1974), Peřina and Peřinová (1976a)].

We point out that the angle brackets in (8.43) and (8.44) also denote the additional average over $\mathcal{K}_j$ or φ_j.

8.5 Photocount statistics

We define the time-dependent photon-number characteristic function $\mathrm{Tr}\,\{\hat{\varrho}(t)\times \times \exp(\mathrm{i}s\hat{n})\} = \mathrm{Tr}\,\{\hat{\varrho}\exp(\mathrm{i}s\hat{n}(t))\}$, where $\hat{n}(t) = \sum_j \hat{a}_j^+(t)\,\hat{a}_j(t)$ is the photon-number operator in the Heisenberg picture [cf. Chapter 3 and (2.30b)]. Further, corresponding to (4.138) we define the integrated intensity probability distributions

$$P_{\mathcal{N},\mathcal{A}}(W, t) = \int \Phi_{\mathcal{N},\mathcal{A}}(\{a_j\}, t)\,\delta(W - \sum_j |\,a_j\,|^2)\,\mathrm{d}^2\{a_j\}, \tag{8.55}$$

and in agreement with (4.137) the integrated intensity is $W = \langle \{a_j\}\,|\,\hat{n}\,|\,\{a_j\}\rangle = \sum_j |\,a_j\,|^2$. By analogy with (4.139) we have for the normal and antinormal characteristic functions, writing the δ-function in (8.55) in the form of the Fourier integral and taking into account that $P_{\mathcal{N},\mathcal{A}}(W, t)$ and $\langle \exp(\mathrm{i}sW)\rangle_{\mathcal{N},\mathcal{A}}$ are related by the Fourier transformations [cf. (3.112) and (5.96)],

$$\langle \exp(\mathrm{i}sW)\rangle_{\mathcal{N},\mathcal{A}} = \int \Phi_{\mathcal{N},\mathcal{A}}(\{a_j\}, t)\exp(\mathrm{i}s\sum_j |\,a_j\,|^2)\,\mathrm{d}^2\{a_j\}. \tag{8.56}$$

If $\Phi_{\mathcal{N}}(\{a_j\}, t)$ does not exist, we calculate $\langle \exp(\mathrm{i}sW)\rangle_{\mathcal{A}}$; however, in order to determine the photocount distribution $p(n, t)$ and its factorial moments $\langle W^k\rangle_{\mathcal{N}}$ with the help of equations (3.102) and (3.103) we need the normal generating function $\langle \exp(\mathrm{i}sW)\rangle_{\mathcal{N}}$, which is determined by the simple substitution (4.191b),

$$\langle \exp(\mathrm{i}sW)\rangle_{\mathcal{N}} = (1 + \mathrm{i}s)^{-M}\left\langle \exp\left(\frac{\mathrm{i}sW}{1 + \mathrm{i}s}\right)\right\rangle_{\mathcal{A}}, \tag{8.57}$$

where M denotes the number of degrees of freedom.

Sometimes we are not interested in the explicit forms of the quasi-distributions. Then the non-existence of $\Phi_{\mathcal{N}}$ may be got round by a direct determination of the generating function $\langle \exp(-\lambda W)\rangle_{\mathcal{N}}$ from the characteristic function $C_{\mathcal{N}}(\{\beta_j\}, t)$ [Rockower and Abraham (1978), Peřina (1979)]. Substituting $\Phi_{\mathcal{N}}(\{a_j\}, t)$ from (8.40) into $\langle \exp(-\lambda W)\rangle_{\mathcal{N}}$ given in (8.56) ($\lambda = -\mathrm{i}s$), changing the order of the integrations and performing the integration over $\{a_j\}$, we arrive at that relation:

$$\langle \exp(-\lambda W)\rangle_{\mathcal{N}} = (\pi\lambda)^{-M}\int \exp\left(-\frac{1}{\lambda}\sum_j |\,\beta_j\,|^2\right) C_{\mathcal{N}}(\{\beta_j\}, t)\,\mathrm{d}^2\{\beta_j\}. \tag{8.58}$$

This relation holds regardless of whether $\Phi_{\mathcal{N}}$ exists or not; this can be proved [Peřina (1979)] using the same relation between $\langle \exp(-\lambda W)\rangle_{\mathcal{A}}$ and $C_{\mathcal{A}}(\{\beta_j\}, t)$, following the steps to (8.58), and making use of (8.57). Further, note that there are always $\lambda(-\mathrm{i}s)$ so small that (8.56) and (8.58) exist [even at $\lambda = 1$, $\langle \exp(-W)\rangle_{\mathcal{N}} = \pi^M\Phi_{\mathcal{A}}(\{0\}, t)$, cf. (4.74) and (4.78)]; then for all values of λ these generating functions are defined by analytic continuation, as discussed in Sec. 3.8 [also Peřina (1979)].

Making use of the quasi-distribution (8.44) in (8.56) we obtain

$$\langle \exp(\mathrm{i}sW)\rangle_{\mathscr{A}} = \left\langle \prod_j (1 - \mathrm{i}sE_j)^{-1/2}(1 - \mathrm{i}sF_j)^{-1/2} \times \right.$$
$$\left. \times \exp\left[\frac{\mathrm{i}sA_{1j}}{1 - \mathrm{i}sE_j} + \frac{\mathrm{i}sA_{2j}}{1 - \mathrm{i}sF_j}\right]\right\rangle \tag{8.59a}$$

and, substituting (8.59a) into (8.57), we have

$$\langle \exp(\mathrm{i}sW)\rangle_{\mathscr{N}} = \left\langle \prod_j [1 - \mathrm{i}s(E_j - 1)]^{-1/2}[1 - \mathrm{i}s(F_j - 1)]^{-1/2} \times \right.$$
$$\left. \times \exp\left[\frac{\mathrm{i}sA_{1j}}{1 - \mathrm{i}s(E_j - 1)} + \frac{\mathrm{i}sA_{2j}}{1 - \mathrm{i}s(F_j - 1)}\right]\right\rangle, \tag{8.59b}$$

where

$$E_j = B_j - |C_j|, \qquad F_j = B_j + |C_j|,$$
$$A_{1,2j} = \frac{1}{2}\left[|\xi_j(t)|^2 \mp \frac{1}{2|C_j|}(\xi_j^2(t)\,C_j^* + \text{c.c.})\right]. \tag{8.60}$$

Of course, the same expressions are obtained from (8.58), where $C_{\mathscr{N}}(\{\beta_j\}, t)$ is given by (8.43) with $B_j(t) \to B_j(t) - 1$ [cf. (4.74)]. The angle brackets in (8.59a, b) denote the average over ξ_j and $\mathscr{K}_j$ with $\Phi_{\mathscr{N}}(\{\xi_j\}, 0)$ and $\bar{P}(\mathscr{K}_j)$. The expressions in the angle brackets of (8.59a, b) are the generating functions for the Laguerre polynomials, as can be seen from (5.64a, b). The quantities $E_j - 1$ and $F_j - 1$ play the role of the „mean numbers of chaotic photons" (the subtraction of unity in the normal generating function eliminates the contribution of the physical vacuum), while the quantities $A_{1,2j}$, related to the incident field, play the role of the „mean numbers of coherent photons".

The deviation of the expression within the angle brackets of (8.59b) from the Poisson generating function reflects the change of the photon statistics caused by the dynamics of the process, whereas the average over the initial complex amplitudes, represented by the angle brackets, leads to an additional change of the photon statistics with respect to the initial state of the field [Srinivas (1978)]. The character of the dynamical change of the photon statistics depends on the sign of the „mean number of chaotic photons" $E_j - 1 = B_j - |C_j| - 1$ ($F_j - 1 = B_j + |C_j| - 1$ is always non-negative). If it is positive, as for instance in the case of propagation of radiation through random media, the uncertainty (bunching effect) of photons is higher than that for the Poisson statistics corresponding to the coherent state. If it is negative, this uncertainty is reduced below that for the coherent state and antibunching of photons and sub-Poissonian statistics may occur, as in nonlinear optical processes (Chapter 9).

The photocount distribution $p(n, t)$ and its factorial moments $\langle W^k\rangle_{\mathscr{N}}$ can be obtained in the same way as in Sec. 5.3 [for example equation (5.103b)] [Peřina, Peřinová and Horák (1973b)]. However, in the following we consider a simplified case in the spirit of the approximate formulae discussed in Sec. 5.3, where all the E_j and F_j are the same, i.e. we assume that all modes are equally damped, only the mean frequency is considered, and all $\mathscr{K}_j$ are assumed to fluctuate uniformly ($\mathscr{K}_j \approx \approx \varepsilon(t)$ in weakly inhomogeneous media). This provides the simplified generating function

$$\langle \exp(\mathrm{i}sW)\rangle_{\mathscr{N}} = \left\langle [1 - \mathrm{i}s(E-1)]^{-M/2} [1 - \mathrm{i}s(F-1)]^{-M/2} \times \right.$$
$$\left. \times \exp\left[\frac{\mathrm{i}sA_1}{1 - \mathrm{i}s(E-1)} + \frac{\mathrm{i}sA_2}{1 - \mathrm{i}s(F-1)}\right]\right\rangle, \qquad (8.61)$$

where

$$E = B - |C|, \qquad F = B + |C|,$$
$$A_{1,2} = \frac{1}{2}\left[\sum_j^M |\xi_j(t)|^2 \mp \frac{1}{2|C|}(C^* \sum_j^M \xi_j^2(t) + \text{c.c.})\right]; \qquad (8.62a)$$

this generating function may be further approximated when the average over the initial random phases is performed, giving approximately

$$A_{1,2} = \frac{1}{2}(|u| \mp |v|)^2 W_0, \qquad (8.62b)$$

where $W_0 = \sum_j |\xi_j|^2$ is the integrated intensity of the incident radiation.

From (8.61) we obtain for the mean number of photons

$$\langle n(t)\rangle = \langle W(t)\rangle_{\mathscr{N}} = \frac{\mathrm{d}}{\mathrm{d}(\mathrm{i}s)} \langle \exp(\mathrm{i}sW)\rangle_{\mathscr{N}}\Big|_{\mathrm{i}s=0} =$$
$$= \langle |u|^2 + |v|^2\rangle \langle W_0\rangle + M\langle |v|^2\rangle + M\langle n_d\rangle (1 - \exp(-\gamma t)); \qquad (8.63)$$

the angle brackets now denote the average over W_0 with weight $P_{\mathscr{N}}(W_0)$, and over $\mathscr{K}$ with $\bar{P}(\mathscr{K})$.

Applying the identities (5.64a, b) to (8.61), we arrive at the following photocount distribution and its factorial moments

$$p(n,t) = \left\langle (EF)^{-M/2} (1 - F^{-1})^n \exp\left(-\frac{A_1}{E} - \frac{A_2}{F}\right) \times \right.$$
$$\times \sum_{k=0}^{n} \frac{1}{\Gamma(k + M/2)\Gamma(n - k + M/2)} \left(\frac{1 - E^{-1}}{1 - F^{-1}}\right)^k \times$$
$$\left. \times L_k^{M/2-1}\left(-\frac{A_1}{E(E-1)}\right) L_{n-k}^{M/2-1}\left(-\frac{A_2}{F(F-1)}\right)\right\rangle, \qquad (8.64a)$$

$$\langle W^k\rangle_{\mathscr{N}} = k! \left\langle (F-1)^k \sum_{l=0}^{k} \frac{1}{\Gamma(l + M/2)\Gamma(k - l + M/2)} \left(\frac{E-1}{F-1}\right)^l \times \right.$$
$$\left. \times L_l^{M/2-1}\left(-\frac{A_1}{E-1}\right) L_{k-l}^{M/2-1}\left(-\frac{A_2}{F-1}\right)\right\rangle, \qquad (8.64b)$$

which are analogous to (5.101c) and (5.104c); here

$$E, F - 1 = |v| (|v| \mp |u|) + \langle n_d\rangle (1 - \exp(-\gamma t)). \qquad (8.65)$$

The corresponding integrated intensity probability distributions $P_{\mathcal{N},\mathcal{A}}(W, t)$ can be calculated as in Sec. 5.3 [Peřina, Peřinová and Horák (1973b)].

The first term in (8.63) represents the response to the incident radiation and it is zero if the incident field is zero, $\langle W_0 \rangle = 0$. The second term represents the self-radiation of the medium and it is non-zero even when $\langle W_0 \rangle = 0$. The third term is the reservoir contribution. If the incident radiation is sufficiently strong, quantum effects may be neglected; then $E - 1$, $F - 1 \approx \langle n_d \rangle (1 - \exp(-\gamma t))$ and we obtain the photocount distribution and its factorial moments as given in (5.101b) and (5.104b) averaged over $\mathcal{K}$, corresponding to the photocount generating function in (5.94) with $\varkappa = 1$; here $\langle n_c \rangle = (|u|^2 + |v|^2) W_0$ and $\langle n_{ch} \rangle = M \langle n_d \rangle \times \times (1 - \exp(-\gamma t))$. If $\langle n_d \rangle = 0$, i.e. the damping mechanism is neglected, we obtain the photodetection equation (8.1), with $\mathcal{K} = |u|^2 + |v|^2$.

As the functions $u(t)$ and $v(t)$ oscillate in the optical region with $t = z/c$, z being the distance travelled by the radiation in the random medium (the mean frequency $\langle \omega \mathcal{K}^{1/2} \rangle$ is about $10^{15}\ \mathrm{s}^{-1}$), some effective values of $u(t)$ and $v(t)$ have to be considered [Peřina, Peřinová and Mišta (1974)],

$$
\begin{aligned}
|u|^2 &= \exp(-\gamma z/c)\left(1 + \frac{1}{2}\,\mathrm{sh}^2\left(\frac{\varphi}{2}\right)\right), \\
|v|^2 &= \frac{1}{2}\exp(-\gamma z/c)\,\mathrm{sh}^2\left(\frac{\varphi}{2}\right), \\
|u|^2 + |v|^2 &= \exp(-\gamma z/c)\,\frac{1 + \mathrm{ch}\,\varphi}{2},
\end{aligned}
\tag{8.66a}
$$

while the maximum deviation of the photocount statistics from those of the incident radiation occurs for

$$
\begin{aligned}
|u|^2 &= \exp(-\gamma z/c)\,\mathrm{ch}^2\left(\frac{\varphi}{2}\right), \\
|v|^2 &= \exp(-\gamma z/c)\,\mathrm{sh}^2\left(\frac{\varphi}{2}\right), \\
|u|^2 + |v|^2 &= \exp(-\gamma z/c)\,\mathrm{ch}\,\varphi.
\end{aligned}
\tag{8.66b}
$$

Neglecting losses, one has in this case

$$|u + v^*|^2 = \exp\varphi = \mathcal{K}. \tag{8.66c}$$

8.6 Diament—Teich and Tatarskii descriptions

The Diament–Teich and Tatarskii descriptions are such phenomenological descriptions which are appropriate for radiation propagating through random media when loss mechanisms and self-radiation of the medium are neglected. Substituting (8.48) with $w_{jl} = z_{jl} = 0$ into (2.4), we may define new effective operators $\hat{a}_j(t)$ as follows

$$\hat{a}_j(t) = [u(t) + v^*(t)]\,\hat{a}_j = u(t)\,\hat{a}_j + [v(t)\,\hat{a}_j^+]^+, \tag{8.67}$$

which compensates self-radiation. This is analogous to the use of complex mode functions in (2.4), which imply that the coefficients of $\hat{a}_j\hat{a}_k$ and $\hat{a}_j^+\hat{a}_k^+$ in the hamiltonian are practically zero when the inhomogeneities are much larger than the wavelength. We then have for the photon number operator, in agreement with (8.66c),

$$\hat{n}(t) = \sum_j \hat{a}_j^+(t)\, \hat{a}_j(t) = \hat{n}_0 \exp \varphi, \tag{8.68a}$$

where $\hat{n}_0 = \sum_j \hat{a}_j^+ \hat{a}_j$ is the photon-number operator of the incident radiation. The average in the coherent state $|\{a_j\}\rangle$ gives

$$W(t) = W_0 \exp \varphi, \tag{8.68b}$$

where $W(t) = \sum_j |a_j(t)|^2$ and $W_0 = \sum_j |a_j|^2$. As a consequence of the energy conservation law, $\langle W(t)\rangle_{\mathcal{N}} = \langle W_0 \rangle$ in agreement with (8.18), and the commutation rules are satisfied in the sense of average values, $\langle[\hat{a}_j(t), \hat{a}_j^+(t)]\rangle = \langle \exp(\varphi)\rangle = 1$.

Thus, substituting the operators (8.67) into the normal characteristic function,

$$C_{\mathcal{N}}(\{\beta_j\}, t) = \mathrm{Tr}\, \{\hat{\varrho}(0) \prod_j \exp[\beta_j \hat{a}_j^+(t)] \prod_k \exp[-\beta_k^* \hat{a}_k(t)]\}, \tag{8.69}$$

we see that it is also normally ordered in the initial operators, and the Fourier transformation (8.40) yields

$$\Phi_{\mathcal{N}}(\{a_j'\}, t) = \langle \prod_j \delta(a_j' - (u + v^*)\, a_j)\rangle. \tag{8.70a}$$

Making use of the substitution (8.55), we have

$$P_{\mathcal{N}}(W, t) = \langle \delta(W - \mathscr{K} W_0)\rangle \tag{8.70b}$$

and, therefore,

$$p(n, t) = \int_0^\infty P_{\mathcal{N}}(W, t) \frac{W^n}{n!} \exp(-W)\, \mathrm{d}W =$$

$$= \iiint_0^\infty P_{\mathcal{N}}(W_0)\, P(\mathscr{K})\, \delta(W - \mathscr{K} W_0) \frac{W^n}{n!} \exp(-W)\, \mathrm{d}W_0\, \mathrm{d}\mathscr{K}\, \mathrm{d}W, \tag{8.71a}$$

which is just the photodetection equation (8.1), with

$$p_0(n, \mathscr{K}) = \int_0^\infty \frac{(\mathscr{K} W_0)^n}{n!} \exp(-\mathscr{K} W_0)\, P_{\mathcal{N}}(W_0)\, \mathrm{d}W_0; \tag{8.71b}$$

thus we have arrived at the Diament–Teich description. From (8.71a) the photon-number generating function can be expressed, by analogy with (2.51e), in the form

$$\langle \exp(\mathrm{i}s\hat{n}(t))\rangle = \langle \exp[(\mathrm{e}^{\mathrm{i}s} - 1)\, W(t)]\rangle_{\mathcal{N}} = \langle \exp[(\mathrm{e}^{\mathrm{i}s} - 1)\, \mathscr{K} W_0]\rangle. \tag{8.72}$$

In this description the effect of the medium is included through the classical relation (8.68b), whereas the quantum relation (8.68a) is appropriate for the Tatarskii description and one has $\langle \hat{n}^k(t) \rangle = \langle \mathscr{K}^k \rangle \langle \hat{n}_0^k \rangle$, and

$$\langle \exp(\mathrm{i}s\hat{n}(t)) \rangle = \langle \exp(\mathrm{i}s\mathscr{K}\hat{n}_0) \rangle = \langle \exp[(\mathrm{e}^{\mathrm{i}s\mathscr{K}} - 1) W_0] \rangle. \tag{8.73}$$

Fourier transformation yields the modified photodetection equation [Peřina (1972a), Peřina, Peřinová, Teich and Diament (1973)]

$$p(n, t) = p_0(0)\,\delta(n) + \sum_{m=1}^{\infty} p_0(m) \frac{\bar{P}(n/m)}{m} = \left\langle \frac{\bar{P}(n/m)}{m} \right\rangle, \tag{8.74}$$

where $p_0(n)$ is the photocount distribution of the incident radiation and $\lim \bar{P}(n/m)/m = \delta(n)$ for $m \to 0$. The first term in (8.74) represents a shot-noise term and is non-zero only when $n = 0$. The second term is well behaved and is comparable

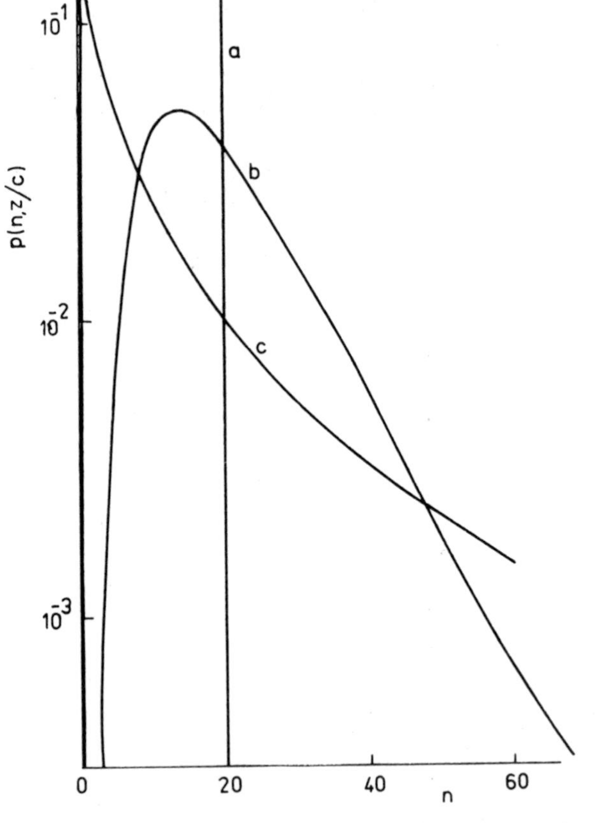

Fig. 8.4 — The photocount distribution for the Fock state $|20\rangle$ and the levels of fluctuations (a) $\sigma = 0$, (b) $\sigma = 1/2$ and (c) $\sigma = 3/2$. Both the Diament—Teich and Tatarskii descriptions give the same results (after Peřina, Peřinová, Teich and Diament, 1973, Phys. Rev. **A7**, 1732).

to (8.71a), being zero when $n = 0$. The shot-noise is related to the quantum relation (8.68a); it is absent when the classical relation (8.68b) is used. Because of the presence of the shot-noise term $p_0(0)\,\delta(n)$ (with $p_0(0) \neq 0$) in the modified photodetection equation (8.74), its regular part is normalized to $1 - p_0(0)$, while $p(n, t)$ given by (8.71a) is normalized to unity. In this way small numerical differences may occur between the photocount distributions in the Diament–Teich and Tatarskii descriptions, which are absent if $p_0(0) = 0$ (e.g. for the Fock state $|N\rangle$). This result may be seen in Fig. 8.4, which shows that both descriptions give the same results when the incident radiation is in the Fock state $|20\rangle$ and for intermediate and saturated levels of turbulence. In this figure the „pure" effect of turbulence on the photocount distribution may be observed as a shift of the peak to a lower number of photoelectrons and as a broadening of the curves with increasing level of turbulence.

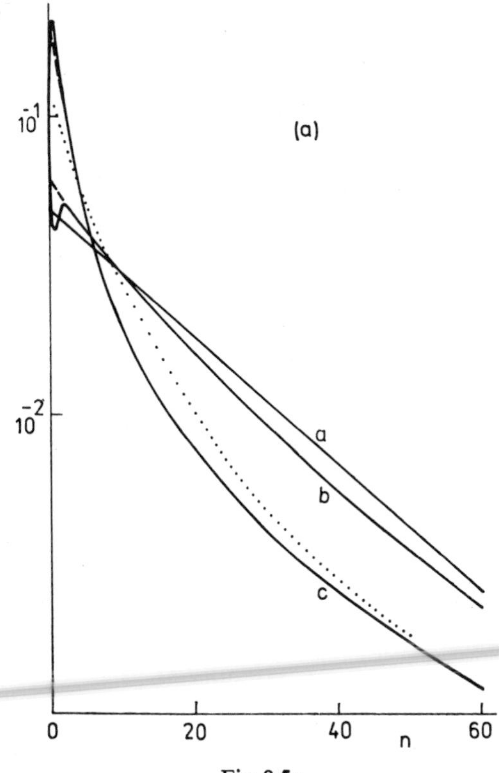

Fig. 8.5a

The similar behaviour of the photocount distribution in the turbulent atmosphere is demonstrated in Fig. 8.5 for initially chaotic and coherent radiation. The dashed curves represent the results according to the Diament–Teich description, while the solid curves are based on the Tatarskii description (the weight value $p_0(0)$ is shown at $n = 0$). The dotted curves were obtained from the quantum description of Sec. 8.5 and are discussed in Sec. 8.7. The photocount statistics of the incident radiation have been described with the help of the formulae of Sec. 5.3 appropriate for the superposition of coherent and chaotic fields. We observe that for non-vacuum states the two descriptions are very similar.

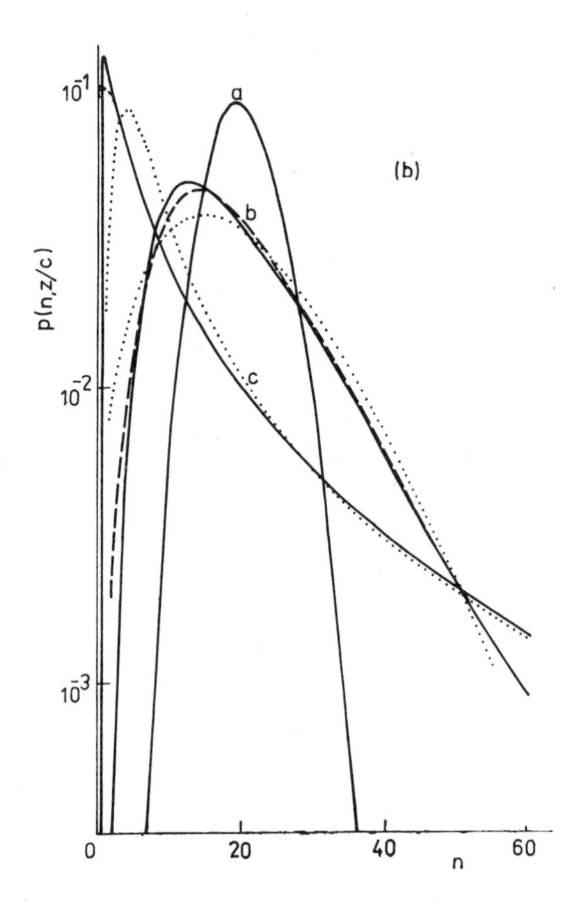

Fig. 8.5 — A comparison of the photocount distributions based on the Tatarskii (solid curves) and the Diament—Teich (dashed curves) descriptions, for $\sigma = 0$ (curve *a*), $\sigma = 1/2$ (curve *b*) and $\sigma = 3/2$ (curve *c*) and (a) $\langle n_c \rangle = 0$, $\langle n_{ch} \rangle = 20$, $\gamma = 0$, (b) $\langle n_c \rangle = 20$, $\langle n_{ch} \rangle = 0$ independently of M and γ. The dotted curves represent values based on the quantum dynamical description. When only a solid curve is shown, all curves coincide (after Peřina, Peřinová, Teich and Diament, 1973, Phys. Rev. **A7**, 1732 and Peřina, Peřinová, Diament and Teich, 1975, Czech. J. Phys. **B25**, 483).

8.7 Comparison of the quantum and phenomenological descriptions

We may now compare the photocount distributions obtained from the phenomenological Diament–Teich and Tatarskii descriptions and from the quantum dynamical description. Substituting (8.66a) into (8.63) and using (8.3) we find that

$$\langle n(z) \rangle = \exp(-\gamma z/c)\left(\frac{\exp(\sigma^2) + 3}{4}\langle W_0 \rangle + \frac{\exp(\sigma^2) - 1}{8} M\right) + \\ + M\langle n_d \rangle (1 - \exp(-\gamma z/c)). \tag{8.75}$$

For $\sigma = 0$, equation (8.75) yields the simple result for the damped radiation field coupled to the reservoir. For the intermediate turbulence levels $\sigma = 1/2$ and 1 and for the saturated level $\sigma = 3/2$, we have

$$\frac{\exp(\sigma^2)+3}{4} = \begin{cases} 1.071, \\ 1.423, \\ 3.122, \end{cases} \quad \frac{\exp(\sigma^2)-1}{8} = \begin{cases} 0.036, & \sigma = 1/2, \\ 0.215, & \sigma = 1, \\ 1.061, & \sigma = 3/2. \end{cases} \tag{8.76}$$

We see that for the intermediate level $\sigma = 1/2$ of fluctuations, the enhancement factor $\exp(\gamma z/c)\langle |u|^2 + |v|^2\rangle$, related to the incident radiation, is small, and the self-radiative quantum-noise contribution may be neglected. For saturated fluctuations the enhancement is more pronounced, but it is diminished by the damping factor $\exp(-\gamma z/c)$. The quantum noise contribution may be neglected if the mean integrated intensity $\langle W_0\rangle$ of the incident radiation is sufficiently high. In this manner, lossy as well as gain media may be treated [Peřina, Peřinová and Mišta (1974), Peřina, Peřinová, Diament and Teich (1975)]. In particular, propagation through weakly inhomogeneous lossless media ($\langle n(z)\rangle = \langle W_0\rangle$) provides a basis for comparing the phenomenological and the quantum descriptions. Such a comparison has been made for a number of initial states of the field, such as the Fock state, for coherent radiation, for chaotic radiation and for their superposition [Peřina, Peřinová, Diament and Teich (1975)]. In Fig. 8.5 this comparison is made for initially chaotic and coherent radiation. Thus the phenomenological descriptions may be applied as effective descriptions to lossless random media.

A behaviour of the photocount distribution for radiation in random media such as is shown in Fig. 8.5 has been obtained also by Estes, Kuppenheimer and Narducci (1970) in a quantum statistical analysis of randomly modulated laser beams, by Bufton, Iyer and Taylor (1977) in connection with the scintillation statistics caused by atmospheric turbulence and speckle in satellite laser ranging and by Churnside and McIntyre (1978a, b) in a study of the signal current probability distribution for optical heterodyne receivers in the turbulent atmosphere.

The conclusions that the peak of the photocount distribution shifts to a lower photoelectron number and that the photocount distribution is broadened with increase of the fluctuations have been experimentally verified by Bluemel, Narducci and Tuft (1972) using the log-normal distribution for the intensity fluctuations of scattered light by a rotating ground glass, by Bertolotti (1974) using nematic liquid crystals with a variable voltage, by Churnside and McIntyre (1978b) who used an optical heterodyne receiver operating in the presence of clear air turbulence over 1.6 km propagation path in the open atmosphere and by Parry (1981) who reported measurements of atmospheric turbulence that induced intensity fluctuations in a laser beam propagating along a 1.125 km path in the atmosphere. In the latest measurements the K-distribution has been adopted [Parry, Pusey, Jakeman and McWhirter (1978)] to describe intensity fluctuations,

$$P_{\mathcal{N}}(W) = \frac{2M^{(M-1)/2}}{\langle W\rangle_{\mathcal{N}}\Gamma(M)}\left(\frac{W}{\langle W\rangle_{\mathcal{N}}}\right)^{(M-1)/2} K_{M-1}\left(2\left[\frac{MW}{\langle W\rangle_{\mathcal{N}}}\right]^{1/2}\right), \tag{8.77a}$$

$$p(n, T) = \left(\frac{M}{\langle n \rangle}\right)^{M/2} \frac{\Gamma(n + M)}{\Gamma(M)} \times$$

$$\times \exp\left(\frac{M}{2\langle n \rangle}\right) W_{-M/2-n,\,(M-1)/2}\left(\frac{M}{\langle n \rangle}\right), \tag{8.77b}$$

where $W_{k,\,l}$ is the Whittaker function again and

$$K_\nu(z) = \frac{\pi/2}{\sin(\pi\nu)}(I_{-\nu}(z) - I_\nu(z)) \tag{8.78}$$

is the McDonald modified Bessel function, $I_\nu(z)$ being the modified Bessel function $J_\nu(z)$ of imaginary argument. Parry has found that the K-distribution is more appropriate for the description of laser beams in the turbulent atmosphere under the conditions of his experiment than is the log-normal distribution. Similar results were obtained by Phillips and Andrews (1981).

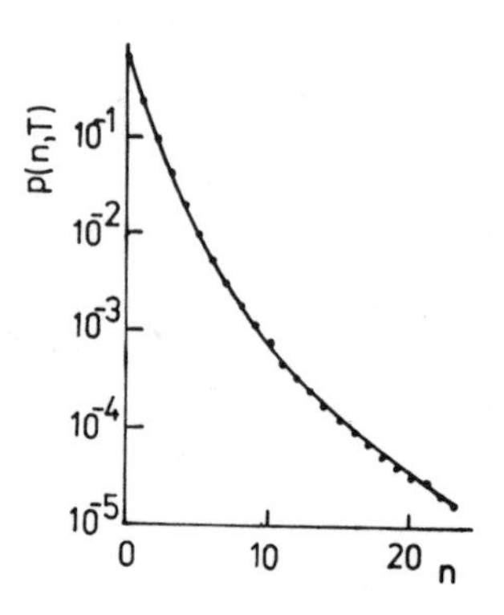

Fig. 8.6 — The photocount distribution of light scattered by rotating coarse-ground glass ($\langle n \rangle = 0.773$, $\langle n^2 \rangle = 2.494$, $T = 10$ μs) (after Bluemel, Narducci and Tuft, 1972, J. Opt. Soc. Am. **62**, 1309).

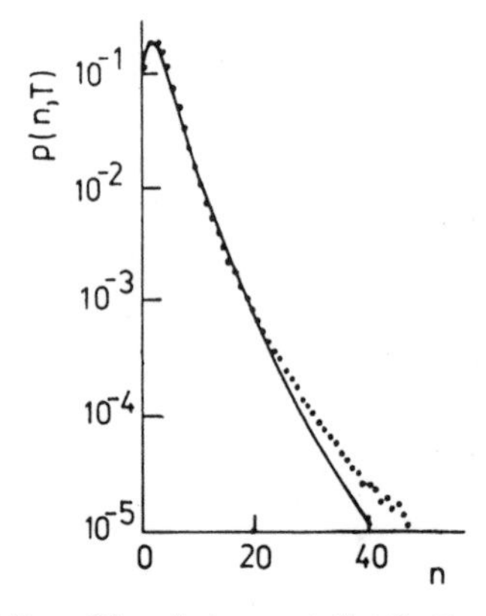

Fig. 8.7 — The photocount distribution for a coarse-ground glass with focused light ($\langle n \rangle =$ $= 3.333$, $\langle n^2 \rangle = 21.454$) (after Bluemel, Narducci and Tuft, 1972, J. Opt. Soc. Am. **62**, 1309).

Fig. 8.6 shows the photocount distribution experimentally observed by Bluemel, Narducci and Tuft (1972), compared with the theoretical one based on the log-normal model ($\langle n \rangle = 0.773$, $\langle n^2 \rangle = 2.494$ are used for the fitting, the average value of the inhomogeneities is 20 μm, and $T = 10$ μs). The results by the same authors when the illuminated pinhole is imaged on the collecting aperture of the detector are shown in Fig. 8.7 ($\langle n \rangle = 3.333$, $\langle n^2 \rangle = 21.454$). The fluctuations obtained are stronger than those corresponding to the Bose–Einstein distribution (line a in Fig. 8.5a).

Fig. 8.8 demonstrates the change of $P_{\mathcal{N}}(I)$, as reported by Bertolotti (1974), when light is scattered by a nematic liquid crystal with increasing applied voltage, so that the number of microturbulences is increasing and finally the negative exponential distribution is reached. A photocount distribution of the form given in Fig. 8.5b has been observed in a turbulent atmosphere by Churnside and McIntyre (1978b); the photocount distribution in the turbulent atmosphere has been also observed by Parry (1981) but, as mentioned above, the theoretically predicted curve was based on the K-distribution.

Measurements of the photon time-interval statistics of laser light scattered by Brownian particles have been reported by Timmermans and Zijlstra (1977), although the interpretation of the experimental data needed some comments [Peřina (1979)]. Measurements of the photocount statistics of scattered light deviated from the Bose–Einstein statistics make it possible to determine small numbers of scattering particles [Koňák et al. (1983)].

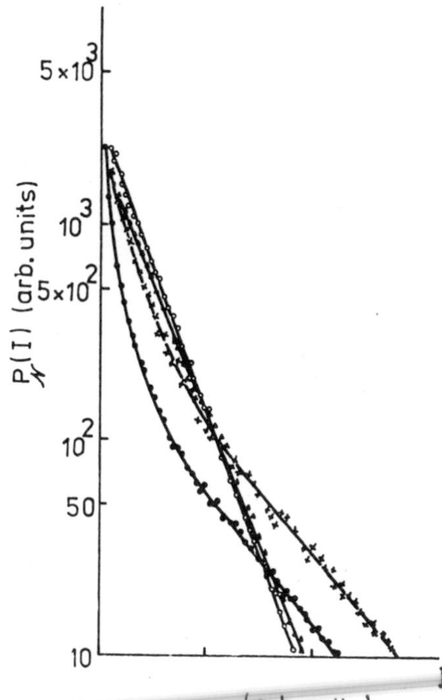

Fig. 8.8 — The intensity probability distribution of laser light scattered in a nematic liquid crystal with variable voltage: 15 V (●), 20 V (×), 30 V (△) and 40—60 V (○) (after Bertolotti, 1974, Photon Correlation and Light Beating Spectroscopy, Plenum Press).

Finally we note that if we neglect losses ($\gamma = \langle n_d \rangle = 0$), then $|u|^2 + |v|^2 = 1 + 2|v|^2$ ($|u|^2 - |v|^2 = 1$ from the commutation rule) and hence the enhancement factor is generally related to the uncertainty (the quantum chaotic component) in the photocount distribution, determined by $M(E - 1 + F - 1)/2 = M|v|^2$, i.e. the quantum noise increases as the amplification is increased [Mollow and Glauber (1967a, b) and Chapter 9].

8.8 Speckle phenomenon

When coherent laser light is reflected from a rough surface or when it propagates through a medium with random refractive index fluctuations, we can observe a typical granular pattern called the speckle phenomenon. This phenomenon depends on the coherence of incident light and the structure of the rough surface or the random medium.

The statistical properties of speckle patterns may be in general very complicated. However, their description is quite analogous to the description of the photocount statistics of radiation, so most of the results from this field can be transferred directly to investigations of the statistical properties of the speckle. If the numbers of scatterers are high, so that the central limit theorem of probability theory can be applied, the negative exponential intensity distribution is appropriate to describe the statistics of the speckle. For smaller numbers of scatterers deviations from Gaussian statistics may occur [Crosignani, Di Porto and Bertolotti (1975)]. For the sum of M speckles the probability distribution (3.139) is obtained. The intensity probability distributions for superposed multifold speckle patterns, in general partially polarized, with a coherent background, are, in fact, given in Sec. 5.3 [equations (5.97a, b), etc.]. The exact formulae have to be based on the normal characteristic function (5.93b) or (5.95a) and require the solution of the Fredholm integral equation (5.42). These questions are discussed in greater detail by Peřina (1977), Saleh and Irshid (1979), Peřina and Horák (1981).

If we define the contrast C of the speckle pattern as $C = \langle(\Delta I)^2\rangle_{\mathcal{N}}^{1/2}/\langle I\rangle_{\mathcal{N}}$, then from (3.141a)

$$C = M^{-1/2}, \tag{8.79}$$

which means that the contrast of the speckle pattern decreases with increase of M. Such a dependence of the intensity probability distribution on M and the contrast C [related by (8.79)], which is similar to the distributions shown in Fig. 8.1a, has been experimentally verified by Ohtsubo and Asakura (1977a). A connection between these quantities and the structure of the rough surface may also be established. For further reading we can recommend reviews of the statistical properties of speckle by Goodman (1975), Dainty (1975, 1976) and Zardecki (1978), where further references may be found (see also references given in the Introduction). Note that the statistics of the speckle fields in terms of the Stokes parameters have been considered by Fercher and Steeger (1981), and the role of speckle in optical fibres has been discussed by Daino, Marchis and Piazzolla (1980).

QUANTUM STATISTICS OF RADIATION IN NONLINEAR MEDIA

We may continue by applying the general methods for investigating the statistical properties of optical fields, to the study of the statistical properties of radiation in nonlinear processes, as discussed in Chapter 6. In particular we consider optical parametric processes, both non-degenerate and degenerate, with classical as well as quantum pumping; Brillouin, Raman and hyper-Raman scattering; multiphoton absorption and emission; and some special nonlinear processes.

9.1 Optical parametric processes with classical pumping

9.1.1 Degenerate case

This is the simplest non-trivial nonlinear case, and it is mathematically fully analogous to one-mode quantum model for the propagation of radiation through random media, discussed in Chapter 8. Assuming the pumping radiation to be so strong that it may be treated as a classical field, the renormalized hamiltonian for second subharmonic generation (degenerate parametric amplification) reads

$$\hat{H}_{rad} = \hbar\omega\hat{a}^+\hat{a} - \frac{1}{2}\hbar g(\hat{a}^2 \exp(\mathrm{i}2\omega t - \mathrm{i}\varphi) + \mathrm{h.c.}); \tag{9.1}$$

here, for simplicity, phase matching is assumed so that the coupling constant $g > 0$ is real, the frequency of pumping is 2ω and its phase is φ (the pumping amplitude is included in the coupling constant g). Losses may be included in the same way as in Chapter 8 [see the hamiltonian (7.1c)]. The corresponding Heisenberg–Langevin equation is

$$\frac{\mathrm{d}\hat{a}}{\mathrm{d}t} = -(\mathrm{i}\omega + \gamma/2)\,\hat{a} + \mathrm{i}g\hat{a}^+ \exp(-\mathrm{i}2\omega t + \mathrm{i}\varphi) + \hat{L}, \tag{9.2}$$

where the Langevin force $\hat{L}$ is given by (7.9) and the generalized Fokker–Planck equation can be written as

$$\frac{\partial \Phi_{\mathscr{A}}}{\partial t} = \left\{(\mathrm{i}\omega + \gamma/2)\frac{\partial}{\partial\alpha}\alpha + \mathrm{c.c.} + \gamma(\langle n_d\rangle + 1)\frac{\partial^2}{\partial\alpha\,\partial\alpha^*} + \right.$$
$$\left. + \left[\mathrm{i}g \exp(\mathrm{i}2\omega t - \mathrm{i}\varphi)\left(\alpha\frac{\partial}{\partial\alpha^*} + \frac{1}{2}\frac{\partial^2}{\partial\alpha^{*2}}\right) + \mathrm{c.c.}\right]\right\}\Phi_{\mathscr{A}}, \tag{9.3a}$$

where α is an eigenvalue of $\hat{a}$ in the coherent state $|\,\alpha\rangle$. The corresponding equation for the characteristic function $C_{\mathscr{A}}(\beta', t)$, $\beta' = \beta \exp(\mathrm{i}\omega t)$, reads

$$\frac{\partial C_{\mathscr{A}}}{\partial t} = \left\{ -\frac{\gamma}{2}\left(\beta' \frac{\partial}{\partial \beta'} + \text{c.c.}\right) - \gamma(\langle n_d \rangle + 1)\,|\beta'|^2 + \right.$$
$$\left. + \mathrm{i}g\left[\exp(-\mathrm{i}\varphi)\,\beta' \frac{\partial}{\partial \beta'^*} - \text{c.c.}\right] + \frac{\mathrm{i}g}{2}\left[\exp(-\mathrm{i}\varphi)\,\beta'^2 - \text{c.c.}\right] \right\} C_{\mathscr{A}}. \tag{9.3b}$$

The solution of the Heisenberg–Langevin equation (9.2), together with its h.c. equation, has the form (8.48) and if losses are neglected

$$u(t) = \operatorname{ch}(gt)\exp(-\mathrm{i}\omega t),$$
$$v(t) = \mathrm{i}\operatorname{sh}(gt)\exp(-i\omega t + \mathrm{i}\varphi). \tag{9.4}$$

The characteristic function $C_{\mathscr{A}}$ and the quasi-distribution $\Phi_{\mathscr{A}}$ are determined by equations (8.43) and (8.44), with the appropriate generating function (8.61), the photocount distribution given by (8.64a) and its factorial moments (8.64b) with $M = 1$ [Mišta, Peřinová, Peřina and Braunerová (1977)], where

$$B(t) = c_1 - c_2 \exp(-\gamma t)\operatorname{ch}(2gt) - c_3 \exp(-\gamma t)\operatorname{sh}(2gt),$$
$$C'(t) = C(t) = \mathrm{i}\exp(-\mathrm{i}2\omega t + \mathrm{i}\varphi)\times$$
$$\times\left[-c_2 \exp(-\gamma t)\operatorname{sh}(2gt) + c_3(1 - \exp(-\gamma t))\operatorname{ch}(2gt)\right],$$
$$c_1 = \frac{\gamma^2(1 + \langle n_d \rangle) - 2g^2}{\gamma^2 - 4g^2}, \qquad c_2 = \frac{\gamma^2\langle n_d \rangle + 2g^2}{\gamma^2 - 4g^2},$$
$$c_3 = \frac{\gamma g(1 + 2\langle n_d \rangle)}{\gamma^2 - 4g^2} \tag{9.5a}$$

for $\gamma/2 \neq g$, while for $\gamma/2 = g$

$$B(t) = \frac{7 + 2\langle n_d \rangle}{8} + \frac{1 + 2\langle n_d \rangle}{4}\gamma t + \frac{1 - 2\langle n_d \rangle}{8}\exp(-2\gamma t),$$
$$C(t) =$$
$$= \mathrm{i}\exp(-\mathrm{i}2\omega t + \mathrm{i}\varphi)\left[\frac{1 + 2\langle n_d \rangle}{4}\gamma t + \frac{1 - 2\langle n_d \rangle}{8}(1 - \exp(-2\gamma t))\right]. \tag{9.5b}$$

Neglecting losses and making the correspondence with equation (4.198) for two-photon coherent states, we see that $u = \mu^*$ and $v = -\nu$. Further, taking into account the relation between the Hermite and Laguerre polynomials

$$H_{2n}(x) = \frac{(-1)^n 2^{2n} n!}{\Gamma(n + 1/2)} L_n^{-1/2}(x^2), \tag{9.6a}$$

the Hermite polynomial $H_n(x)$ being defined as

$$H_n(x) = (-1)^n \exp(x^2)\,\frac{\mathrm{d}^n \exp(-x^2)}{\mathrm{d}x^n}, \tag{9.6b}$$

we can prove the following identity by simple operator algebra [Mišta, Peřinová, Peřina and Braunerová (1977)]:

$$\sum_{k=0}^{\infty} \frac{(-1)^k}{\Gamma(k+1/2)\,\Gamma(n-k+1/2)} L_k^{-1/2}(x^2)\, L_{n-k}^{-1/2}(y^2) =$$

$$= \frac{1}{2^n n!} H_n\left(\frac{x+y}{2^{1/2}}\right) H_n\left(\frac{x-y}{2^{1/2}}\right). \tag{9.7}$$

This identity enables us to transform the expressions (8.64a) and (8.59b) for the photocount distribution and for the photocount generating function to the forms (4.206) and (4.207) respectively ($B = |u|^2$, $C = uv$), which have been obtained for the two-photon coherent states by Yuen (1976). This demonstrates the full mathematical equivalence of the two-photon coherent state technique with the results obtained for the one-mode hamiltonian (8.28) or for the hamiltonian (9.1). From this it is evident that the two-photon coherent states are generated rather by the degenerate parametric amplification process with classical pumping, than by two-photon stimulated emission, unless the atom variables are described classically.

Since $(B-1)^2 - |C|^2 = (\mathrm{ch}^2 gt - 1)^2 - \mathrm{ch}^2 gt\, \mathrm{sh}^2 gt = -\mathrm{sh}^2 gt < 0$, which is involved in the Fourier transformation to obtain $\Phi_{\mathcal{N}}(\alpha, t)$, this quasi-distribution does not exist at all; on the other hand, $\Phi_{\mathcal{A}}(\alpha, t)$ always exists, since $B^2 - |C|^2 = \mathrm{ch}^2 gt > 0$.

For the mean number of photons and for the variance of the integrated intensity we obtain [Stoler (1974), Mišta and Peřina (1977b, c)]

$$\langle n(t)\rangle = \langle W(t)\rangle_{\mathcal{N}} = \langle \hat{a}^+(t)\,\hat{a}(t)\rangle = \frac{\mathrm{d}}{\mathrm{d(is)}} \langle \exp(\mathrm{i}sW)\rangle_{\mathcal{N}}\Big|_{\mathrm{i}s=0} =$$

$$= [\mathrm{ch}\,(2gt) + \mathrm{sh}\,(2gt)\sin(2\vartheta - \varphi)]\,|\xi|^2 + \mathrm{sh}^2(gt), \tag{9.8a}$$

$$\langle (\Delta W(t))^2\rangle_{\mathcal{N}} = \langle \hat{a}^{+2}(t)\,\hat{a}^2(t)\rangle - \langle \hat{a}^+(t)\,\hat{a}(t)\rangle^2 =$$

$$= \frac{\mathrm{d}^2}{\mathrm{d(is)}^2} \langle \exp(\mathrm{i}sW)\rangle_{\mathcal{N}}\Big|_{\mathrm{i}s=0} - \left[\frac{\mathrm{d}}{\mathrm{d(is)}} \langle \exp(\mathrm{i}sW)\rangle_{\mathcal{N}}\Big|_{\mathrm{i}s=0}\right]^2 =$$

$$= \mathrm{sh}\,(gt)\left[\frac{\mathrm{sh}\,(3gt) - \mathrm{sh}\,(gt)}{2} + 2|\xi|^2 \mathrm{sh}\,(3gt) + \right.$$

$$\left. + 2|\xi|^2\, \mathrm{ch}\,(3gt)\sin(2\vartheta - \varphi)\right], \tag{9.8b}$$

where, for simplicity, we do not consider the reservoir, and ϑ is the phase of the initial complex amplitude of the incident coherent signal field. Alternatively, these expressions can be obtained if we substitute the solution for $\hat{a}(t)$ and perform the normal ordering in the initial operators. The last term in (9.8a) and the first term in (9.8b) represent the quantum noise contributions (from the physical vacuum), since they are independent of the incident field. It has been shown by Stoler (1974) that if the phases are related so that $2\vartheta - \varphi = -\pi/2$, maximal antibunching of photons occurs, i.e. $\langle (\Delta W(t))^2\rangle_{\mathcal{N}} < 0$. In this case the photocount distribution $p(n, t)$ is narrower than the Poisson distribution corresponding to the coherent state and we have sub-Poissonian radiation. This holds for a certain time interval after the switching on of the interaction. In Fig. 9.1 we can see the time behaviour of the mean intensity (mean number of photons) including this sub-Poissonian regime. The attenuation of the incident field initially in the coherent state stops at $t \approx 6 \times 10^{-5}$ s, in approximate agreement with the region of antibunching, seen in Fig. 9.2, where $\langle W^2 \rangle_{\mathcal{N}} / \langle W \rangle_{\mathcal{N}}^2 - 1 < 0$ (more complicated relations between the attenuation of the field and the photon antibunching can be demonstrated in Brillouin scattering, see Sec. 9.4). In this figure the reduced factorial moments $\langle W^k \rangle_{\mathcal{N}} / \langle W \rangle_{\mathcal{N}}^k - 1$ are shown for $k = 2 - 5$ (solid curves); the dotted curve represents the quantity $\langle (\Delta W)^2 \rangle_{\mathcal{N}} / \langle W \rangle_{\mathcal{N}}$, with its minimum about -0.4 [it is equal to -1 for the Fock state, equation (3.153)]. Antibunching manifests itself also in the higher-order moments. However,

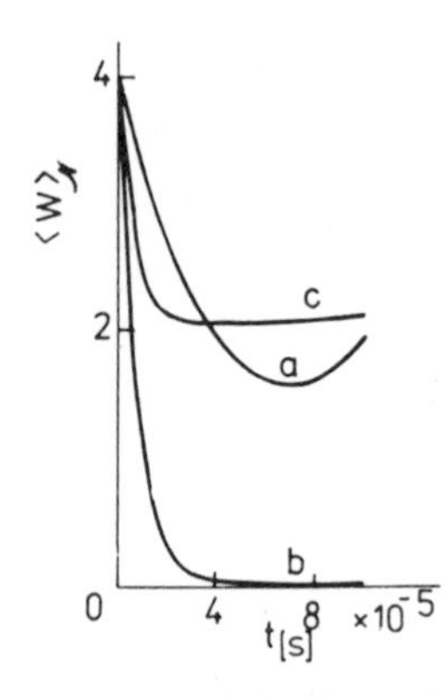

Fig. 9.1 — The time behaviour of the mean integrated intensity for $|\xi| = 2$, $2\vartheta - \varphi = -\pi/2$, $g = 10^4\ \mathrm{s}^{-1}$; curves a—c are for $\gamma = \langle n_d \rangle = 0$; $\gamma = 10^5\ \mathrm{s}^{-1}$, $\langle n_d \rangle = 0$ and $\gamma = 10^5\ \mathrm{s}^{-1}$, $\langle n_d \rangle = 2$ respectively (after Mišta, Peřinová, Peřina and Braunerová, 1977, Acta Phys. Pol. **A51**, 739).

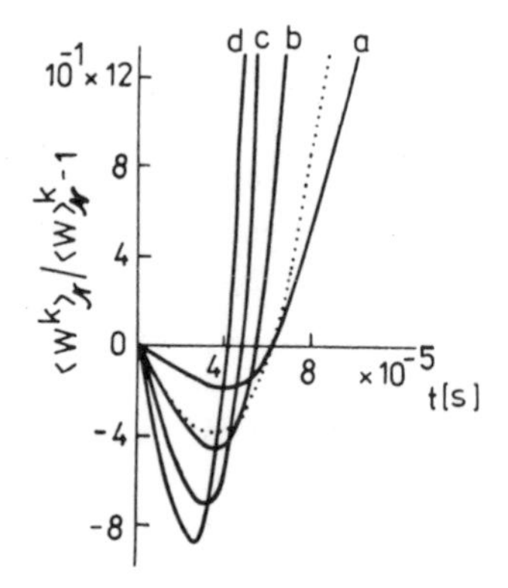

Fig. 9.2 — The reduced factorial moments $\langle W^k \rangle_{\mathcal{N}} / \langle W \rangle_{\mathcal{N}}^k - 1$, $k = 2$—5 (solid curves a—d respectively) for $|\xi| = 2$, $2\vartheta - \varphi = -\pi/2$, $g = 10^4\ \mathrm{s}^{-1}$ and $\gamma = \langle n_d \rangle = 0$. The dotted curve represents the quantity $\langle (\Delta W)^2 \rangle_{\mathcal{N}} / \langle W \rangle_{\mathcal{N}}$ (after Mišta, Peřinová, Peřina and Braunerová, 1977, Acta Phys. Pol. **A51**, 739).

the curves corresponding to various values of k have not a common point of intersection with the t-axis and thus the initially coherent field cannot be coherent again at any later time (this may happen in Brillouin scattering, Sec. 9.4). However, at a certain instant the field may be coherent in fourth order. This is a typical property of fields having no classical analogues (for classical fields coherence to all orders follows from second- and fourth-order coherence [Titulaer and Glauber (1965), Sec. 3.3]. Further we observe that the amplification is related, as before, to an increase in quantum noise, and the decrease of the fluctuations in n below the level appropriate to the coherent state is usually related to the attenuation of the field, although this is not necessary (Sec. 9.4). The asymptotic values of the curves shown are $\langle W^k \rangle_{\mathcal{N}} / \langle W \rangle_{\mathcal{N}}^k - 1 = (2k - 1)!! - 1$ ($= 2$ for $k = 2$), i.e. chaotic fluctuations are enhanced for long-time intervals [cf. (3.131)]. Note that the entropy properties of „superchaotic" fields have been investigated by Sotskii and Glazatchev (1981).

In Fig. 9.3 we see the time development of the photocount distribution including the sub-Poissonian regime ($2\vartheta - \varphi = -\pi/2$) for $g = 10^4\ \mathrm{s}^{-1}$, neglecting losses ($\gamma = \langle n_d \rangle = 0$), and with $|\xi|^2 = 4$ for the initial intensity and $t = 2 \times 10^{-5}$ s (*a*), 3×10^{-5} s (*b*), 6×10^{-5} s (*c*) and 8×10^{-5} s (*d*). The full curves show the actual

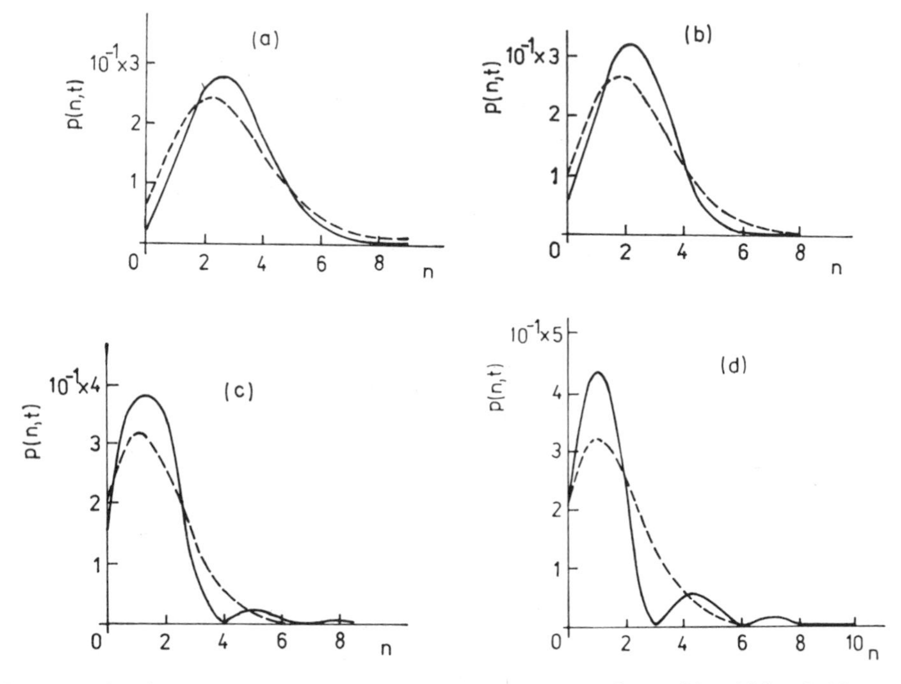

Fig. 9.3 — The photocount distribution for a degenerate parametric amplifier, $|\xi| = 2$, $2\vartheta - \varphi = -\pi/2$, $g = 10^4\ \mathrm{s}^{-1}$, $\gamma = \langle n_d \rangle = 0$ and (a) $t = 2\times10^{-5}$ s, (b) $t = 3\times10^{-5}$ s, (c) $t = 6\times10^{-5}$ s and (d) $t = 8\times10^{-5}$ s. The full curves represent the actual distributions, the dashed curves are the corresponding Poisson distributions (after Mišta, Peřinová, Peřina and Braunerová, 1977, Acta Phys. Pol. **A51**, 739).

distribution, the dashed curves represent the Poisson distribution with the same $\langle W(t) \rangle_{\mathcal{N}}$ appropriate to the coherent state. Figures (*a*) and (*b*) clearly demonstrate sub-Poissonian behaviour when the actual distributions are narrower than the corresponding Poisson distributions. For later times a typical oscillating behaviour is reached, reflecting the appearance of competition between states. Sub-Poissonian behaviour can also occur if losses are included, but with increasing $\langle n_d \rangle$ it is quickly smoothed out [Mišta, Peřinová, Peřina and Braunerová (1977)].

Some efforts to enhance photon antibunching by using interference have been made by Bandilla and Ritze (1979).

A method of producing photon antibunching based on degenerate parametric amplification with classical pumping, has been proposed by Stoler (1974). In one nonlinear crystal radiation of frequency 2ω is generated, and it is used as the pumping radiation in a second crystal, where the fundamental wave of frequency ω is introduced as the initial signal. If $2\vartheta - \varphi = -\pi/2$, the initial signal is attenuated at the beginning of the interaction and antibunching has to occur in the signal mode.

9.1.2 Non-degenerate case

The non-degenerate version of the parametric generation ($\xi_1 = \xi_2 = 0$) or amplification ($\xi_1 \neq 0$, $\xi_2 = 0$) process with classical pumping is described by the renormalized hamiltonian

$$\hat{H}_{\text{rad}} = \hbar\omega_1 \hat{a}_1^+ \hat{a}_1 + \hbar\omega_2 \hat{a}_2^+ \hat{a}_2 - \hbar g(\hat{a}_1 \hat{a}_2 \exp(\mathrm{i}\omega t - \mathrm{i}\varphi) + \text{h.c.}), \tag{9.9}$$

assuming phase matching again (the coupling constant g is real) and $\omega = \omega_1 + \omega_2$. The corresponding Heisenberg–Langevin equations are

$$\frac{\mathrm{d}\hat{a}_1}{\mathrm{d}t} = -\mathrm{i}\omega_1 \hat{a}_1 + \mathrm{i}g\hat{a}_2^+ \exp(-\mathrm{i}\omega t + \mathrm{i}\varphi) + \hat{L}_1,$$

$$\frac{\mathrm{d}\hat{a}_2}{\mathrm{d}t} = -\mathrm{i}\omega_2 \hat{a}_2 + \mathrm{i}g\hat{a}_1^+ \exp(-\mathrm{i}\omega t + \mathrm{i}\varphi) + \hat{L}_2, \tag{9.10}$$

where the Langevin forces are given by (7.9) for every mode. If one neglects losses, the solution of (9.10) has the form

$$\hat{a}_1(t) = \exp(-\mathrm{i}\omega_1 t)\left[\hat{a}_1 \operatorname{ch}(gt) + \mathrm{i}\hat{a}_2^+ \operatorname{sh}(gt) \exp(\mathrm{i}\varphi)\right],$$

$$\hat{a}_2(t) = \exp(-\mathrm{i}\omega_2 t)\left[\hat{a}_2 \operatorname{ch}(gt) + \mathrm{i}\hat{a}_1^+ \operatorname{sh}(gt) \exp(\mathrm{i}\varphi)\right]. \tag{9.11}$$

The generalized Fokker–Planck equation for this process reads [Graham (1968a, b), Mišta and Peřina (1978)]

$$\frac{\partial \Phi_{\mathscr{A}}}{\partial t} = \left\{ \sum_{j=1}^{2} \left[(\mathrm{i}\omega_j + \gamma_j/2) \frac{\partial}{\partial \alpha_j} \alpha_j + \text{c.c.} + \gamma_j(\langle n_{dj} \rangle + 1) \frac{\partial^2}{\partial \alpha_j \, \partial \alpha_j^*} \right] + \right.$$
$$\left. + \left[\mathrm{i}g \exp(\mathrm{i}\omega t - \mathrm{i}\varphi) \left(\alpha_1 \frac{\partial}{\partial \alpha_2^*} + \alpha_2 \frac{\partial}{\partial \alpha_1^*} + \frac{\partial^2}{\partial \alpha_1^* \, \partial \alpha_2^*} \right) + \text{c.c.} \right] \right\} \Phi_{\mathscr{A}}. \tag{9.12a}$$

Making use of the transformation $\alpha_j = \alpha_j' \exp(-\mathrm{i}\omega_j t + \mathrm{i}\varphi/2)$ and performing the Fourier transformation, we have for the characteristic function

$$\frac{\partial C_{\mathscr{A}}(\beta_1', \beta_2', t)}{\partial t} =$$
$$= \left\{ \sum_{j=1}^{2} \left[-\frac{\gamma_j}{2} \left(\beta_j' \frac{\partial}{\partial \beta_j'} + \beta_j'^* \frac{\partial}{\partial \beta_j'^*} \right) - \gamma_j(\langle n_{dj} \rangle + 1) \, |\beta_j'|^2 \right] + \right.$$
$$\left. + \mathrm{i}g \left[\left(\beta_2' \frac{\partial}{\partial \beta_1'^*} + \beta_1' \frac{\partial}{\partial \beta_2'^*} + \beta_1' \beta_2' \right) - \text{c.c.} \right] \right\} C_{\mathscr{A}}(\beta_1', \beta_2', t); \tag{9.12b}$$

its solution is of the form, returning to the original variables,

$$C_{\mathscr{A}}(\beta_1, \beta_2, t) = \\ = \langle \exp \{ - \sum_{j=1}^{2} [|\beta_j|^2 B_j(t) + \beta_j \xi_j^*(t) - \beta_j^* \xi_j(t)] + \\ + (D_{12}^*(t) \beta_1 \beta_2 + \text{c.c.})\} \rangle. \tag{9.13a}$$

Then Fourier transformation provides the quasi-distribution

$$\Phi_{\mathscr{A}}(\alpha_1, \alpha_2, t) = \left\langle (\pi^2 K(t))^{-1} \exp \left\{ - \frac{B_2(t)}{K(t)} |\alpha_1 - \xi_1(t)|^2 - \right.\right. \\ \left.\left. - \frac{B_1(t)}{K(t)} |\alpha_2 - \xi_2(t)|^2 - \left[\frac{D_{12}^*(t)}{K(t)} (\alpha_1 - \xi_1(t))(\alpha_2 - \xi_2(t)) + \text{c.c.} \right] \right\} \right\rangle, \tag{9.13b}$$

where [Mišta and Peřina (1978)]

$$B_1(t) = B_1(\langle n_{d1} \rangle, \langle n_{d2} \rangle, t) = \\ = \frac{\gamma(\langle n_{d1} \rangle + 1)}{2} \left[\frac{\gamma}{\gamma^2 - 4g^2} (1 - \text{ch}(2gt) \exp(-\gamma t)) + \right. \\ \left. + \frac{1}{\gamma} (1 - \exp(-\gamma t)) - \frac{2g}{\gamma^2 - 4g^2} \text{sh}(2gt) \exp(-\gamma t) \right] + \\ + \frac{\gamma(\langle n_{d2} \rangle + 1)}{2} \left[\frac{\gamma}{\gamma^2 - 4g^2} (1 - \text{ch}(2gt) \exp(-\gamma t)) - \right. \\ \left. - \frac{1}{\gamma} (1 - \exp(-\gamma t)) - \frac{2g}{\gamma^2 - 4g^2} \text{sh}(2gt) \exp(-\gamma t) \right] + \\ + g \left[\frac{2g}{\gamma^2 - 4g^2} (\text{ch}(2gt) \exp(-\gamma t) - 1) + \right. \\ \left. + \frac{\gamma}{\gamma^2 - 4g^2} \text{sh}(2gt) \exp(-\gamma t) \right] + \text{ch}(2gt) \exp(-\gamma t),$$

$$B_2(t) = B_1(\langle n_{d2} \rangle, \langle n_{d1} \rangle, t),$$

$$D_{12}(t) = \mathrm{i} \exp(-\mathrm{i}\omega t + \mathrm{i}\varphi) \left\{ \frac{\gamma(\langle n_{d1} \rangle + \langle n_{d2} \rangle + 2)}{2} \times \right. \\ \times \left[\frac{2g}{\gamma^2 - 4g^2} (1 - \text{ch}(2gt) \exp(-\gamma t)) - \frac{\gamma}{\gamma^2 - 4g^2} \text{sh}(2gt) \exp(-\gamma t) \right] - \\ - g \left[\frac{\gamma}{\gamma^2 - 4g^2} (1 - \text{ch}(2gt) \exp(-\gamma t)) - \frac{2g^2}{\gamma^2 - 4g^2} \text{sh}(2gt) \exp(-\gamma t) \right] + \\ \left. + \text{sh}(2gt) \exp(-\gamma t) \right\},$$

$$\xi_1(t) = \exp(-\mathrm{i}\omega_1 t - \gamma t/2) [\xi_1 \text{ch}(gt) + \mathrm{i}\xi_2^* \exp(\mathrm{i}\varphi) \text{sh}(gt)], \\ \xi_2(t) = \exp(-\mathrm{i}\omega_2 t - \gamma t/2) [\xi_2 \text{ch}(gt) + \mathrm{i}\xi_1^* \exp(\mathrm{i}\varphi) \text{sh}(gt)], \\ K(t) = B_1(t) B_2(t) - |D_{12}(t)|^2; \tag{9.14a}$$

the expressions for $B_{1,2}(t)$ and $D_{12}(t)$ hold if $\gamma/2 \neq g$ ($\gamma_1 = \gamma_2 = \gamma$), whereas for $\gamma/2 = g$

$$\begin{aligned}
B_1(t) &= B_1(\langle n_{d1}\rangle, \langle n_{d2}\rangle, t) = \\
&= \frac{\gamma(\langle n_{d1}\rangle + 1)}{2}\left[\frac{t}{2} + \frac{1 - \exp(-\gamma t)}{\gamma} + \frac{1 - \exp(-2\gamma t)}{4\gamma}\right] + \\
&+ \frac{\gamma(\langle n_{d2}\rangle + 1)}{2}\left[\frac{t}{2} + \frac{\exp(-\gamma t) - 1}{\gamma} + \frac{1 - \exp(-2\gamma t)}{4\gamma}\right] - \\
&- \frac{\gamma}{4}\left[t + \frac{\exp(-2\gamma t) - 1}{2\gamma}\right] + \frac{1}{2}(1 + \exp(-2\gamma t)), \\
B_2(t) &= B_1(\langle n_{d2}\rangle, \langle n_{d1}, t), \\
D_{12}(t) &= \\
&= \mathrm{i}\exp(-\mathrm{i}\omega t + \mathrm{i}\varphi)\left\{\frac{\gamma(\langle n_{d1}\rangle + \langle n_{d2}\rangle + 2)}{4}\left[t + \frac{\exp(-2\gamma t) - 1}{2\gamma}\right] - \right. \\
&\left. - \frac{\gamma}{4}\left[t + \frac{1 - \exp(-2\gamma t)}{2}\right] + \frac{1}{2}(1 - \exp(-2\gamma t))\right\}; \qquad (9.14b)
\end{aligned}$$

while, neglecting losses, $\gamma = \langle n_{d1}\rangle = \langle n_{d2}\rangle = 0$ and

$$\begin{aligned}
B_1(t) &= B_2(t) = K(t) = \mathrm{ch}^2\, gt, \\
D_{12}(t) &= \frac{\mathrm{i}}{2}\,\mathrm{sh}\,(2gt)\exp(-\mathrm{i}\omega t + \mathrm{i}\varphi). \qquad (9.14c)
\end{aligned}$$

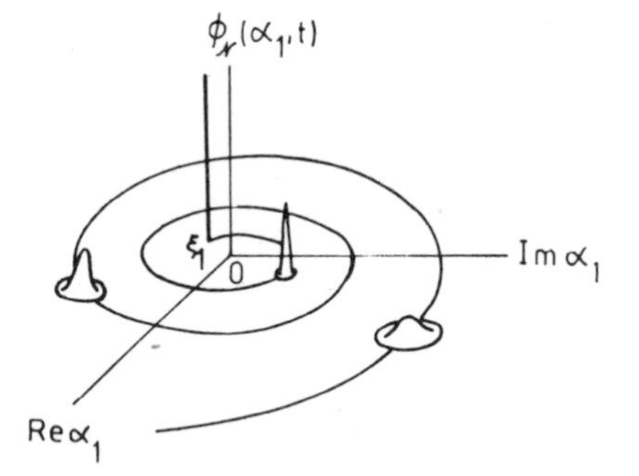

Fig. 9.4 — A schematic description of the time evolution of $\Phi_{\mathcal{N}}(\alpha_1, t)$ for the signal mode of the parametric amplifier, which is initially in the coherent state $|\,\xi_1\rangle$ and the idler mode starts from the vacuum.

Calculating $C_{\mathcal{N}}(\beta_1, 0, t)$ and $\Phi_{\mathcal{N}}(\alpha_1, t)$ with the use of (9.11), we find that the initial field $|\,\xi_1, 0\rangle$ is transformed by this process to the superposition of a signal $\xi_1(t) = \xi_1\,\mathrm{ch}\,(gt)\exp(-\mathrm{i}\omega_1 t)$ and noise $B_1(t) = 1 + \mathrm{sh}^2(gt)$, with $\Phi_{\mathscr{A}}$ given by (8.54), that is

$$\Phi_{\mathcal{N}}(\alpha_1, t) = [\pi(B_1(t) - 1)]^{-1}\exp\left[-\frac{|\,\alpha_1 - \xi_1(t)\,|^2}{B_1(t) - 1}\right], \qquad (9.15)$$

and the peak of this Gaussian distribution, whose standard deviation $\mathrm{sh}^2\, gt$ increases with time, moves along the spiral $\xi_1(t) = \xi_1 \operatorname{ch}(gt) \exp(-i\omega_1 t)$ [Mollow and Glauber (1967a, b), Mišta (1969), Mišta and Peřina (1978)], as illustrated in Fig. 9.4. For the mean number of photons in the first signal mode we obtain the expression

$$\langle \hat{a}_1^+(t)\, \hat{a}_1(t) \rangle = \langle \hat{a}_1^+ \hat{a}_1 \rangle \operatorname{ch}^2(gt) + \langle \hat{a}_2^+ \hat{a}_2 \rangle \operatorname{sh}^2(gt) + \\ + \frac{i}{2} \operatorname{sh}(2gt) \left(\langle \hat{a}_1^+ \hat{a}_2^+ \rangle \exp(i\varphi) - \text{c.c.} \right) \underset{t\to\infty}{\simeq} \\ \underset{t\to\infty}{\simeq} \frac{1}{4} (\langle n_1 \rangle + \langle n_2 \rangle + 1) \exp(2gt), \tag{9.16}$$

provided that the initial phases are uncertain; this is in agreement with (6.9) [$\langle n_2 \rangle = 0$, $\langle n_1 \rangle \gg 1$, so that the quantum noise $\exp(2gt)/4$ can be neglected]. We observe again that the amplification of the field is related to an increase of uncertainty.

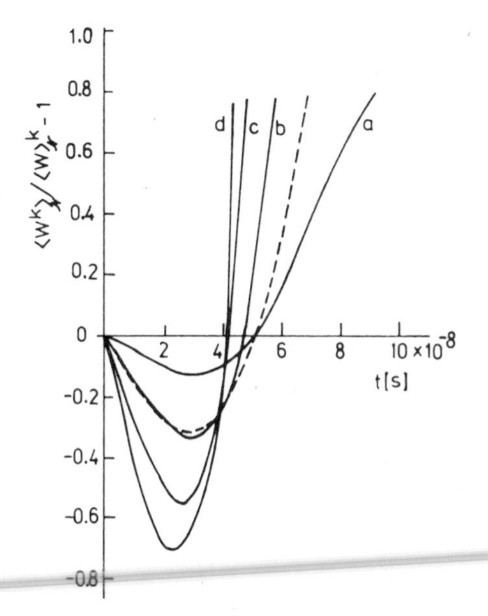

Fig. 9.5 — The time behaviour of the reduced factorial moments $\langle W^k \rangle_{\mathcal{N}} / \langle W \rangle_{\mathcal{N}}^k - 1$ for the parametric amplification process, $\varphi_1 + \varphi_2 - \varphi = -\pi/2$, $|\xi|^2 = 2$, $|g| = 10^7\ \text{s}^{-1}$, $\gamma = \langle n_d \rangle = 0$ for $k = 2$—5 (curves *a*—*d* respectively); the dashed curve represents the quantity $\langle (\Delta W)^2 \rangle_{\mathcal{N}} / \langle W \rangle_{\mathcal{N}}$ (after Peřinová and Peřina, 1981, Opt. Acta **28**, 769).

The solution of equation (9.12a) has again the form of a multidimensional superposition of coherent and chaotic radiation, averaged over the initial amplitudes, and the normal generating function, the photocount distribution and its factorial moments are expressed by equations (8.61) and (8.64a, b) with $M = 2$. The mean intensity equals

$$\langle W(t) \rangle_{\mathcal{N}} = \left. \frac{\partial^2 C_{\mathcal{N}}}{\partial \beta_1\, \partial(-\beta_1^*)} \right|_{\beta_1=\beta_2=0} + \left. \frac{\partial^2 C_{\mathcal{N}}}{\partial \beta_2 (-\partial \beta_2^*)} \right|_{\beta_1=\beta_2=0} = \\ = B_1(t) + B_2(t) + |\xi_1(t)|^2 + |\xi_2(t)|^2, \tag{9.17}$$

where the first two terms represent quantum noise related to the physical vacuum. The non-existence of the Glauber–Sudarshan quasi-distribution $\Phi_{\mathcal{N}}(\alpha_1, \alpha_2, t)$ in this process [Mollow and Glauber (1967b)] is related to possible anticorrelation between the signal and idler modes 1 and 2, $\langle \Delta W_1 \Delta W_2 \rangle_{\mathcal{N}} < 0$, as follows from the

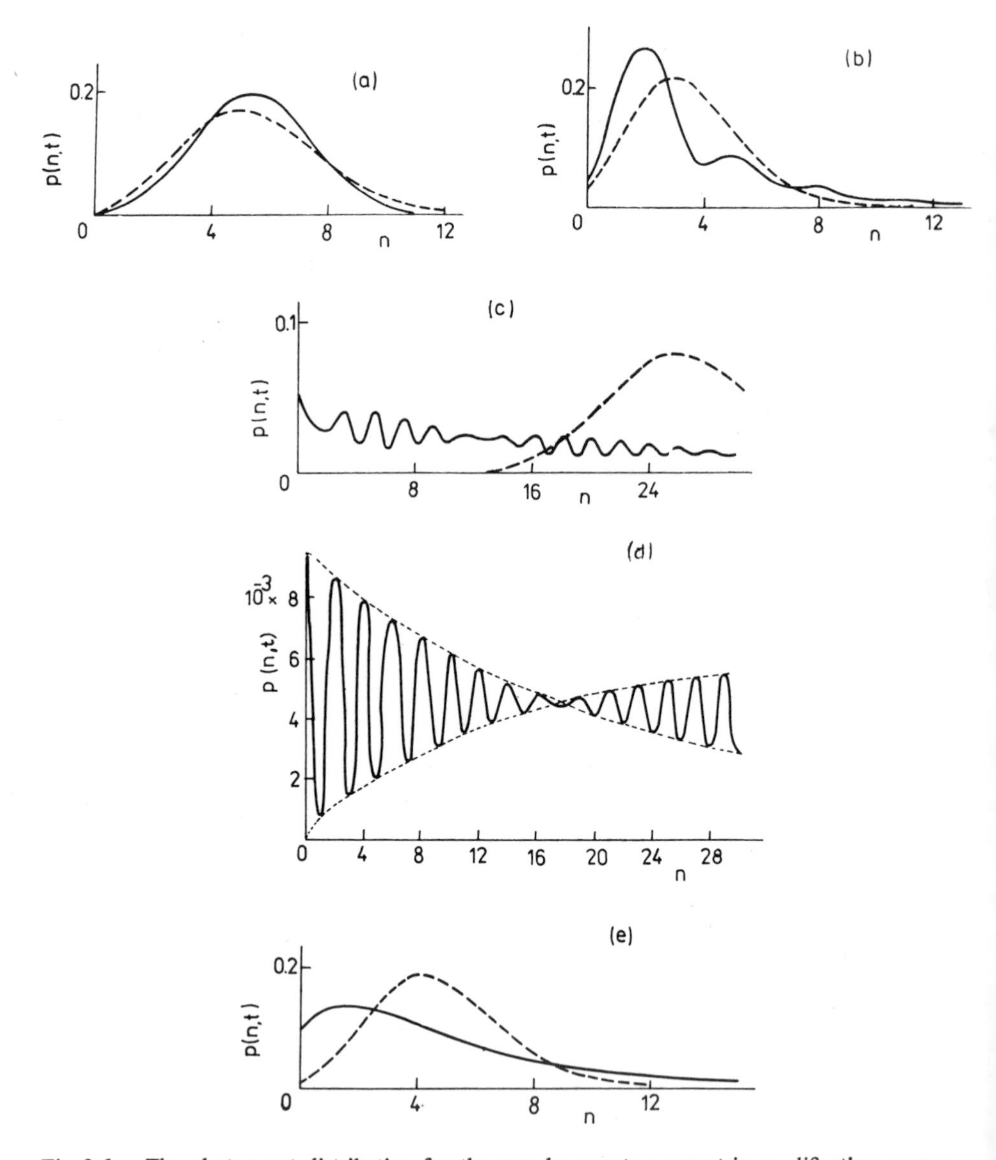

Fig. 9.6 — The photocount distribution for the non-degenerate parametric amplification process, $\varphi_1 + \varphi_2 - \varphi = -\pi/2$, $|\xi_1| = |\xi_2| = 2$, $g = 10^4\ \mathrm{s}^{-1}$ and (a) $\gamma = \gamma_1 = \gamma_2 = 0$, $\langle n_{d1} \rangle = \langle n_{d2} \rangle = 0$, $t = 2 \times 10^{-5}$ s, (b) as in (a), $t = 9 \times 10^{-5}$ s, (c) as in (a), $t = 2 \times 10^{-4}$ s, (d) as in (a), $t = 3 \times 10^{-4}$ s and (e) $\gamma = 2 \times 10^4\ \mathrm{s}^{-1}$, $\langle n_{d1} \rangle = \langle n_{d2} \rangle = 2$, $t = 7 \times 10^{-5}$ s. The solid curves represent the actual distributions, the dashed curves are the corresponding Poisson distributions; in (d) the Poisson distribution is not shown (after Mišta and Peřina, 1978, Czech. J. Phys. **B28**, 392).

following expression [Mišta and Peřina (1977b, c)]

$$\langle(\Delta W)^2\rangle_{\mathcal{N}} = \langle(\Delta W_1)^2\rangle_{\mathcal{N}} + \langle(\Delta W_2)^2\rangle_{\mathcal{N}} + 2\langle\Delta W_1 \Delta W_2\rangle_{\mathcal{N}} =$$
$$= 2\,\mathrm{sh}\,(gt)\left[\frac{\mathrm{sh}\,(3gt) - \mathrm{sh}\,(gt)}{2} + \mathrm{sh}\,(3gt)\,(|\xi_1|^2 + |\xi_2|^2) + \right.$$
$$\left. + 2\,\mathrm{ch}\,(3gt)\,|\xi_1|\,|\xi_2|\sin(\varphi_1 + \varphi_2 - \varphi)\right]. \tag{9.18}$$

Here φ_j are the phases of the initial complex amplitudes ξ_j of the coherent field, $\langle(\Delta W)^2\rangle_{\mathcal{N}} = \langle\mathcal{N}\hat{n}^2(t)\rangle - \langle\hat{n}(t)\rangle^2$, and $\hat{n}(t) = \hat{a}_1^+(t)\,\hat{a}_1(t) + \hat{a}_2^+(t)\,\hat{a}_2(t)$, $\mathcal{N}$ being the normal ordering operator. Expression (9.18) generalizes the Stoler expression (9.8b) for the case of second subharmonic generation, that is obtained by putting $\xi_1 = \xi_2 = \xi$ and dividing the expression by 2; here the simultaneous detection of both modes is assumed. Antibunching is maximal when $\varphi_1 + \varphi_2 - \varphi = -\pi/2$ and is due to the coupling of modes. It is a purely quantum phenomenon, as discussed by Paul and Brunner (1981). The time behaviour of the reduced factorial moments $\langle W^k\rangle_{\mathcal{N}}/\langle W\rangle_{\mathcal{N}}^k - 1$ has been found to be quite similar to that in the above degenerate case, as shown in Fig. 9.5. The time behaviour of $p(n, t)$ is demonstrated in Fig. 9.6.

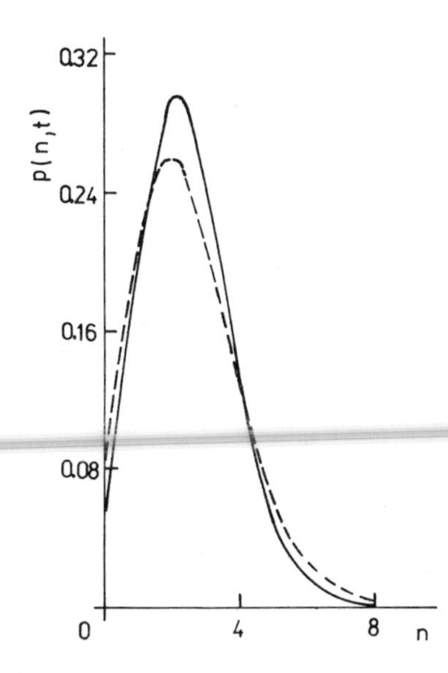

Fig. 9.7 — The photocount distribution giving evidence for sub-Poissonian radiation in parametric amplification process with losses, $\varphi_1 + \varphi_2 - \varphi = -\pi/2$, $|\xi|^2 = \langle n_d\rangle = 2$, $|g| = 10^7\,\mathrm{s}^{-1}$, $\gamma = 10^6\,\mathrm{s}^{-1}$, $t = 3\times10^{-8}$ s. The solid curve represents the actual distribution, the dashed one is the Poisson distribution with the same $\langle W\rangle_{\mathcal{N}}$ (after Peřinová and Peřina, 1981, Opt. Acta **28**, 769).

The solid curves represent the actual distribution and the dashed ones again represent the Poisson distribution with the same $\langle W(t)\rangle_{\mathcal{N}}$. In Fig. 9.6a sub-Poissonian behaviour takes place, while in Fig. 9.6e the photocount distribution of the reservoir is reached (the Mandel–Rice formula with $M = 2$ is appropriate). At the beginning of the interaction the actual photocount distribution is indeed narrower than the corresponding Poisson distribution. In the generating function (8.59b) this is reflected by a negative „mean number of chaotic photons" $E_j - 1 = B_j - |C_j| - 1$. Also Fig. 9.7 shows evidence for sub-Poissonian radiation even if losses are included ($\langle n_d \rangle = 2$, $\gamma = 10^6\ \mathrm{s}^{-1}$). The effect of pumping fluctuations has been investigated by Peřinová and Peřina (1981), and is shown in Fig. 9.8; Fig. 9.8a represents the normalized variance $\langle (\Delta W)^2 \rangle_{\mathcal{N}} = \langle (\Delta W_1)^2 \rangle_{\mathcal{N}} + \langle (\Delta W_2)^2 \rangle_{\mathcal{N}} + 2\langle \Delta W_1 \Delta W_2 \rangle_{\mathcal{N}}$, whereas the normalized correlations between fluctuations $\langle \Delta W_1 \Delta W_2 \rangle_{\mathcal{N}}$ in different modes, for various levels of fluctuation, are shown in Fig. 9.8b. We observe that sub-Poissonian behaviour is progressively eliminated as the level of fluctuations in the pumping radiation increases. Here $g = g_0 + \varepsilon g_1$ has been assumed, ε being a real fluctuating quantity obeying a Gaussian distribution with standard deviation σ and $\langle g \rangle = g_0$.

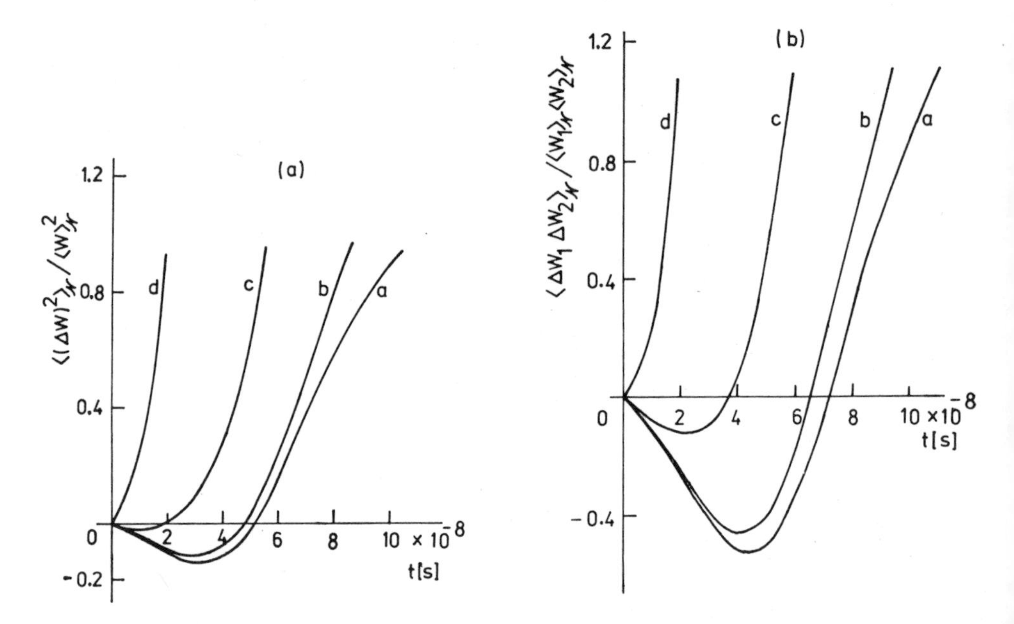

Fig. 9.8 — The time behaviour of the normalized variance $\langle (\Delta W)^2 \rangle_{\mathcal{N}} / \langle W \rangle_{\mathcal{N}}^2$ (a) and the normalized correlation of fluctuations $\langle \Delta W_1 \Delta W_2 \rangle_{\mathcal{N}} / \langle W_1 \rangle_{\mathcal{N}} \langle W_2 \rangle_{\mathcal{N}}$ (b) for parametric amplification process, showing the smoothing of antibunching and anticorrelation with increasing level of fluctuations of the pumping field; $\varphi_1 + \varphi_2 - \varphi = -\pi/2$, $|\xi|^2 = 2$, $\gamma = \langle n_d \rangle = 0$, $g_0 = 10^7$, curves *a*—*d* are given for $g_1 = 0$; $g_1 = 10^6\ \mathrm{s}^{-1}$, $\sigma \doteq 10$; $g_1 = 10^7\ \mathrm{s}^{-1}$, $\sigma = 1$ and $g_1 = 10^7\ \mathrm{s}^{-1}$, $\sigma = 10$ respectively (after Peřinová and Peřina, 1981, Opt. Acta **28**, 769).

The renormalized hamiltonian for up-frequency conversion ($\xi_1 = 0$) and down-frequency conversion ($\xi_2 = 0$) can be written as

$$\hat{H}_{\mathrm{rad}} = \hbar\omega_1 \hat{a}_1^+ \hat{a}_1 + \hbar\omega_2 \hat{a}_2^+ \hat{a}_2 + \hbar(g\hat{a}_1 \hat{a}_2^+ \exp(\mathrm{i}\omega t - \mathrm{i}\varphi) + \mathrm{h.c.}),$$
$$\omega_1 = \omega_2 + \omega, \tag{9.19}$$

leading to the Heisenberg equations

$$\frac{d\hat{a}_1}{dt} = -i\omega_1\hat{a}_1 - ig^*\hat{a}_2 \exp(-i\omega t + i\varphi),$$

$$\frac{d\hat{a}_2}{dt} = -i\omega_2\hat{a}_2 - ig\hat{a}_1 \exp(i\omega t - i\varphi), \tag{9.20}$$

whose solution is

$$\hat{a}_1(t) = \exp(-i\omega_1 t)\left[\hat{a}_1 \cos(|g|t) - i\frac{g^*}{|g|}\exp(i\varphi)\hat{a}_2 \sin(|g|t)\right],$$

$$\hat{a}_2(t) = \exp(-i\omega_2 t)\left[\hat{a}_2 \cos(|g|t) - i\frac{g}{|g|}\exp(-i\varphi)\hat{a}_1 \sin(|g|t)\right]. \tag{9.21}$$

Losses can be taken into account in the same way as before. Thus, compared to the parametric generation and amplification processes, the solution is now periodic. If, for instance, the second mode starts from the vacuum, then

$$\langle\hat{a}_1^+(t)\,\hat{a}_1(t)\rangle = \langle n_1\rangle \cos^2(|g|t), \tag{9.22}$$

$$\langle\hat{a}_2^+(t)\,\hat{a}_2(t)\rangle = \langle n_1\rangle \sin^2(|g|t),$$

$\langle\hat{a}_1^+(t)\,\hat{a}_1(t)\rangle + \langle\hat{a}_2^+(t)\,\hat{a}_2(t)\rangle = \langle n_1\rangle$, $\langle n_1\rangle$ being the initial mean number of photons in mode 1; this agrees with (6.11). Since in (9.21) only the initial annihilation operators are present, the normal characteristic function $C_{\mathcal{N}}(\beta_1, \beta_2, t)$ is also normally ordered in the initial operators, and therefore $\Phi_{\mathcal{N}}(\alpha_1, \alpha_2, t)$ has the form of a δ-function for initially coherent light and the initial statistics are conserved by this process [Mišta (1969)]. Assuming g real (the phase-matching condition is fulfilled), we can obtain anticorrelation of modes 1 and 2 from (9.21), $\langle\Delta W_1 \Delta W_2\rangle_{\mathcal{N}} = -|\xi_1|^2|\xi_2|^2 \times \sin^2(2gt)/2$, regardless of the intensity level of the field, provided that the field is initially in the coherent state $|\xi_1, \xi_2\rangle$ and the pumping phase φ is uniformly distributed in the interval $(0, 2\pi)$. However, $\langle(\Delta W)^2\rangle_{\mathcal{N}} = \langle(\Delta W_1)^2\rangle_{\mathcal{N}} + \langle(\Delta W_2)^2\rangle_{\mathcal{N}} + 2\langle\Delta W_1 \Delta W_2\rangle_{\mathcal{N}} = 0$, that is no bunching or antibunching (which is the purely quantum effect) can occur. If $g(t)$ is the coupling stochastic function [Kryszewski and Chrostowski (1977), Mielniczuk (1979), Srinivasan and Udayabaskaran (1979), Mielniczuk and Chrostowski (1981)], then antibunching can occur [$g(t) = g_0 + \varepsilon g_1(t)$, $\langle g_1^*(t)\, g_1(t')\rangle = 2D\delta(t - t')$]. If the stochastic perturbation is weak, the perturbation method of solving the equations of motion provides [Srinivasan and Udayabaskaran (1979)], for the variation of the mean numbers of photons in modes 1 and 2 and for the normalized fourth-order correlation function of mode 1, $\gamma^{(2,2)}_{\mathcal{N};11}(t_1, t_2, t_1, t_2) = \langle\hat{a}_1^+(t_1)\,\hat{a}_1^+(t_2)\,\hat{a}_1(t_1)\,\hat{a}_1(t_2)\rangle/\langle\hat{a}_1^+(t_1)\hat{a}_1(t_1)\rangle\,\langle\hat{a}_1^+(t_2)\,\hat{a}_1(t_2)\rangle$, which depends on t_1 if t_2 is fixed, the results which are shown in Fig. 9.9. In Fig. 9.9a mode 1 is represented by the full curve and starts from the vacuum, mode 2 is shown by the dashed curve, and the dot-dashed line represents the behaviour of the sum of the mean numbers. Values of $\gamma^{(2,2)}_{\mathcal{N};11}$ lower than unity exhibit the existence of photon antibunching in the generated mode 1. If the stochastic perturbation is strong, the oscillating behaviour vanishes, nevertheless antibunching still occurs [Mielniczuk (1979), Mielniczuk and Chrostowski (1981)].

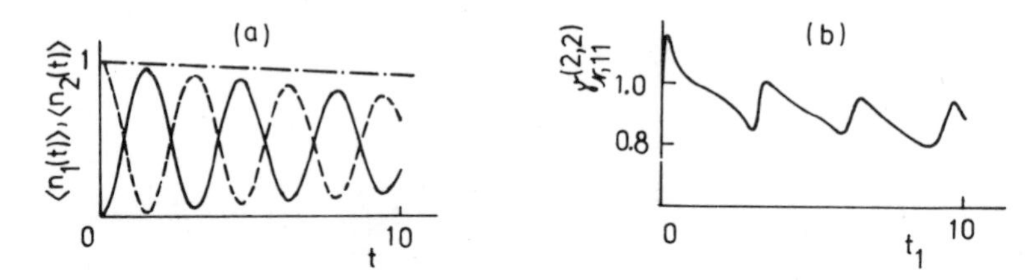

Fig. 9.9 — The mean numbers $\langle n_1(t)\rangle$ and $\langle n_2(t)\rangle$ (a) and the fourth-order degree of coherence $\gamma^{(2,2)}_{\mathcal{N},11}$ (b) as functions of time for a frequency converter with stochastic pumping; $\langle n_1(t)\rangle$ is shown by the solid curve, $\langle n_2(t)\rangle$ by the dashed curve and the dot-dashed line represents their sum; in (b) t_2 is fixed. Mode 2 is assumed to be initially coherent (after Srinivasan and Udayabaskaran, 1979, Opt. Acta **26**, 1535).

The optical parametric processes treated above may be unified, and, including losses, we have

$$\hat{H} = \sum_{j=1}^{2} \hbar\omega_j \hat{a}_j^+ \hat{a}_j - \hbar[g_{11}^*(t)\,\hat{a}_1^2 + g_{22}^*(t)\,\hat{a}_2^2 + g_{12}^*(t)\,\hat{a}_1\hat{a}_2 +$$
$$+ g_{12}'(t)\,\hat{a}_1^+\hat{a}_2 + \text{h.c.}] + \sum_{j=1}^{2}\sum_{l} \hbar(\psi_l^{(j)}\hat{b}_l^{(j)+}\hat{b}_l^{(j)} +$$
$$+ \varkappa_{jl}\hat{a}_j^+\hat{b}_l^{(j)} + \varkappa_{jl}^*\hat{a}_j\hat{b}_l^{(j)+}), \tag{9.23}$$

with the Heisenberg–Langevin equations

$$\frac{d\hat{a}_1}{dt} = -(i\omega_1 + \gamma_1/2)\,\hat{a}_1 + i(2g_{11}\hat{a}_1^+ + g_{12}'\hat{a}_2 + g_{12}\hat{a}_2^+) + \hat{L}_1,$$
$$\frac{d\hat{a}_2}{dt} = -(i\omega_2 + \gamma_2/2)\,\hat{a}_2 + i(g_{12}'^*\hat{a}_1 + g_{12}\hat{a}_1^+ + 2g_{22}\hat{a}_2^+) + \hat{L}_2. \tag{9.24}$$

The corresponding Fokker–Planck equation reads

$$\frac{\partial \Phi_{\mathscr{A}}}{\partial t} = \left\{ \sum_{j=1}^{2} \left[(i\omega_j + \gamma_j/2)\frac{\partial}{\partial \alpha_j}\alpha_j + \text{c.c.} \right] + \right.$$
$$+ \left[i(2g_{11}^*\alpha_1 + g_{12}^*\alpha_2 + g_{12}'^*\alpha_2^*)\frac{\partial}{\partial \alpha_1^*} + (2g_{22}^*\alpha_2 + g_{12}^*\alpha_1 + g_{12}'\alpha_1^*)\times \right.$$
$$\left. \times \frac{\partial}{\partial \alpha_2^*} - g_{11}\frac{\partial^2}{\partial \alpha_1^2} - g_{22}\frac{\partial^2}{\partial \alpha_2^2} - g_{12}\frac{\partial^2}{\partial \alpha_1\,\partial \alpha_2}\Big) + \text{c.c.} \right] +$$
$$\left. + \sum_{j=1}^{2} \gamma_j(\langle n_{dj}\rangle + 1)\frac{\partial^2}{\partial \alpha_j\,\partial \alpha_j^*} \right\} \Phi_{\mathscr{A}}, \tag{9.25a}$$

$$\frac{\partial C_{\mathcal{N}\mathscr{A}}}{\partial t} = \left\{ \left[((-i\omega_1 - \gamma_1/2)\,\beta_1^* + i(2g_{11}^*\beta_1 + g_{12}^*\beta_2 + g_{12}'^*\beta_2^*))\frac{\partial}{\partial \beta_1^*} + \right.\right.$$

$$+((-i\omega_2 - \gamma_2/2)\beta_2^* + i(g_{12}^*\beta_1 + g_{12}'\beta_1^* + 2g_{22}^*\beta_2))\times$$
$$\times\frac{\partial}{\partial\beta_2^*} + (-1)^\varepsilon i(g_{11}\beta_1^{*2} + g_{22}\beta_2^{*2} + g_{12}\beta_1^*\beta_2^*) + \text{c.c.}\Bigg] -$$
$$-\sum_{j=1}^{2}\gamma_j(\langle n_{dj}\rangle + \varepsilon)|\beta_j|^2\Bigg\} C_{\mathcal{N},\mathcal{A}}, \tag{9.25b}$$

where $\varepsilon = 0$ for $C_\mathcal{N}$ and $\varepsilon = 1$ for $C_\mathcal{A}$. The solution for $C_\mathcal{N}$ has the form

$$C_\mathcal{N}(\beta_1, \beta_2, t) = \Bigg\langle \exp\Bigg\{-\sum_{j=1}^{2}(B_j(t) - 1)|\beta_j|^2 +$$
$$+\left[\frac{1}{2}\sum_{j=1}^{2}C_j^*(t)\beta_j^2 + D_{12}^*(t)\beta_1\beta_2 + \bar{D}_{12}(t)\beta_1\beta_2^* + \text{c.c.}\right] +$$
$$+\sum_{j=1}^{2}(\beta_j\xi_j^*(t) - \beta_j^*\xi_j(t))\Bigg\}\Bigg\rangle; \tag{9.26}$$

the corresponding photocount distribution and its factorial moments are expressed n the next section [the fourth-fold versions of (9.35) and (9.36a, b)].

9.2 Interaction of three one-mode boson quantum fields

If we consider the third pumping mode in the hamiltonian (9.9) to be also a quantum mode and if we remove the assumption of phase matching, we may write the hamiltonian for the three-mode interaction in the general form [Graham (1968a, b, 1970)]

$$\hat{H}_{rad} = \sum_{j=1}^{3}\hbar\omega_j\hat{a}_j^+\hat{a}_j - \hbar(g\hat{a}_1\hat{a}_2\hat{a}_3^+ + \text{h.c.}), \tag{9.27}$$

where $\omega_3 = \omega_1 + \omega_2$. This hamiltonian describes, in a fully quantum manner, sum and differency generation, frequency conversion as well as superradiative emission, as represented by hamiltonians (4.212), (4.213) and (4.216) [Walls and Barakat (1970), Nussenzveig (1973)]. The Heisenberg–Langevin equations are

$$\frac{d\hat{a}_1}{dt} = -(i\omega_1 + \gamma_1/2)\hat{a}_1 + ig^*\hat{a}_2^+\hat{a}_3 + \hat{L}_1,$$
$$\frac{d\hat{a}_2}{dt} = -(i\omega_2 + \gamma_2/2)\hat{a}_2 + ig^*\hat{a}_1^+\hat{a}_3 + \hat{L}_2,$$
$$\frac{d\hat{a}_3}{dt} = -(i\omega_3 + \gamma_3/2)\hat{a}_3 + ig\hat{a}_1\hat{a}_2 + \hat{L}_3, \tag{9.28}$$

with the corresponding Fokker–Planck equation

$$\frac{\partial\Phi_\mathcal{A}}{\partial t} = \Bigg\{\sum_{j=1}^{3}\left[(i\omega_j + \gamma_j/2)\frac{\partial}{\partial\alpha_j}\alpha_j + \text{c.c.} + \gamma_j(\langle n_{dj}\rangle + 1)\frac{\partial^2}{\partial\alpha_j\,\partial\alpha_j^*}\right] +$$
$$+\left[ig^*\left(\alpha_1^*\alpha_2^*\frac{\partial}{\partial\alpha_3^*} - \alpha_1^*\alpha_3\frac{\partial}{\partial\alpha_2} - \alpha_2^*\alpha_3\frac{\partial}{\partial\alpha_1} - \alpha_3\frac{\partial^2}{\partial\alpha_1\,\partial\alpha_2}\right) + \text{c.c.}\right]\Bigg\}\Phi_\mathcal{A}; \tag{9.29a}$$

introducing the new variables $\alpha'_j = \alpha_j \exp(\mathrm{i}\omega_j t)$, $\beta'_j = \beta_j \exp(\mathrm{i}\omega_j t)$, which exclude the high frequency harmonic oscillations, we have for $C_{\mathscr{A}}$

$$\frac{\partial C_{\mathscr{A}}}{\partial t} = \left\{ -\sum_{j=1}^{3} \left[\frac{\gamma_j}{2} \left(\beta'_j \frac{\partial}{\partial \beta'_j} + \text{c.c.} \right) + \gamma_j(\langle n_{dj} \rangle + 1) | \beta'_j |^2 \right] + \right.$$
$$+ \left[\mathrm{i}g \left(\beta'_1 \beta'_2 \frac{\partial}{\partial \beta'_3} - \beta'^*_3 \frac{\partial^2}{\partial \beta'^*_1 \partial \beta'^*_2} + \beta'_2 \frac{\partial^2}{\partial \beta'^*_1 \partial \beta'_3} + \right. \right.$$
$$\left. \left. + \beta'_1 \frac{\partial^2}{\partial \beta'^*_2 \partial \beta'_3} \right) - \text{c.c.} \right] \Bigg\} C_{\mathscr{A}}. \tag{9.29b}$$

From (9.28) the following identities are obtained

$$\frac{\mathrm{d}}{\mathrm{d}t}(\langle \hat{a}_1^+ \hat{a}_1 \rangle + \langle \hat{a}_3^+ \hat{a}_3 \rangle) = -\gamma_1 \langle \hat{a}_1^+ \hat{a}_1 \rangle - \gamma_3 \langle \hat{a}_3^+ \hat{a}_3 \rangle + \gamma_1 \langle n_{d1} \rangle + \gamma_2 \langle n_{d2} \rangle,$$

$$\frac{\mathrm{d}}{\mathrm{d}t}(\langle \hat{a}_2^+ \hat{a}_2 \rangle + \langle \hat{a}_3^+ \hat{a}_3 \rangle) = -\gamma_2 \langle \hat{a}_2^+ \hat{a}_2 \rangle - \gamma_3 \langle \hat{a}_3^+ \hat{a}_3 \rangle + \gamma_2 \langle n_{d2} \rangle + \gamma_3 \langle n_{d3} \rangle,$$

$$\frac{\mathrm{d}}{\mathrm{d}t}(\langle \hat{a}_1^+ \hat{a}_1 \rangle - \langle \hat{a}_2^+ \hat{a}_2 \rangle) = -\gamma_1 \langle \hat{a}_1^+ \hat{a}_1 \rangle + \gamma_2 \langle \hat{a}_2^+ \hat{a}_2 \rangle + \gamma_1 \langle n_{d1} \rangle - \gamma_2 \langle n_{d2} \rangle,$$
$$\tag{9.30}$$

where the identity (7.17) has been applied. The identities in (9.30) correspond to the conservation laws (6.7), which are obtained for $\gamma_1 = \gamma_2 = \gamma_3 = \langle n_{d1} \rangle = \langle n_{d2} \rangle = \langle n_{d3} \rangle = 0$.

The short-time solutions (up to $| gt |^2$) of (9.28) are easily obtained, taking into account that $\hat{a}_j(t) \approx \hat{a}_j(0) + \hat{a}'_j(0)\, t + \hat{a}''_j(0)\, t^2/2$, giving

$$\hat{a}_1(t) = \exp(-\mathrm{i}\omega_1 t - \gamma_1 t/2) \left\{ \hat{a}_1 + \mathrm{i}t(g^* \hat{a}_2^+ \hat{a}_3 - \sum_l \varkappa_{1l} \hat{b}_l^{(1)}) - \right.$$
$$\left. - \frac{|g|^2 t^2}{2} \left[\hat{a}_1(\hat{a}_2^+ \hat{a}_2 - \hat{a}_3^+ \hat{a}_3) - \mathrm{i} \frac{\gamma_1 - \gamma_2 - \gamma_3}{2g} \hat{a}_2^+ \hat{a}_3 \right] \right\},$$

$$\hat{a}_2(t) = \exp(-\mathrm{i}\omega_2 t - \gamma_2 t/2) \left\{ \hat{a}_2 + \mathrm{i}t(g^* \hat{a}_1^+ \hat{a}_3 - \sum_l \varkappa_{2l} \hat{b}_l^{(2)}) - \right.$$
$$\left. - \frac{|g|^2 t^2}{2} \left[\hat{a}_2(\hat{a}_1^+ \hat{a}_1 - \hat{a}_3^+ \hat{a}_3) - \mathrm{i} \frac{\gamma_2 - \gamma_1 - \gamma_3}{2g} \hat{a}_1^+ \hat{a}_3 \right] \right\},$$

$$\hat{a}_3(t) = \exp(-\mathrm{i}\omega_3 t - \gamma_3 t/2) \left\{ \hat{a}_3 + \mathrm{i}t(g \hat{a}_1 \hat{a}_2 - \sum_l \varkappa_{3l} \hat{b}_l^{(3)}) - \right.$$
$$\left. - \frac{|g|^2 t^2}{2} \left[\hat{a}_3(\hat{a}_1 \hat{a}_1^+ + \hat{a}_2^+ \hat{a}_2) - \mathrm{i} \frac{\gamma_3 - \gamma_1 - \gamma_2}{2g^*} \hat{a}_1 \hat{a}_2 \right] \right\}, \tag{9.31}$$

where $\hat{a}_j \equiv \hat{a}_j(0)$. One can verify that $[\hat{a}_j(t), \hat{a}_k^+(t)] \approx \hat{1}\delta_{jk}$ up to $(| g | t)^2$. Here we have neglected the second-order reservoir terms containing $(| \varkappa | t)^2$, which give no contributions to the calculated quantities. One can easily derive expressions for the time dependent mean photon numbers $\langle \hat{a}_j^+(t)\, \hat{a}_j(t) \rangle$ up to $(| g | t)^2$ [Peřina (1976)].

We may now apply the general methods of determining the quantum statistical properties, as explained in Chapter 7. Either we can substitute the solutions (9.31) in the characteristic functions $C_{\mathcal{N}}$ or $C_{\mathcal{A}}$ and use the Baker–Hausdorff identity (the assumptions for its application are fulfilled up to $(|g|t)^2$) to obtain them in the normal form with respect to the initial operators, or we can directly solve the equation (9.29b) for the characteristic function, for which a special iterative procedure has been developed [Peřinová and Peřina (1978b)], which is convergent if $|gt\xi| < 1$. Thus we arrive at

$$\begin{aligned} C_{\mathcal{N}}(\{\beta_j\}, t) = \Big\langle \exp\Big\{ & A(t)\,\beta_1\beta_2\beta_3^* - \text{c.c.} - \sum_{j=1}^{3} B_j(t)\,|\beta_j|^2 + \\ & + \sum_{j=1}^{3}\left[\frac{1}{2}\,C_j^*(t)\,\beta_j^2 + \text{c.c.}\right] + \\ & + [D_{12}(t)\,\beta_1^*\beta_2^* + D_{13}(t)\,\beta_1^*\beta_3^* + D_{23}(t)\,\beta_2^*\beta_3^* + \bar{D}_{13}(t)\,\beta_1\beta_3^* + \\ & + \bar{D}_{23}(t)\,\beta_2\beta_3^* + \text{c.c.}] + \sum_{j=1}^{3}[\beta_j\xi_j^*(t) - \beta_j^*\xi_j(t)]\Big\rangle\Big\}, \end{aligned} \qquad (9.32)$$

where

$$A(t) = \frac{\mathrm{i}}{2}\,g\gamma_3\langle n_{d3}\rangle\,t^2\exp\left(-\frac{\gamma_1+\gamma_2+\gamma_3}{2}\,t\right),$$

$$\begin{aligned} B_j(t) &= \langle n_{dj}\rangle\,(1-\exp(-\gamma_j t)) + \\ &\quad + (1-\delta_{j3})\,|g|^2t^2\,|\xi_3|^2\exp(-\gamma_j t), \qquad j = 1, 2, 3, \end{aligned}$$

$$C_j(t) = 0, \qquad j = 1, 2, 3,$$

$$\begin{aligned} D_{12}(t) = \Big\{ & \mathrm{i}g^*\xi_3\left[t + \left(\frac{\gamma_1+\gamma_2-\gamma_3}{2} + \gamma_1\langle n_{d1}\rangle + \gamma_2\langle n_{d2}\rangle\right)\frac{t^2}{2}\right] - \\ & - \frac{1}{2}\,|g|^2t^2\xi_1\xi_2\Big\}\exp\left(-\mathrm{i}\omega_3 t - \frac{\gamma_1+\gamma_2}{2}\,t\right), \end{aligned}$$

$$D_{j3}(t) = -\frac{1}{2}\,|g|^2t^2\xi_j\xi_3\exp\left[-\mathrm{i}(\omega_j+\omega_3)\,t - \frac{\gamma_j+\gamma_3}{2}\,t\right], \qquad j = 1, 2,$$

$$\bar{D}_{13}(t) = \frac{\mathrm{i}}{2}\,g(\gamma_3\langle n_{d3}\rangle - \gamma_1\langle n_{d1}\rangle)\,t^2\xi_2\exp\left(-\mathrm{i}\omega_2 t - \frac{\gamma_1+\gamma_3}{2}\,t\right),$$

$$\bar{D}_{23}(t) = \bar{D}_{13}(1 \leftrightarrow 2), \qquad (9.33a)$$

$$\begin{aligned} \xi_1(t) &= \exp(-\mathrm{i}\omega_1 t - \gamma_1 t/2)\,\times \\ &\times\left[\xi_1 + \mathrm{i}g^*\left(t + \frac{\gamma_1-\gamma_2-\gamma_3}{4}\,t^2\right)\xi_2^*\xi_3 + \frac{1}{2}\,|g|^2t^2\xi_1(|\xi_3|^2 - |\xi_2|^2)\right], \end{aligned}$$

$$\begin{aligned} \xi_2(t) &= \exp(-\mathrm{i}\omega_2 t - \gamma_2 t/2)\,\times \\ &\times\left[\xi_2 + \mathrm{i}g^*\left(t + \frac{\gamma_2-\gamma_1-\gamma_3}{4}\,t^2\right)\xi_1^*\xi_3 + \frac{1}{2}\,|g|^2t^2\xi_2(|\xi_3|^2 - |\xi_1|^2)\right], \end{aligned}$$

$$\xi_3(t) = \exp(-i\omega_3 t - \gamma_3 t/2) \times$$
$$\times \left[\xi_3 + ig\left(t + \frac{\gamma_3 - \gamma_1 - \gamma_2}{4}t^2\right)\xi_1\xi_2 - \frac{1}{2}|g|^2 t^2 \xi_3 (1 + |\xi_1|^2 + |\xi_2|^2)\right]. \tag{9.33b}$$

If losses are neglected, $A(t) = 0$ and $\Phi_{\mathscr{A}}$ has again the form of the averaged, shifted, multidimensional Gaussian distribution [Peřinová and Peřina (1978c)]

$$\begin{aligned}\Phi_{\mathscr{A}}(\{\alpha_j\}, t) = \Big\langle \frac{c_0^{1/2}}{\pi^3} \exp\{-\sum_{j=1}^{3} A_j |\alpha_j - \xi_j(t)|^2 + \\ &+ [D(\alpha_1 - \xi_1(t))(\alpha_2 - \xi_2(t)) + \\ &+ E(\alpha_1 - \xi_1(t))(\alpha_3 - \xi_3(t)) + F(\alpha_1 - \xi_1(t))(\alpha_3^* - \xi_3^*(t)) + \\ &+ G(\alpha_2 - \xi_2(t))(\alpha_3 - \xi_3(t))' + \\ &+ H(\alpha_2 - \xi_2(t))(\alpha_3^* - \xi_3^*(t)) + \text{c.c.}]\}\Big\rangle,\end{aligned} \tag{9.34}$$

c_0 being the normalization constant and A_j, D, E, F, G and H being certain functions of time. In separate modes $\Phi_{\mathscr{A}}(\alpha_j, t)$ has the form (8.44). The generating function can be transformed to the form

$$\langle \exp(isW)\rangle_{\mathscr{N}} = \left\langle \prod_{j=1}^{6} (1 - is\lambda_j)^{-1/2} \exp\left(\frac{isA_j}{1 - is\lambda_j}\right)\right\rangle, \tag{9.35}$$

with the photocount distribution and its factorial moments in the forms

$$p(n, t) = \left\langle \prod_{j=1}^{6} (1 + \lambda_j)^{-1/2} \exp\left(-\frac{A_j}{1 + \lambda_j}\right) \sum_{\sum_{j=1}^{6} k_j = n} \prod_{j=1}^{6} \left(\frac{\lambda_j}{1 + \lambda_j}\right)^{k_j} \times \right.$$
$$\left. \times \frac{1}{\Gamma(k_j + 1/2)} L_{k_j}^{-1/2}\left(-\frac{A_j}{\lambda_j(\lambda_j + 1)}\right)\right\rangle, \tag{9.36a}$$

$$\langle W^k \rangle_{\mathscr{N}} = \left\langle k! \sum_{\sum_{j=1}^{6} k_j = k} \prod_{j=1}^{6} \frac{(\lambda_j)^{k_j}}{\Gamma(k_j + 1/2)} L_{k_j}^{-1/2}\left(-\frac{A_j}{\lambda_j}\right)\right\rangle; \tag{9.36b}$$

the quantities A_j and λ_j play again the role of the „mean numbers of coherent photons" and „mean numbers of chaotic photons" respectively.

Alternative methods of dealing with the impossibility of solving equations (9.28) and (9.29a, b) in closed form are based on numerical calculation [Walls and Barakat (1970), Walls and Tindle (1972), Mostowski and Rzażewski (1978)], making use of the commutativity of the free and the interaction hamiltonians. Some operator linearization methods have been suggested [Katriel and Hummer (1981)], which make it possible to obtain linearized closed form solutions for a class of nonlinear optical processes, including four-wave mixing. Nevertheless, the above approach is simple and enables one to derive explicitly and in a systematic way short-time quantum statistics within the interaction time $t = z/c$ (z being the distance travelled by the beam in the active crystal [Shen (1967)]; t is indeed very small in practical situations for a propagating wave). Such solutions provide information for more detailed computations and make it possible to determine, in a straightforward way, all the possible states of the field exhibiting anticorrelation and antibunching, correlation and bunching and full coherence.

We can demonstrate the region of validity of the short-time approximation given by $|gt\xi_{\text{laser}}| < 1$ [Peřinová and Peřina (1978b)]; we obtain a maximal photon flux 10^{26} photons/s (corresponding to power about 10 MW) if we choose $g = 100\ \text{s}^{-1}$ and $t = 10^{-10}$ s for a crystal of linear dimensions in centimeters, so that the number of photons in the crystal is restricted as $|\xi|^2 < 10^{16}$. If the photoefficiency $\eta = 10^{-1}$, and the detection time interval is $T = 10^{-8}$ s, and the detection area $S = 10^{-4}\ \text{cm}^2$, then the maximum accessible mean photoelectron number is $\langle n \rangle = \eta TS |\xi|^2 = 10^3$. It has been shown by Kozierowski, Tanaś and Kielich (1978) [see also Kielich (1981)] that the validity of the short-time approximation is also determined by the ratio of the mean numbers of photons in the second-harmonic mode and the fundamental mode, which has to be small. Thus the short-time (short-length) approximation seems to be sufficient to describe the propagation of optical waves through nonlinear media.

Fluctuations in individual modes and correlations between fluctuations in different modes are easily obtained from the normal characteristic function [see (9.32)],

$$\begin{aligned}
\langle (\Delta W_j)^2 \rangle_{\mathcal{N}} &= \langle \hat{a}_j^{+2}(t)\, \hat{a}_j^2(t) \rangle - \langle \hat{a}_j^+(t)\, \hat{a}_j(t) \rangle^2 = \\
&= \left. \frac{\partial^4 C_{\mathcal{N}}}{\partial \beta_j^2\, \partial(-\beta_j^*)^2} \right|_{\{\beta_j\}=0} - \left[\left. \frac{\partial^2 C_{\mathcal{N}}}{\partial \beta_j\, \partial(-\beta_j^*)} \right|_{\{\beta_j\}=0} \right]^2 = \\
&= \langle B_j^2 + |C_j|^2 + 2B_j |\xi_j(t)|^2 + [C_j \xi_j^{*2}(t) + \text{c.c.}] \rangle + \\
&\quad + \langle (B_j + |\xi_j(t)|^2)^2 \rangle - \langle B_j + |\xi_j(t)|^2 \rangle^2, \qquad (9.37\text{a})
\end{aligned}$$

$$\begin{aligned}
\langle \Delta W_j \Delta W_k \rangle_{\mathcal{N}} &= \langle \hat{a}_j^+(t)\, \hat{a}_k^+(t)\, \hat{a}_j(t)\, \hat{a}_k(t) \rangle - \langle \hat{a}_j^+(t)\, \hat{a}_j(t) \rangle \langle \hat{a}_k^+(t)\, \hat{a}_k(t) \rangle = \\
&= \left. \frac{\partial^4 C_{\mathcal{N}}}{\partial \beta_j\, \partial(-\beta_j^*)\, \partial \beta_k\, \partial(-\beta_k^*)} \right|_{\{\beta_j\}=0} - \left. \frac{\partial^2 C_{\mathcal{N}}}{\partial \beta_j\, \partial(-\beta_j^*)} \frac{\partial^2 C_{\mathcal{N}}}{\partial \beta_k\, \partial(-\beta_k^*)} \right|_{\{\beta_j\}=0} = \\
&= \langle |D_{jk}|^2 + |\bar{D}_{jk}|^2 + [D_{jk} \xi_j^*(t)\, \xi_k^*(t) - \bar{D}_{jk} \xi_j(t)\, \xi_k^*(t) + \text{c.c.}] \rangle + \\
&\quad + \langle (B_j + |\xi_j(t)|^2)(B_k + |\xi_k(t)|^2) \rangle - \langle B_j + |\xi_j(t)|^2 \rangle \langle B_k + |\xi_k(t)|^2 \rangle, \\
&\qquad j \neq k. \qquad (9.37\text{b})
\end{aligned}$$

If the field is initially coherent, the last two terms in (9.37a, b) are cancelled.

Applying this to (9.32), we arrive at

$$\begin{aligned}
\langle (\Delta W_j)^2 \rangle_{\mathcal{N}} &= 2|g|^2 t^2 |\xi_j|^2 |\xi_3|^2 + 2\gamma_j t \langle n_{dj} \rangle \times \\
&\times \left[\left(1 - \frac{3}{2}\gamma_j t\right) |\xi_j|^2 + (\mathrm{i} g^* t \xi_1^* \xi_2^* \xi_3 + \text{c.c.}) \right] + \gamma_j^2 t^2 \langle n_{dj} \rangle^2, \\
&\qquad j = 1, 2,
\end{aligned}$$

$$\langle(\Delta W_3)^2\rangle_{\mathcal{N}} = 2\gamma_3 t\langle n_{d3}\rangle\left[\left(1 - \frac{3}{2}\gamma_3 t\right)|\xi_3|^2 + (\mathrm{i}gt\xi_1\xi_2\xi_3^* + \mathrm{c.c.})\right] + \\ + \gamma_3^2 t^2\langle n_{d3}\rangle^2,$$

$$\langle\Delta W_1 \Delta W_2\rangle_{\mathcal{N}} = \\ = \mathrm{i}g^*\xi_1^*\xi_2^*\xi_3\left[t - \left(\frac{3\gamma_1 + 3\gamma_2 + \gamma_3}{2} - \gamma_1\langle n_{d1}\rangle - \gamma_2\langle n_{d2}\rangle\right)\frac{t^2}{2}\right] + \\ + \mathrm{c.c.} + |g|^2 t^2[(1 + 2|\xi_1|^2 + 2|\xi_2|^2)|\xi_3|^2 - |\xi_1|^2|\xi_2|^2],$$

$$\langle\Delta W_j \Delta W_3\rangle_{\mathcal{N}} = -|g|^2 t^2|\xi_j|^2|\xi_3|^2 + \\ + \left[\frac{\mathrm{i}}{2}g(\gamma_j\langle n_{dj}\rangle - \gamma_3\langle n_{d3}\rangle)t^2\xi_1\xi_2\xi_3^* + \mathrm{c.c.}\right], \quad j = 1, 2 \tag{9.38a}$$

and

$$\langle(\Delta W)^2\rangle_{\mathcal{N}} = \langle(\Delta W_1)^2\rangle_{\mathcal{N}} + \langle(\Delta W_2)^2\rangle_{\mathcal{N}} + 2\langle\Delta W_1\Delta W_2\rangle_{\mathcal{N}} = \\ = 4|g|t|\xi_1||\xi_2||\xi_3|\sin(\varphi_1 + \varphi_2 + \psi - \varphi_3) + \\ + 2|g|^2 t^2[(3(|\xi_1|^2 + |\xi_2|^2) + 1)|\xi_3|^2 - |\xi_1|^2|\xi_2|^2], \tag{9.38b}$$

provided that the field is initially in the coherent state $|\xi_1, \xi_2, \xi_3\rangle$; in (9.38b) losses are neglected. In parametric amplification $\xi_2 = 0$ and $\langle(\Delta W)^2\rangle_{\mathcal{N}} = 6|g|^2 t^2 \times \times(|\xi_1|^2 + 1/3)|\xi_3|^2 > 0$ and in parametric generation $\xi_1 = \xi_2 = 0$ and $\langle(\Delta W)^2\rangle_{\mathcal{N}} = = 2|g|^2 t^2|\xi_3|^2 > 0$. From (9.38a) we see that, neglecting losses ($\gamma_j = \langle n_{dj}\rangle = 0$), the subfrequency modes are subject to quantum noise (photon bunching) in the interaction, while the sum-frequency mode is coherent in this approximation. If losses are included, then all $\langle(\Delta W_j)^2\rangle_{\mathcal{N}} \approx 2\gamma_j t\langle n_{dj}\rangle > 0$ $(j = 1, 2, 3)$. Further, there is phase-dependent anticorrelation between the subfrequency signal and idler modes 1 and 2, which is maximal if $\varphi_1 + \varphi_2 + \psi - \varphi_3 = -\pi/2$, where φ_j are again phases of ξ_j and ψ is the phase of g. This phase condition is in agreement with that of Sec. 9.1, if the pumping is classical. There always exists anticorrelation (antibunching) in the course of sum-frequency generation with $\xi_3 = 0$; it is given by $\langle\Delta W_1\Delta W_2\rangle_{\mathcal{N}} = -|g|^2 t^2|\xi_1|^2|\xi_2|^2$, independent of the initial phases. This is a typical higher-order quantum term in the intensities, which cannot be obtained if the pumping mode is treated classically. We also see from (9.38a) that anticorrelation always occurs between the subfrequency and sum-frequency modes at the beginning of the interaction, if losses are neglected. It is a consequence of the intermode coupling. Proceeding to higher powers of t, one can show that the sum-frequency mode 3 exhibits antibunching

$$\langle(\Delta W_3)^2\rangle_{\mathcal{N}} = -\frac{2}{3}|g|^6 t^6|\xi_1|^4|\xi_2|^4, \tag{9.39}$$

provided that $\xi_3 = 0$, in close analogy to the similar result [see equation (9.52d)] of Kozierowski and Tanaś (1977) and Kielich, Kozierowski and Tanaś (1978) for second harmonic generation. If losses are included, the anticorrelation between modes 1 and 3 or 2 and 3 may be smoothed or reinforced by the reservoir terms

$ig(\gamma_j\langle n_{dj}\rangle - \gamma_3\langle n_{d3}\rangle)\, t^2\xi_1\xi_2\xi_3^*/2$ + c.c., depending on the initial phase relation and the relation between $\gamma_j\langle n_{dj}\rangle$ and $\gamma_3\langle n_{d3}\rangle$ ($j = 1, 2$). For instance, it is reinforced if $\gamma_3\langle n_{d3}\rangle > \gamma_j\langle n_{dj}\rangle$ and $\varphi_1 + \varphi_2 + \psi - \varphi_3 = -\pi/2$. In the course of parametric amplification ($\xi_2 = 0$), there is anticorrelation $\langle \Delta W_1 \Delta W_3\rangle_{\mathcal{N}} = -|g|^2 t^2 |\xi_1|^2 |\xi_3|^2$ and $\langle \Delta W_2 \Delta W_3\rangle_{\mathcal{N}} = 0$. The anticorrelation may also be reinforced if some modes are initially chaotic. For example, if the idler mode 2 is initially chaotic (it may represent an effective mode of molecular and atomic vibrations in Raman scattering), with the mean number of particles $\langle n_{ch2}\rangle$, we obtain, in the case of sum-frequency generation ($\xi_3 = 0$), enhanced anticorrelation [Trung and Schütte (1978), Peřinová and Peřina (1978b)]

$$\langle \Delta W_1 \Delta W_2\rangle_{\mathcal{N}} = -|g|^2 t^2 |\xi_1|^2 \langle n_{ch2}\rangle (1 + \langle n_{ch2}\rangle), \tag{9.40a}$$

while in the case of subfrequency generation ($\xi_1 = 0$)

$$\langle \Delta W_2 \Delta W_3\rangle_{\mathcal{N}} = -|g|^2 t^2 \langle n_{ch2}\rangle |\xi_3|^2 (1 + \langle n_{ch2}\rangle); \tag{9.40b}$$

also in general

$$\langle \Delta W_1 \Delta W_3\rangle_{\mathcal{N}} = -|g|^2 t^2 |\xi_1|^2 |\xi_3|^2 (1 + 2\langle n_{ch2}\rangle). \tag{9.40c}$$

If both the signal and idler are initially chaotic,

$$\langle \Delta W_1 \Delta W_2\rangle_{\mathcal{N}} = -|g|^2 t^2 \langle n_{ch1}\rangle \langle n_{ch2}\rangle (1 + \langle n_{ch1}\rangle + \langle n_{ch2}\rangle) \tag{9.41a}$$

for sum-frequency generation ($\xi_3 = 0$) and

$$\langle \Delta W_1 \Delta W_3\rangle_{\mathcal{N}} = -|g|^2 t^2 \langle n_{ch1}\rangle |\xi_3|^2 (1 + \langle n_{ch1}\rangle) \tag{9.41b}$$

if $\langle n_{ch2}\rangle = 0$; and if all modes are initially chaotic and moreover $\langle n_{ch3}\rangle = 0$, (9.41a) holds. It should be noted that a random mode may create an enhancement of anticorrelation between different modes because fluctuations in them may be opposite, as a consequence of the randomness; however $\langle (\Delta W_j)^2\rangle_{\mathcal{N}} < 0$ may occur only in consequence of quantum noise (from terms arising from the commutators) [Paul and Brunner (1981)]. An example is the frequency converter with classical stochastic pumping.

Note that questions of saturation, which is lost in the short-time approximation treatment, have been discussed by Graham (1968a, b, 1970, 1973), Graham and Haken (1968), Oliver and Bendjaballah (1980) and Abraham (1980). Methods of possibly enhancing antibunching in the non-degenerate parametric process (with classical pumping), using interference, have been suggested by Bandilla and Ritze (1980a). The statistical properties of parametrically excited spin waves have been investigated by Mikhailov (1976). Also, higher-order parametric processes have been treated [Peřinová, Peřina and Knesel (1977), Graham (1973)], but the results obtained are more complex.

Proposals for generating quantum fields exhibiting photon antibunching and sub-Poissonian radiation have been given by Stoler (1974), based on the degenerate parametric amplifier with classical pumping and by Paul and Brunner (1980) and

Chmela, Horák and Peřina (1981), based on the non-degenerate parametric amplifier with classical pumping. Paul and Brunner also investigated memory effects caused by the dispersion of the nonlinear medium. However, they neglected the initial correlations of radiation modes in the second step of the device, which were included by Chmela et al. (1981). Paul and Brunner proposed to use a c.w. picosecond laser emitting light of frequency ω_3, which enters a nonlinear crystal in which both signal and idler waves are generated, both with low efficiency. Subsequently, the latter waves are attenuated and the phase difference is changed by π. Then in the second crystal sum-frequency radiation is generated (the incident signal and idler waves are attenuated) and antibunching of photons may be registered by means of the Hanbury Brown–Twiss coincidence technique; antibunching would then be demonstrated by an excess of delayed coincidences, or by photocount measurements [$p(n)$ being narrower than the Poisson distribution]. Chmela et al. suggested using two powerful lasers as sources of intense coherent subfrequency radiations at ω_1 and ω_2, respectively. Each of these subfrequency light beams is optically divided into two beams, one of these being strong and the other weak. In the first step the intense sum-frequency wave at ω_3 is generated by the two subfrequency waves at ω_1 and ω_2 ($\omega_1 + \omega_2 = \omega_3$); in the second step a nonlinear interaction between the strong sum-frequency wave at ω_3 (pumping wave) and the two weak subfrequency waves at ω_1 and ω_2, which were not employed in the nonlinear process in the first step, takes place. The optical paths of both the strong and weak waves at ω_j ($j = 1, 2$), interacting in the first and the second steps, respectively, must be equal to satisfy the initial condition for sum-frequency generation and for antibunching in subfrequency modes ($\varphi_1 + \varphi_2 - \varphi = -\pi/2$) in the second step of the experiment. In this way the initial uncorrelation of the beams at the second step is achieved.

The effect of pump coherence on frequency conversion and parametric amplification has been investigated by Crosignani, Di Porto, Ganiel, Solimeno and Yariv (1972). A quantum theory of light propagation in amplifying media has been considered by Foerster and Glauber (1971).

9.3 Second and higher harmonic and subharmonic generation

If we cannot distinguish between the signal and idler modes, the degenerate version of the hamiltonian (9.27) is appropriate,

$$\hat{H}_{\text{rad}} = \hbar\omega_1\hat{a}_1^+\hat{a}_1 + \hbar\omega_2\hat{a}_2^+\hat{a}_2 - \hbar(g\hat{a}_1^2\hat{a}_2^+ + \text{h.c.}), \tag{9.42}$$

where $\omega_2 = 2\omega_1$. Rather as in Sec. 9.2 the Heisenberg–Langevin and Fokker–Planck equations may be obtained. The Heisenberg–Langevin equations, by analogy with (9.28), are

$$\begin{aligned} \frac{\mathrm{d}\hat{a}_1}{\mathrm{d}t} &= -(\mathrm{i}\omega_1 + \gamma_1/2)\,\hat{a}_1 + \mathrm{i}\,2g^*\hat{a}_1^+\hat{a}_2 + \hat{L}_1, \\ \frac{\mathrm{d}\hat{a}_2}{\mathrm{d}t} &= -(\mathrm{i}\omega_2 + \gamma_2/2)\,\hat{a}_2 + \mathrm{i}g\hat{a}_1^2 + \hat{L}_2, \end{aligned} \tag{9.43}$$

from which, using the identity (7.17),

$$\frac{\mathrm{d}}{\mathrm{d}t}(\langle \hat{a}_1^+\hat{a}_1\rangle + 2\langle \hat{a}_2^+\hat{a}_2\rangle) =$$
$$= -\gamma_1\langle \hat{a}_1^+\hat{a}_1\rangle - 2\gamma_2\langle \hat{a}_2^+\hat{a}_2\rangle + \gamma_1\langle n_{d1}\rangle + 2\gamma_2\langle n_{d2}\rangle. \tag{9.44}$$

The short-time solution of (9.43) is

$$\hat{a}_1(t) = \exp(-\mathrm{i}\omega_1 t - \gamma_1 t/2)\Big[\hat{a}_1 + \mathrm{i}g^* t\Big(2\hat{a}_1^+\hat{a}_2 - \frac{1}{g^*}\sum_l \varkappa_{1l}\hat{b}_l^{(1)}\Big) +$$
$$+ |g|^2 t^2\Big(2\hat{a}_1\hat{a}_2^+\hat{a}_2 - \hat{a}_1^+\hat{a}_1^2 + \frac{\mathrm{i}\gamma_2}{2g}\hat{a}_1^+\hat{a}_2\Big)\Big],$$
$$\hat{a}_2(t) = \exp(-\mathrm{i}\omega_2 t - \gamma_2 t/2)\Big\{\hat{a}_2 + \mathrm{i}gt\Big(\hat{a}_1^2 - \frac{1}{g}\sum_l \varkappa_{2l}\hat{b}_l^{(2)}\Big) -$$
$$- |g|^2 t^2\Big[(2\hat{a}_1^+\hat{a}_1 + 1)\hat{a}_2 - \frac{\mathrm{i}}{2g^*}\Big(\gamma_1 - \frac{\gamma_2}{2}\Big)\hat{a}_1^2\Big]\Big\}. \tag{9.45}$$

One can also obtain expressions for $\langle \hat{a}_j^+(t)\,\hat{a}_j(t)\rangle$, $j = 1, 2$ [see Peřina (1976)].

The corresponding Fokker–Planck equation reads

$$\frac{\partial \Phi_{\mathscr{A}}}{\partial t} = \Big\{\frac{\partial}{\partial \alpha_1}[(\mathrm{i}\omega_1 + \gamma_1/2)\alpha_1 - \mathrm{i}\,2g^*\alpha_1^*\alpha_2] +$$
$$+ \frac{\partial}{\partial \alpha_2}[(\mathrm{i}\omega_2 + \gamma_2/2)\alpha_2 - \mathrm{i}g\alpha_1^2] + \text{c.c.} +$$
$$+ \gamma_1(\langle n_{d1}\rangle + 1)\frac{\partial^2}{\partial\alpha_1\,\partial\alpha_1^*} + \gamma_2(\langle n_{d2}\rangle + 1)\frac{\partial^2}{\partial\alpha_2\,\partial\alpha_2^*} +$$
$$+ \Big(\mathrm{i}g\alpha_2^*\frac{\partial^2}{\partial\alpha_1^{*2}} + \text{c.c.}\Big)\Big\}\Phi_{\mathscr{A}} \tag{9.46a}$$

and

$$\frac{\partial C_{\mathscr{N},\mathscr{A}}}{\partial t} = \Big\{(\mathrm{i}\omega_1 - \gamma_1/2)\beta_1\frac{\partial}{\partial\beta_1} + \text{c.c.} +$$
$$+ [(\mathrm{i}\omega_2 - \gamma_2/2)\beta_2 \mp \mathrm{i}g\beta_1^2]\frac{\partial}{\partial\beta_2} + [-(\mathrm{i}\omega_2 + \gamma_2/2)\beta_2^* \mp$$
$$\mp \mathrm{i}g^*\beta_1^{*2}]\frac{\partial}{\partial\beta_2^*} - \gamma_1\langle n_{d1}\rangle|\beta_1|^2 - \gamma_2\langle n_{d2}\rangle|\beta_2|^2 +$$
$$+ \Big[\Big(\mathrm{i}2g\beta_1\frac{\partial^2}{\partial\beta_1^*\,\partial\beta_2} - \mathrm{i}g\beta_2^*\frac{\partial^2}{\partial\beta_1^{*2}}\Big) - \text{c.c.}\Big]\Big\}C_{\mathscr{N},\mathscr{A}}. \tag{9.46b}$$

In the same way as the characteristic function (9.32) was obtained, we have in the degenerate case [Peřinová and Peřina (1978a)]

$$C_{\mathscr{N}}(\beta_1, \beta_2, t) = \Big\langle \exp\Big[A(t)\,\beta_1^2\beta_2^* - \text{c.c.} - B_1(t)|\beta_1|^2 -$$

$$- B_2(t) | \beta_2 |^2 + \left(\frac{1}{2} C_1^*(t) \beta_1^2 + \frac{1}{2} C_2^*(t) \beta_2^2 + D_{12}^*(t) \beta_1 \beta_2 + \right.$$

$$\left. + \bar{D}_{12}(t) \beta_1 \beta_2^* + \text{c.c.} \right) + (\beta_1 \xi_1^*(t) + \beta_2 \xi_2^*(t) - \text{c.c.}) \Big] \Big\rangle, \quad (9.47)$$

where

$$A(t) = \frac{\mathrm{i}}{2} g \gamma_2 t^2 \langle n_{d2} \rangle,$$

$$B_j(t) = \langle n_{dj} \rangle (1 - \exp(-\gamma_j t)) + \delta_{j1} 4 | g |^2 t^2 | \xi_2 |^2 \exp(-\gamma_1 t),$$

$$C_1(t) = -\left[\mathrm{i}2g^* t \xi_2 + | g |^2 t^2 | \xi_1 |^2 + \mathrm{i}g^* \left(\gamma_1 - \frac{\gamma_2}{2} + 2\gamma_1 \langle n_{d1} \rangle \right) t^2 \xi_2 \right] \times$$

$$\times \exp(-\mathrm{i}2\omega_1 t - \gamma_1 t),$$

$$C_2(t) = 0,$$

$$D_{12}(t) = -2 | g |^2 t^2 \xi_1 \xi_2 \exp(-\mathrm{i}\omega_1 t - \mathrm{i}\omega_2 t - \gamma_1 t/2 - \gamma_2 t/2), \quad (9.48a)$$

$$\bar{D}_{12}(t) = \mathrm{i}g(\gamma_2 \langle n_{d2} \rangle - \gamma_1 \langle n_{d1} \rangle) t^2 \xi_1 \exp(-\mathrm{i}\omega_1 t - \gamma_1 t/2 - \gamma_2 t/2),$$

$$\xi_1(t) = \exp(-\mathrm{i}\omega_1 t - \gamma_1 t/2) \times$$

$$\times \left[\xi_1 + \mathrm{i}2g^* t \xi_1^* \xi_2 - | g |^2 t^2 \xi_1 (| \xi_1 |^2 - 2 | \xi_2 |^2) - \frac{\mathrm{i}g^*}{2} \gamma_2 t^2 \xi_1^* \xi_2 \right],$$

$$\xi_2(t) = \exp(-\mathrm{i}\omega_2 t - \gamma_2 t/2) \times \quad (9.48b)$$

$$\times \left[\xi_2 + \mathrm{i}gt \xi_1^2 - | g |^2 t^2 \xi_2 (1 + 2 | \xi_1 |^2) - \frac{\mathrm{i}g}{2} \left(\gamma_1 - \frac{\gamma_2}{2} \right) t^2 \xi_1^2 \right].$$

Neglecting losses, $\Phi_{\mathscr{A}}(\alpha_j, t)$ in separate modes is of the form (8.44); $C_{\mathscr{A}}$ can be written in real variables in the form

$$C_{\mathscr{A}}(\beta_1, \beta_2, t) = \langle \exp(\hat{\beta}^+ \hat{A} \hat{\beta} + \mathrm{i} \hat{\beta}^+ \hat{\xi}) \rangle, \quad (9.49a)$$

where

$$\hat{\beta} = \begin{pmatrix} \beta_{1x} \\ \beta_{1y} \\ \beta_{2x} \\ \beta_{2y} \end{pmatrix}, \qquad \hat{\xi} = \begin{pmatrix} -2\xi_{1y}(t) \\ 2\xi_{1x}(t) \\ -2\xi_{2y}(t) \\ 2\xi_{2x}(t) \end{pmatrix},$$

$$\hat{A} = \begin{pmatrix} -B - 1 + C_x & C_y & D_x & D_y \\ C_y & -B - 1 - C_x & D_y & -D_x \\ D_x & D_y & -1 & 0 \\ D_y & -D_x & 0 & -1 \end{pmatrix}, \quad (9.49b)$$

and $\beta_j = \beta_{jx} + \mathrm{i}\beta_{jy}$, $\xi_j = \xi_{jx} + \mathrm{i}\xi_{jy}$, $j = 1, 2$, $B \equiv B_1(t)$, $C = C_x + \mathrm{i}C_y \equiv C_1(t)$, $D = D_x + \mathrm{i}D_y \equiv D_{12}(t)$. We can diagonalize the matrix $\hat{A}$ with the use of a unitary transformation matrix $\hat{U}$ to $\hat{\Lambda} = \hat{U}^+ \hat{A} \hat{U}$; then, introducing new variables $\hat{\gamma} = \hat{U}^+ \hat{\beta}$ and $\hat{\eta} = \hat{U}^+ \hat{\xi}$, performing the Fourier transformation and returning to the original variables, we arrive at the shifted multidimensional Gaussian distribution

$$\Phi_{\mathscr{A}}(\hat{\alpha}, t) = \left\langle \frac{1}{\pi^2(\mathrm{Det}\,\hat{A})^{1/2}} \exp\left[\frac{1}{4}(\hat{\alpha} - \hat{\xi})^+ \hat{A}^{-1}(\hat{\alpha} - \hat{\xi})\right]\right\rangle, \tag{9.50}$$

$$\hat{\alpha} = (\alpha_{1x}, \alpha_{1y}, \alpha_{2x}, \alpha_{2y})^+, \alpha_j = \alpha_{jx} + \mathrm{i}\alpha_{jy}, \qquad j = 1, 2.$$

Applying (8.57), the normal generating function can be obtained, together with the photocount distribution and its factorial moments, in the forms of the four-fold versions of the expressions (5.92), (5.98a) and (5.103a) ($\varkappa = 1$).

Alternatively, we can write in the complex form

$$C_{\mathscr{A}}(\beta_1, \beta_2, t) = \langle \exp(\hat{\beta}^T \hat{A} \hat{\beta} + \hat{\beta}^T \hat{\xi})\rangle, \tag{9.51a}$$

where T denotes the transposition and

$$\hat{\beta} = \begin{pmatrix} \beta_1 \\ \beta_1^* \\ \beta_2 \\ \beta_2^* \end{pmatrix}, \qquad \hat{\xi} = \begin{pmatrix} \xi_1^*(t) \\ -\xi_1(t) \\ \xi_2^*(t) \\ -\xi_2(t) \end{pmatrix},$$

$$\hat{A} = \frac{1}{2}\begin{pmatrix} C^* & -B-1 & D^* & 0 \\ -B-1 & C & 0 & D \\ D^* & 0 & 0 & -1 \\ 0 & D & -1 & 0 \end{pmatrix}; \tag{9.51b}$$

the matrix $\hat{A}$ can be diagonalized to the block form

$$\hat{A} = \frac{1}{2}\begin{pmatrix} C_1'^* & -B_1'-1 & 0 & 0 \\ -B_1'-1 & C_1' & 0 & 0 \\ 0 & 0 & C_2'^* & -B_2'-1 \\ 0 & 0 & -B_2'-1 & C_2' \end{pmatrix}; \tag{9.51c}$$

this leads, by means of the Fourier transformation, to (8.44) for $\Phi_{\mathscr{A}}$ and consequently to the two-fold generating function (8.59b) [or (9.35)], with the corresponding photocount distribution and its factorial moments given in (9.36a, b). If $B_1' \approx B_2'$ and $C_1' \approx C_2'$, then equations (8.64a, b) with $M = 2$ are appropriate. We note again that anticorrelation and antibunching are related to the negative chaotic component $E_j - 1$, which causes the actual photocount distribution to be narrower than the corresponding Poisson distribution for the coherent state.

For fluctuations in the subharmonic and second harmonic modes and for the correlation of fluctuations between them, we obtain for the initially coherent field, using (9.37a, b),

$$\begin{aligned}\langle(\Delta W_1)^2\rangle_{\mathscr{N}} &= \mathrm{i}2g^*\xi_1^{*2}\xi_2\left[t + t^2\left(3\gamma_1\langle n_{d1}\rangle - \frac{3}{2}\gamma_1 - \frac{1}{4}\gamma_2\right)\right] + \\ &+ \text{c.c.} + 2|g|^2 t^2(2|\xi_2|^2 + 12|\xi_1|^2|\xi_2|^2 - |\xi_1|^4) + \\ &+ 2|\xi_1|^2\langle n_{d1}\rangle\gamma_1 t\left(1 - \frac{3}{2}\gamma_1 t\right) + \gamma_1^2 t^2\langle n_{d1}\rangle^2,\end{aligned}$$

$$\langle(\Delta W_2)^2\rangle_{\mathcal{N}} = i2g^*\gamma_2 t^2\langle n_{d2}\rangle \xi_1^{*2}\xi_2 + \text{c.c.} + \gamma_2 t\langle n_{d2}\rangle\times$$
$$\times(2 - 3\gamma_2 t)\,|\,\xi_2\,|^2 + \gamma_2^2 t^2\langle n_{d2}\rangle^2,$$
$$\langle \Delta W_1 \Delta W_2\rangle_{\mathcal{N}} = -4\,|\,g\,|^2\,t^2\,|\,\xi_1\,|^2\,|\,\xi_2\,|^2 + i(\gamma_1\langle n_{d1}\rangle - \gamma_2\langle n_{d2}\rangle)\,t^2\times$$
$$\times(g\xi_1^2\xi_2^* - \text{c.c.}). \tag{9.52a}$$

Thus, neglecting losses, the subharmonic mode 1 exhibits antibunching and sub-Poissonian behaviour if $2\varphi_1 - \varphi_2 + \psi = -\pi/2$, in agreement with the results of Stoler (1974) and Sec. 9.1. For second harmonic generation ($\xi_2 = 0$) we have antibunching independent of the initial phases, $\langle(\Delta W_1)^2\rangle_{\mathcal{N}} = -2\,|\,g\,|^2\,t^2\,|\,\xi_1\,|^4$, which needs the quantum description of both the modes, being of higher order in the intensity. This led Chmela et al. (1981) to propose a cascade experiment to observe antibunching in the fundamental mode with successive filtering of the second harmonic mode; for this, optical fibres may be suitable. More generally for the kth harmonics

$$\langle(\Delta W_1)^2\rangle_{\mathcal{N}} = -k(k-1)\,|\,g\,|^2\,t^2\,|\,\xi_1\,|^{2k}. \tag{9.52b}$$

In the course of the second subharmonic generation ($\xi_1 = 0$) we have photon bunching in the fundamental mode, $\langle(\Delta W_1)^2\rangle_{\mathcal{N}} = \langle(\Delta W)^2\rangle_{\mathcal{N}} = 4\,|\,g\,|^2\,t^2\,|\,\xi_2\,|^2 > 0$.

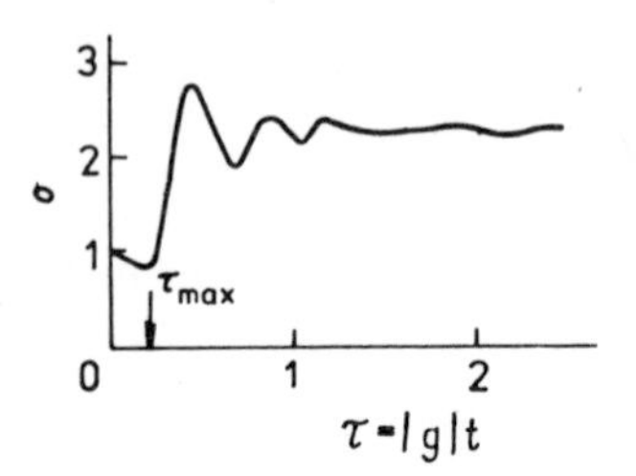

Fig. 9.10 — The time evolution of $\sigma = [\langle(\Delta n_2)^2\rangle/\langle n_2\rangle]^{1/2}$ for the second harmonic generation, $\langle(\Delta n_2)^2\rangle = \langle n_2\rangle + \langle(\Delta W_2)^2\rangle_{\mathcal{N}}$, $\langle n_1\rangle = 50$, $\langle n_2\rangle = 0$ at $t = 0$ (after Walls and Tindle, 1972, J. Phys. **A5**, 534).

Between modes 1 and 2 there is in general anticorrelation at the beginning of the interaction ($\gamma_j = \langle n_{dj}\rangle = 0$), whereas in the course of the second harmonic or subharmonic generation they are uncorrelated in this approximation. More generally for the kth-order process [Hofman (1980)]

$$\langle \Delta W_1 \Delta W_2\rangle_{\mathcal{N}} = -(k!)^2\,|\,g\,|^2\,t^2\,|\,\xi_1\,|^2\,|\,\xi_2\,|^2\sum_{s=0}^{k-2}\frac{|\,\xi_1\,|^{2s}}{s!(k-s-1)!\,(s+1)!}\,. \tag{9.52c}$$

The second or the kth harmonic mode is coherent in this approximation ($\gamma_j = \langle n_{dj}\rangle = 0$). This has been verified experimentally by Clark, Estes and Narducci (1970), Akhmanov, Tchirkin and Tunkin (1970) and Akhmanov and Tchirkin (1971). However, in higher powers of t, antibunching occurs in the course of second harmonic generation ($\xi_2 = 0$) [Kozierowski, and Tanaś (1977), Kielich, Kozierowski and Tanaś (1978)],

$$\langle(\Delta W_2)^2\rangle_{\mathcal{N}} = -\frac{8}{3}|g|^6 t^6 |\xi_1|^8. \tag{9.52d}$$

This result is in agreement with the numerical solution of Walls and Tindle (1972), as shown in Fig. 9.10. Antibunching may be attainable by present experimental devices up to $\tau_{\max}$. If the pump exceeds some critical value, self-pulsing behaviour can occur, as shown in Fig. 9.11 [McNeil, Drummond and Walls (1978)], which can be a basis for optical turbulence and noise [Savage and Walls (1983)].

If losses are included, then $\langle(\Delta W_{1,2})^2\rangle_{\mathcal{N}} > 0$. However, anticorrelation between modes 1 and 2 may be enhanced by the reservoir terms $\mathrm{i}(\gamma_1\langle n_{d1}\rangle - \gamma_2\langle n_{d2}\rangle)\, gt^2\xi_1^{2}\xi_1^{2}\xi_2^* +$ + c.c., provided that $\gamma_2\langle n_{d2}\rangle > \gamma_1\langle n_{d1}\rangle$ and $2\varphi_1 - \varphi_2 + \psi = -\pi/2$, by analogy with the similar result given in Sec. 9.2 for the non-degenerate case.

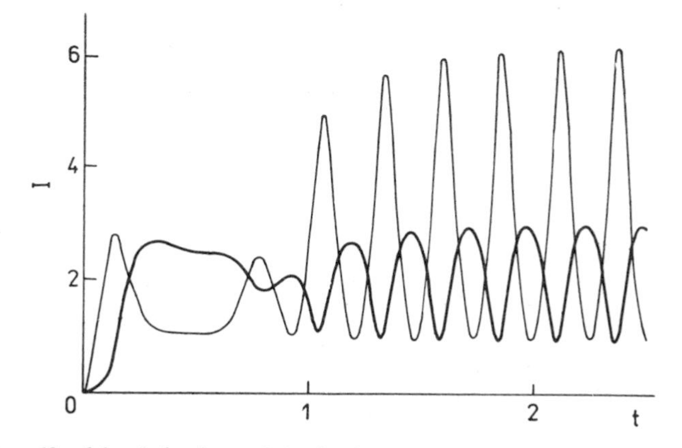

Fig. 9.11 — The self-pulsing behaviour of the fundamental intensity (light curve) and of the second harmonic intensity (heavy curve) (after McNeil, Drummond and Walls, 1978, Opt. Comm. **27**, 292).

Further investigation of the interaction between the fundamental and second harmonic waves inside a Fabry–Perot cavity with external coherent driving fields was carried out by Drummond, McNeil and Walls (1979, 1980b, 1981), who found that antibunching of the fundamental wave in the steady state may occur and some bistable operation is possible. A review has been given by Walls, Drummond and McNeil (1981). Modes 1 and 2 have been coupled to the external pumping fields $p_j(t)\exp(-\mathrm{i}\omega_j t)$, $j = 1, 2$; then by applying the method of small fluctuations around the steady state in the Langevin equations, which are obtained by replacing the operators in the Heisenberg–Langevin equations by classical complex amplitudes α_1 and α_2 and putting $\mathrm{d}\alpha_1/\mathrm{d}t = \mathrm{d}\alpha_2/\mathrm{d}t = 0$, the bistability shown in Fig. 9.12a has been found. In a similar way optical bistability in the three-mode interaction has been discussed by Mišta (1981) and Schütte, Germey, Tiebel and Worlitzer (1981a, b). Optical bistability (or multistability) is also evident from Fig. 9.6, where oscillations in $p(n, t)$ show evidence for competition states [Lugiato, Mandel, Dembiński and Kossakowski (1978)]. Fig. 9.12b shows how the intensity varies as a function of pumping p in the laser containing the saturable absorber [P. Mandel (1978)]. Starting from a situation where the laser does not oscillate, we increase p to reach the bistable domain in which the intensity remains zero until we reach the second boundary, where a sudden jump occurs and the laser begins to operate. For higher values of p we are on the upper branch. Decreasing the value of p, we remain on the upper branch until we reach the critical value, where the laser stops its operation. Thus a hysteresis cycle is obtained and the value of the intensity in the bistable regime

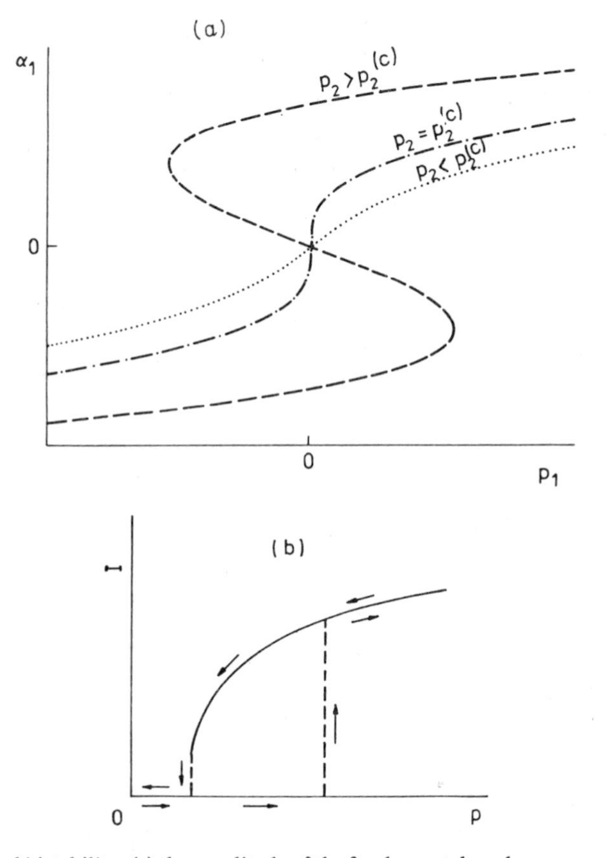

Fig. 9.12 — Optical bistability: (a) the amplitude of the fundamental mode versus pump parameter p_1 for p_2 below, at and above threshold value $p_2^{(c)}$ (after Drummond, McNeil and Walls, 1979, Opt. Comm. **28**, 255); (b) schematic intensity behaviour of the laser with a saturable absorber, as function of the pumping p.

depends upon the path followed to reach this value. This is true also for other non-linear optical phenomena. In a bistable state, the photocount distribution exhibits two peaks, reflecting the competition of two states. At the beginning of the interaction one of them is dominant, while in the development of the system the other may increase and the former decrease.

Recently considerable work has been done to investigate bistability by Walls and his collaborators [Hassan, Drummond and Walls (1978), Walls (1980), Walls and Zoller (1980), Drummond and Walls (1980, 1981), Drummond, McNeil and Walls (1980a, b, 1981), Drummond, Gardiner and Walls (1981), Walls, Zoller, Steÿn–Ross (1981), Walls, Kunasz, Drummond and Zoller (1981), Reid, McNeil and Walls (1981), Walls, Drummond and McNeil (1981)] and by other authors [Agrawal and Carmichael (1980), Lugiato (1981) and references therein, Graham and Schenzle (1981), Zardecki (1981), Chrostowski, Zardecki and Delisle (1981), Hioe and Singh (1981), Schütte et al. (1983)].

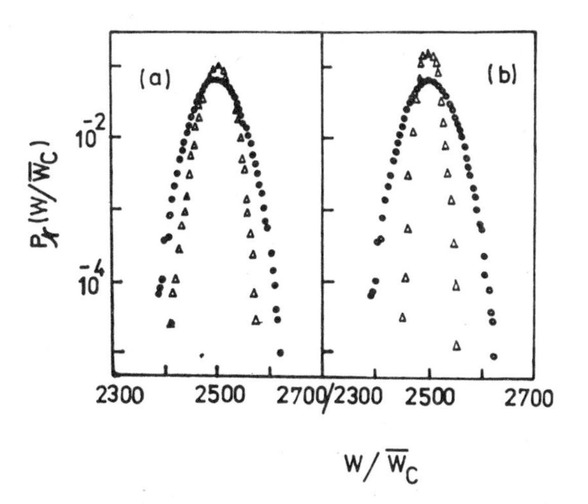

Fig. 9.13 — Measured intensity-probability distributions giving evidence for sub-Poissonian behaviour (a) in the fundamental wave, (b) in the second harmonic wave; the initial distributions are represented by circles and the final distributions by triangles; $\overline{W}_c$ is the mean intensity within one coherence area (after Wagner, Kurowski and Martienssen, 1979, Z. Phys. **B33**, 391).

Till now no experiment has been realized using nonlinear optical phenomena to detect photon antibunching as a basic property of quantum fields having no classical analogues. However, Wagner, Kurowski and Martienssen (1979) performed a simulation experiment using an analogue device and a computer, employing an electro-optical filter. Fig. 9.13 shows the logarithmic probability $P_{\mathcal{N}}(W)$, which is equal to $p(n)$ for strong fields, (*a*) in the fundamental component and (*b*) in the second harmonic wave. Their measurement provides an analogue demonstration of sub--Poissonian behaviour in second harmonic generation.

Higher-harmonic and subharmonic generations have also been discussed [Peřinová, Peřina and Knesel (1977), Peřina, Peřinová and Knesel (1977), Kielich, Kozierowski and Tanaś (1978), Kielich (1981)], using the coherent-state technique.

Using the equations of motion for the number operators, derived from (9.43) with $\gamma_j = L_j = 0$, Crosignani, Di Porto and Solimeno (1972) and Chmela (1973) investigated the second-harmonic generation process with initial coherent and chaotic light, respectively. Crosignani et al. demonstrated that complete depletion of the fundamental wave is impossible in the process if the quantum approach is adopted, and Chmela demonstrated explicitly that this process is more effective with a chaotic fundamental wave than with a coherent one, because of bunching of chaotic photons.

A correction to antibunching in second subharmonic generation with classical pumping, derived from a perturbative quantum treatment of the pumping field, has been obtained by Neumann and Haug (1979). Photon statistics of the second-harmonic generation in gaseous systems have been discussed by Nayak and Mohanty (1977), who adopted the Fock-state technique in the density matrix treatment and took into account atomic motions. Simple models for sub-Poissonian radiation fields have been proposed by Baltes, Quattropani and Schwendimann (1978) and Quattropani et al. (1980). Reviews of results concerning the antibunching phenomenon have been published recently by Loudon (1980), Kozierowski (1981), Kielich (1981) and Paul (1982). Methods used for describing the interaction of N two-level atoms with the electromagnetic field have been exploited by Orszag (1979) to find also solutions for a parametric amplifier, a frequency converter, and Brillouin and Raman scattering. Semiclassical phenomenological methods for describing radiation statistics and coherence properties in nonlinear optical processes have been developed and applied by Tunkin and Tchirkin (1970), Akhmanov and Tchirkin (1971) and Akhmanov, Dyakov and Tchirkin (1981).

Further results concerning the quantum statistical and coherence properties of radiation in optical parametric processes are available in the literature [Mollow and Glauber (1967a, b), Mišta (1969), Tucker and Walls (1969), Mishkin and Walls (1969), Smither and Lu (1974), Graham (1968a, b), Graham and Haken (1968), Walls and Barakat (1970), Echtermeyer (1971), Dewael (1975), Trung and Schütte (1977, 1978)].

9.4 Raman, Brillouin and hyper-Raman scattering

9.4.1 Reservoir phonon system

First we consider the Raman scattering of intense classical laser light with complex amplitude e_L by an infinite Markoffian system of phonons, as expressed by the renormalized hamiltonian [Walls (1973), Peřina (1981a, b)]

$$\hat{H} = \hbar\omega_L \hat{a}_L^+ \hat{a}_L + \hbar\omega_S \hat{a}_S^+ \hat{a}_S + \hbar\omega_A \hat{a}_A^+ \hat{a}_A + \sum_l \hbar\omega_{Vl} \hat{a}_{Vl}^+ \hat{a}_{Vl} - \\ - \sum_l (\hbar g_l e_L \hat{a}_S^+ \hat{a}_{Vl}^+ + \hbar\varkappa_l^* e_L \hat{a}_{Vl} \hat{a}_A^+ + \text{h.c.}), \tag{9.53}$$

where the first four terms represent the free hamiltonians of the laser (L), Stokes (S), anti-Stokes (A) and vibration phonon (V) modes, with the corresponding annihilation operators $\hat{a}_L$, $\hat{a}_S$, $\hat{a}_A$, $\hat{a}_{Vl}$ and with frequencies ω_L, ω_S, ω_A, ω_{Vl} respectively; the remaining terms are the interaction terms between laser photons, scattered Stokes and anti-Stokes photons and phonons, g_l and $\varkappa_l$ being the Stokes and anti-Stokes coupling constants. The Heisenberg equations read

$$\frac{\mathrm{d}\hat{a}_S}{\mathrm{d}t} = -\mathrm{i}\omega_S\hat{a}_S + \mathrm{i}\sum_l g_l e_L \hat{a}_{Vl}^+,$$
$$\frac{\mathrm{d}\hat{a}_A}{\mathrm{d}t} = -\mathrm{i}\omega_A\hat{a}_A + \mathrm{i}\sum_l \varkappa_l^* e_L \hat{a}_{Vl},$$
$$\frac{\mathrm{d}\hat{a}_{Vl}}{\mathrm{d}t} = -\mathrm{i}\omega_{Vl}\hat{a}_{Vl} + \mathrm{i}g_l e_L \hat{a}_S^+ + \mathrm{i}\varkappa_l e_L^* \hat{a}_A. \tag{9.54}$$

Introducing the frequency mismatch $\Delta = \omega_L - (\omega_S + \omega_A)/2$ and defining new operators by the relation

$$\hat{a}_j(t) = \hat{A}_j(t)\exp(-\mathrm{i}\omega_j t - \mathrm{i}\Delta t), \qquad j = S, A, \tag{9.55}$$

we obtain, after eliminating the chaotic phonon operators $\hat{a}_{Vl}$ in the Wigner–Weisskopf approximation as we did in Chapter 7, the following set of equations of motion:

$$\frac{\mathrm{d}}{\mathrm{d}t}\begin{pmatrix}\hat{A}_S\\ \hat{A}_A^+\end{pmatrix} = \begin{pmatrix}\frac{1}{2}\gamma_S|E_L|^2 + \mathrm{i}\Delta & \frac{1}{2}\gamma_{SA}E_L^2\\ -\frac{1}{2}\gamma_{SA}^* E_L^{*2} & -\frac{1}{2}\gamma_A|E_L|^2 - \mathrm{i}\Delta\end{pmatrix}\begin{pmatrix}\hat{A}_S\\ \hat{A}_A^+\end{pmatrix} + \begin{pmatrix}E_L\hat{L}_S\\ E_L^*\hat{L}_A^+\end{pmatrix}. \tag{9.56}$$

Here $\gamma_S = 2\pi|g|^2\varrho$, $\gamma_A = 2\pi|\varkappa|^2\varrho$, $\gamma_{SA} = 2\pi g\varkappa^*\varrho$ are the gain, the damping, and the mutual damping constants of modes S and A ($|\gamma_{SA}|^2 = \gamma_S\gamma_A$), ϱ being the phonon density function. At resonance, $\Delta = 0$ and those ω_{Vl} are strongly coupled to S and A for which $\mp\omega_{Vl} \simeq \omega_{S,A} - \omega_L$, that is

$$2\omega_L \simeq \omega_S + \omega_A, \tag{9.57}$$

in agreement with (6.13); further, $E_L = e_L\exp(\mathrm{i}\omega_L t)$ and the Langevin forces

$$\hat{L}_S(t) = \mathrm{i}\sum_l g_l\hat{a}_{Vl}^+(0)\exp\left[\mathrm{i}\left(\frac{\omega_S - \omega_A}{2} + \omega_{Vl}\right)t\right],$$
$$\hat{L}_A(t) = \mathrm{i}\sum_l \varkappa_l^*\hat{a}_{Vl}(0)\exp\left[-\mathrm{i}\left(\frac{\omega_S - \omega_A}{2} + \omega_{Vl}\right)t\right] \tag{9.58}$$

satisfy

$$\langle\hat{L}_S(t)\,\hat{L}_S^+(t')\rangle_R = \gamma_S\langle n_V\rangle\,\delta(t - t'),$$
$$\langle\hat{L}_A(t)\,\hat{L}_A^+(t')\rangle_R = \gamma_A(\langle n_V\rangle + 1)\,\delta(t - t'),$$
$$\frac{1}{2}\langle\hat{L}_S(t)\,\hat{L}_A(t') + \hat{L}_A(t')\,\hat{L}_S(t)\rangle_R = -\gamma_{SA}\left(\langle n_V\rangle + \frac{1}{2}\right)\delta(t - t'),$$
$$\langle\hat{L}_S(t)\,\hat{L}_A^+(t')\rangle_R = \langle\hat{L}_S^+(t)\,\hat{L}_A(t')\rangle_R = 0, \tag{9.59}$$

where $\langle\,\rangle_R$ denotes the average over the phonon reservoir system and $\langle n_V\rangle$ is the mean number of phonons. However, in this treatment the dynamics of the phonon system is lost, compared to other treatments presented in this chapter. The corresponding generalized Fokker–Planck equation is derived directly from the Heisenberg–Langevin equations (9.56) and the properties of the Langevin forces given in (9.59) in the standard way, as discussed in Chapter 7; alternatively using the master equation (7.36) with $N_1 = \langle n_V \rangle + 1$, $N_2 = \langle n_V \rangle$, $\hat{O}^{(n)} = (2\pi\varrho)^{1/2} (gE_L \hat{A}_S^+ + \varkappa E_L^* \hat{A}_A)$ [Peřina (1981a)],

$$\begin{aligned}
\frac{\partial \Phi_{\mathscr{A}}}{\mathrm{d}t} = &-\left[\left(\frac{\gamma_S}{2} |E_L|^2 + \mathrm{i}\Delta\right) \frac{\partial}{\partial \alpha_S} (\alpha_S \Phi_{\mathscr{A}}) + \text{c.c.}\right] + \\
&+ \left[\left(\frac{\gamma_A}{2} |E_L|^2 - \mathrm{i}\Delta\right) \frac{\partial}{\partial \alpha_A} (\alpha_A \Phi_{\mathscr{A}}) + \text{c.c.}\right] - \\
&- \left[\frac{\gamma_{SA}}{2} E_L^2 \left(\alpha_A^* \frac{\partial \Phi_{\mathscr{A}}}{\partial \alpha_S} - \alpha_S^* \frac{\partial \Phi_{\mathscr{A}}}{\partial \alpha_A}\right) + \text{c.c.}\right] + \\
&+ \gamma_S |E_L|^2 \langle n_V \rangle \frac{\partial^2 \Phi_{\mathscr{A}}}{\partial \alpha_S \partial \alpha_S^*} + \gamma_A |E_L|^2 (\langle n_V \rangle + 1) \frac{\partial^2 \Phi_{\mathscr{A}}}{\partial \alpha_A \partial \alpha_A^*} - \\
&- \left[\gamma_{SA} E_L^2 \left(\langle n_V \rangle + \frac{1}{2}\right) \frac{\partial^2 \Phi_{\mathscr{A}}}{\partial \alpha_S \partial \alpha_A} + \text{c.c.}\right],
\end{aligned} \tag{9.60}$$

where α_j are eigenvalues of $\hat{A}_j$ $(j = S, A)$ in the coherent state $|\alpha_S, \alpha_A\rangle$. The corresponding equations of motion for the characteristic functions $C_{\mathscr{N},\mathscr{A}}$ are

$$\begin{aligned}
\frac{\partial C_{\mathscr{N},\mathscr{A}}}{\partial t} = &\left\{\left[\left(\frac{\gamma_S}{2} |E_L|^2 - \mathrm{i}\Delta\right) \beta_S \frac{\partial}{\partial \beta_S} + \text{c.c.}\right] - \right. \\
&- \left[\left(\frac{\gamma_A}{2} |E_L|^2 + \mathrm{i}\Delta\right) \beta_A \frac{\partial}{\partial \beta_A} + \text{c.c.}\right] + \\
&+ \left[\frac{\gamma_{SA}}{2} E_L^2 \left(\beta_A^* \frac{\partial}{\partial \beta_S} - \beta_S^* \frac{\partial}{\partial \beta_A}\right) + \text{c.c.}\right] - \\
&- \gamma_S |E_L|^2 (\langle n_V \rangle + 1 - \varepsilon) |\beta_S|^2 - \gamma_A |E_L|^2 (\langle n_V \rangle + \varepsilon) |\beta_A|^2 - \\
&\left. - \left[\gamma_{SA}^* E_L^{*2} \left(\langle n_V \rangle + \frac{1}{2}\right) \beta_S \beta_A + \text{c.c.}\right]\right\} C_{\mathscr{N},\mathscr{A}},
\end{aligned} \tag{9.61}$$

where $\varepsilon = 0$ for $C_{\mathscr{N}}$ and $\varepsilon = 1$ for $C_{\mathscr{A}}$. Here and in the earlier transformations of the Fokker–Planck equation to the corresponding equations of motion for the characteristic functions the following property

$$|C_{\mathscr{A}}(\{\beta_j\}, t)| \leqq \prod_j \exp\left(-\frac{1}{2} |\beta_j|^2\right) \to 0, \qquad \text{for } \{\beta_j\} \to \infty \tag{9.62}$$

has been employed when integrating by parts [Peřina and Peřinová (1975)].

Note that from (9.56) it follows that

$$\begin{aligned}
&\frac{\mathrm{d}}{\mathrm{d}t} \langle \hat{A}_S^+(t) \hat{A}_S(t) + \hat{A}_A^+(t) \hat{A}_A(t) \rangle_R = \\
&= |E_L|^2 [\gamma_S(\langle \hat{A}_S^+(t) \hat{A}_S(t) \rangle_R + \langle n_V \rangle + 1) - \gamma_A(\langle \hat{A}_A^+(t) \hat{A}_A(t) \rangle_R - \langle n_V \rangle)].
\end{aligned} \tag{9.63}$$

The operator solutions of (9.56) can be written in the form

$$\hat{A}_S(t) = U_S(t)\,\hat{a}_S + V_S(t)\,\hat{a}_A^+ + \sum_l W_{Sl}(t)\,\hat{a}_{Vl}^+(0),$$
$$\hat{A}_A(t) = U_A(t)\,\hat{a}_A + V_A(t)\,\hat{a}_S^+ + \sum_l W_{Al}(t)\,\hat{a}_{Vl}(0) \tag{9.64}$$

and $\hat{a}_j = \hat{a}_j(0) = \hat{A}_j(0)$. The time-dependent functions U_S, U_A, V_S, V_A, W_{Sl} and W_{Al} satisfy the following identities

$$|U_S(t)|^2 - |V_S(t)|^2 - \sum_l |W_{Sl}(t)|^2 = 1,$$
$$|U_A(t)|^2 - |V_A(t)|^2 + \sum_l |W_{Al}(t)|^2 = 1,$$
$$U_S(t)\,V_A(t) - V_S(t)\,U_A(t) - \sum_l W_{Sl}(t)\,W_{Al}(t) = 0, \tag{9.65a}$$

etc., which are a consequence of the commutation rules

$$[\hat{A}_S(t), \hat{A}_S^+(t)] = [\hat{A}_A(t), \hat{A}_A^+(t)] = \hat{1},$$
$$[\hat{A}_S(t), \hat{A}_A(t)] = \hat{0}, \tag{9.65b}$$

respectively. The normal characteristic function has the form

$$C_{\mathcal{N}}(\beta_S, \beta_A, t) = \langle \exp\{-B_S(t)\,|\beta_S|^2 - B_A(t)\,|\beta_A|^2 + \tag{9.66a}$$
$$+ [D_{SA}(t)\,\beta_S^*\beta_A^* + \text{c.c.}] + [\beta_S\xi_S^*(t) + \beta_A\xi_A^*(t) - \text{c.c.}]\}\rangle,$$

where, assuming for simplicity $\Delta = 0$ $(2\omega_L = \omega_S + \omega_A)$

$$\xi_S(t) = U_S(t)\,\xi_S + V_S(t)\,\xi_A^*,$$
$$\xi_A(t) = U_A(t)\,\xi_A + V_A(t)\,\xi_S^*,$$
$$U_S(t) = \frac{1}{\gamma_S - \gamma_A}(\gamma_S \exp(xt) - \gamma_A) \geqq 0,$$
$$U_A(t) = \frac{1}{\gamma_S - \gamma_A}(\gamma_S - \gamma_A \exp(xt)),$$
$$V_S(t) = -V_A(t) = \frac{(\gamma_S\gamma_A)^{1/2}}{\gamma_S - \gamma_A}(\exp(xt) - 1)\exp(\mathrm{i}2\varphi_L + \mathrm{i}\psi_S - \mathrm{i}\psi_A),$$
$$x = \frac{1}{2}(\gamma_S - \gamma_A)\,|E_L|^2,$$
$$B_S(t) = |V_S(t)|^2 + (\langle n_V\rangle + 1)\sum_l |W_{Sl}(t)|^2 =$$
$$= \frac{1}{(\gamma_S - \gamma_A)^2}\{\gamma_S^2(\exp(2xt) - 1) + 2\gamma_S\gamma_A(1 - \exp(xt))\} +$$
$$+ \frac{\gamma_S\langle n_V\rangle}{\gamma_S - \gamma_A}(\exp(2xt) - 1) \geqq 0,$$

$$B_A(t) = |V_A(t)|^2 + \langle n_V \rangle \sum_l |W_{Al}(t)|^2 =$$

$$= \frac{\gamma_S \gamma_A}{(\gamma_S - \gamma_A)^2} (\exp(\varkappa t) - 1)^2 + \frac{\gamma_A \langle n_V \rangle}{\gamma_A - \gamma_S} (1 - \exp(2\varkappa t)) \geqq 0,$$

$$D_{SA}(t) = V_S(t)\, U_A(t) + (\langle n_V \rangle + 1) \sum_l W_{Sl}(t)\, W_{Al}(t) =$$

$$= \frac{(\gamma_S \gamma_A)^{1/2}}{\gamma_A - \gamma_S} \left\{ \frac{(\exp(\varkappa t) - 1)(\gamma_A - \gamma_S \exp(\varkappa t))}{\gamma_A - \gamma_S} + \right.$$

$$\left. + \langle n_V \rangle (\exp(2\varkappa t) - 1) \right\} \exp(\mathrm{i}2\varphi_L + \mathrm{i}\psi_S - \mathrm{i}\psi_A), \qquad (9.66b)$$

where φ_L, ψ_S and ψ_A are the phases of E_L, g and $\varkappa$ respectively. [The functions $W_{Sl}(t)$ and $W_{Al}(t)$ are not explicitly needed and the expressions derived from (9.65a) are sufficient.] The angle brackets in (9.66a) denote the average over the initial complex amplitudes ξ_S and ξ_A, with the probability function $\Phi_{\mathcal{N}}(\xi_S, \xi_A, 0)$. If the Raman dispersion effect is neglected, $\gamma_S = \gamma_A = \gamma$ and

$$U_S(t) = 1 + \frac{\gamma}{2} |E_L|^2 t,$$

$$U_A(t) = 1 - \frac{\gamma}{2} |E_L|^2 t,$$

$$V_S(t) = -V_A(t) = \frac{\gamma}{2} |E_L|^2 t \exp(\mathrm{i}2\varphi_L),$$

$$B_S(t) = \frac{\gamma^2}{4} |E_L|^4 t^2 + \gamma |E_L|^2 t(\langle n_V \rangle + 1),$$

$$B_A(t) = \frac{\gamma^2}{4} |E_L|^4 t^2 + \gamma |E_L|^2 t \langle n_V \rangle,$$

$$D_{SA}(t) = -\left[\frac{\gamma^2}{4} |E_L|^4 t^2 + \gamma |E_L|^2 t \left(\langle n_V \rangle + \frac{1}{2} \right) \right] \exp(\mathrm{i}2\varphi_L). \qquad (9.66c)$$

If $\Delta \neq 0$,

$$U_S(t) = \frac{1}{p_1 - p_2} \left[\left(p_1 + \frac{\gamma_A}{2} |E_L|^2 + \mathrm{i}\Delta \right) \exp(p_1 t) - \right.$$

$$\left. - \left(p_2 + \frac{\gamma_A}{2} |E_L|^2 + \mathrm{i}\Delta \right) \exp(p_2 t) \right],$$

$$U_A(t) = \frac{1}{p_1^* - p_2^*} \left[\left(p_1^* - \frac{\gamma_S}{2} |E_L|^2 + \mathrm{i}\Delta \right) \exp(p_1^* t) - \right.$$

$$\left. - \left(p_2^* - \frac{\gamma_S}{2} |E_L|^2 + \mathrm{i}\Delta \right) \exp(p_2^* t) \right],$$

$$V_S(t) = \frac{1}{2} \gamma_{SA} E_L^2 \frac{\exp(p_1 t) - \exp(p_2 t)}{p_1 - p_2},$$

$$V_A(t) = -\frac{1}{2}\gamma_{SA}E_L^2 \frac{\exp(p_1^* t) - \exp(p_2^* t)}{p_1^* - p_2^*},$$

$$p_{1,2} = \frac{1}{2}\left\{\frac{1}{2}(\gamma_S - \gamma_A)\,|\,E_L\,|^2 \pm \right. \tag{9.66d}$$

$$\left. \pm \left[\frac{1}{4}(\gamma_S - \gamma_A)^2\,|\,E_L\,|^4 - 4\left(\Delta^2 - \frac{i}{2}\Delta(\gamma_S + \gamma_A)\,|\,E_L\,|^2\right)\right]^{1/2}\right\}.$$

The functions $B_S(t)$, $B_A(t)$ and $D_{SA}(t)$ for $\Delta \neq 0$ are more complicated [Peřina (1981b)].

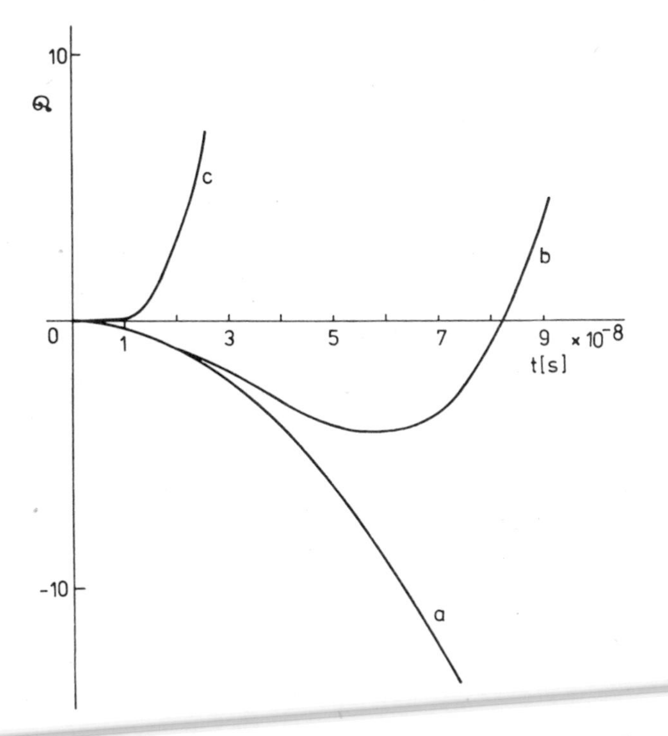

Fig. 9.14 — The time evolution of the determinant $\mathscr{D}$ related to the existence of $\Phi_{\mathscr{N}}$ for $\gamma_S = \gamma_A = 10^4\ \text{s}^{-1}$, $|\,E_L\,|^2 = 10^4$, $|\,\Delta\,| = 10^6\ \text{s}^{-1}$ and $\langle n_V \rangle = 0$ (curve *a*), $\langle n_V \rangle = 10$ (curve *b*) and $\langle n_V \rangle = 100$ (curve *c*) (after Peřina, 1981b, Opt. Acta **28**, 1529).

One can show that

$$\mathscr{D} = B_S B_A - |\,D_{SA}\,|^2 = -\frac{\gamma_S\gamma_A}{(\gamma_S - \gamma_A)^2}(\exp(xt) - 1)^2 < 0, \tag{9.67}$$

provided that $\Delta = 0$, that is the quasi-distribution $\Phi_{\mathscr{N}}(\alpha_S, \alpha_A, t)$ never exists for $t > 0$, if the fields are in a coherent state at $t = 0$. If $\Delta \neq 0$, the analytic expression for $\mathscr{D}$ is complicated; however, in Fig. 9.14 we can see the time dependence of $\mathscr{D}$ for particular cases. In general, for small Δ the determinant $\mathscr{D} < 0$ is practically independent of $\langle n_V \rangle$; for high values of Δ and small $\langle n_V \rangle$, $\mathscr{D} < 0$ and $\Phi_{\mathscr{N}}$ does not exist, whereas it exists if Δ, $\langle n_V \rangle \gg 1$. For $\Phi_{\mathscr{A}}$ we obtain

$$\Phi_{\mathscr{A}}(\alpha_S, \alpha_A, t) = \left\langle (\pi^2 \mathscr{D}')^{-1} \exp\left[-\frac{|\alpha_S - \xi_S(t)|^2}{B_S'(t)} - \right.\right.$$
$$\left.\left. - \frac{|B_S'^{1/2}(t)(\alpha_A - \xi_A(t)) - B_S'^{-1/2}(t) D_{SA}(t)(\alpha_S^* - \xi_S^*(t))|^2}{\mathscr{D}'} \right] \right\rangle, \tag{9.68}$$

where $\mathscr{D}' = B_S' B_A' - |D_{SA}|^2 > 0$, $B_{S,A}' = B_{S,A} + 1$. If the field is initially coherent, $\Phi_{\mathscr{A}}$ has the form of a shifted multidimensional Gaussian distribution involving the intermode correlation and the correlation of real and imaginary parts of the complex amplitudes α_S and α_A. Therefore the statistics of the superposition of signal and noise in the generalized quantum form are appropriate for the statistical description of the scattering process under discussion. If the coupling between the Stokes and anti-Stokes modes is neglected, then $D_{SA} = 0$, and $\Phi_{\mathscr{A}}(\alpha_S, \alpha_A, t)$ is the product of independent distributions for α_S and α_A of the form (8.54).

The normal generating function can be obtained either by using (8.57) after determining the antinormal generating function $\langle \exp(isW) \rangle_{\mathscr{A}}$ with the help of (9.68), or using (8.58), which leads to

$$\langle \exp(-\lambda W) \rangle_{\mathscr{N}} = \left(1 - \frac{\lambda}{\lambda_1}\right)^{-1} \left(1 - \frac{\lambda}{\lambda_2}\right)^{-1} \times$$
$$\times \exp\left(\frac{\lambda \mathscr{A}}{1 - \dfrac{\lambda}{\lambda_1}} + \frac{\lambda \mathscr{B}}{1 - \dfrac{\lambda}{\lambda_2}} \right), \tag{9.69a}$$

where

$$\lambda_{1,2} \quad \frac{-(B_S + B_A) \pm [(B_S + B_A)^2 - 4\mathscr{D}]^{1/2}}{2\mathscr{D}},$$

$$\mathscr{A} = \frac{1}{\mathscr{D}(\lambda_1 - \lambda_2)} \left[|\xi_S(t)|^2 \left(B_A + \frac{1}{\lambda_1}\right) + \right.$$
$$\left. + |\xi_A(t)|^2 \left(B_S + \frac{1}{\lambda_1}\right) - (D_{SA} \xi_S^*(t) \xi_A^*(t) + \text{c.c.}) \right],$$

$$\mathscr{B} = \frac{1}{\mathscr{D}(\lambda_2 - \lambda_1)} \left[|\xi_S(t)|^2 \left(B_A + \frac{1}{\lambda_2}\right) + \right.$$
$$\left. + |\xi_A(t)|^2 \left(B_S + \frac{1}{\lambda_2}\right) - (D_{SA} \xi_S^*(t) \xi_A^*(t) + \text{c.c.}) \right], \tag{9.69b}$$

provided that the Stokes and anti-Stokes radiation modes are initially coherent. Further, it is the case that $\lambda_1 < 0$, $\lambda_2 > 0$ if $\mathscr{D} < 0$; if $\mathscr{D} > 0$, then $\lambda_1, \lambda_2 < 0$. Since $(-\lambda_1^{-1})$ and $(-\lambda_2^{-1})$ play the role of the "mean numbers of chaotic photons", this provides an insight into the occurrence of photon antibunching $(-\lambda_2 < 0)$, sub-Poissonian behaviour and the non-existence of $\Phi_{\mathscr{N}}(\mathscr{D} < 0)$.

Applying (5.64a, b) to (9.69a) we arrive at the photocount distribution and its factorial moments:

$$p(n, t) = \frac{p(0, t)}{(1 - \lambda_1)^n} \sum_{m=0}^{n} \frac{1}{m!(n-m)!} \left(\frac{1-\lambda_1}{1-\lambda_2}\right)^m \times$$

$$\times L_m^0\left(-\frac{\mathscr{B}\lambda_2^2}{\lambda_2 - 1}\right) L_{n-m}^0\left(-\frac{\mathscr{A}\lambda_1^2}{\lambda_1 - 1}\right), \tag{9.70a}$$

$$\langle W^k \rangle_{\mathscr{N}} = \frac{(-1)^k k!}{\lambda_2^k} \sum_{l=0}^{k} \frac{(\lambda_2/\lambda_1)^l}{l!(k-l)!} L_l^0(-\mathscr{A}\lambda_1) L_{k-l}^0(-\mathscr{B}\lambda_2), \tag{9.70b}$$

where

$$p(0, t) = \langle \exp(-W) \rangle_{\mathscr{N}} = \left(1 - \frac{1}{\lambda_1}\right)^{-1} \left(1 - \frac{1}{\lambda_2}\right)^{-1} \times$$

$$\times \exp\left(\frac{\mathscr{A}}{1 - \frac{1}{\lambda_1}} + \frac{\mathscr{B}}{1 - \frac{1}{\lambda_2}}\right). \tag{9.70c}$$

For the mean integrated intensity and the variance of the integrated intensity we obtain

$$\langle W \rangle_{\mathscr{N}} = \langle W_S \rangle_{\mathscr{N}} + \langle W_A \rangle_{\mathscr{N}} = -(\mathscr{A} + \mathscr{B}) - \left(\frac{1}{\lambda_1} + \frac{1}{\lambda_2}\right) =$$

$$= |\xi_S(t)|^2 + |\xi_A(t)|^2 + B_S + B_A, \tag{9.71a}$$

$$\langle (\Delta W)^2 \rangle_{\mathscr{N}} = \langle (\Delta W_S)^2 \rangle_{\mathscr{N}} + \langle (\Delta W_A)^2 \rangle_{\mathscr{N}} + 2\langle \Delta W_S \Delta W_A \rangle_{\mathscr{N}} =$$

$$= \frac{2\mathscr{A}}{\lambda_1} + \frac{2\mathscr{B}}{\lambda_2} + \frac{1}{\lambda_1^2} + \frac{1}{\lambda_2^2}, \tag{9.71b}$$

where

$$\langle (\Delta W_S)^2 \rangle_{\mathscr{N}} = B_S^2 + 2B_S |\xi_S(t)|^2,$$

$$\langle (\Delta W_A)^2 \rangle_{\mathscr{N}} = B_A^2 + 2B_A |\xi_A(t)|^2,$$

$$\langle \Delta W_S \Delta W_A \rangle_{\mathscr{N}} = |D_{SA}|^2 + (D_{SA}\xi_S^*(t)\xi_A^*(t) + \text{c.c.}) \tag{9.71c}$$

are fluctuations in the Stokes and anti-Stokes modes and the fluctuation-correlation between these modes. The terms B_S^2, B_A^2 and $|D_{SA}|^2$ in (9.71c) are independent of the initial field and represent quantum noise arising from vacuum fluctuations, due to the commutators.

It is evident from (9.66a) that separately both the Stokes and anti-Stokes modes behave like a superposition of the signal $\xi_j(t)$ and the noise $B_j(t)$ ($j = S, A$). If moreover $\varkappa = 0$, i.e. $\gamma_A = 0$ (only the Stokes interaction is switched on) or $g = 0$, i.e. $\gamma_S = 0$ (only the anti-Stokes interaction is switched on), then $D_{SA} = 0$ and from (9.66b)

$$B_S(t) = [\exp(\gamma_S |E_L|^2 t) - 1](\langle n_V \rangle + 1),$$

$$\xi_S(t) = \xi_S \exp\left(\frac{\gamma_S}{2} |E_L|^2 t\right), \tag{9.72a}$$

or

$$B_A(t) = [1 - \exp(-\gamma_A | E_L|^2 t)] \langle n_V \rangle,$$
$$\xi_A(t) = \xi_A \exp\left(-\frac{\gamma_A}{2} | E_L |^2 t\right), \quad (9.72b)$$

so the Stokes mode is amplified and the quantum noise increases exponentially with time, whereas the anti-Stokes mode is attenuated and the quantum noise is saturated by the value $\langle n_V \rangle$ in the course of stimulated scattering, in agreement with the results of Chapter 6.

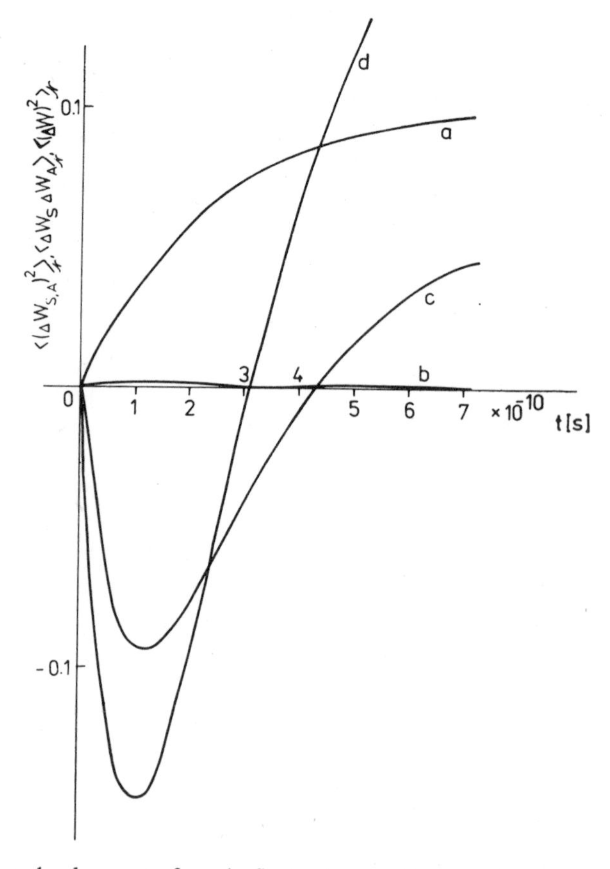

Fig. 9.15 — The time development of mode fluctuations $\langle(\Delta W_S)^2\rangle_{\mathcal{N}}$ (curve *a*) and $\langle(\Delta W_A)^2\rangle_{\mathcal{N}}$ (curve *b*) and their correlation $\langle \Delta W_S \Delta W_A \rangle_{\mathcal{N}}$ (curve *c*) and of the field fluctuations $\langle(\Delta W)^2\rangle_{\mathcal{N}} = \langle(\Delta W_S)^2\rangle_{\mathcal{N}} + \langle(\Delta W_A)^2\rangle_{\mathcal{N}} + 2\langle \Delta W_S \Delta W_A \rangle_{\mathcal{N}}$ (curve *d*) for $\gamma_S = 100\ \mathrm{s}^{-1}$, $\gamma_A = 10^4\ \mathrm{s}^{-1}$, $|E_L|^2 = 10^6$, $|\xi_S| = |\xi_A| = 2^{1/2}$, $|\Delta| = 1 - 10^6$ (all the curves coincide), $\langle n_V \rangle = 0$, under the initial phase condition (9.74) (after Peřina, 1981b, Opt. Acta **28**, 1529).

In the short-time approximation we obtain from (9.71b)

$$\langle(\Delta W)^2\rangle_{\mathcal{N}} \cong 2\gamma_S^{1/2} | E_L|^2 t | \xi_S | (\gamma_S^{1/2} | \xi_S | - \gamma_A^{1/2} | \xi_A |) + \\ + 2 | E_L|^2 t \langle n_V \rangle (\gamma_S^{1/2} | \xi_S | - \gamma_A^{1/2} | \xi_A |)^2, \quad (9.73)$$

provided that the phase conditions for maximal anticorrelation are fulfilled, that is if

$$\Delta\varphi = 2\varphi_L - \varphi_S - \varphi_A + \psi_S - \psi_A = 2n\pi, \qquad n = 0, 1, \ldots, \tag{9.74}$$

φ_S and φ_A being the phases of ξ_S and ξ_A respectively. From (9.73) it is obvious that sub-Poissonian behaviour is possible in the stimulated scattering after the switching on of the interaction, particularly if $\langle n_V \rangle \approx 0$ and $\gamma_A^{1/2} \mid \xi_A \mid > \gamma_S^{1/2} \mid \xi_S \mid$.

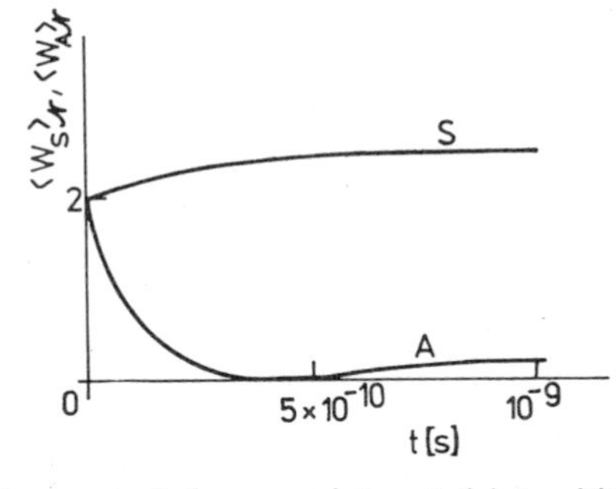

Fig. 9.16 — The time development of the mean integrated intensities $\langle W_S \rangle_{\mathcal{N}}$ and $\langle W_A \rangle_{\mathcal{N}}$ in the Stokes and anti-Stokes modes for the same parameter values as in Fig. 9.15 (after Peřina, 1981a, Opt. Acta **28**, 325).

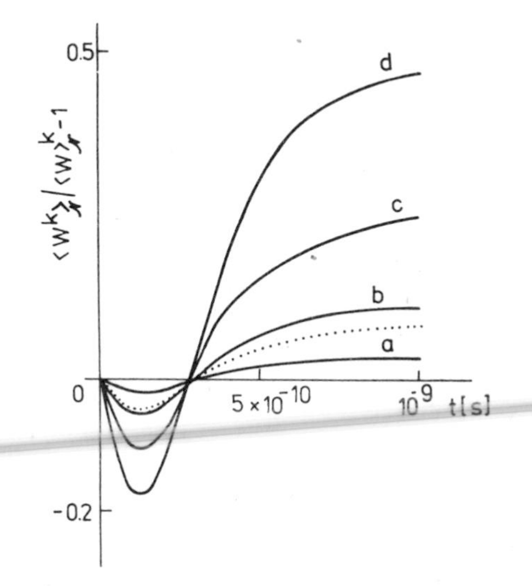

Fig. 9.17 — The time behaviour of the reduced factorial moments $\langle W^k \rangle_{\mathcal{N}} / \langle W \rangle_{\mathcal{N}}^k - 1$ for Raman scattering, curves a—d correspond to $k = 2 - 5$ respectively, for the same parameter values as in Fig. 9.15; the dotted curve represents the quantity $\langle (\Delta W)^2 \rangle_{\mathcal{N}} / \langle W \rangle_{\mathcal{N}}$ (after Peřina, 1981a, Opt. Acta **28**, 325).

In Fig. 9.15 we show the time behaviour of the fluctuations $\langle (\Delta W_{S,A})^2 \rangle_{\mathcal{N}}$, $\langle \Delta W_S \Delta W_A \rangle_{\mathcal{N}}$ and $\langle (\Delta W)^2 \rangle_{\mathcal{N}}$ if the initial phase condition (9.74) is satisfied for the given values of γ_S, γ_A, $| E_L |^2$, $| \xi_S |$, $| \xi_A |$, Δ and $\langle n_V \rangle$. The anticorrelation of modes S and A appears in the competition between these modes, as demonstrated in the time dependence of the integrated intensities $\langle W_S \rangle_{\mathcal{N}}$ and $\langle W_A \rangle_{\mathcal{N}}$ in Fig. 9.16. The reduced factorial moments $\langle W^k \rangle_{\mathcal{N}} / \langle W \rangle_{\mathcal{N}}^k - 1$ for this case are shown in Fig. 9.17, $k = 2 - 5$; the dotted curve represents the quantity $\langle (\Delta W)^2 \rangle_{\mathcal{N}} / \langle W \rangle_{\mathcal{N}}$, which is equal to -1 for the Fock state. It is evident that the effect is smoothed out when the

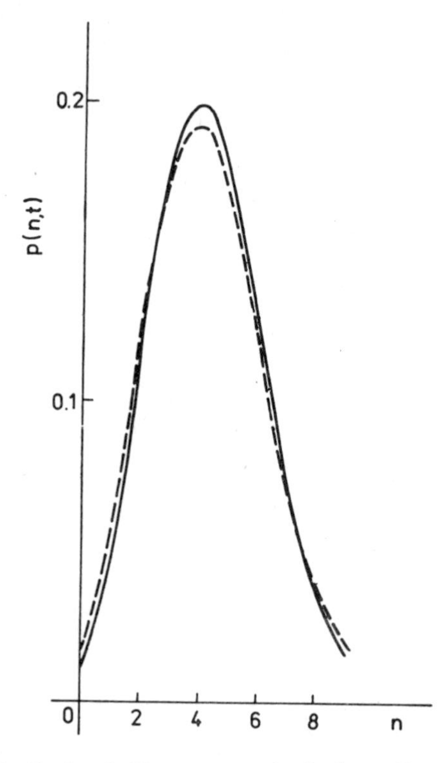

Fig. 9.18 — Photocount distribution for Raman scattering in the antibunching regime ((9.74) holds) for $\gamma_S = 1\,\text{s}^{-1}$, $\gamma_A = 2\,\text{s}^{-1}$, $|E_L|^2 = 10^5$, $|\xi_S| = 1$, $|\xi_A| = 2$, $|\Delta| = 10^5\,\text{s}^{-1}$, $\langle n_V \rangle = 0$ and $t = 1.6 \times 10^{-6}$ s. The full curve represents the actual distribution and the dashed curve is the corresponding Poisson distribution (after Peřina, 1981b, Opt. Acta **28**, 1529).

fields are initially chaotic, or if the initial phase of the laser field is random. Antibunching is also rapidly eliminated if $\langle n_V \rangle$ increases, as is obvious from (9.73). The photocount distribution giving evidence for antibunching and sub-Poissonian behaviour is shown in Fig. 9.18 for $t = 1.6 \times 10^{-6}$ s. The full curve represents the actual distribution, the dashed one is the Poisson distribution with the same $\langle W(t) \rangle_{\mathcal{N}}$, corresponding to the field in the coherent state (as before the photocount distributions are shown by continuous curves for convenience, although they are defined for integer n only). In this case the "chaotic mean photon number" $(-1/\lambda_2)$ is large and negative, which causes narrowing of the photocount distribution compared to the Poisson distribution and reduction of the level of fluctuations in n below the level of these fluctuations in the physical vacuum, associated with the coherent state.

An analogous quantum description of Raman scattering with damping has been proposed by Germey, Schütte and Tiebel (1981).

9.4.2 Dynamics of photon and phonon modes

If we wish to retain also the dynamics of the phonons, the simplest way is to consider an effective boson one-mode field $\hat{a}_V$ for phonons. If the laser field is again assumed to be strong, we have then effective interaction hamiltonian

$$\hat{H}_{\text{int}} = -(\hbar g \hat{a}_S^+ \hat{a}_V^+ \exp(-\mathrm{i}\omega_L t + \mathrm{i}\varphi_L) + \\ + \hbar \varkappa^* \hat{a}_V \hat{a}_A^+ \exp(-\mathrm{i}\omega_L t + \mathrm{i}\varphi_L) + \text{h.c.}). \tag{9.75}$$

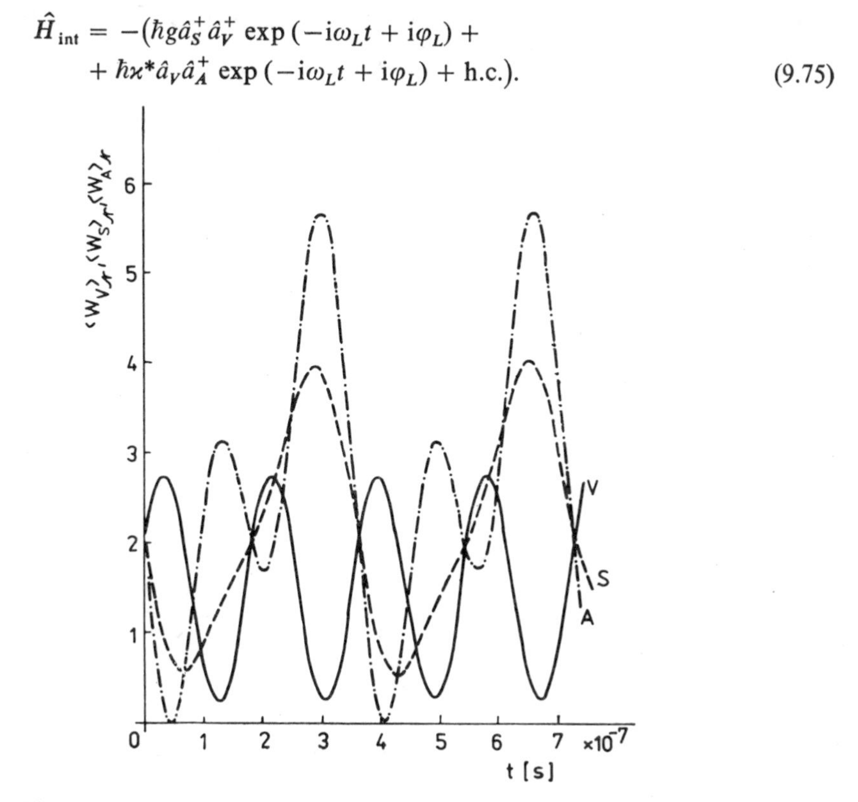

Fig. 9.19 — The time behaviour of the mean integrated intensities $\langle W_S\rangle_{\mathcal{N}}$, $\langle W_A\rangle_{\mathcal{N}}$ and $\langle W_V\rangle_{\mathcal{N}}$ under the phase conditions related to (9.77), $g = 10^7\ \text{s}^{-1}$, $\varkappa = 2\times 10^7\ \text{s}^{-1}$, $|\xi_S|^2 = |\xi_A|^2 = |\xi_V|^2 = 2$ (after Pieczonková, 1982b, Opt. Acta **29**, 1509).

Hence, the Stokes interaction corresponds to the parametric amplification process and the anti-Stokes interaction to the frequency conversion process. The damping of modes may be included in the same way as above. Having eliminated the reservoir variables in the Wigner–Weisskopf approximation, we have

$$\frac{\mathrm{d}\hat{a}_S}{\mathrm{d}t} = -\left(\mathrm{i}\omega_S + \frac{\gamma_S}{2}\right)\hat{a}_S + \mathrm{i}g\hat{a}_V^+ \exp(-\mathrm{i}\omega_L t + \mathrm{i}\varphi_L) + \hat{L}_S,$$
$$\frac{\mathrm{d}\hat{a}_A}{\mathrm{d}t} = -\left(\mathrm{i}\omega_A + \frac{\gamma_A}{2}\right)\hat{a}_A + \mathrm{i}\varkappa^*\hat{a}_A \exp(-\mathrm{i}\omega_L t + \mathrm{i}\varphi_L) + \hat{L}_A,$$
$$\frac{\mathrm{d}\hat{a}_V}{\mathrm{d}t} = -\left(\mathrm{i}\omega_V + \frac{\gamma_V}{2}\right)\hat{a}_V + \mathrm{i}g\hat{a}_S^+ \exp(-\mathrm{i}\omega_L t + \mathrm{i}\varphi_L) + \\ + \mathrm{i}\varkappa\hat{a}_A \exp(\mathrm{i}\omega_L t - \mathrm{i}\varphi_L) + \hat{L}_V, \tag{9.76}$$

together with the Hermitian conjugate equations, where γ_S, γ_A, γ_V are now the damping constants of the corresponding modes and the Langevin forces are defined by (7.9) separately for modes S, A and V.

This system of coupled equations may be solved by Laplace transformation [Pieczonková and Peřina (1981)] in closed form, like that obtained in (9.64), but with more complicated coefficients. The following analysis is the same as that given in the previous section, i.e. the complete normal characteristic function involving all modes S, A, V is determined, and from it the fluctuations in separate modes,

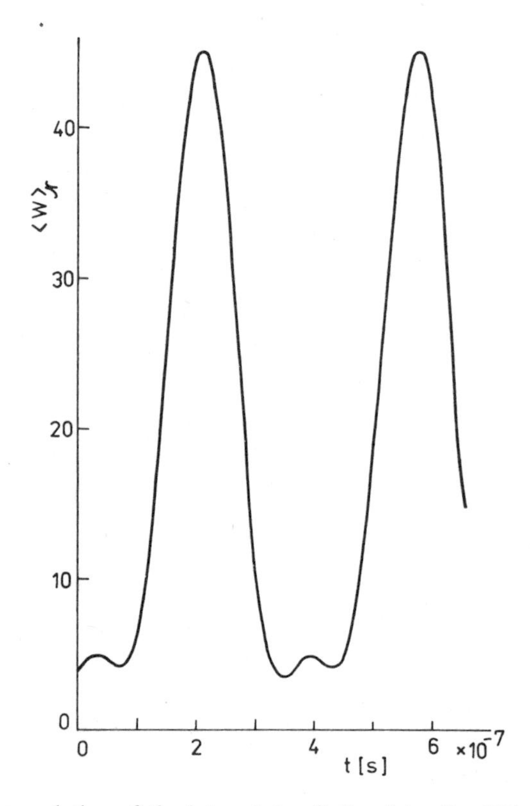

Fig. 9.20 — The time evolution of the integrated radiation intensity $\langle W\rangle_{\mathcal{N}} = \langle W_S\rangle_{\mathcal{N}} + \langle W_A\rangle_{\mathcal{N}}$ under the phase conditions related to (9.78) for the same parameter values as in Fig. 9.19 (after Pieczonková, 1982a, Czech. J. Phys. **B32, 831**).

including the phonon mode, and their correlations are obtained in the standard way provided that the photon and phonon modes are initially coherent (Brillouin scattering) and under various initial phase conditions. Particularly, it has been found in the subsequent studies [Pieczonková (1982a, b)] that anticorrelation, antibunching and sub-Poissonian behaviour may be periodic phenomena and that the initially coherent field may return to the coherent state at later times, provided that the anti-Stokes interaction predominates. In the latter papers the photon and phonon statistics have been completely determined and the generating function, the pho-

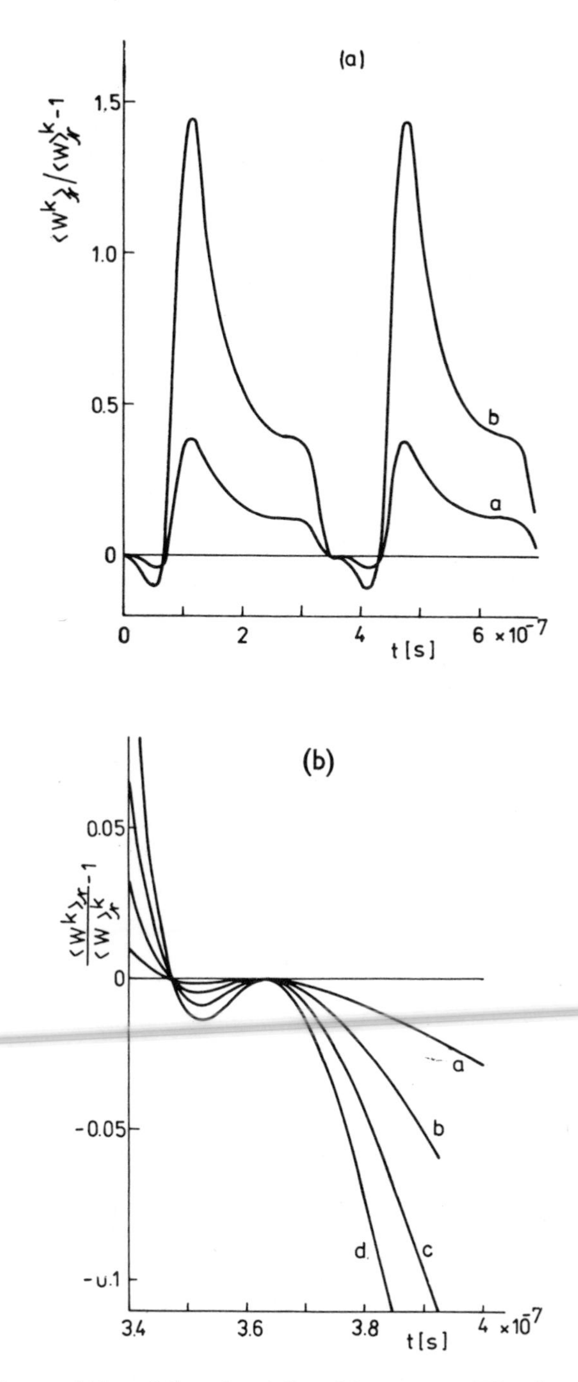

Fig. 9.21 — The time evolution of the reduced factorial moments $\langle W^k \rangle_{\mathcal{N}} / \langle W \rangle^k_{\mathcal{N}} - 1$ for the same parameter values as in Fig. 9.19; curves a, b in (a) are for $k = 2, 3$, curves a—d in (b) are for $k = 2 - 5$ respectively and they provide detailed time behaviour (after Pieczonková, 1982a, Czech. J. Phys. **B32**, 831).

tocount distribution and its factorial moments are obtained in the forms (9.69) and (9.70), in some cases with an additional coherent component [cf. expressions (5.95), (5.101c) and (5.104c)]. All these quantities are computed and some of the results are demonstrated in the following figures.

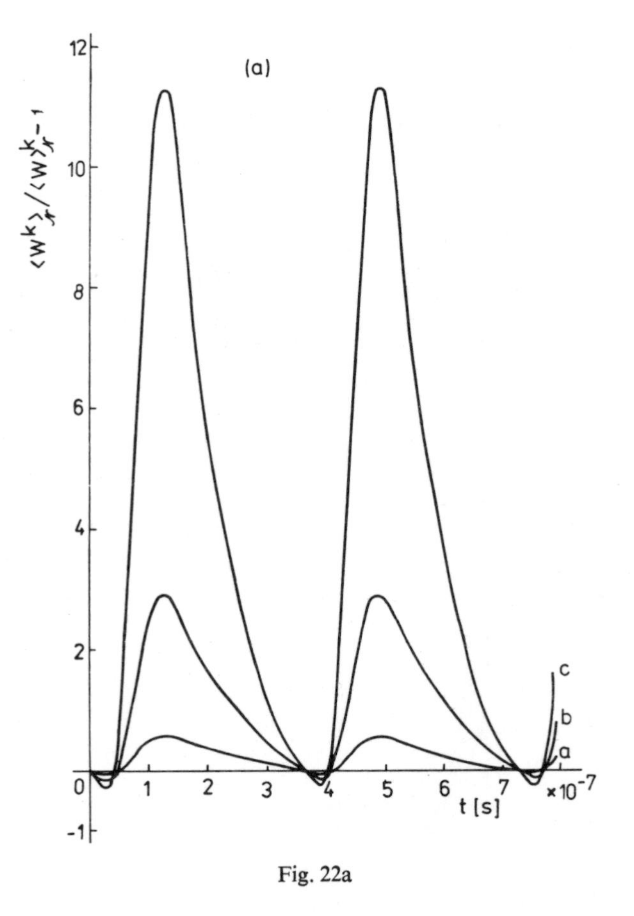

Fig. 22a

Fig. 9.19 demonstrates the time dependence of the mean integrated intensities in the Stokes, anti-Stokes and vibration phonon modes under initial phase conditions corresponding to

$$\begin{aligned} &\sin(\varphi_S + \varphi_V - \varphi_L) = -1, \\ &\cos(2\varphi_L - \varphi_S - \varphi_A) = -1, \\ &(\sin(\varphi_L - \varphi_A + \varphi_V) = 1), \end{aligned} \tag{9.77}$$

when antibunching is maximal (if $\varphi_S + \varphi_V - \varphi_L = -\pi/2$ and $2\varphi_L - \varphi_S - \varphi_A = \pi$, then $\varphi_L - \varphi_A + \varphi_V = \pi/2$; all modes are initially coherent and $g = 10^7\,\mathrm{s}^{-1}$, $\varkappa = 2\times 10^7\,\mathrm{s}^{-1}$, $|\xi_S|^2 = |\xi_A|^2 = |\xi_V|^2 = 2$). In Fig. 9.20 we see the time-dependence of the integrated intensity $\langle W\rangle_{\mathcal{N}} = \langle W_S\rangle_{\mathcal{N}} + \langle W_A\rangle_{\mathcal{N}}$ for the same values of the parameters as in Fig. 9.19 under phase conditions corresponding to

$$\begin{aligned}&\sin(\varphi_S + \varphi_V - \varphi_L) = -1,\\ &\cos(2\varphi_L - \varphi_S - \varphi_A) = 1,\\ &(\sin(\varphi_L - \varphi_A + \varphi_V) = -1),\end{aligned} \tag{9.78}$$

that is, for instance, $\varphi_S + \varphi_V - \varphi_L = -\pi/2$, $2\varphi_L - \varphi_S - \varphi_A = 0$ and consequentl $\varphi_L - \varphi_A + \varphi_V = -\pi/2$. The time behaviour of the reduced factorial moments $\langle W^k\rangle_{\mathcal{N}}/\langle W\rangle^k_{\mathcal{N}} - 1$ is given in Figs. 9.21a, b for the same values of the parameters

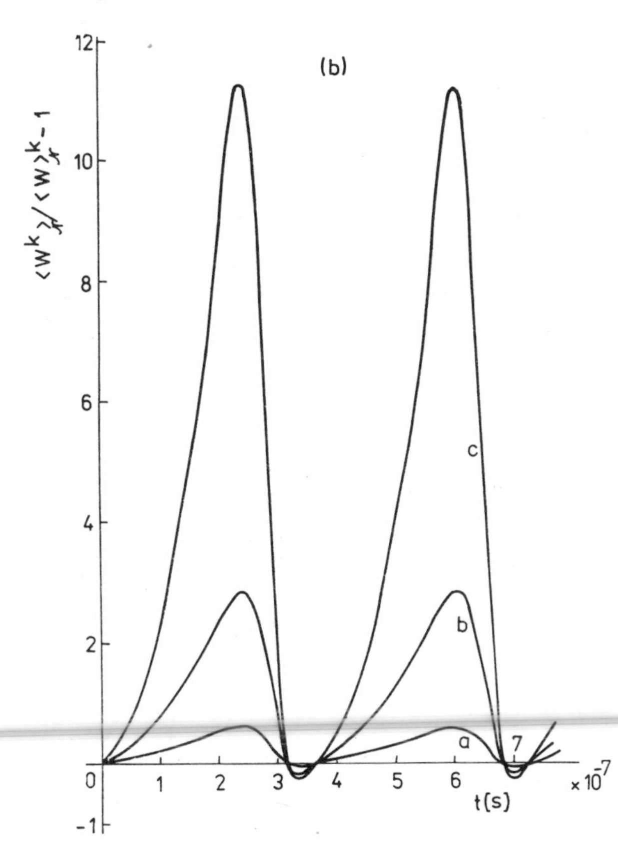

Fig. 9.22 — The time evolution of the reduced factorial moments $\langle W^k\rangle_{\mathcal{N}}/\langle W\rangle^k_{\mathcal{N}} - 1$, $\langle W\rangle_{\mathcal{N}} = \langle W_S\rangle_{\mathcal{N}} + \langle W_A\rangle_{\mathcal{N}} + \langle W_V\rangle_{\mathcal{N}}$, $k = 2-4$ (curves *a*—*c* respectively) under the initial phase conditions related to (a) (9.77), (b) (9.79) (after Pieczonková, 1982b, Opt. Acta **29**, 1509).

and $k = 2 - 5$. The more detailed behaviour revealed by Fig. 9.21b demonstrates the interesting possibility that the field may return to the coherent state at the corresponding times. Moreover, returning to Fig. 9.20 we see that sub-Poissonian behaviour need not neccessarily be related to the attenuation of the field (photons with short time delays are still absent from the beam, but the number of photons with higher time delays may increase, so that the beam may be amplified), as was the

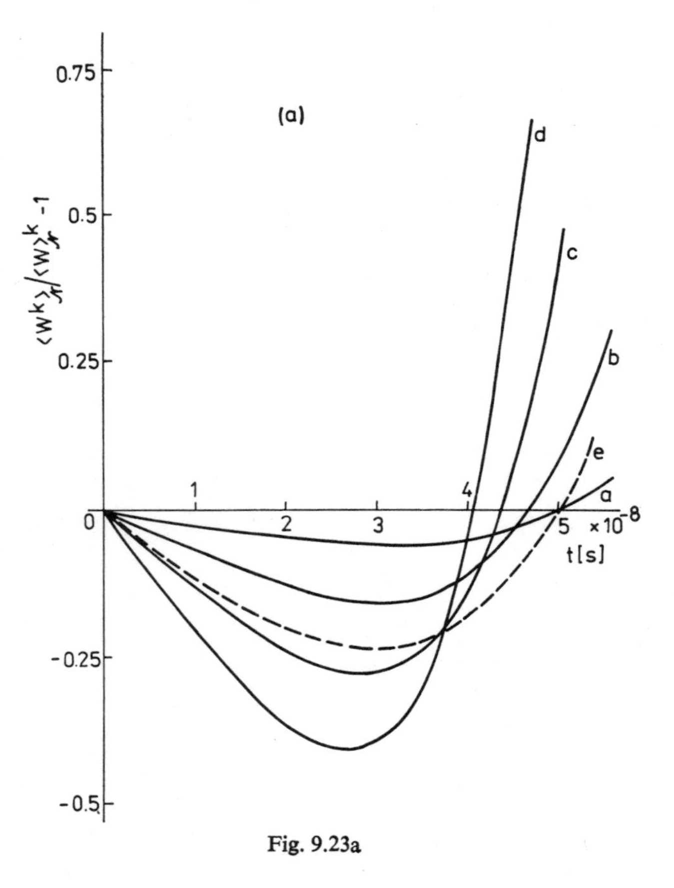

Fig. 9.23a

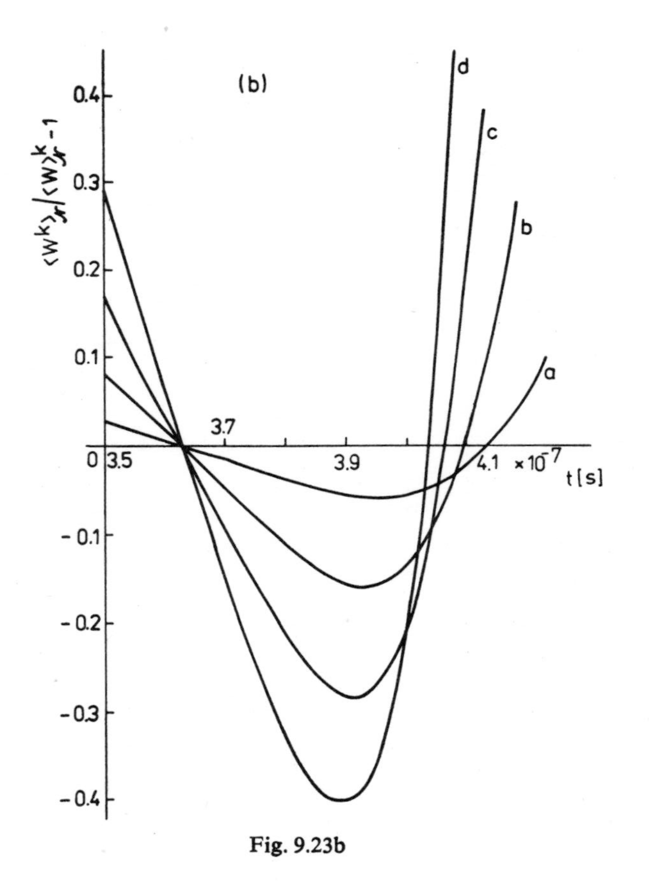

Fig. 9.23b

case, for example, in Figs. 9.1 and 9.2. In the antibunching regime the actual photocount distribution has again been shown to be narrower than the corresponding Poisson distribution; for later times typical oscillations of $p(n, t)$ occurred showing competition between states, as in optical parametric processes. Variations in the periodical antibunching effects have been found for modes S, A and S, V, with the possibility that the field may return to the coherent state at later times; however, no antibunching has been found with modes A, V, regardless of the initial phase conditions, in agreement with the short-time approximation involving classical laser light.

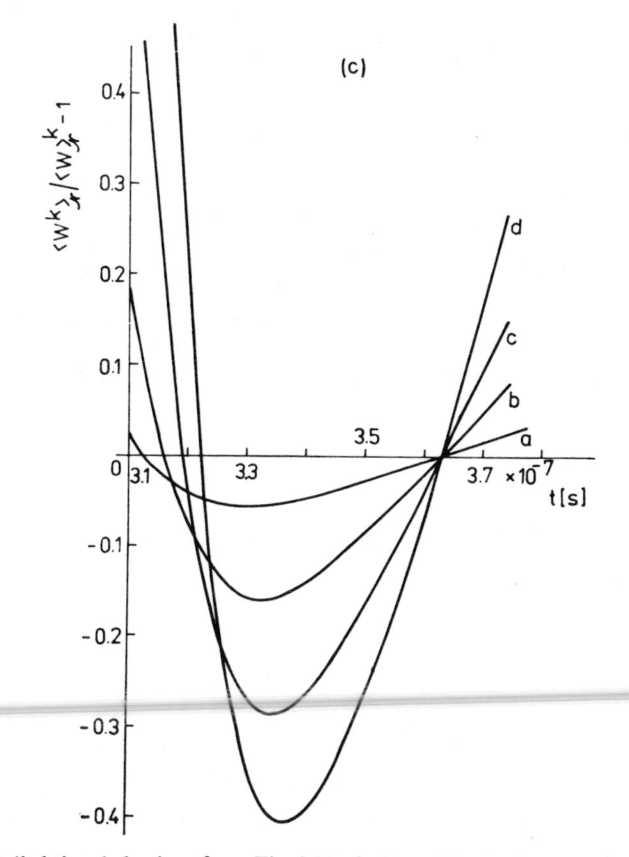

Fig. 9.23 — Detailed time behaviour from Fig. 9.22a (a, b) and from Fig. 9.22b (c), curves a—c are for $k = 2 — 5$, e is $\langle(\Delta W)^2\rangle_{\mathcal{N}}/\langle W\rangle_{\mathcal{N}}$ (after Pieczonková, 1982b, Opt. Acta **29**, 1509).

Figs. 9.22a and b demonstrate the time behaviour of the reduced factorial moments $\langle W^k\rangle_{\mathcal{N}}/\langle W\rangle^k_{\mathcal{N}} - 1$, $\langle W\rangle_{\mathcal{N}} = \langle W_S\rangle_{\mathcal{N}} + \langle W_A\rangle_{\mathcal{N}} + \langle W_V\rangle_{\mathcal{N}}$, under the initial phase conditions (9.77) and

$$\begin{aligned} &\sin(\varphi_S + \varphi_V - \varphi_L) = 1, \\ &\cos(2\varphi_L - \varphi_S - \varphi_A) = -1, \\ &(\sin(\varphi_L - \varphi_A + \varphi_V) = -1), \end{aligned} \tag{9.79}$$

respectively, that is, for example, $\varphi_S + \varphi_V - \varphi_L = \pi/2$, $2\varphi_L - \varphi_S - \varphi_A = -\pi$ and consequently $\varphi_L - \varphi_A + \varphi_V = -\pi/2$. We have again chosen $g = 10^7\ \mathrm{s}^{-1}$, $\varkappa = 2\times 10^7\ \mathrm{s}^{-1}$, $|\xi_S|^2 = |\xi_A|^2 = |\xi_V|^2 = 2$. One can see from Fig. 9.22b that sub-Poissonian behaviour does not occur immediately after the switching on of the interaction under conditions (9.79). More detail of the behaviour of the reduced factorial moments shown in Fig. 9.22a is given in Figs. 9.23a, b and a detail of Fig. 9.22b, demonstrating sub-Poissonian and coherent-state behaviour for later times, is shown in Fig. 9.23c. The photocount distributions narrower than the Poisson distribution with the same $\langle W(t)\rangle_{\mathcal{N}}$ correspond again to antibunching regimes. Note again that all these effects are purely quantal.

9.4.3 Completely quantum description

Now we remove the assumption of classical coherent light and all the modes will be dealt with as fully quantum modes. Then the hamiltonian of the Raman scattering process may be written as follows [Walls (1970)]

$$\hat{H} = \sum_{j=L,S,A,V} \hbar\omega_j \hat{a}_j^+ \hat{a}_j - (\hbar g \hat{a}_L \hat{a}_S^+ \hat{a}_V^+ + \hbar\varkappa^* \hat{a}_L \hat{a}_V \hat{a}_A^+ + \text{h.c.}). \tag{9.80}$$

From the hamiltonian the following Heisenberg equations are derived

$$\begin{aligned}
\frac{d\hat{a}_L}{dt} &= -i\omega_L \hat{a}_L + ig^* \hat{a}_S \hat{a}_V + i\varkappa \hat{a}_A \hat{a}_V^+, \\
\frac{d\hat{a}_S}{dt} &= -i\omega_S \hat{a}_S + ig \hat{a}_L \hat{a}_V^+, \\
\frac{d\hat{a}_A}{dt} &= -i\omega_A \hat{a}_A + i\varkappa^* \hat{a}_L \hat{a}_V, \\
\frac{d\hat{a}_V}{dt} &= -i\omega_V \hat{a}_V + ig \hat{a}_L \hat{a}_S^+ + i\varkappa \hat{a}_L^+ \hat{a}_A;
\end{aligned} \tag{9.81}$$

in this section we neglect losses, for simplicity. An exact closed solution cannot be obtained even for this set of equations (together with the Hermitian conjugate equations). However, they may be solved approximately in powers of t, as in the case of optical parametric processes [Szlachetka, Kielich, Peřina and Peřinová (1979, 1980b)],

$$\begin{aligned}
\hat{a}_L(t) = \exp(-i\omega_L t)\Big\{&\hat{a}_L + ig^* t \hat{a}_S \hat{a}_V + i\varkappa t \hat{a}_V^+ \hat{a}_A - \frac{1}{2}|g|^2 t^2 \times \\
&\times \hat{a}_L(\hat{a}_S^+ \hat{a}_S + \hat{a}_V^+ \hat{a}_V + 1) + \frac{1}{2}|\varkappa|^2 t^2 \hat{a}_L(\hat{a}_A^+ \hat{a}_A - \hat{a}_V^+ \hat{a}_V)\Big\},
\end{aligned}$$

$$\begin{aligned}
\hat{a}_S(t) = \exp(-i\omega_S t)\Big\{&\hat{a}_S + igt \hat{a}_L \hat{a}_V^+ + \frac{1}{2}|g|^2 t^2 \hat{a}_S(\hat{a}_L^+ \hat{a}_L - \hat{a}_V^+ \hat{a}_V) - \\
&- \frac{1}{2} g\varkappa t^2 \hat{a}_V^{+2} \hat{a}_A + \frac{1}{2} g\varkappa^* t^2 \hat{a}_L^2 \hat{a}_A^+\Big\},
\end{aligned}$$

$$\hat{a}_A(t) = \exp(-\mathrm{i}\omega_A t)\Big\{\hat{a}_A + \mathrm{i}\varkappa^* t\hat{a}_L\hat{a}_V - \frac{1}{2}|\varkappa|^2 t^2\hat{a}_A(\hat{a}_L^+\hat{a}_L + \hat{a}_V^+\hat{a}_V + 1) -$$
$$- \frac{1}{2} g^*\varkappa^* t^2\hat{a}_S\hat{a}_V^2 - \frac{1}{2} g\varkappa^* t^2\hat{a}_S^+\hat{a}_L^2\Big\},$$
$$\hat{a}_V(t) = \exp(-\mathrm{i}\omega_V t)\Big\{\hat{a}_V + \mathrm{i}gt\hat{a}_L\hat{a}_S^+ + \mathrm{i}\varkappa t\hat{a}_L^+\hat{a}_A - \frac{1}{2}|g|^2 t^2 \times$$
$$\times \hat{a}_V(\hat{a}_S^+\hat{a}_S - \hat{a}_L^+\hat{a}_L) - \frac{1}{2}|\varkappa|^2 t^2\hat{a}_V(\hat{a}_L^+\hat{a}_L - \hat{a}_A^+\hat{a}_A)\Big\}, \tag{9.82}$$

where $\hat{a}_j \equiv \hat{a}_j(0)$ again. We can calculate also the expressions for time-dependent mean number-operators $\langle \hat{a}_j^+(t)\,\hat{a}_j(t)\rangle$, $j = L, S, A, V$ up to t^2 and it appears that $(\mathrm{d}/\mathrm{d}t)\,(\hat{n}_L(t) + \hat{n}_S(t) + \hat{n}_A(t)) = \hat{0}$ and $(\mathrm{d}/\mathrm{d}t)\,(\hat{n}_V(t) + \hat{n}_A(t) - \hat{n}_S(t)) = \hat{0}$, i.e. these combinations of $\hat{n}_j$ are constants of motion. In the same way as for the optical parametric processes we obtain

$$C_{\mathcal{N}}(\beta_L, \beta_S, \beta_A, \beta_V, t) = \Big\langle \exp\Big\{\sum_{j=L,S,A,V}\Big[-B_j(t)\,|\beta_j|^2 +$$
$$+ \left(\frac{1}{2}C_j^*(t)\,\beta_j^2 + \text{c.c.}\right) + \beta_j\xi_j^*(t) - \beta_j^*\xi_j(t)\Big] +$$
$$+ \sum_{j<k}\left(D_{jk}(t)\,\beta_j^*\beta_k^* + \bar{D}_{jk}(t)\,\beta_j\beta_k^* + \text{c.c.}\right)\Big\}\Big\rangle, \tag{9.83}$$

where the set (L, S, A, V) is assumed to be ordered. Here

$$B_L(t) = |\varkappa|^2 t^2\,|\xi_A|^2, \qquad C_L(t) = -g^*\varkappa t^2\xi_S\xi_A\exp(-\mathrm{i}2\omega_L t),$$
$$B_S(t) = |g|^2 t^2\,|\xi_L|^2, \qquad C_S(t) = 0,$$
$$B_A(t) = 0, \qquad C_A(t) = 0,$$
$$B_V(t) = |g|^2 t^2\,|\xi_L|^2 + |\varkappa|^2 t^2\,|\xi_A|^2,$$
$$C_V(t) = -g\varkappa t^2\xi_S^*\xi_A\exp(-\mathrm{i}2\omega_V t),$$
$$D_{LS}(t) = -\frac{1}{2}|g|^2 t^2\xi_L\xi_S\exp[-\mathrm{i}(\omega_L + \omega_S)\,t],$$
$$D_{LA}(t) = -\frac{1}{2}|\varkappa|^2 t^2\xi_L\xi_A\exp[-\mathrm{i}(\omega_L + \omega_A)\,t],$$
$$D_{LV}(t) = \left[\mathrm{i}\varkappa t\xi_A - (|g|^2 + |\varkappa|^2)\frac{t^2}{2}\xi_L\xi_V\right]\exp[-\mathrm{i}(\omega_L + \omega_V)\,t],$$
$$D_{SA}(t) = -\frac{1}{2}g\varkappa^* t^2\xi_L^2\exp[-\mathrm{i}(\omega_S + \omega_A)\,t],$$
$$D_{SV}(t) = \left(\mathrm{i}gt\xi_L - \frac{1}{2}|g|^2 t^2\xi_S\xi_V - g\varkappa t^2\xi_A\xi_V^*\right)\exp[-\mathrm{i}(\omega_S + \omega_V)\,t],$$
$$D_{AV}(t) = -\frac{1}{2}|\varkappa|^2 t^2\xi_A\xi_V\exp[-\mathrm{i}(\omega_A + \omega_V)\,t],$$
$$\bar{D}_{LS}(t) = -g\varkappa^* t^2\xi_L\xi_A^*\exp[\mathrm{i}(\omega_L - \omega_S)\,t], \tag{9.84}$$

all other $\bar{D}_{jk}(t) = 0$; $\xi_j, j = L, S, A, V$ are the initial complex amplitudes over which the average in (9.83) is carried out. Expressions (9.37a, b) then provide the quantities $\langle(\Delta W_j)^2\rangle_{\mathcal{N}}$ and $\langle\Delta W_j \Delta W_k\rangle_{\mathcal{N}}$.

Assume first that all photon and phonon modes are initially coherent, i.e. the complete boson field consisting of photons and phonons is in the coherent state $|\xi_L, \xi_S, \xi_A, \xi_V\rangle$ for $t = 0$. This means that the coherent radiation is scattered by coherent acoustical phonons (stimulated Brillouin scattering). Then (9.37a) provides

$$\begin{aligned}
\langle(\Delta W_L)^2\rangle_{\mathcal{N}} &= 2|\varkappa|^2 t^2 |\xi_L|^2 |\xi_A|^2 - (Rt^2 + \text{c.c.}),\\
\langle(\Delta W_S)^2\rangle_{\mathcal{N}} &= 2|g|^2 t^2 |\xi_L|^2 |\xi_S|^2,\\
\langle(\Delta W_A)^2\rangle_{\mathcal{N}} &= 0,\\
\langle(\Delta W_V)^2\rangle_{\mathcal{N}} &= 2(|g|^2 t^2 |\xi_L|^2 + |\varkappa|^2 t^2 |\xi_A|^2)|\xi_V|^2 - (Ut^2 + \text{c.c.}),
\end{aligned} \tag{9.85}$$

where $R = g\varkappa^* \xi_L^2 \xi_S^* \xi_A^*$ and $U = g\varkappa \xi_S^* \xi_A \xi_V^{*2}$. Thus the anti-Stokes mode is coherent up to t^2 and the Stokes and phonon modes always exhibit bunching (in practice $|\xi_L|^2 \gg |\xi_V|^2 \gg |\xi_{S,A}|^2$); in the laser mode photon bunching as well as antibunching may occur [Tänzler and Schütte (1981a)]. If $g = \varkappa$, one has

$$\langle(\Delta W_L)^2\rangle_{\mathcal{N}} = 2|g|^2 t^2 |\xi_L|^2 |\xi_A|^2 \left[1 - \frac{|\xi_S|}{|\xi_A|} \cos(2\varphi_L - \varphi_S - \varphi_A)\right]. \tag{9.86}$$

The maximal antibunching arises if the initial phase condition (9.74) is satisfied and $|\xi_S| > |\xi_A|$ holds in stimulated scattering.

Similarly (9.37b) leads to

$$\begin{aligned}
\langle\Delta W_L \Delta W_S\rangle_{\mathcal{N}} &= -|g|^2 t^2 |\xi_L|^2 |\xi_S|^2 + (Rt^2 + \text{c.c.}),\\
\langle\Delta W_L \Delta W_A\rangle_{\mathcal{N}} &= -|\varkappa|^2 t^2 |\xi_L|^2 |\xi_A|^2,\\
\langle\Delta W_L \Delta W_V\rangle_{\mathcal{N}} &= \mathrm{i}\varkappa t \xi_A \xi_L^* \xi_V^* + \text{c.c.} - |g|^2 t^2 |\xi_L|^2 |\xi_V|^2 +\\
&\quad + |\varkappa|^2 t^2 [|\xi_A|^2 (1 + 2|\xi_V|^2 + 2|\xi_L|^2) -\\
&\quad - |\xi_L|^2 |\xi_V|^2] + (Rt^2 + Ut^2 + \text{c.c.}),\\
\langle\Delta W_S \Delta W_A\rangle_{\mathcal{N}} &= -\frac{1}{2}(Rt^2 + \text{c.c.}),\\
\langle\Delta W_S \Delta W_V\rangle_{\mathcal{N}} &= \mathrm{i}gt \xi_L \xi_S^* \xi_V^* +\\
&\quad + \text{c.c.} + |g|^2 t^2 [|\xi_L|^2 (1 + 2|\xi_V|^2 + 2|\xi_S|^2) -\\
&\quad - |\xi_S|^2 |\xi_V|^2] + (Rt^2 - Ut^2 + \text{c.c.}),\\
\langle\Delta W_A \Delta W_V\rangle_{\mathcal{N}} &= -|\varkappa|^2 t^2 |\xi_A|^2 |\xi_V|^2.
\end{aligned} \tag{9.87}$$

Thus, anticorrelation between the laser and the Stokes modes may occur in the course of spontaneous scattering in the anti-Stokes mode ($\xi_A = 0$) or, for instance, in stimulated scattering if the interference terms Rt^2 + c.c. are zero because of $\Delta\varphi = \pi/2$; the anticorrelation is maximal when $\Delta\varphi = \pi$. There is always anticorrelation of a fully quantum nature between the laser and anti-Stokes modes, and between the anti-Stokes and phonon modes. After the switching on of the interaction phase-dependent anticorrelation may occur, if $\varphi_L + \varphi_V - \varphi_A - \psi_A = -\pi/2$, between the laser and phonon modes; in this case also phase independent anticorrelation occurs in the course of spontaneous scattering ($\xi_S = \xi_A = 0$),

$$\langle \Delta W_L \Delta W_V \rangle_{\mathcal{N}} = -(|g|^2 + |\varkappa|^2) t^2 |\xi_L|^2 |\xi_V|^2. \tag{9.88}$$

Between the Stokes and anti-Stokes modes anticorrelation is also possible if the initial phase condition (9.74) is satisfied, in agreement with the earlier results. Between the Stokes and phonon modes phase-dependent anticorrelation may occur if $\varphi_S + \varphi_V - \varphi_L - \psi_S = -\pi/2$.

We can easily calculate the quantities $\langle(\Delta W_j + \Delta W_k)^2\rangle_{\mathcal{N}} = \langle(\Delta W_j)^2\rangle_{\mathcal{N}} + \langle(\Delta W_k)^2\rangle_{\mathcal{N}} + 2\langle\Delta W_j \Delta W_k\rangle_{\mathcal{N}}$, $j \neq k$. For example $\langle(\Delta W_L + \Delta W_V)^2\rangle_{\mathcal{N}} = -2|\varkappa|^2 t^2 |\xi_L|^2 |\xi_V|^2$ in spontaneous scattering, etc.

We can now consider scattering of coherent light by chaotic phonons. From (9.37a, b) we obtain in the same way

$$\begin{aligned}\langle(\Delta W_L)^2\rangle_{\mathcal{N}} = {} & 2|g|^2 t^2 |\xi_L|^2 |\xi_S|^2 \langle n_V\rangle + \\ & + 2|\varkappa|^2 t^2 |\xi_L|^2 |\xi_A|^2 (\langle n_V\rangle + 1) - \\ & - [Rt^2(1 + 2\langle n_V\rangle) + \text{c.c.}],\end{aligned}$$

$$\langle(\Delta W_S)^2\rangle_{\mathcal{N}} = 2|g|^2 t^2 |\xi_L|^2 |\xi_S|^2 (\langle n_V\rangle + 1),$$

$$\langle(\Delta W_A)^2\rangle_{\mathcal{N}} = 2|\varkappa|^2 t^2 |\xi_L|^2 |\xi_A|^2 \langle n_V\rangle,$$

$$\begin{aligned}\langle(\Delta W_V)^2\rangle_{\mathcal{N}} = {} & \langle n_V\rangle^2 + 2|g|^2 t^2 [|\xi_L|^2 \langle n_V\rangle (|\xi_S|^2 + 1) + \\ & + \langle n_V\rangle^2 (|\xi_L|^2 - |\xi_S|^2)] + 2|\varkappa|^2 t^2 [|\xi_A|^2 \langle n_V\rangle (|\xi_L|^2 + 1) + \\ & + \langle n_V\rangle^2 (|\xi_A|^2 - |\xi_L|^2)] + (2Rt^2 \langle n_V\rangle + \text{c.c.}),\end{aligned}$$

$$\begin{aligned}\langle\Delta W_L \Delta W_S\rangle_{\mathcal{N}} = {} & -|g|^2 t^2 |\xi_L|^2 |\xi_S|^2 (2\langle n_V\rangle + 1) + \\ & + [Rt^2(\langle n_V\rangle + 1) + \text{c.c.}],\end{aligned}$$

$$\begin{aligned}\langle\Delta W_L \Delta W_A\rangle_{\mathcal{N}} = {} & -|\varkappa|^2 t^2 |\xi_L|^2 |\xi_A|^2 (2\langle n_V\rangle + 1) + \\ & + [Rt^2 \langle n_V\rangle + \text{c.c.}],\end{aligned}$$

$$\begin{aligned}\langle\Delta W_L \Delta W_V\rangle_{\mathcal{N}} = {} & |g|^2 t^2 [\langle n_V\rangle^2 (|\xi_S|^2 - |\xi_L|^2) - \langle n_V\rangle |\xi_L|^2 (2|\xi_S|^2 + \\ & + 1)] + |\varkappa|^2 t^2 [\langle n_V\rangle^2 (|\xi_A|^2 - |\xi_L|^2) + \\ & + \langle n_V\rangle |\xi_L|^2 (2|\xi_A|^2 - 1) + \\ & + |\xi_A|^2 (1 + 2|\xi_L|^2 + 2\langle n_V\rangle)] + (Rt^2 + \text{c.c.}),\end{aligned}$$

$$\langle\Delta W_S \Delta W_A\rangle_{\mathcal{N}} = -\frac{1}{2}[Rt^2(2\langle n_V\rangle + 1) + \text{c.c.}],$$

$$\begin{aligned}\langle\Delta W_S \Delta W_V\rangle_{\mathcal{N}} = {} & |g|^2 t^2 [\langle n_V\rangle^2 (|\xi_L|^2 - |\xi_S|^2) + \\ & + |\xi_S|^2 \langle n_V\rangle (2|\xi_L|^2 - 1) + |\xi_L|^2 (1 + 2|\xi_S|^2 + \\ & + 2\langle n_V\rangle)] + [Rt^2(1 + \langle n_V\rangle) + \text{c.c.}],\end{aligned}$$

$$\langle \Delta W_A \Delta W_V \rangle_{\mathcal{N}} = -|\varkappa|^2 t^2 [|\xi_A|^2 \langle n_V \rangle (2|\xi_L|^2 + 1) + \langle n_V \rangle^2 (|\xi_A|^2 - |\xi_L|^2)] - (R t^2 \langle n_V \rangle + \text{c.c.}). \quad (9.89)$$

We see that in this case no antibunching is possible in separate modes ($|\xi_L|^2 \gg \langle n_V \rangle \gg |\xi_{S,A}|^2$), since the averaging procedure involving the chaotic phonons leads to the smoothing of the effect. For spontaneous scattering $\langle (\Delta W_L)^2 \rangle_{\mathcal{N}} = \langle (\Delta W_S)^2 \rangle_{\mathcal{N}} = \langle (\Delta W_A)^2 \rangle_{\mathcal{N}} = 0$. However, in this case there is enhancement of the anticorrelation between modes L and S, modes L and A and modes S and A, under the same conditions as before; for example, corresponding to (9.88) for spontaneous scattering,

$$\langle \Delta W_L \Delta W_V \rangle_{\mathcal{N}} = -(|g|^2 + |\varkappa|^2)\, t^2 |\xi_L|^2 \langle n_V \rangle (\langle n_V \rangle + 1), \quad (9.90)$$

which also holds approximately for stimulated scattering if $|\xi_L|^2 \gg \langle n_V \rangle \gg |\xi_{S,A}|^2$. Enhanced anticorrelation occurs between modes L and S (L and A) in stimulated Stokes (anti-Stokes) and spontaneous anti-Stokes (Stokes) scattering. Further $\langle \Delta W_{S,A} \Delta W_V \rangle_{\mathcal{N}} \geqq 0$ is always true.

If, in addition, the Stokes and anti-Stokes modes are initially chaotic, with the mean numbers of photons $\langle n_S \rangle$ and $\langle n_A \rangle$ respectively, and the conditions $|\xi_L|^2 \gg \langle n_V \rangle \gg \langle n_S \rangle, \langle n_A \rangle$ are assumed, then the modes L and S, modes L and A and modes L and V are anticorrelated as above (the phase-dependent terms being cancelled in this case), modes S and A are independent and $\langle \Delta W_{S,A} \Delta W_V \rangle_{\mathcal{N}} \geqq 0$.

Further, even for $\langle n_V \rangle = 0$, bunching of laser and Stokes photons occurs and the anti-Stokes mode tends to be coherent, $\langle (\Delta W_A)^2 \rangle_{\mathcal{N}} = 0$; anticorrelation may occur between the laser and Stokes modes and Stokes and anti-Stokes modes and there is anticorrelation between the laser and anti-Stokes modes, $\langle \Delta W_L \Delta W_A \rangle_{\mathcal{N}} = -|\varkappa|^2 t^2 |\xi_L|^2 |\xi_A|^2$, as a result of phonon vacuum fluctuations.

Thus anticorrelations between incident laser photons and scattered photons can arise only in stimulated Raman scattering. In spontaneous scattering the incident laser photons and scattered photons are not correlated. Anticorrelation between scattered photons and phonons occurs neither in spontaneous nor in stimulated Raman scattering. Photon–phonon anticorrelation takes place between incident laser photons and phonons in spontaneous Raman scattering.

As mentioned above the conditions $|g\xi_L t| < 1$ and $|\varkappa \xi_L t| < 1$ determine the validity of the short-time approximation, which permits laser pulses of about 10 MW to be used. Taking the length of the scattering medium as of the order of centimeters, then $t \approx 10^{-10}$ s and $|gt|$ or $|\varkappa t| \approx 10^{-8}$. Further $\langle n_V \rangle \approx 10^{12}\ \text{cm}^{-3}$ for hydrogen gas at 10 atm. pressure and a temperature of 300 K, and consequently

$$\frac{\langle \Delta W_L \Delta W_S \rangle_{\mathcal{N}}}{\langle W_L \rangle_{\mathcal{N}} \langle W_S \rangle_{\mathcal{N}}} \approx \frac{\langle \Delta W_L \Delta W_A \rangle_{\mathcal{N}}}{\langle W_L \rangle_{\mathcal{N}} \langle W_A \rangle_{\mathcal{N}}} \approx -|g|^2 t^2 (2\langle n_V \rangle + 1) \approx \approx -|\varkappa|^2 t^2 (2\langle n_V \rangle + 1) \approx -2 \times 10^{-4}; \quad (9.91a)$$

further it holds that (g or $\varkappa = 0$)

$$\frac{\langle \Delta W_L \Delta W_{S,A} \rangle_{\mathcal{N}}}{[\langle (\Delta W_L)^2 \rangle_{\mathcal{N}} \langle (\Delta W_{S,A})^2 \rangle_{\mathcal{N}}]^{1/2}} \approx -1. \tag{9.91b}$$

We point out again that all these effects are intrinsically quantum mechanical.

Further results are available in papers by Szlachetka et al. (1979, 1980a, b), particularly when the Stokes and anti-Stokes modes are intially coherent and chaotic. In the paper by Szlachetka et al. (1980b) interesting tables are given, showing the strong dependence of the correlation properties of Raman and hyper-Raman scattering on the initial photon statistics.

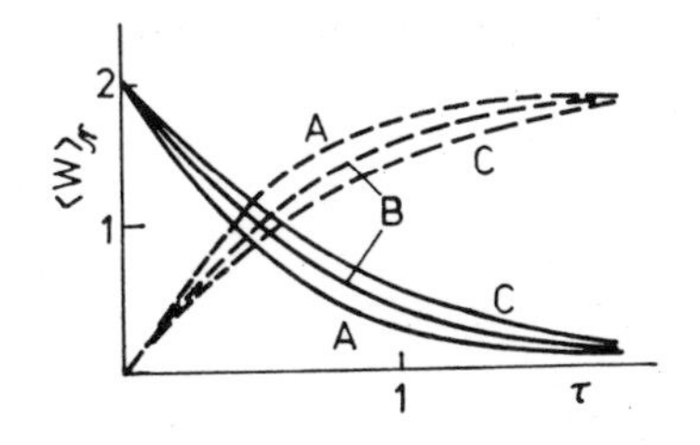

Fig. 9.24 — The time development of the incident integrated intensity $\langle W(t) \rangle_{\mathcal{N}}$ (solid curves) and Stokes scattered light $\langle W_S(t) \rangle_{\mathcal{N}}$ (dashed curves) for initially chaotic radiation (A), coherent radiation (B) and a Fock state (C), $\langle |\xi|^2 \rangle = 2$, $\langle |\xi_S|^2 \rangle = 0$, $\tau = |g| t$ (after Simaan, 1975, J. Phys. **A8**, 1620).

It is a consequence of the form (9.83) for the characteristic function, that the quasi-distribution $\Phi_{\mathcal{A}}$, the generating function and the measurable photocount distribution $p(n, t)$ and its factorial moments $\langle W^k \rangle_{\mathcal{N}}$ may be determined in the forms (8.44), (8.59b) and (8.64a, b).

An alternative approach to Raman scattering has been developed in which atomic variables are eliminated using the master equation and the Fock states, or the rate equations for the conditional photon number probabilities [Loudon (1973), McNeil and Walls (1974), Simaan (1975)]. A similar approach to stimulated Raman scattering has been developed by Gupta and Mohanty (1980) and Mohanty and Gupta (1981a, b), using the density matrix formalism in the Fock states and including effects of atomic motion, detuning, populations of lower and upper levels, and higher-order non-linearities. They have shown that the coherence properties of the Stokes beam are improved with increasing amplification and with decreasing detuning. In Fig. 9.24 we demonstrate the time evolution of the incident radiation and of the scattered Stokes radiation for initially chaotic radiation, for coherent radiation, and for the Fock state [Simaan (1975)]. This figure shows that Raman scattering is more effective for chaotic light than for coherent light. Similar conclusions apply for multiphoton absorption [see Chapter 6 and Schubert and Wilhelmi (1980), Bandilla (1978)] and for second-harmonic generation [Crosignani, Di Porto and Solimeno (1972), Chmela (1973)].

A more general connection can be found [Chmela (1979a–e, 1981a, b)] between the efficiency of nonlinear optical processes and intermodal correlations. For instance, the correlation between subfrequency modes supports sum-frequency generation; but it acts against subfrequency generation, while the anticorrelation between them acts in the opposite way. The correlation between the subfrequency and sum-frequency modes supports difference-frequency generation, while the anticorrelation between them supports sum-frequency generation.

In Fig. 9.25 the normalized correlation function $\langle \hat{a}_L^+(t)\, \hat{a}_S^+(t)\, \hat{a}_L(t)\, \hat{a}_S(t)\rangle / \langle \hat{a}_L^+(t)\, \hat{a}_L(t)\rangle \langle \hat{a}_S^+(t)\, \hat{a}_S(t)\rangle$ is shown for initially chaotic radiation, for coherent radiation and for the Fock state again [Simaan (1975)]. When it is less than unity, anticorrelation between the laser (randomized laser or Fock state) and scattered Stokes modes occurs, in agrement with the above discussion.

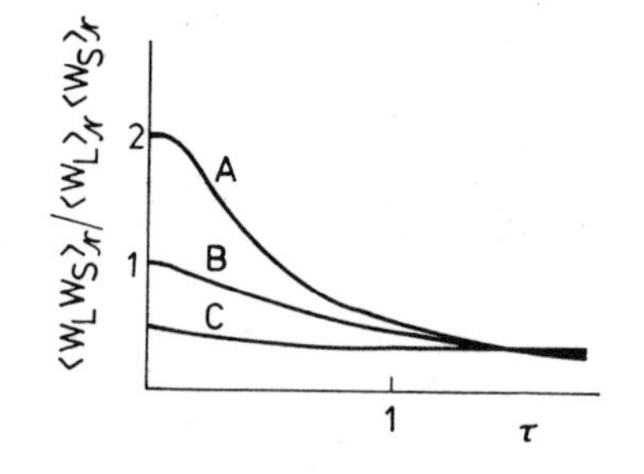

Fig. 9.25 — The normalized correlation between incident (laser) and Stokes photons for the same cases as in Fig. 9.24 (after Simaan, 1975, J. Phys. **A8**, 1620).

A number of experimental investigations have been carried out, particularly for Rayleigh scattering [Arecchi, Giglio and Tartari (1967), Arecchi and Degiorgio (1972), Pusey, Schaefer and Koppel (1974), Pike and Jakeman (1974), Aoki, Okabe and Sakurai (1974), Aoki, Endo, Takayanagi and Sakurai (1976), Timmermans and Zijlstra (1977), Aoki and Sakurai (1979, 1980), Yoshikawa, N. Suzuki and T. Suzuki (1981)], where measurements of the photocount statistics have been reported. Particularly, Aoki et al. (1976) verified the validity of expressions (5.108) for the factorial moments $\langle W^2\rangle_{\mathcal{N}}$ and $\langle W^3\rangle_{\mathcal{N}}$ $(P = 1, \varphi = 0)$, and Aoki and Sakurai (1979) verified the validity of the expressions for the photocount statistics for partially polarized light.

Quantum statistics of anti-Stokes luminescence, including the cooperative mechanism, has been discussed by Gonzáles–Díaz (1978) and the quantum theory of the stimulated Raman effect in a fluctuating medium has been considered by Iwasawa (1976). Entropy considerations for scattered electromagnetic fields have been given by Crosignani and Tedeschi (1976), and the role of Raman scattering in optical fibres has been investigated by Crosignani, Di Porto and Solimeno (1980). The first observation of macroscopic quantum fluctuations in stimulated Raman scattering has been reported by Walmsley and Raymer (1983).

9.4.4 Hyper-Raman scattering

Consider first non-degenerate hyper-Raman scattering. Denote by 1 and 2 two modes of laser light incident onto the scattering medium. If this light is sufficiently intense, hyper-Raman scattering may take place, characterized by the effective hamiltonian

$$\hat{H} = \sum_{j=1,2,S,A,V} \hbar\omega_j \hat{a}_j^+ \hat{a}_j - \hbar(g\hat{a}_1\hat{a}_2\hat{a}_S^+\hat{a}_V^+ + \varkappa^*\hat{a}_1\hat{a}_2\hat{a}_A^+\hat{a}_V + \text{h.c.}), \tag{9.92}$$

where $\omega_{S,A} = \omega_1 + \omega_2 \mp \omega_V$. In the standard way we can derive the Heisenberg equations of motion for the photon and phonon operators $\hat{a}_j(t)$, from which the following conservation laws are obtained:

$$\frac{\mathrm{d}}{\mathrm{d}t}[\hat{n}_1(t) - \hat{n}_2(t)] = \hat{0},$$

$$\frac{\mathrm{d}}{\mathrm{d}t}[\hat{n}_1(t) + \hat{n}_S(t) + \hat{n}_A(t)] = \hat{0},$$

$$\frac{\mathrm{d}}{\mathrm{d}t}[\hat{n}_2(t) + \hat{n}_S(t) + \hat{n}_A(t)] = \hat{0},$$

$$\frac{\mathrm{d}}{\mathrm{d}t}[\hat{n}_V(t) + \hat{n}_A(t) - \hat{n}_S(t)] = \hat{0}; \tag{9.93}$$

they generalize the corresponding conservation laws for Raman scattering. The solution of the Heisenberg equations up to t^2 is more complex [Peřinová et al. (1979a), Szlachetka et al. (1980)], although it naturally generalizes the solution for Raman scattering. Further, we can determine the normal characteristic function in a form similar to that given by (9.83); however, additional higher-order terms now arise, such as $\bar{D}_{jkk}\beta_j \mid \beta_k \mid^2$, $\bar{D}_{jkj} \mid \beta_j \mid^2 \beta_k$ and their complex conjugates, and $\bar{D}_{jkjk} \mid \beta_j \mid^2 \mid \beta_k \mid^2$ giving the additional contribution $\langle(\bar{D}_{jkk}\xi_j(t) + \bar{D}_{jkj}\xi_k(t) + \text{c.c.}) + \bar{D}_{jkjk}\rangle$ in (9.37b).

The behaviour of the fluctuations in separate modes and of their correlations is quite similar to that for Raman scattering, for instance

$$\langle(\Delta W_S)^2\rangle_{\mathcal{N}} = 2 \mid g \mid^2 t^2 \mid \xi_1 \mid^2 \mid \xi_2 \mid^2 \mid \xi_S \mid^2 (\langle n_V \rangle + 1), \tag{9.94a}$$

$$\langle(\Delta W_A)^2\rangle_{\mathcal{N}} = 2 \mid \varkappa \mid^2 t^2 \mid \xi_1 \mid^2 \mid \xi_2 \mid^2 \mid \xi_A \mid^2 \langle n_V \rangle, \tag{9.94b}$$

$$\langle \Delta W_{1,2} \Delta W_S \rangle_{\mathcal{N}} = - \mid g \mid^2 t^2 \mid \xi_{1,2} \mid^2 \mid \xi_S \mid^2 [(2\langle n_V \rangle + 1) \mid \xi_{2,1} \mid^2 + \langle n_V \rangle] + \\ + (Rt^2(\langle n_V \rangle + 1) + \text{c.c.}), \tag{9.94c}$$

$$\langle \Delta W_{1,2} \Delta W_A \rangle_{\mathcal{N}} = - \mid \varkappa \mid^2 t^2 \mid \xi_{1,2} \mid^2 \mid \xi_A \mid^2 [(2\langle n_V \rangle + 1) \mid \xi_{2,1} \mid^2 + \\ + \langle n_V \rangle + 1] + (Rt^2 \langle n_V \rangle + \text{c.c.}), \tag{9.94d}$$

assuming that the phonon mode is initially chaotic, while all the radiation modes are initially coherent and $R = g\varkappa^*\xi_1^2\xi_2^2\xi_S^*\xi_A^*$. Thus between the laser and scattered modes anticorrelation is possible (for instance if $R = 0$). In the course of spontaneous scattering ($\xi_S = \xi_A = 0$) anticorrelation takes place between the laser mode 1 (2) and the phonon mode V,

$$\langle \Delta W_{1,2} \Delta W_V \rangle_{\mathcal{N}} = -(|g|^2 + |\varkappa|^2)\, t^2 |\xi_1|^2 |\xi_2|^2 \langle n_V \rangle (\langle n_V \rangle + 1) \quad (9.95)$$

by analogy with (9.90) for Raman scattering; this is also approximately true in stimulated scattering, provided that $|\xi_{1,2}|^2 \gg \langle n_V \rangle \gg |\xi_{S,A}|^2$, as is usual in practical situations. The correlation properties of the modes S and A are expressed in (9.89) with R given above and $\langle \Delta W_{S,A} \Delta W_V \rangle_{\mathcal{N}} \geqq 0$. Nevertheless, there is one difference compared to Raman scattering, namely anticorrelation occurs between the laser modes as a consequence of their coupling in spontaneous scattering,

$$\langle \Delta W_1 \Delta W_2 \rangle_{\mathcal{N}} = -|g|^2 t^2 |\xi_1|^2 |\xi_2|^2 (\langle n_V \rangle + 1) - \\ - |\varkappa|^2 t^2 |\xi_1|^2 |\xi_2|^2 \langle n_V \rangle; \quad (9.96)$$

in this case antibunching of incident laser photons arises, since $\langle (\Delta W_1 + \Delta W_2)^2 \rangle_{\mathcal{N}} = 2\langle \Delta W_1 \Delta W_2 \rangle_{\mathcal{N}} < 0$ [$\langle (\Delta W_1)^2 \rangle_{\mathcal{N}} = \langle (\Delta W_2)^2 \rangle_{\mathcal{N}} = 0$ in spontaneous scattering]. Quite similar behaviour of the fluctuations and their correlations can be deduced if, in addition, the Stokes and anti-Stokes photons are initially chaotic. However, in this case the modes S and A are uncorrelated. Anticorrelation (9.96) and the corresponding antibunching of laser photons just correspond to the case of double-beam two-photon absorption [Simaan and Loudon (1975)]; this means that pairs of photons in the laser beam are absorbed, producing photon-anticorrelation in the beam. Because of the presence of higher-order terms in the characteristic function, an explicit determination of the photocount statistics is more complicated in the present case.

Consider now degenerate hyper-Raman scattering [Peřinová et al. (1979b)]. In this case we may write the following effective hamiltonian

$$\hat{H} = \sum_{j=L,S,A,V} \hbar\omega_j \hat{a}_j^+ \hat{a}_j - \hbar(g\hat{a}_L^2 \hat{a}_S^+ \hat{a}_V^+ + \varkappa^* \hat{a}_L^2 \hat{a}_A^+ \hat{a}_V + \text{h.c.}) \quad (9.97)$$

and $\omega_{S,A} = 2\omega_L \mp \omega_V$. In the standard way one can derive the Heisenberg equations and obtain approximate solutions up to t^2, and the other quantities specifying the quantum statistics. In this interaction the combinations $\hat{n}_L(t) + 2\hat{n}_S(t) + 2\hat{n}_A(t)$ and $\hat{n}_V(t) + \hat{n}_A(t) - \hat{n}_S(t)$ are constants of motion. In the normal characteristic function (9.83) one now has additional higher-order terms $\bar{D}_{LLL}\beta_L |\beta_L|^2$ and the complex conjugate and $\bar{D}_{LLLL}|\beta_L|^4$, giving the additional terms $\langle 4(\bar{D}_{LLL}\xi_L(t) + \text{c.c.}) + 4\bar{D}_{LLLL} \rangle$ in (9.37a).

Assuming initially chaotic phonons and coherent photons again the behaviour of the fluctuations and of their correlations is quite similar to that in non-degenerate scattering. For example

$$\langle (\Delta W_S)^2 \rangle_{\mathcal{N}} = 2|g|^2 t^2 |\xi_L|^4 |\xi_S|^2 (\langle n_V \rangle + 1), \quad (9.98a)$$

$$\langle (\Delta W_A)^2 \rangle_{\mathcal{N}} = 2|\varkappa|^2 t^2 |\xi_L|^4 |\xi_A|^2 \langle n_V \rangle; \quad (9.98b)$$

the Stokes and anti-Stokes photons may be anticorrelated in stimulated scattering if, as above, $(Rt^2 + \text{c.c.}) \leqq 0$, $R = g\varkappa^* \xi_L^4 \xi_S^* \xi_A^*$; further

$$\langle \Delta W_L \Delta W_S \rangle_{\mathcal{N}} = \\ = -2 | g |^2 t^2 | \xi_L |^2 | \xi_S |^2 [| \xi_L |^2 (2\langle n_V \rangle + 1) + 2\langle n_V \rangle] + \\ + [2Rt^2(\langle n_V \rangle + 1) + \text{c.c.}], \tag{9.98c}$$

$$\langle \Delta W_L \Delta W_A \rangle_{\mathcal{N}} = \\ = -2 | \varkappa |^2 t^2 | \xi_L |^2 | \xi_A |^2 [| \xi_L |^2 (2\langle n_V \rangle + 1) + 2(\langle n_V \rangle + 1)] + \\ + [2Rt^2 \langle n_V \rangle + \text{c.c.}], \tag{9.98d}$$

and for the correlation of the fluctuations in the laser and phonon modes during spontaneous scattering (or if $| \xi_L |^2 \gg \langle n_V \rangle \gg | \xi_S |^2, | \xi_A |^2$ in stimulated scattering) we have

$$\langle \Delta W_L \Delta W_V \rangle_{\mathcal{N}} = -2(| g |^2 + | \varkappa |^2) t^2 | \xi_L |^4 \langle n_V \rangle (\langle n_V \rangle + 1), \tag{9.98e}$$

giving anticorrelation of these modes. These phenomena can also occur as a result of phonon vacuum fluctuations ($\langle n_V \rangle = 0$), but $\langle (\Delta W_A)^2 \rangle_{\mathcal{N}} = \langle \Delta W_L \Delta W_V \rangle_{\mathcal{N}} = 0$ in this case. Corresponding to the anticorrelation (9.96) between the laser modes in non-degenerate scattering, antibunching arises in the laser mode in the course of spontaneous scattering

$$\langle (\Delta W_L)^2 \rangle_{\mathcal{N}} = -2 | g |^2 t^2 | \xi_L |^4 (\langle n_V \rangle + 1) - 2 | \varkappa |^2 t^2 | \xi_L |^4 \langle n_V \rangle; \tag{9.99}$$

this effect is analogous to one-beam two-photon absorption [Simaan and Loudon (1975)]. This antibunching is also produced by phonon vacuum fluctuations ($\langle n_V \rangle = 0$). While modes S and A may be phase anticorrelated, as shown in (9.89), they are uncorrelated if, in addition, these modes are initially chaotic.

Further results concerning the statistical properties of radiation in hyper-Raman scattering are available in the literature [Simaan (1978), Peřinová et al. (1979a, b), Szlachetka et al. (1980a, b, 1980), Tänzler and Schütte (1981a, b)]. Particularly Szlachetka et al. (1980b) provided a systematic treatment of the photon and phonon correlation properties in multiphoton Raman processes, including their dependence on the initial statistics of the modes and classification of the nature of the correlations (for bunching, antibunching, correlation, anticorrelation and uncorrelation in modes); the behaviour of the quasi-distribution $\Phi_{\mathcal{N}}$ has been also investigated for these cases. The quantum statistics of stimulated hyper-Raman scattering in an inhomogeneously broadened system have been dealt with by Mohanty and Gupta (1981c).

Thus, we can conclude that scattering processes provide a number of possibilities of observing radiation that exhibits antibunching and anticorrelation phenomena.

9.5 Multiphoton absorption

Multiphoton absorption is described by the master equation (7.36) if $N_2 = 0$, $m = 0$ and $n = k$. Thus for the k-photon absorption of single-mode radiation we have the master equation for the reduced density matrix in the form

$$\frac{\partial \hat{\varrho}}{\partial t} = -\frac{\mu}{2}[\hat{a}^{+k}\hat{a}^{k}\hat{\varrho} - 2\hat{a}^{k}\hat{\varrho}\hat{a}^{+k} + \hat{\varrho}\hat{a}^{+k}\hat{a}^{k}], \tag{9.100}$$

where $\mu = 2KN_1 > 0$. If we multiply this equation by the number operator $\hat{a}^+\hat{a}$ and take the trace, we obtain

$$\frac{\mathrm{d}}{\mathrm{d}t}\langle \hat{a}^+\hat{a}\rangle = -k\mu\langle \hat{a}^{+k}\hat{a}^{k}\rangle, \tag{9.101}$$

which points out the strong dependence of the multiphoton absorption on the statistical properties of the light. If $k = 1$, then $\langle \hat{a}^+(t)\,\hat{a}(t)\rangle = \exp(-\mu t)\,\langle \hat{a}^+(0)\,\hat{a}(0)\rangle$, which is the standard result. For radiation in the coherent state

$$\frac{\mathrm{d}}{\mathrm{d}t}|\alpha|^2 = -k\mu(|\alpha|^2)^k \tag{9.102a}$$

and for chaotic radiation

$$\frac{\mathrm{d}}{\mathrm{d}t}\langle \hat{a}^+\hat{a}\rangle = -kk!\,\mu\langle \hat{a}^+\hat{a}\rangle^k, \tag{9.102b}$$

reflecting the fact that this process is $k!$-times more effective for chaotic light than for coherent light. For the photon-number distribution $p(n, t) = \langle n\,|\,\hat{\varrho}(t)\,|\,n\rangle$ we derive from (9.100)

$$\frac{\mathrm{d}p(n, t)}{\mathrm{d}t} = -\mu\left[\frac{n!}{(n-k)!}\,p(n, t) - \frac{(n+k)!}{n!}\,p(n+k, t)\right] \tag{9.103a}$$

and for the moments we obtain

$$\frac{\mathrm{d}\langle n^l\rangle}{\mathrm{d}t} = -\mu\left\langle \frac{n!}{(n-k)!}[n^l - (n-k)^l]\right\rangle. \tag{9.103b}$$

The generalized Fokker–Planck equation is much more complicated for the process of multiphoton absorption, either for $\Phi_{\mathscr{A}}$ or for $C_{\mathscr{A}}$ ($C_{\mathscr{N}}$). The most convenient way of solving (9.103a) is to use the generating function (3.64),

$$G(\lambda, t) = \sum_{n=0}^{\infty} p(n, t)\,(1-\lambda)^n, \tag{9.104}$$

so that (9.103a) reduces to

$$\frac{\partial G}{\partial t} = -\mu(-1)^k\left[(1-\lambda)^k\frac{\partial^k G}{\partial \lambda^k} - \frac{\partial^k G}{\partial \lambda^k}\right]; \tag{9.105a}$$

for one-photon absorption we have

$$\frac{\partial G}{\partial t} = -\mu\lambda\frac{\partial G}{\partial \lambda}, \tag{9.105b}$$

having the solution $G(\lambda, t) = G(\lambda \exp(-\mu t), 0)$ $(G(0, t) = \sum_{n=0}^{\infty} p(n, t) = 1)$; for two-photon absorption

$$\frac{\partial G}{\partial t} = -\mu\lambda(\lambda - 2)\frac{\partial^2 G}{\partial \lambda^2}. \tag{9.105c}$$

The solution of this equation can be expressed in terms of the Gegenbauer polynomials $C_n^{(k)}(x)$ [Tornau and Bach (1974), Bandilla and Ritze (1976a)], and the photon-number distribution and its factorial moments have the following forms

$$p(n, t) = \frac{2^n \Gamma(n - 1/2)}{n!\,\Gamma(-1/2)} \sum_{m=n}^{\infty} b_m C_{m-n}^{(n-1/2)}(0) \exp\left[-m(m-1)\mu t\right], \tag{9.106a}$$

$$\langle W^k \rangle_{\mathcal{N}} = 2^k \frac{\Gamma(k - 1/2)}{\Gamma(-1/2)} \sum_{n=k}^{\infty} b_n C_{n-k}^{(k-1/2)}(1) \exp\left[-n(n-1)\mu t\right], \tag{9.106b}$$

where

$$C_{m-n}^{(n-1/2)}(0) = \begin{cases} \dfrac{(-1)^{(m-n)/2}\Gamma((n+m-1)/2)}{\Gamma(n-1/2)\,\Gamma(m/2 - n/2 + 1)}, & m - n \text{ even}, \\ 0 & m - n \text{ odd}, \end{cases} \tag{9.107a}$$

$$C_{n-k}^{(k-1/2)}(1) = \binom{n + k - 2}{n - k} \tag{9.107b}$$

and

$$b_0 = \sum_{n \text{ even}} p(n, t),$$

$$b_1 = -\sum_{n \text{ odd}} p(n, t),$$

$$b_n = \left(n - \frac{1}{2}\right) \int_{-1}^{+1} \left[g(x, 0) - b_0 + b_1 x\right] C_{n-2}^{(3/2)}(x)\,dx, \qquad n \geq 2, \tag{9.107c}$$

with $g(x, \mu t) = G(\lambda, t)$, $x = 1 - \lambda$.

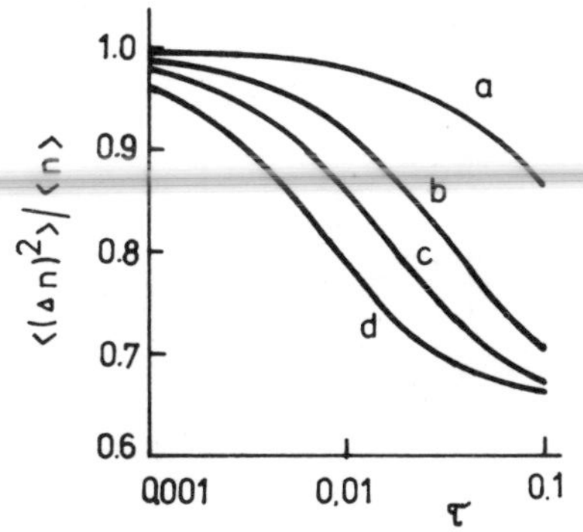

Fig. 9.26 — The fluctuation quantity $\langle(\Delta n)^2\rangle/\langle n\rangle$ versus $\tau = \mu t$ for coherent light and (a) $\langle n(0)\rangle = 1$, (b) $\langle n(0)\rangle = 5$, (c) $\langle n(0)\rangle = 10$ and (d) $\langle n(0)\rangle = 20$ (after Tornau and Bach, 1974, Opt. Comm. **11**, 46).

In Fig. 9.26 the time-dependence of the quantity $\langle(\Delta n)^2\rangle/\langle n\rangle$ is shown for initially coherent light and various initial photon numbers (Tornau and Bach (1974), see also Schubert and Wilhelmi (1976)). This quantity is always less than unity, which demonstrates antibunching and sub-Poissonian behaviour in two-photon absorption. This phenomenon is easily explained taking into account that pairs of photons are more probably absorbed, which produces "holes" in the beam, leading to photon antibunching.

The photon statistics of higher-order absorption processes have been discussed by Paul, Mohr and Brunner (1976), Mohr and Paul (1978), Voigt, Bandilla and Ritze (1980), Zubairy and Yeh (1980), Voigt and Bandilla (1981) and Mohr (1981). It has been particularly shown that [Paul, Mohr and Brunner (1976)] (in approximation $\langle n(t)\rangle \gg 1$, $\langle(\Delta n(t))^2\rangle/\langle n(t)\rangle \ll \langle n(t)\rangle$)

$$\frac{\langle(\Delta n(t))^2\rangle}{\langle n(t)\rangle} = \left[\frac{\langle(\Delta n(0))^2\rangle}{\langle n(0)\rangle} - \frac{k}{2k-1}\right]\left[\frac{\langle n(t)\rangle}{\langle n(0)\rangle}\right]^{2k-1} + \frac{k}{2k-1}, \tag{9.108}$$

and for initially coherent light $\langle(\Delta n(0))^2\rangle/\langle n(0)\rangle = 1$. For higher values of t the first term tends to zero and consequently (9.108) tends to the value $k/(2k-1)$, and 2/3 for $k = 2$, i.e. $\langle(\Delta W)^2\rangle_{\mathcal{N}}/\langle W\rangle_{\mathcal{N}} = -1/3$ in that limit; the ultimate value for $t \to \infty$ is $-1/2$. The asymptotic photon-number distribution can be expressed in the form [Mohr and Paul (1978)]

$$p(n) = \left(\frac{2k-1}{k}\right)^{1/2} (2\pi\langle n\rangle)^{-1/2} \exp\left[-\frac{2k-1}{k}\frac{(n-\langle n\rangle)^2}{2\langle n\rangle}\right]. \tag{9.109}$$

The dependence of the quantity $\langle(\Delta n(t))^2\rangle/\langle n(t)\rangle$ on $\langle n(t)\rangle/\langle n(0)\rangle$ for $k = 1, 2, 3, 5$ and 10 is shown in Fig. 9.27, showing photon antibunching and sub-Poissonian statistics also for higher-order absorption. For $t \to \infty$, $\langle(\Delta n(t))^2\rangle/\langle n(t)\rangle \to (k + 1)/6$. The sub-Poissonian distributions for two- and three-photon absorbers in the asymptotic state are illustrated in Fig. 9.28 [Voigt, Bandilla and Ritze (1980)].

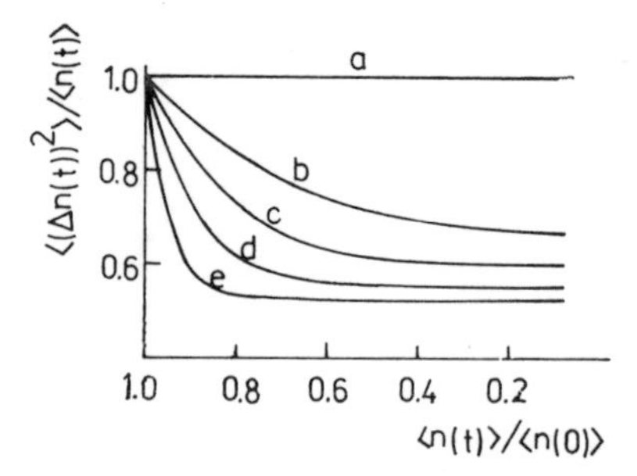

Fig. 9.27 — The change of the quantity $\langle(\Delta n(t))^2\rangle/\langle n(t)\rangle$ as a function of $\langle n(t)\rangle/\langle n(0)\rangle$ due to the k-photon absorption for $k = 1, 2, 3, 5$ and 10 (curves a—e respectively); for $k = 1$ the statistics are conserved (after Mohr and Paul, 1978, Ann. Physik **35**, 461).

General solutions of the master equation for multiphoton absorption have been found by Zubairy and Yeh (1980). In Figs. 9.29a, b we show the transient behaviour of the photon-number distribution $p(n, t)$ for $\mu t = 0, 0.1, 0.2$ and ∞ for two-photon absorption and initially coherent light (Fig. 9.29a) and chaotic light (Fig. 9.29b) if

$\langle n(0)\rangle = 10$ [Zubairy and Yeh (1980)]. Very similar results have been obtained for three- and four-photon absorption. The transient photon-number distribution for two-photon absorption has been discussed also by Simaan and Loudon (1978).

Using the master equation approach single- as well as multimode multiphoton absorption has been investigated in a number of papers [McNeil and Walls (1974),

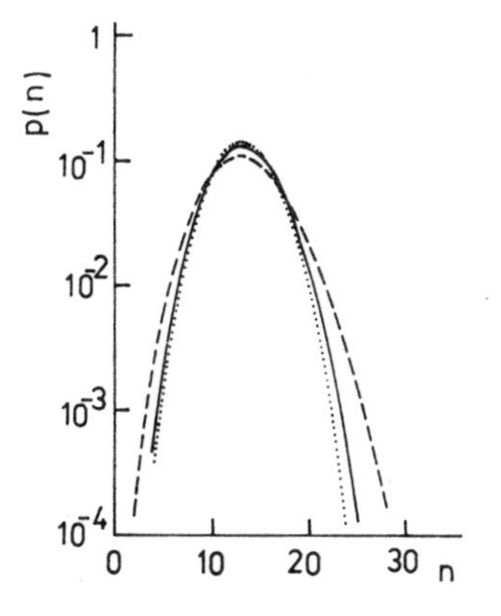

Fig. 9.28 — The photon-number distribution for coherent light (Poisson distribution, (- - -), two-photon absorbed light (———) and three-photon absorbed light (......) — asymptotic states (after Voigt, Bandilla and Ritze, 1980, Z. Phys. **B36**, 295).

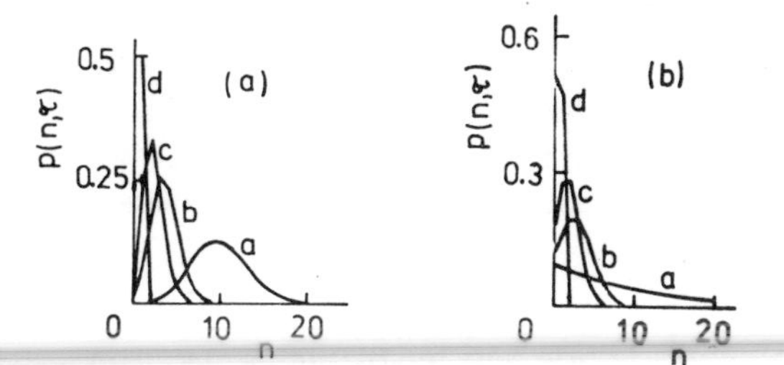

Fig. 9.29 — The photon-number distribution for (a) two-photon absorption of initially coherent light, (b) two-photon absorption of initially chaotic light; $\langle n(0)\rangle = 10$, curves *a—d* are for $\tau = \mu t = 0, 0.1, 0.2$ and ∞ respectively (cf. Fig. 6.7) (after Zubairy and Yeh, 1980, Phys. Rev. **A21**, 1624).

Simaan and Loudon (1975, 1978), Simaan (1979)]. The effect of partial coherence of driving fields in two-photon absorbers in a resonator has been discussed by Chaturvedi, Drummond and Walls (1977). They have obtained antibunching for initially coherent light of amount $\gamma_{\mathcal{N}}^{(2,2)}(0) = \langle|\alpha(0)|^4\rangle_{\mathcal{N}}/\langle|\alpha(0)|^2\rangle_{\mathcal{N}}^2 = 1 - 1/3\langle n\rangle$, which is rapidly smoothed out if the field is initially partially coherent. The use of interference to enhance photon antibunching in two-photon absorbed beams has been suggested by Bandilla and Ritze (1980b, 1981). Photon statistics for a laser with an intercavity two-photon absorber have been reported by Bandilla and Ritze (1976b), Bandilla (1978) and Hildred (1980). Some memory effects and spatial and temporal behaviour of optical beams during two-photon absorption, including photon antibunching, have been treated by Mohr and Paul (1979). Further discussions of antibunching have been given by Hildred and Hall (1978) and Quattropani, Schwendimann and Baltes (1980). Interactions of coherent and incoherent light pulses with matter have been considered by Carusotto and Strati (1973) and Carusotto (1974).

9.6 Multiphoton emission

The process of multiphoton emission is described again by the master equation (7.36), where $N_1 = 0$, $n = k$ and $m = k$, i.e. for k-photon emission we have

$$\frac{\partial \hat{\varrho}}{\partial t} = -\frac{\mu}{2}[\hat{a}^k \hat{a}^{+k} \hat{\varrho} - 2\hat{a}^{+k} \hat{\varrho} \hat{a}^k + \hat{\varrho} \hat{a}^k \hat{a}^{+k}] \tag{9.110a}$$

and for $k = 2$

$$\frac{\partial p(n, \tau)}{\partial \tau} = n(n-1)\, p(n-2, \tau) - (n+2)(n+1)\, p(n, \tau), \tag{9.110b}$$

where $\tau = \mu t$, $\mu = 2KN_2 > 0$. By analogy with (9.101), which is valid for k-photon absorption, we obtain for k-photon emission

$$\frac{\mathrm{d}}{\mathrm{d}t} \langle \hat{a}\hat{a}^+ \rangle = k\mu \langle \hat{a}^k \hat{a}^{+k} \rangle, \tag{9.111}$$

showing that antinormal ordering of field operators is appropriate to describe emission [see equations (2.14a, b) and sections 3.1 and 3.4]; it is also evident that multiphoton emission is strongly dependent on the statistical properties of the light. For $k = 1$, $\langle \hat{a}(t)\, \hat{a}^+(t) \rangle = \exp(\mu t) \langle \hat{a}(0)\, \hat{a}^+(0) \rangle$.

The generalized Fokker–Planck equation is simpler for multiphoton emission than for multiphoton absorption and for two-photon emission it can be reduced to the following equation of motion for the antinormal characteristic function [Peřina and Peřinová (1983)]

$$\frac{\partial C_{\mathscr{A}}}{\partial t} = -\frac{\mu}{2}\left(\beta^2 \frac{\partial^2}{\partial \beta^2} + 2\beta \frac{\partial^2}{\partial \beta^2} \frac{\partial}{\partial \beta^*} + \text{c.c.}\right) C_{\mathscr{A}}. \tag{9.112}$$

For two-mode two-photon emission we have [McNeil and Walls (1974)]

$$\frac{\partial \hat{\varrho}}{\partial t} = -\frac{\mu}{2}[\hat{a}_1 \hat{a}_2 \hat{a}_1^+ \hat{a}_2^+ \hat{\varrho} - 2\hat{a}_1^+ \hat{a}_2^+ \hat{\varrho} \hat{a}_1 \hat{a}_2 + \hat{\varrho} \hat{a}_1 \hat{a}_2 \hat{a}_1^+ \hat{a}_2^+] \tag{9.113a}$$

and

$$\frac{\partial p(n_1, n_2, \tau)}{\partial \tau} = n_1 n_2 p(n_1 - 1, n_2 - 1, \tau) - (n_1 + 1)(n_2 + 1)\, p(n_1, n_2, \tau). \tag{9.113b}$$

For the antinormal characteristic function we have [Sibilia and Bertolotti (1981)]

$$\frac{\partial C_{\mathscr{A}}}{\partial t} = -\frac{\mu}{2}\left(\beta_1\beta_2 \frac{\partial^2}{\partial\beta_1\,\partial\beta_2} + \beta_1 \frac{\partial^3}{\partial\beta_1\,\partial\beta_2\,\partial\beta_2^*} + \right.$$
$$\left. + \beta_2 \frac{\partial^3}{\partial\beta_1\,\partial\beta_1^*\,\partial\beta_2} + \text{c.c.}\right) C_{\mathscr{A}}; \tag{9.114}$$

here and in (9.112) the time dependence $\exp(-\mathrm{i}\omega_j t)$ has been eliminated.

The solution of (9.114) has been obtained by Sibilia and Bertolotti (1981) up to (μt) and by Peřinová, Peřina, Bertolotti and Sibilia (1982) up to $(\mu t)^2$. Returning to the original variables, we can write for the normal characteristic function

$$\begin{aligned} C_{\mathscr{N}}(\beta_1, \beta_2, t) = \exp\{&\bar{D}_{1212}(t)\,|\,\beta_1\,|^2\,|\,\beta_2\,|^2 - \\ &- [\bar{D}_{121}(t)\,|\,\beta_1\,|^2\beta_2 + \bar{D}_{122}(t)\beta_1\,|\,\beta_2\,|^2 - \text{c.c.}] - \\ &- B_1(t)\,|\,\beta_1\,|^2 - B_2(t)\,|\,\beta_2\,|^2 + [D_{12}(t)\,\beta_1^*\beta_2^* + \text{c.c.}] + \\ &+ [\bar{D}_{12}(t)\,\beta_1\beta_2^* + \text{c.c.}] + \\ &+ [\beta_1\xi_1^*(t) + \beta_2\xi_2^*(t) - \text{c.c.}]\}, \end{aligned} \tag{9.115}$$

provided that the field is initially coherent, and it holds that

$$\begin{aligned} \bar{D}_{1212}(t) &= \mu t, \\ \bar{D}_{122}(t) &= \mu t \xi_1^* \exp(\mathrm{i}\omega_1 t), \\ \bar{D}_{121}(t) &= \mu t \xi_2^* \exp(\mathrm{i}\omega_2 t), \\ B_1(t) &= \mu t(|\,\xi_2\,|^2 + 1), \\ B_2(t) &= \mu t(|\,\xi_1\,|^2 + 1), \\ D_{12}(t) &= \frac{1}{2}\mu t \xi_1\xi_2 \exp[-\mathrm{i}(\omega_1 + \omega_2)\,t], \\ \bar{D}_{12}(t) &= -\mu t \xi_1^*\xi_2 \exp[\mathrm{i}(\omega_1 - \omega_2)\,t], \\ \xi_1(t) &= \xi_1\left[1 + \frac{\mu t}{2}(1 + |\,\xi_2\,|^2)\right]\exp(-\mathrm{i}\omega_1 t), \\ \xi_2(t) &= \xi_2\left[1 + \frac{\mu t}{2}(1 + |\,\xi_1\,|^2)\right]\exp(-\mathrm{i}\omega_2 t). \end{aligned} \tag{9.116}$$

Using (9.37a, b) with the additional higher-order terms as in hyper-Raman scattering we arrive at

$$\begin{aligned} \langle W_j\rangle_{\mathscr{N}} &= |\,\xi_j\,|^2 + \mu t(1 + |\,\xi_1\,|^2)(1 + |\,\xi_2\,|^2), \qquad j = 1, 2, \\ \langle(\Delta W_1)^2\rangle_{\mathscr{N}} &= 2\mu t\,|\,\xi_1\,|^2(1 + |\,\xi_2\,|^2) > 0, \\ \langle(\Delta W_2)^2\rangle_{\mathscr{N}} &= 2\mu t\,|\,\xi_2\,|^2(1 + |\,\xi_1\,|^2) > 0, \\ \langle\Delta W_1\Delta W_2\rangle_{\mathscr{N}} &= \mu t[1 + 2(|\,\xi_1\,|^2 + |\,\xi_2\,|^2) + 3\,|\,\xi_1\,|^2\,|\,\xi_2\,|^2] > 0, \\ \langle(\Delta W)^2\rangle_{\mathscr{N}} &= 2\mu t[1 + 3(|\,\xi_1\,|^2 + |\,\xi_2\,|^2) + 5\,|\,\xi_1\,|^2\,|\,\xi_2\,|^2] > 0. \end{aligned} \tag{9.117}$$

Consequently, in two-photon stimulated emission only photon bunching can occur. The same is true up to $(\mu t)^2$ [Peřinová et al. (1982)].

Similarly in the degenerate one-mode case we obtain

$$C_{\mathcal{N}}(\beta, t) = \exp\Big\{\bar{D}_{LLLL}(t)\,|\,\beta\,|^4 - [\bar{D}_{LLL}(t)\,|\,\beta\,|^2\beta - \text{c.c.}] -$$

$$- B_L(t)\,|\,\beta\,|^2 + \left[\frac{1}{2}C_L(t)\,\beta^{*2} + \text{c.c.}\right] + [\beta\xi_L^*(t) - \text{c.c.}]\Big\}, \quad (9.118)$$

where

$$\begin{aligned}
\bar{D}_{LLLL}(t) &= \mu t,\\
\bar{D}_{LLL}(t) &= 2\mu t\xi_L^* \exp(\mathrm{i}\omega_L t),\\
B_L(t) &= 4\mu t(1 + |\,\xi_L\,|^2),\\
C_L(t) &= \mu t\xi_L^2 \exp(-\mathrm{i}2\omega_L t),\\
\xi_L(t) &= \xi_L[1 + \mu t(2 + |\,\xi_L\,|^2)]\exp(-\mathrm{i}\omega_L t)
\end{aligned} \quad (9.119)$$

and consequently

$$\langle(\Delta W)^2\rangle_{\mathcal{N}} = 2\mu t[2(1 - |\,\xi_L\,|^2)^2 + 3\,|\,\xi_L\,|^2] > 0 \quad (9.120)$$

and only photon bunching is possible. The same conclusion has been obtained by Schubert and Vogel (1981). However, one can see from (9.117) that the classical inequality $\langle\Delta W_1\Delta W_2\rangle_{\mathcal{N}}^2 \leqq \langle(\Delta W_1)^2\rangle_{\mathcal{N}}\langle(\Delta W_2)^2\rangle_{\mathcal{N}}$ is violated so that photon bunching expressed by $\langle(\Delta W)^2\rangle_{\mathcal{N}} = \langle(\Delta W_1)^2\rangle_{\mathcal{N}} + \langle(\Delta W_2)^2\rangle_{\mathcal{N}} + 2\langle\Delta W_1\Delta W_2\rangle_{\mathcal{N}}$ is supported [for antibunching $\langle(\Delta W)^2\rangle_{\mathcal{N}} < 0$, the above inequality is also violated, but $\langle\Delta W_1\Delta W_2\rangle_{\mathcal{N}} < 0$].

Nevertheless, Carusotto (1980) has shown that if the initial field is coherent and less than one-sixth of the total number of atoms are maintained in the excited state, then photon antibunching can arise. On the other hand, if more than one-sixth of the total number of atoms are maintained in the upper state, enhanced bunching occurs. Enhanced fluctuations and enhanced photon bunching (compared to chaotic radiation) can also be obtained by two-photon spontaneous emission [McNeil and Walls (1975c)]. The photon distribution of spontaneous two-photon emission has been derived by Hirota and Ikehara (1976). However, McNeil and Walls (1975a, b), Nayak and Mohanty (1979) and Gupta and Mohanty (1981) demonstrated, using the master equation approach in terms of the Fock states, that if the pumping and losses are included the two- and three-photon laser may provide sharper photon-number distributions than the one-photon laser. This is shown in Figs. 9.30a, b for single and multimode two- and three-photon lasers [the normalization is always $\sum_{n=0}^{\infty}\varrho(n, n) = 1$; however in the one-mode case only the values of $\varrho(n, n)$ for $n = 0, 2, 4, \ldots$ or $0, 3, 6, \ldots$ are non-zero in two-photon or three-photon process]. The photon-number distribution is shown by full curves for convenience. Recently Bandilla and Voigt (1982) showed that saturated k-photon emission leads to a Poisson-like photon-number distribution independently of the initial photon statistics; in the stationary regime a two-photon laser provides amplitude-stabilized light with small bunching of photons. If $\langle n\rangle \gg 1$ and $\langle(\Delta n)^2\rangle \ll \langle n\rangle^2$, then antibunching is possible in multiphoton laser if losses prevail and bunching occurs if multiphoton emission dominates; in detailed balance fluctuations are comparable with one-photon laser [Herzog (1983)].

The two-photon coherent states, discussed in Sec. 4.10, were introduced by Yuen (1976) just in connection with the process of two-photon emission [see the hamiltonian (7.34) with $n = 2$ and $m = 2$]. However, this is correct only if the atomic variables are treated as classical ones, otherwise two-photon stimulated emission does not lead to the two-photon coherent states [Golubev (1979), Schubert and Vogel (1981)]

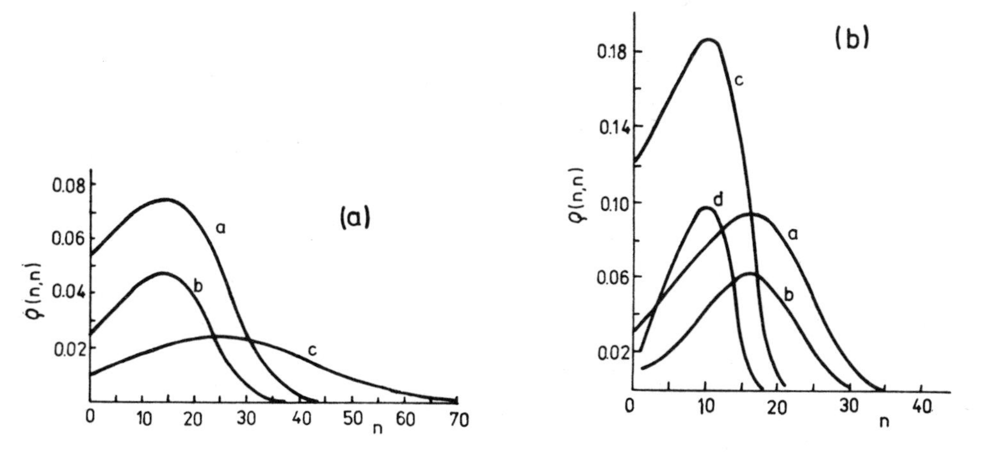

Fig. 9.30 — The photon-number distributions for (a) two-photon, (b) two- and three-photon lasers. In (a) the distributions are shown for single mode two-photon (curve *a*), two-mode two-photon (curve *b*) and one-photon (curve *c*) lasers. In (b) the distributions for single mode two-photon (curve *a*), multimode two-photon (curve *b*), single mode three-photon (curve *c*) and multimode three-photon (curve *d*) lasers are shown (after Nayak and Mohanty, 1979, Phys. Rev. **A19,** 1204 and Gupta and Mohanty, 1981, Opt. Acta **28**, 521).

(which are generated rather by the degenerate amplification process with strong pumping under the appropriate initial phase conditions). This is also indicated by the fact that the radiation in two-photon coherent states can exhibit photon antibunching, whereas, as shown above, two-photon emission provides radiation with bunching.

General solutions of the master equation for multiphoton emission have been derived by Zubairy and Yeh (1980). The effect of cooperative atomic interactions on the photon statistics in a two-photon laser has been investigated by Sharma and Brescausin (1981). Reid et al. (1981) suggested a unified approach to multiphoton lasers and multiphoton bistability. Bistability in multiphoton lasers was also discussed by Sczaniecki (1980, 1982).

9.7 Resonance fluorescence

Although resonance fluorescence is a one-photon phenomenon, it is of increasing theoretical and experimental importance, since it was used by Kimble, Dagenais and Mandel (1977) to observe photon antibunching for the first time. Later more precise experiments were performed by Kimble, Dagenais and Mandel (1978), Dagenais and Mandel (1978) and by Leuchs, Rateike and Walther (1979) [see Walls (1979)].

Carmichael and Walls (1976), Kimble and Mandel (1976) and Cohen–Tannoudji (1977) predicted that the fourth-order correlation function of light emitted by a single atom undergoing resonance fluorescence exhibits photon antibunching. The normal and time ordered intensity correlation function can be written in the factorized form

$$\langle \mathscr{T}\mathscr{N}\hat{I}(t)\,\hat{I}(t+\tau)\rangle = \langle \hat{I}\rangle \langle \hat{I}(\tau)\rangle_G, \tag{9.121}$$

where $\mathscr{T}$ and $\mathscr{N}$ are operators of time and normal ordering, $\hat{I} = \hat{A}^{(-)}\hat{A}^{(+)}$ is the intensity operator, $\langle \hat{I}\rangle$ is the steady-state mean intensity and $\langle \hat{I}(\tau)\rangle_G$ is the mean intensity of light that is radiated by an atom driven by an external field at time τ if it starts in the ground state at $t = 0$. As an atom cannot radiate in its ground state, it follows that $\langle \hat{I}(\tau)\rangle_G$ always starts from zero at $\tau = 0$ and then grows with τ and reaches its steady-state value $\langle \hat{I}\rangle$ after a time, longer than the natural life time.

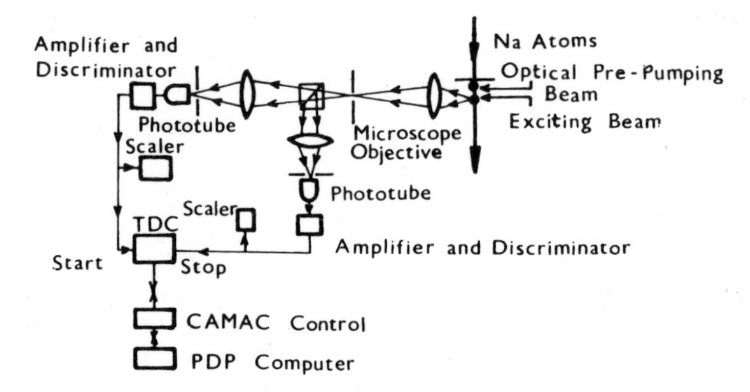

Fig. 9.31 — An outline of the device for measuring antibunching in resonance fluorescence radiation (after Dagenais and Mandel, 1978, Phys. Rev. **A18**, 2217).

Denoting the Rabi frequency as $\Omega = 2\mathscr{P}\mathscr{E}/h$ ($\mathscr{P}$ being the atomic dipole matrix element and $\mathscr{E}$ the driving field amplitude) and writing β for the half of the Einstein A-coefficient for the transition, then [Carmichael and Walls (1976), Kimble and Mandel (1976, 1977), Cohen–Tannoudji (1977)]

$$\langle \hat{I}(\tau)\rangle_G = \langle \hat{I}\rangle\,[1 + \lambda(\tau)], \tag{9.122}$$

where $\lambda(\tau)$ represents the normalized correlation function of fluctuations from both the detectors of the Hanbury Brown–Twiss correlation arrangement, and

$$\lambda(\tau) = \frac{\langle \mathscr{T}\mathscr{N}\hat{I}_1(t)\,\hat{I}_2(t+\tau)\rangle}{\langle \hat{I}_1(t)\rangle\,\langle \hat{I}_2(t+\tau)\rangle} - 1 \tag{9.123a}$$

and

$$1 + \lambda(\tau) = 1 - \exp\left(-\frac{3}{2}\beta\tau\right)[\cos(\Omega'\beta\tau) + (3/2\Omega')\sin(\Omega'\beta\tau)],$$

$$\Omega' = \left(\frac{\Omega^2}{\beta^2} - \frac{1}{4} \right)^{1/2} . \tag{9.123b}$$

A sketch of the experimental arrangement of Kimble, Dagenais and Mandel (1977) and Dagenais and Mandel (1978) is given in Fig. 9.31. They used an atomic beam of sodium atoms optically pumped in order to prepare a pure two-level system. The atomic beam was irradiated at right angles with a highly stabilized dye laser tuned on resonance with the $3^2S_{1/2}$, $F = 2$, $m_F = 2$ to $3^2P_{3/2}$, $F = 3$, $m_F = 3$ transition in sodium. The intensity of the atomic beam was reduced so that on the average no more than one atom is present in the observation region at a time. The fluorescent light from a small observation volume is obtained in a direction orthogonal to both the atomic and laser beams. Further the usual Hanbury Brown–Twiss coincidence arrangement is used, i.e. the fluorescent light is divided by a splitter and the arrival of photons in each beam is detected by two photomultipliers. The pulses from the two detectors are fed to the start and stop inputs of a time to digital converter (TDC) where the time intervals τ between the start and stop pulses are digitized in units of 0.5 ns and stored. The number of events $n(\tau)$ stored at address τ is a measure of the joint probability density $\eta^2 \gamma_{\mathcal{N}}^{(2,2)}(\tau)$ of separation of photons by τ seconds (η being the photodetector efficiency).

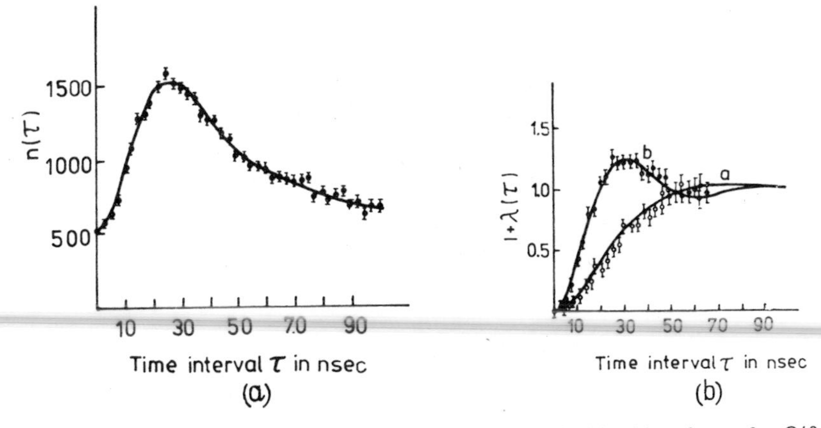

Fig. 9.32 — Comparison of (a) unnormalized measured values of $n(\tau)$ with a theory for $\Omega/\beta = 3.3$; (b) normalized correlation function $1 + \lambda(\tau)$ for $\Omega/\beta = 1.4$ (curve *a*) and 3.3 (curve *b*) (after Dagenais and Mandel, 1978, Phys. Rev. A18, 2217).

The results of the measurements of Dagenais and Mandel (1978) are shown in Figs. 9.32a, b for $\Omega/\beta = 3.3$ and 1.4. Fig. 9.32a represents the unnormalized data and we see that the increasing part of the curve violates the classical inequality $\gamma^{(2,2)}(\tau) \leqq \gamma^{(2,2)}(0)$ $(\Gamma^{(2,2)^2}(\mathbf{x}_1, \mathbf{x}_2, \mathbf{x}_2, \mathbf{x}_1, t, t+\tau, t+\tau, t) = \langle I_1(t) I_2(t+\tau) \rangle^2 \leqq$ $\leqq \langle I_1^2(t) \rangle \langle I_2^2(t) \rangle = \Gamma^{(2,2)}(\mathbf{x}_1, \mathbf{x}_1, \mathbf{x}_1, \mathbf{x}_1, t, t, t, t)\, \Gamma^{(2,2)}(\mathbf{x}_2, \mathbf{x}_2, \mathbf{x}_2, \mathbf{x}_2, t, t, t, t))$ and demonstrates the antibunching phenomenon in resonance fluorescence radiation. If the results are normalized (and also scattered light is eliminated and transit time effects and atom number fluctuations are taken into account), we obtain Fig. 9.32b giving clear evidence for photon antibunching in resonance fluorescence [the quantity (9.123a) equals -1 for $\tau = 0$, so that $1 + \lambda(0)$ is zero, whereas for classical systems it is maximal].

Similar experimental results have been obtained by Leuchs, Rateike and Walther (1979), and Cresser et al. (1982), for $\gamma_{\mathcal{N}}^{(2,2)}(\tau)$ (Fig. 9.33), demonstrating antibunching in the vicinity of $\tau = 0$, where

$$\gamma_{\mathcal{N}}^{(2,2)}(\tau) < 1 \qquad \text{and} \qquad \gamma_{\mathcal{N}}^{(2,2)}(0) = 0.$$

Thus it has been experimentally demonstrated that fluorescent photons from one atom exhibit antibunching in time, which may be regarded as a reflection of the fact that the atom makes a quantum jump to the ground state in the process of emitting a photon and is unable to radiate again immediately afterwards. No classical system can exhibit such a behaviour. Thus the photon coincidence technique makes it possible to test the behaviour of the atom and to observe the antibunching effect which is a direct manifestation of the quantum nature of light, in agreement with predictions of quantum electrodynamics.

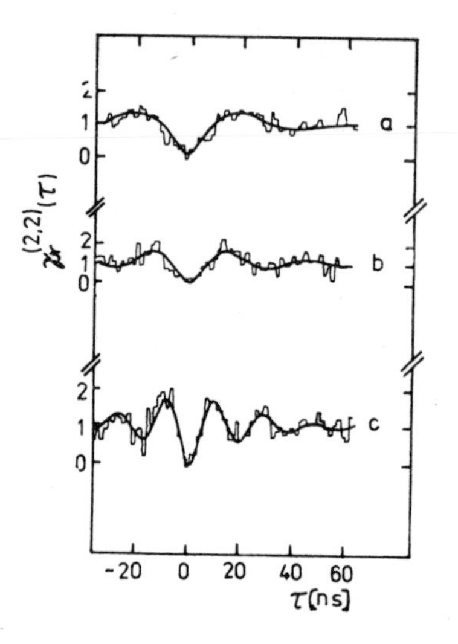

Fig. 9.33 — Photon correlation measurements of fluorescent light from sodium; the experimental points were obtained by Leuchs, Rateike and Walther (1979), the theoretical predictions are shown by full curves; $\Omega/\beta = 4.4$ (curve a), 8.6 (curve b) and 11.4 (curve c) (after Walls (1979), reprinted by permission from Nature, Vol. 280, No. 5722, pp. 451—454; Copyright © 1979 Macmillan Journals Limited).

The photon-number distribution of resonance fluorescence radiation has been discussed by Mandel (1979), Cook (1980, 1981) and Smirnov and Troshin (1981). Mandel (1979) and Smirnov and Troshin (1981) demonstrated that the photon-number distribution $p(n, t)$ for fluorescent photons from one atom is indeed narrower than the corresponding Poisson distribution, in analogy to the results above for optical parametric processes and Raman and Brillouin scattering. Mandel (1979) showed that the quantity $[\langle(\Delta n)^2\rangle - \langle n\rangle]/\langle n\rangle$ is negative with the absolute maximum value 3/4. Sub-Poissonian photon statistics in resonance fluorescence of one atom have been observed by Short and Mandel (1983).

With respect to the basic importance of these experiments an extensive discussion occurred in the literature concerning the effect of atomic fluctuations, which smooth out the antibunching phenomenon [Jakeman, Pike, Pusey and Vaugham (1977), Carmichael, Drummond, Meystre and Walls (1978), Carmichael, Drummond, Walls and Meystre (1980), Schubert, Süsse, Vogel, Welsch and Wilhelmi (1982)].

Further discussion has been devoted to photon antibunching in resonance fluorescence from impurity atoms in solids and from diatomic molecules, including the behaviour of the transient non-stationary spectra [Süsse, Vogel, Welsch and Wilhelmi (1979), Süsse, Vogel and Welsch (1980a, b, 1981)]. The effect of amplitude and phase fluctuations of the laser pump has been investigated by Schubert, Süsse, Vogel and Welsch (1981).

Several papers have been devoted to various questions concerning the statistical and correlation properties of resonance-fluorescence radiation. Two-time photon correlations in resonance fluorescence have been discussed by Mandel (1981b) and the polarization properties of resonance fluorescence by Mandel (1981c). The intensity correlations in resonance fluorescence excited by intense incoherent light have been considered by Georges (1981). General correlation properties of resonantly scattered light have been investigated by Kazantsev, Smirnov and Sokolov (1980) and Kazantsev, Smirnov, Sokolov and Tumajkin (1981). Radiation properties of systems of three and four continuously pumped atoms have been derived by Steudel (1981). Entropy considerations for resonance-fluorescence radiation have been given by Walls, Carmichael, Gragg and Schieve (1978). The use of reduced quantum fluctuations in resonance fluorescence to produce squeezed states for the detection of gravitational waves has been suggested by Walls and Zoller (1981). A review of the theory of intensity-dependent resonance light scattering and resonance fluorescence has been prepared by Mollow (1981).

A general approach to radiation statistics including the time evolution of atomic variables has been proposed by Meystre, Geneux, Quattropani and Faist (1975), Baltes, Meystre and Quattropani (1976) and Baltes, Quattropani and Meystre (1976).

9.8 Other interesting nonlinear phenomena

9.8.1 Coherent γ-emission by stimulated annihilation of electron–positron pairs

The process of coherent γ-emission by stimulated annihilation of electron–positron pairs has been discussed by Bertolotti and Sibilia (1979), Sibilia and Bertolotti (1981) and Peřinová, Peřina, Bertolotti and Sibilia (1982). As shown by Sibilia and Bertolotti (1981), this process is described in the same manner as the process of two-photon stimulated emission, discussed above, only the coupling constant μ is negative, as a consequence of the presence of the particle and the anti-particle in the interaction.

Thus in this case, as follows from the expressions (9.117) and $\mu < 0$, anticorrelation and antibunching of γ-photons always occur. The second-order corrections, which are always positive, have also been derived [Peřinová et al. (1982)], and they have been shown to be negligible in the annihilation of electron–positron pairs. Therefore the magnitudes of fluctuations and their correlations given in (9.117) are relevant for the description of the process.

9.8.2 A solvable model for light scattering

Let us consider now a light scattering model related to free electron lasers assuming that the scattering medium is treated as a two-level system (one state before the act of scattering, the other after the scattering) and light is described in the standard quantal way. Consider the scattering of two laser beams 1 and 2, described by the annihilation and creation operators $\hat{a}_1, \hat{a}_2$ and $\hat{a}_1^+, \hat{a}_2^+$ respectively, by a particle (electron) beam, which is described by annihilation and creation operators $\hat{c}_1$, $\hat{c}_2$ and $\hat{c}_1^+$, $\hat{c}_2^+$, respectively [Peřina, Peřinová, Křepelka, Lukš, Sibilia and Bertolotti (1983)]. The interaction hamiltonian of this process can be written in the form (7.34), where $\hat{O}^{(2)} = \hat{O} = \hat{O}^+ = (\hat{a}_1^+ + \hat{a}_2^+)(\hat{a}_1 + \hat{a}_2)$, that means it is uncertain in which beam the annihilation or the creation of a photon occurs (interference fringes arise, Sec. 4.9), caused by the presence of the virtual terms $\hat{a}_1^+\hat{a}_1$ and $\hat{a}_2^+\hat{a}_2$. The master equation (7.36) reads

$$\frac{\partial \hat{\varrho}}{\partial t} = K\left\{\hat{O}\hat{\varrho}\hat{O} - \frac{1}{2}\hat{O}^2\hat{\varrho} - \frac{1}{2}\hat{\varrho}\hat{O}^2\right\}, \tag{9.124}$$

where $K = 2|\mu^{(2)}|^2 (N_1 + N_2)$. Introducing the new variables

$$\hat{A}_1 = 2^{-1/2}(\hat{a}_1 + \hat{a}_2), \quad \hat{A}_2 = 2^{-1/2}(\hat{a}_1 - \hat{a}_2), \tag{9.125}$$

which also satisfy the ordinary commutation rules, $[\hat{A}_j, \hat{A}_j^+] = \hat{1}$, etc., we can obtain for the $\Phi_{\mathscr{A}}$ in the standard way

$$\frac{\partial \Phi_{\mathscr{A}}}{\partial t} = -2K\left\{A_1\frac{\partial}{\partial A_1} - |A_1|^2\frac{\partial^2}{\partial A_1\,\partial A_1^*} + A_1^2\frac{\partial^2}{\partial A_1^2} + \text{c.c.}\right\}\Phi_{\mathscr{A}}, \tag{9.126a}$$

where $A_1 = (\alpha_1 + \alpha_2)/2^{1/2}$, $A_2 = (\alpha_1 - \alpha_2)/2^{1/2}$, if α_1 and α_2 are eigenvalues of $\hat{a}_1$ and $\hat{a}_2$ in the coherent state $|\alpha_1, \alpha_2\rangle$. The equation of motion for the antinormal characteristic function is of the same form as the Fokker–Planck equation (9.126a),

$$\frac{\partial C_{\mathscr{A}}}{\partial t} = -2K\left\{\gamma_1\frac{\partial}{\partial \gamma_1} - |\gamma_1|^2\frac{\partial^2}{\partial \gamma_1\,\partial \gamma_1^*} + \gamma_1^2\frac{\partial^2}{\partial \gamma_1^2} + \text{c.c.}\right\}C_{\mathscr{A}}, \tag{9.126b}$$

where γ_j are the variables related to A_j ($\gamma_1 = (\beta_1 + \beta_2)/2^{1/2}$, $\gamma_2 = (\beta_1 - \beta_2)/2^{1/2}$, β_j being related to α_j, see (8.40)). Introducing the polar variables ϱ and φ, $\gamma_1 = \varrho \exp(\mathrm{i}\varphi)$, $\gamma_1^* = \varrho \exp(-\mathrm{i}\varphi)$ ($\varrho = (\gamma_1\gamma_1^*)^{1/2}$, $\varphi = \log(\gamma_1/\gamma_1^*)/2\mathrm{i}$), we obtain the diffusion equation from (9.126b),

$$\frac{\partial C_{\mathscr{A}}}{\partial t} = 2K \frac{\partial^2 C_{\mathscr{A}}}{\partial \varphi^2}. \tag{9.127}$$

It is easy to verify that a solution of the equation (9.126b) can be written in the form

$$C_{\mathscr{A}}(\gamma_1, \gamma_2, t) = \sum_{n=0}^{\infty} \sum_{m=0}^{\infty} c_{nm}(\gamma_2) \gamma_1^n \gamma_1^{*m} \exp[-2K(n-m)^2 t], \tag{9.128}$$

where $c_{nm}(\gamma_2)$ are arbitrary functions of γ_2 (determined by the initial conditions). A solution of the equation (9.127) can be written in the form

$$C_{\mathscr{A}}(\gamma_1, \gamma_2, t) = \int_{-\infty}^{+\infty} C_{\mathscr{A}}(\varrho, \bar{\varphi}, \gamma_2, 0)(8\pi K t)^{-1/2} \exp\left[-\frac{(\varphi - \bar{\varphi})^2}{8Kt}\right] d\bar{\varphi}. \tag{9.129}$$

Assuming the initial field to be in the coherent state $|\xi_1\rangle|\xi_2\rangle$, we have the final form of the characteristic functions

$$\begin{aligned} C_{\mathscr{N}}(\beta_1, \beta_2, t) &= C_{\mathscr{A}}(\beta_1, \beta_2, t) \exp(|\beta_1|^2 + |\beta_2|^2) = \\ &= \exp\left[\frac{1}{2}(\beta_1 - \beta_2)(\xi_1^* - \xi_2^*) - \text{c.c.}\right] \sum_{n,m=0}^{\infty} \frac{(-1)^m}{2^{n+m} n! \, m!} \times \\ &\times (\beta_1 + \beta_2)^n (\beta_1^* + \beta_2^*)^m (\xi_1^* + \xi_2^*)^n (\xi_1 + \xi_2)^m \exp[-2Kt(n-m)^2], \end{aligned} \tag{9.130a}$$

with the corresponding quasi-distribution

$$\begin{aligned} \Phi_{\mathscr{A}}(\alpha_1, \alpha_2, t) &= \pi^{-2} \exp\left\{-|\alpha_1|^2 - |\alpha_2|^2 - |\xi_1|^2 - |\xi_2|^2 + \right. \\ &+ \left.\left[\frac{1}{2}(\alpha_1 - \alpha_2)(\xi_1^* - \xi_2^*) + \text{c.c.}\right]\right\} \sum_{n,m=0}^{\infty} \frac{1}{2^{n+m} n! \, m!} \times \\ &\times (\alpha_1 + \alpha_2)^m (\alpha_1^* + \alpha_2^*)^n (\xi_1^* + \xi_2^*)^n (\xi_1 + \xi_2)^m \exp[-2Kt(n-m)^2], \end{aligned} \tag{9.130b}$$

which represents the solution of the equation (9.126a) and the inverse Fourier transform of $C_{\mathscr{A}}(\beta_1, \beta_2, t)$ given in (9.130a) as well.

In the usual way, evaluating derivatives at zero of the characteristic function (9.130a), we arrive at the mean integrated intensities and their variances in separate modes and the correlation of fluctuations between modes,

$$\langle W_1 \rangle_{\mathscr{N}} = \frac{1}{2}(|\xi_1|^2 + |\xi_2|^2) + \frac{1}{2}(|\xi_1|^2 - |\xi_2|^2) \exp(-2Kt),$$

$$\langle W_2 \rangle_{\mathscr{N}} = \frac{1}{2}(|\xi_1|^2 + |\xi_2|^2) + \frac{1}{2}(|\xi_2|^2 - |\xi_1|^2) \exp(-2Kt),$$

$$\begin{aligned} \langle (\Delta W_1)^2 \rangle_{\mathscr{N}} &= \langle (\Delta W_2)^2 \rangle_{\mathscr{N}} = -\langle \Delta W_1 \Delta W_2 \rangle_{\mathscr{N}} = \\ &= \frac{1}{8}[|\xi_1|^4 + |\xi_2|^4 - (\xi_1^{*2} \xi_2^2 + \text{c.c.})] - \end{aligned}$$

$$-\frac{1}{4}(|\xi_1|^2 - |\xi_2|^2)^2 \exp(-4Kt) +$$

$$+\frac{1}{8}[(|\xi_1|^2 - |\xi_2|^2)^2 + (\xi_1^*\xi_2 - \text{c.c.})^2]\exp(-8Kt) \tag{9.131a}$$

and consequently

$$\langle(\Delta W)^2\rangle_{\mathcal{N}} = \langle(\Delta W_1)^2\rangle_{\mathcal{N}} + \langle(\Delta W_2)^2\rangle_{\mathcal{N}} + 2\langle\Delta W_1 \Delta W_2\rangle_{\mathcal{N}} = 0, \tag{9.131b}$$

that is the whole field is coherent for all times. The increase of noise in separate modes is compensated by the anticorrelation between modes. In (9.131a) the fourth-order moments are phase dependent, depending on terms $2|\xi_1|^2|\xi_2|^2 \cos(2\varphi_1 - 2\varphi_2)$, φ_j being the phases of ξ_j.

Applying (8.58) we obtain for the normal generating function

$$\langle\exp(-\lambda W)\rangle_{\mathcal{N}} = \exp[-\lambda(|\xi_1|^2 + |\xi_2|^2)], \tag{9.132}$$

that is, in agreement with (9.131b), the photon-number distribution is Poissonian with $\langle n\rangle = \langle W\rangle_{\mathcal{N}} = \langle W_1\rangle_{\mathcal{N}} + \langle W_2\rangle_{\mathcal{N}} = |\xi_1|^2 + |\xi_2|^2$ independently of t. Hence, the whole field is coherent for all times, as may be expected from the form of the hamiltonian for this process. In separate modes the generating functions read

$$\langle\exp(-\lambda W_{1,2})\rangle_{\mathcal{N}} = \exp\left[-\frac{\lambda}{2}(|\xi_1|^2 + |\xi_2|^2)\right]\times$$

$$\times\sum_{l,j=0}^{\infty}\frac{1}{l!j!}\left[\mp\frac{\lambda}{4}(\xi_1^* + \xi_2^*)(\xi_1 - \xi_2)\right]^l\times$$

$$\times\left[\mp\frac{\lambda}{4}(\xi_1 + \xi_2)(\xi_1^* - \xi_2^*)\right]^j \exp[-2Kt(l-j)^2] \tag{9.133}$$

and the photon-number distributions and its factorial moments $p_{1,2}(n, t)$ and $\langle W_{1,2}^k\rangle_{\mathcal{N}}$ respectively can be derived in the standard ways by means of derivatives [equations (3.102) and (3.103) with $\eta = 1$ and $\lambda = -is$].

It is easy to see that if the mean integrated intensity of one of the modes increases, the other decreases and vice versa. For example, if $|\xi_1| > |\xi_2|$, then $\langle W_1\rangle_{\mathcal{N}}$ decreases from $|\xi_1|^2$ to $(|\xi_1|^2 + |\xi_2|^2)/2$ and $\langle W_2\rangle_{\mathcal{N}}$ increases from $|\xi_2|^2$ to the same limit for $t \to \infty$. If $|\xi_1| = |\xi_2| = |\xi|$, then $\langle W_1\rangle_{\mathcal{N}} = \langle W_2\rangle_{\mathcal{N}} = |\xi|^2$ and $\langle(\Delta W_1)^2\rangle_{\mathcal{N}} = \langle(\Delta W_2)^2\rangle_{\mathcal{N}} = \langle\Delta W_1 \Delta W_2\rangle_{\mathcal{N}} = 0$ independently of t, i.e. both the modes are Poissonian and uncorrelated. If $\xi_2 = 0$, we have

$$\langle W_1\rangle_{\mathcal{N}} = \frac{1}{2}|\xi_1|^2[1 + \exp(-2Kt)],$$

$$\langle W_2\rangle_{\mathcal{N}} = \frac{1}{2}|\xi_1|^2[1 - \exp(-2Kt)]. \tag{9.134a}$$

Further, one can prove that in general $\langle(\Delta W_1)^2\rangle_{\mathcal{N}} = \langle(\Delta W_2)^2\rangle_{\mathcal{N}} \geqq 0$ and $\langle\Delta W_1 \Delta W_2\rangle_{\mathcal{N}} \leqq 0$. The quantities $\langle(\Delta W_{1,2})^2\rangle_{\mathcal{N}}$ and $-\langle\Delta W_1 \Delta W_2\rangle_{\mathcal{N}}$ are maximal if $\varphi_1 - \varphi_2 = \pi/2$, minimal if $\varphi_1 - \varphi_2 = 0$ (or $n\pi$), and intermediate if $\varphi_1 - \varphi_2 = \pi/4$. If $\xi_2 = 0$

$$\langle(\Delta W_1)^2\rangle_{\mathcal{N}} = \langle(\Delta W_2)^2\rangle_{\mathcal{N}} = -\langle\Delta W_1 \Delta W_2\rangle_{\mathcal{N}} = \frac{1}{8}|\xi_1|^4 (1 - \exp(-4Kt))^2. \tag{9.134b}$$

The saturation values are

$$\langle W_1\rangle_{\mathcal{N}} = \langle W_2\rangle_{\mathcal{N}} = \frac{1}{2}(|\xi_1|^2 + |\xi_2|^2), \tag{9.135a}$$

$$\langle(\Delta W_1)^2\rangle_{\mathcal{N}} = \langle(\Delta W_2)^2\rangle_{\mathcal{N}} = -\langle\Delta W_1 \Delta W_2\rangle_{\mathcal{N}} = \frac{1}{8}[|\xi_1|^4 + |\xi_2|^4 - 2|\xi_1|^2|\xi_2|^2 \cos(2\varphi_1 - 2\varphi_2)]. \tag{9.135b}$$

Hence, we see that the properties of modes 1 and 2 vary from the Poisson statistics with $\langle(\Delta W_{1,2})^2\rangle_{\mathcal{N}} = 0$ to the asymptotic value (9.135b), which is maximal with the magnitude $(|\xi_1|^2 + |\xi_2|^2)^2/8$ for $\varphi_1 - \varphi_2 = \pi/2$ and minimal with the magnitude $(|\xi_1|^2 - |\xi_2|^2)^2/8$ if $\varphi_1 = \varphi_2$ and thus it is always less than the value $\langle(\Delta W_1)^2\rangle_{\mathcal{N}} = \langle(\Delta W_2)^2\rangle_{\mathcal{N}} = \langle W_1\rangle^2_{\mathcal{N}} = \langle W_2\rangle^2_{\mathcal{N}} = (|\xi_1|^2 + |\xi_2|^2)^2/4$ appropriate for Gaussian radiation (the maximal value of the saturated variance is just one half of its Gaussian value).

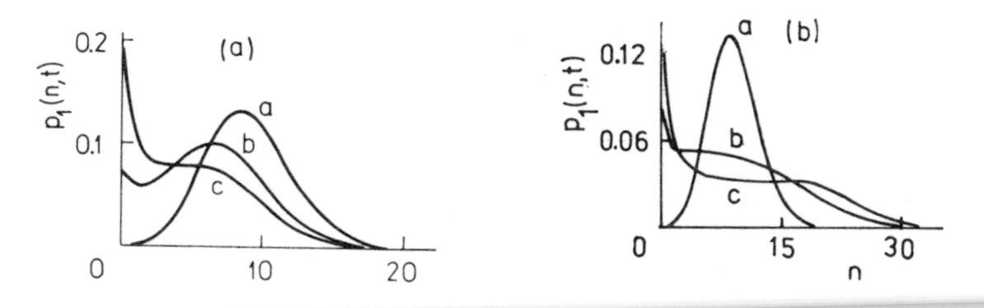

Fig. 9.34 — The photon-number distributions in mode 1 for interfering double-beam radiation scattered by an electron beam for (a) $\xi_1 = 3$, $\xi_2 = 0$, $\varphi_1 = \varphi_2$ and $2Kt = 0$ (Poisson distribution, curve *a*), $2Kt = 1$ (curve *b*) and $2Kt = 5$ (curve *c*); (b) $\xi_1 = 3i$, $\xi_2 = (13.5)^{1/2}$, $\varphi_1 - \varphi_2 = \pi/2$ and $2Kt = 0$ (Poisson distribution, *a*), $2Kt = 0.25$ (curve *b*) and $2Kt = 2$ (curve *c*) (after Peřina et al. 1983, Opt. Acta **30**, **959**).

In Fig. 9.34a we show the photon-number distribution $p_1(n, t)$ for $\zeta_1 = 3$, $\xi_2 = 0$ and $\varphi_1 = \varphi_2$ for (a) $2Kt = 0$ (the Poissonian distribution), (b) $2Kt = 1$ and (c) $2Kt = 5$, and in Fig. 9.34b for $\xi_1 = 3i$, $\zeta_2 = (13.5)^{1/2}$, $\varphi_1 - \varphi_2 = \pi/2$ and (a) $2Kt = 0$, (b) $2Kt = 0.25$ and (c) $2Kt = 2$. In the former case $|\xi_1| > |\xi_2|$ and the distribution is shifted to lower n, whereas in the latter case $|\xi_1| < |\xi_2|$ and a large plateau arises. The two-peak behaviour of the distribution gives evidence for the existence of bistable states.

9.9 Phase-transition analogies

In connection with the statistical properties of nonlinear optical processes we have mentioned a number of cases where bistable behaviour and phase transitions are possible. Now we may finally follow some analogies in the behaviour of radiation in nonlinear optical processes and the laser and in the behaviour of superconductors, superfluids and ferromagnets. In all these cases a cooperative behaviour can occur as a consequence of the broken symmetry induced by an external field (for example, random trains of optical radiation from a traditional source are ordered to a periodical behaviour of the electric field vector appropriate for a coherent state, or random spins are ordered in one direction in a ferromagnet). Such cooperative phenomena may be treated as second-order phase transitions in the spirit of the Landau–Ginzburg theory [Sargent and Scully (1972), Sargent, Scully and Lamb (1974), Haken (1978)].

When describing superconductivity, superfluidity or ferromagnetism, we may start with the distribution function

$$f = \exp\left(-\frac{F}{KT}\right), \tag{9.136a}$$

where the free energy is chosen as

$$F = F_0(T) + \frac{\alpha(T)}{2}\psi^2 + \frac{\beta(T)}{4}\psi^4, \tag{9.137a}$$

with some coefficients $\alpha(T)$ and $\beta(T) > 0$. The corresponding equation of motion for the wave function ψ of an electron pair $(\mathbf{k}, -\mathbf{k})$ (for ferromagnets $\psi \to$ magnetization $\boldsymbol{M}$), which plays the role of an ordering parameter, is

$$\dot{\psi} = -\frac{\partial F}{\partial \psi} + L = -\alpha\psi - \beta\psi^3 + L; \tag{9.138a}$$

here the Langevin force L is determined by thermal fluctuations and the coefficient $\alpha(T)$ can be written in the form

$$\alpha(T) = a(T - T_c), \tag{9.139a}$$

where a is another constant and T_c is the critical temperature. In a disordered state, the entropy $S = -\partial F/\partial T = S_0$ and only the minimum value at $\psi = 0$ of F given by (9.137a) is admissible; in this case $T > T_c$ and $\alpha > 0$. On the other hand, an ordered state occurs if $T < T_c$ and $\alpha < 0$ and F has two non-zero minima at $\psi = \pm(-\alpha/\beta)^{1/2}$ and

$$S = S_0 + \frac{a^2}{2\beta}(T - T_c); \tag{9.140a}$$

the entropy S is continuous at $T = T_c$. The specific heat $c = T\,\partial S/\partial T$ at the disordered state is

$$c = T\frac{\partial S_0}{\partial T}, \tag{9.141a}$$

whereas in the ordered state

$$c = T\frac{\partial S_0}{\partial T} + \frac{a^2}{2\beta}T \tag{9.142a}$$

and so it is discontinuous at $T = T_c$.

The same scheme may be developed, with small deviations, for the laser and for those nonlinear optical processes with a threshold. All the following equations are just analogous to those given above. For the distribution function we write

$$f = \exp\left(-\frac{B}{Q}\right), \tag{9.136b}$$

where Q is the quantum gain, proportional to KT in the case of thermal equilibrium, and we have for the potential

$$B = B_0(d) + \frac{\bar{\alpha}(d)}{2}E^2 + \frac{\bar{\beta}(d)}{4}E^4; \tag{9.137b}$$

d is the population inversion. The equation of motion for the wave field E, playing the role of the ordering parameter, reads [cf. (5.74)]

$$\dot{E} = -\bar{\alpha}E - \bar{\beta}E^3 + L. \tag{9.138b}$$

The Langevin force L describes the spontaneous emission. The parameter $\bar{\alpha}$ is expressed as

$$\bar{\alpha}(d) = \bar{a}(d_c - d), \tag{9.139b}$$

where d_c is the threshold value of the population inversion and the correspondence $-d_c \leftrightarrow T_c$ is appropriate. Thus, for $\bar{S} = \partial B/\partial d$ we have $\bar{S} = \bar{S}_0$ below threshold, where $d < d_c$ and $\bar{\alpha} > 0$ and the potential B has its minimum value at $E = 0$. Above threshold $d > d_c$ and $\bar{\alpha} < 0$ and

$$\bar{S} = \bar{S}_0 + \frac{\bar{a}^2}{2\bar{\beta}}(d_c - d), \tag{9.140b}$$

with the non-zero minima of B at $E = \pm(-\bar{\alpha}/\bar{\beta})^{1/2}$; S is continuous at $d = d_c$. However, $\bar{c} = d\,\partial\bar{S}/\partial d$ is again discontinuous at threshold $d = d_c$, being

$$\bar{c} = d\frac{\partial \bar{S}_0}{\partial d} \tag{9.141b}$$

in disordered state below threshold, and

$$\bar{c} = d\frac{\partial \bar{S}_0}{\partial d} - \frac{\bar{a}^2}{2\bar{\beta}}d \tag{9.142b}$$

in the ordered state above threshold; thus $\bar{c}$ has discontinuity at $d = d_c$.

Consequently an ordered cooperative state, described by a coherent state, arises in these phenomena from a chaotic state far from thermal equilibrium if the ordering parameter exceeds some critical threshold value (for parameters $T < T_c$ or $d > d_c$). This is demonstrated in Fig. 5.1 or 5.5, showing the change of the photon statistics of the laser from the region below threshold, where Bose–Einstein statistics are appropriate, to the region well above threshold, where Poissonian statistics are correct. This conclusion is true for other nonlinear optical phenomena as well [Graham (1973, 1974)]. However, there are other phenomena in chemistry, biology and other sciences, as treated in synergetics [Haken (1978)], which may be described in analogous ways. A single biological cell represents a cooperative system operating in several modes in a coherent way.

The spectral linewidth of laser noise and of fluctuations near the second-order phase transition has been discussed by Jakeman and Pike (1971). The coherent-state technique has been applied to solid states, superfluidity, ferromagnetism, etc. by Carruthers and Dy (1966), López (1967), Langer (1968, 1969), Rezende and Zagury (1969) [see Man'ko (1972)] and Malkin and Man'ko (1979), among others.

CHAPTER 10

CONCLUSIONS

In this monograph we have investigated in a systematic way the quantum statistical properties of optical fields, with particular attention to nonlinear optical phenomena, using the coherent-state technique. Chapters 2–4 are based on the material of the previous monograph on Coherence of Light by this author [Van Nostrand Reinhold Company, London (1972)] in order to have this monograph self-contained. These chapters provide thorough information on the Fock and the coherent-state descriptions of radiation and on quantum correlation theory, which are necessary for any quantum statistical treatment. Applications of the general coherent-state methods to particular kinds of fields important in practice, such as chaotic fields, laser fields and their superposition, appropriate to optical communication, are made in Chapter 5. The other chapters of the book deal mostly with nonlinear phenomena, reviewed in a traditional way in Chapter 6. The use of the coherent-state technique for the description of optical fields interacting with matter is outlined, in the framework of the Heisenberg and Schrödinger pictures, in Chapter 7. Chapter 8 is devoted to the development of the quantum dynamical theory of the statistical properties of radiation in random media, including self-radiation. Using the generalized Fokker–Planck equation and the Heisenberg–Langevin equations together with the quantum characteristic functions, the most important nonlinear optical phenomena, such as optical parametric processes, scattering, multiphoton absorption and emission, etc., are investigated in Chapter 9. Particular attention is devoted to anticorrelation and antibunching effects and to sub-Poissonian behaviour in optical fields having no classical analogues.

Although the subject of this monograph may be considered to be to some extent closed, from a theoretical point of view, there are still various problems to be solved, particularly in the theory of the quantum statistical properties of nonlinear optical processes. For instance,

- stronger mathematical methods are needed for solving the operator Heisenberg–Langevin equations for nonlinear optical processes and for solving the generalized Fokker–Planck equations for these processes,
- still closer analogies may be followed among various physical as well as non-physical phenomena, on the basis of synergetics, with a deeper understanding of the transition from chaos to an ordered cooperative behaviour,
- a more thorough theory of chaos should be developed, with exact stochastic descriptions in terms of the correlation functions and quasi-distributions,
- further research on optical bistability may have interesting applications,
- soliton propagation problems should be investigated in greater detail.

However, from the experimental point of view, in the field of nonlinear optical phenomena a period of interesting experiments begins, following the pioneering experiments of Mandel and his collaborators and Walther and his collaborators, on the measurement of antibunching and sub-Poissonian statistics and the production of fields having no classical analogues including squeezing states. Such optical fields might be used in optical communication [Jodoin and Mandel (1971), Mandel (1976b), Yuen and Shapiro (1978a), Shapiro et al. (1979), Helstrom (1979), Shapiro (1980)] and in the detection of gravitational waves using optical interferometers [Caves (1981), Walls and Zoller (1981), Loudon (1981), Walls and Milburn (1981)].

Although we intended to include possibly all the references related to the subject of this book, this was found to be impossible so we included only those references which were most important for the text of the monograph.

REFERENCES

Abate, J. A., H. J. Kimble and L. Mandel, 1976, *Phys. Rev.* **A14,** 788.

Ablekov, V. K., V. S. Avdyrevskii, Yu. N. Babaev, S. A. Koljadin, A. V. Frolov and V. A. Fulov, 1980, *Proc. Acad. Sci. USSR* **251,** 1098.

Ablekov, V. K., Yu. N. Babaev, S. A. Koljadin and A. V. Frolov, 1980, *Proc. Acad. Sci. USSR* **250,** 90.

Abraham, N. B., 1980, *Phys. Rev.* **A21,** 1595.

Abraham, N. B., J. C. Huang, D. A. Kranz and E. B. Rockower, 1981, *Phys. Rev.* **A24,** 2556.

Abraham, N. B. and S. R. Smith, 1977, *Phys. Rev.* **A15,** 421.

Abraham, N. B. and S. R. Smith, 1981, *Opt. Comm.* **38,** 372.

Adaptive Optics, 1977, *J. Opt. Soc. Am.* **67,** No 3.

Agarwal, G. S., 1969, *Phys. Rev.* **178,** 2025.

Agarwal, G. S., 1970, *Phys. Rev.* **A1,** 1445.

Agarwal, G. S., 1973, Progress in Optics, Vol. 11, ed. E. Wolf, North-Holland, Amsterdam, p. 1.

Agarwal, G. S., 1975, *Z. Physik* **B22,** 207.

Agarwal, G. S. and E. Wolf, 1968a, *Phys. Lett.* **26A,** 485.

Agarwal, G. S. and E. Wolf, 1968b, *Phys. Rev. Lett.* **21,** 180.

Agarwal, G. S. and E. Wolf, 1968c, *Phys. Rev. Lett.* **21,** 656(E).

Agarwal, G. S. and E. Wolf, 1970a, *Phys. Rev.* **D2,** 2161.

Agarwal, G. S. and E. Wolf, 1970b, *Phys. Rev.* **D2,** 2187.

Agarwal, G. S. and E. Wolf, 1970c, *Phys. Rev.* **D2,** 2206.

Agrawal, G. P., 1978, *J. Opt. Soc. Am.* (*Lett.*) **68,** 1135.

Agrawal, G. P. and H. J. Carmichael, 1980, *Opt. Acta* **27,** 651.

Agrawal, G. P. and C. L. Mehta, 1974, *J. Phys.* **A7,** 607.

Akhiezer, N. I., 1970, Classical Moment Problem and some Related Questions, Oliver and Boyd, Edinburgh, Chap. I.

Akhiezer, A. I. and V. B. Berestetsky, 1965, Quantum Electrodynamics, Interscience, New York.

Akhmanov, S. A., Yu. E. Dyakov and A. S. Tchirkin, 1981, Introduction to Statistical Radiophysics and Optics, Nauka, Moscow (in Russian).

Akhmanov, S. A., R. V. Khokhlov and A. P. Sukhorukov, 1972, Laser Handbook, Vol. 2, eds. F. T. Arecchi and E. O. Schulz-Dubois, North-Holland, Amsterdam, p. 1151.

Akhmanov, S. A. and A. S. Tchirkin, 1971, Statistical Phenomena in Nonlinear Optics, Lomonosov Univ., Moscow (in Russian).

Akhmanov, S. A., A. S. Tchirkin and V. G. Tunkin, 1970, *Opto-Electronics* **2,** No 2.

Aldridge, M. D., 1969, *J. Appl. Phys.* **40,** 1720.

Allen, L. and J. H. Eberly, 1975, Optical Resonance and Two-Level Atoms, J. Wiley, New York.

Anisimov, V. Ya. and B. A. Sotskii, 1977a, *Opt. Spectr.* (USSR) **42,** 563.

Anisimov, V. Ya. and B. A. Sotskii, 1977b, *Opt. Spectr.* (USSR) **43,** 125.

Aoki, T., 1977, *Phys. Rev.* **A16,** 2432.

Aoki, T., Y. Endo, H. Takayanagi and K. Sakurai, 1976, *Phys. Rev.* **A13,** 853.

Aoki, T., Y. Okabe and K. Sakurai, 1974, *Phys. Rev.* **A10,** 259.

Aoki, T. and K. Sakurai, 1979, *Phys. Rev.* **A20,** 1593.

Aoki, T. and K. Sakurai, 1970, *Phys. Rev.* **A22,** 684.

Arecchi, F. T., 1965, *Phys. Rev. Lett.* **15,** 912.
Arecchi, F. T., 1969, Quantum Optics, ed. R. J. Glauber, Acad. Press, New York.
Arecchi, F. T., M. Asdente and A. M. Ricca, 1976, *Phys. Rev.* **14,** 383.
Arecchi, F. T., A. Berné and P. Burlamacchi, 1966, *Phys. Rev. Lett.* **16,** 32.
Arecchi, F. T., A. Berné and A. Sona, 1966, *Phys. Rev. Lett.* **17,** 260.
Arecchi, F. T., A. Berné, A. Sona and P. Burlamacchi, 1966, *IEEE J. Quant. Electr.* **QE-2,** 341.
Arecchi, F. T., E. Courtens, R. Gilmore and H. Thomas, 1972, *Phys. Rev.* **A6,** 2211.
Arecchi, F. T., E. Courtens, R. Gilmore and H. Thomas, 1973, Coherence and Quantum Optics, eds. L. Mandel and E. Wolf, Plenum Press, New York, p. 191.
Arecchi, F. T. and V. Degiorgio, 1971, *Phys. Rev.* **A3,** 1108.
Arecchi, F. T. and V. Degiorgio, 1972, Laser Handbook, Vol. 1, eds. F. T. Arecchi and E. O. Schulz-Dubois, North-Holland, Amsterdam, p. 191.
Arecchi, F. T., V. Degiorgio and B. Querzola, 1967, *Phys. Rev. Lett.* **19,** 1168.
Arecchi, F. T., E. Gatti and A. Sona, 1966, *Phys. Lett.* **20,** 27.
Arecchi, F. T., M. Giglio and A. Sona, 1967, *Phys. Lett.* **25A,** 341.
Arecchi, F. T., M. Giglio and U. Tartari, 1967, *Phys. Rev.* **163,** 186.
Arecchi, F. T., G. L. Masserini and P. Schwendimann, 1969, *Rivista Nuovo Cim.* **1,** 181.
Arecchi, F. T. and A. Politi, 1980, *Phys. Rev. Lett.* **45,** 1219.
Arecchi, F. T., A. Politi and L. Ulivi, 1982, *Phys. Lett.* **87A,** 333.
Arecchi, F. T. and A. M. Ricca, 1977, *Phys. Rev.* **A15,** 308.
Arecchi, F. T., G. S. Rodari and A. Sona, 1967, *Phys. Lett.* **25A,** 59.
Armstrong, J. A. and A. W. Smith, 1967, Progress in Optics, Vol. 6, ed. E. Wolf, North-Holland, Amsterdam, p. 211.
Astafunov, V. G. and G. H. Glazov, 1980, *Opt. Spectr.* (USSR) **48,** 568.
Baldwin, G. C., 1969, An Introduction to Nonlinear Optics, Plenum Press, New York.
Baltes, H. P., 1976, Progress in Optics, Vol. 13, ed. E. Wolf, North-Holland, Amsterdam, p. 1.
Baltes, H. P., 1977, *Appl. Phys.* **12,** 221.
Baltes, H. P., ed., 1978, Inverse Source Problems, Springer, Berlin.
Baltes, H. P., H. A. Ferwerda, A. S. Glass and B. Steinle, 1981, *Opt. Acta* **28,** 11.
Baltes, H. P., J. Geist and A. Walther, 1978, Inverse Source Problems, ed. H. P. Baltes, Springer, Berlin, p. 119.
Baltes, H. P. and E. R. Hilf, 1976, Spectra of Finite Systems, BI-Wissenschaftsverlag, Bibl. Inst. Mannheim.
Baltes, H. P., P. Meystre and A. Quattropani, 1976, *Nuovo Cim.* **32B,** 303.
Baltes, H. P., A. Quattropani and P. Meystre, 1976, *Infrared Phys.* **16,** 9.
Baltes, H. P., A. Quattropani and P. Schwendimann, 1978, *Helv. Phys. Acta* **51,** 534.
Baltes, H. P., B. Steinle and G. Antes, 1976, *Opt. Comm.* **18,** 242.
Baltes, H. P., B. Steinle and M. Pabst, 1976, *Phys. Rev.* **A13,** 1866.
Bandilla, A., 1977, *Opt. Comm.* **23,** 299.
Bandilla, A., 1978, Quantum Optics, eds. J. Heldt and J. Czub, Univ. Toruń and Gdańsk, p. 20.
Bandilla, A. and H. H. Ritze, 1976a, *Ann. Physik* **33,** 207.
Bandilla, A. and H. H. Ritze, 1976b, *Opt. Comm.* **19,** 169.
Bandilla, A. and H. H. Ritze, 1979, *Opt. Comm.* **28,** 126.
Bandilla, A. and H. H. Ritze, 1980a, *Opt. Comm.* **34,** 190.
Bandilla, A. and H. H. Ritze, 1980b, *Opt. Comm.* **32,** 195.
Bandilla, A. and H. H. Ritze, 1981, *Ann. Physik* **38,** 123.
Bandilla, A. and H. Voigt, 1982, Quantum statistics of light after saturated two-photon emission process and the photon statistics of a two-photon laser, preprint, ZOS-Berlin.
Barakat, R., 1976, *J. Opt. Soc. Am.* **66,** 211.
Barakat, R., 1977, *Opt. Comm.* **23,** 147.
Barakat, R., 1980, *J. Opt. Soc. Am.* **70,** 688.

Barakat, R., 1981, *J. Opt. Soc. Am.* **71,** 86.
Barakat, R. and J. Blake, 1976, *Phys. Rev.* **A13,** 1122.
Barakat, R. and J. Blake, 1980, *Phys. Rep.* **60,** 225.
Barashev, P. P., 1970a, *J. Exp. Theor. Phys.* (USSR) **59,** 1318.
Barashev, P. P., 1970b, *Phys. Lett.* **32A,** 291.
Barashev, P. P., 1971, *Phys. Lett.* **36A,** 205.
Barashev, P. P., 1976, *Opt. Spectr.* (USSR) **40,** 349.
Bark, A. and S. R. Smith, 1977, *Phys. Rev.* **A15,** 269.
Bartolino, R., M. Bertolotti, F. Scudieri and D. Sette, 1973, *Appl. Opt.* **12,** 2917.
Bastiaans, M. J., 1977, *Opt. Acta* **24,** 261.
Bastiaans, M. J., 1981, *Opt. Acta* **28,** 1215.
Becker, W., M. O. Scully and M. S. Zubairy, 1982, *Phys. Rev. Lett.* **48,** 475.
Bédard, G., 1966a, *Phys. Rev.* **151,** 1038.
Bédard, G., 1966b, *Phys. Lett.* **21,** 32.
Bédard, G., 1967a, *J. Opt. Soc. Am.* **57,** 1201.
Bédard, G., 1967b, *Phys. Lett.* **24A,** 613.
Bédard, G., 1967c, *Proc. Phys. Soc.* **90,** 131.
Bédard, G., 1967d, *Phys. Rev.* **161,** 1304.
Bédard, G., J. C. Chang and L. Mandel, 1967, *Phys. Rev.* **160,** 1496.
Bénard, C. M., 1969, *C. R. Acad. Sci. Paris* **268,** 1504.
Bénard, C. M., 1970a, Quantum Optics, eds. S. M. Kay and A. Maitland, Academic Press, London, p. 535.
Bénard, C., 1970b, *Phys. Rev.* **A2,** 2140.
Bénard, C., 1975, *J. Math. Phys.* **16,** 710.
Bendjaballah, C., 1969, *C. R. Acad. Sci. Paris* **268,** 1719.
Bendjaballah, C., 1971, *C. R. Acad. Sci. Paris* **272,** 1244.
Bendjaballah, C., 1973, *J. Phys.* **A6,** 837.
Bendjaballah, C., 1975, *Opt. Comm.* **14,** 153.
Bendjaballah, C., 1979, *J. Appl. Phys.* **50,** 62.
Bendjaballah, C., 1980, *Opt. Comm.* **34,** 164.
Bendjaballah, C. and F. Perrot, 1971, *Opt. Comm.* **3,** 21.
Bendjaballah, C. and F. Perrot, 1973, *J. Appl. Phys.* **44,** 5130.
Beran, M., J. De Velis and G. Parrent, 1967, *Phys. Rev.* **154,** 1224.
Beran, M. J. and T. L. Ho, 1969, *J. Opt. Soc. Am.* **59,** 1134.
Beran, M. and G. B. Parrent, 1964, Theory of Partial Coherence, Prentice-Hall, Englewood Cliffs, New Jersey.
Berezansky, J. M., 1968, Expansions in Eigenfunctions of Selfconjugate Operators, American Math. Soc., New York.
Bertolotti, M., 1974, Photon Correlation and Light Beating Spectroscopy, eds. H. Z. Cummins and E. R. Pike, Plenum Press, New York.
Bertolotti, M., B. Crosignani and P. Di Porto, 1970, *J. Phys.* **A3,** L 37.
Bertolotti, M., B. Crosignani, P. Di Porto and D. Sette, 1966, *Phys. Rev.* **150,** 1054.
Bertolotti, M., B. Crosignani, P. Di Porto and D. Sette, 1967, *Z. Phys.* **205,** 129.
Bertolotti, M., S. Martellucci, F. Scudieri and R. Bartolino, 1973, Coherence and Quantum Optics, eds. L. Mandel and E. Wolf, Plenum Press, New York, p. 449.
Bertolotti, M. and C. Sibilia, 1979, *Appl. Phys.* **19,** 127.
Bertolotti, M. and C. Sibilia, 1980, *Phys. Rev.* **A21,** 234.
Bertrand, P. P. and E. A. Mishkin, 1967, *Phys. Lett.* **25A,** 204.
Bespalov, V. I., 1979, Inversion of Wave Front of Optical Radiation in Nonlinear Media, Acad. Sci. USSR, Gorkii (in Russian).
Bialynicka-Birula, Z., 1968, *Phys. Rev.* **173,** 1207.

Blake, J. and R. Barakat, 1972, *Opt. Comm.* **6,** 278.
Blake, J. and R. Barakat, 1973, *J. Phys.* **A6,** 1196.
Blake, J. and R. Barakat, 1976, *Opt. Comm.* **16,** 303.
Blake, J. and R. Barakat, 1977, *Opt. Comm.* **20,** 10.
Blažek, M., 1979, *Acta Phys. Slov.* **29,** 3.
Bloembergen, N., 1965, Nonlinear Optics, W. A. Benjamin, New York.
Bluemel, V., L. M. Narducci and R. A. Tuft, 1972, *J. Opt. Soc. Am.* **62,** 1309.
Bogolyubov, N. N. and D. V. Shirkov, 1959, Introduction to the Theory of Quantized Fields, Interscience, New York.
Bojcov, V. F., Yu. E. Murachver and S. G. Sljusarev, 1973, *Opt. Spectr.* (USSR) **35,** 708.
Bonifacio, R., 1980, *Opt. Comm.* **32,** 440.
Bonifacio, R., L. M. Narducci and E. Montaldi, 1966, *Phys. Rev. Lett.* **16,** 1125.
Born, M. and E. Wolf, 1965, Principles of Optics, 3rd ed., Pergamon, Oxford.
Bothe, W., 1927, *Z. Phys.* **41,** 345.
Bourret, R. C., 1960, *Nuovo Cim.* **18,** 347.
Brand, H., R. Graham and A. Schenzle, 1980, *Opt. Comm.* **32,** 359.
Brannen, E., H. I. S. Ferguson and W. Wehlau, 1958, *Can. J. Phys.* **36,** 871.
Brevik, I. and E. Suhonen, 1968, *Phys. Norveg.* **3,** 135.
Brevik, I. and E. Suhonen, 1970, *Nuovo Cim.* **65B,** 187.
Brown, R. Hanbury, 1964, *Sky and Telesc.* **28,** 64.
Brown, R. Hanbury and R. Q. Twiss, 1956a, *Nature* (London) **177,** 27.
Brown, R. Hanbury and R. Q. Twiss, 1956b, *Nature* (London) **178,** 1046.
Brown, R. Hanbury and R. Q. Twiss, 1956c, *Nature* (London) **178,** 1447.
Brown, R. Hanbury and R. Q. Twiss, 1957a, *Proc. Roy. Soc.* (*A*) **242,** 300.
Brown, R. Hanbury and R. Q. Twiss, 1957b, *Proc. Roy. Soc.* (*A*) **243,** 291.
Brown, R. Hanbury and R. Q. Twiss, 1958, *Proc. Roy. Soc.* (*A*) **248,** 199 and 222.
Bufton, J. L., R. S. Iyer and L. S. Taylor, 1977, *Appl. Opt.* **16,** 2408.
Bureš, J., C. Delisle and A. Zardecki, 1971, *Can. J. Phys.* **49,** 3064.
Bureš, J., C. Delisle and A. Zardecki, 1972a, *Can. J. Phys.* **50,** 760.
Bureš, J., C. Delisle and A. Zardecki, 1972b, *Can. J. Phys.* **50,** 1307.
Bureš, J., C. Delisle and A. Zardecki, 1972c, *Phys. Rev.* **A6,** 2237.
Burge, R. E., M. A. Fiddy, A. H. Greenaway and G. Ross, 1974, *J. Phys.* **D7,** L 65.
Burge, R. E., M. A. Fiddy, A. H. Greenaway and G. Ross, 1976, *Proc. Roy. Soc. Lond.* **A350,** 191.
Butylkin, V. S., A. E. Kaplan, Yu. G. Khronopulo and E. I. Yakubovitch, 1977, Resonant Interactions of Light with Matter, Nauka, Moscow (in Russian).
Cahill, K. E., 1965, *Phys. Rev.* **138,** B 1566.
Cahill, K. E., 1969, *Phys. Rev.* **180,** 1244.
Cahill, K. E. and R. J. Glauber, 1969a, *Phys. Rev.* **177,** 1857.
Cahill, K. E. and R. J. Glauber, 1969b, *Phys. Rev.* **177,** 1882.
Campagnoli, G. and G. Zambotti, 1968, *Nuovo Cim.* **57A,** 468.
Cantor, B. I. and M. C. Teich, 1975, *J. Opt. Soc. Am.* **65,** 786.
Cantrell, C. D., 1969, *Phys. Lett.* **29A,** 469.
Cantrell, C. D., 1970, *Phys. Rev.* **A1,** 672.
Cantrell, C. D., 1971, *Phys. Rev.* **A3,** 728.
Cantrell, C. D. and J. R. Fields, 1973, *Phys. Rev.* **A7,** 2063.
Cantrell, C. D., M. Lax and W. A. Smith, 1973, Coherence and Quantum Optics, eds. L. Mandel and E. Wolf, Plenum Press, New York, p. 785.
Cantrell, C. D. and W. A. Smith, 1971, *Phys. Lett.* **37A,** 167.
Carmichael, H. J., P. Drummond, P. Meystre and D. F. Walls, 1978, *J. Phys.* **A11,** L 121.
Carmichael, H. J., P. D. Drummond, D. F. Walls and P. Meystre, 1980, *Opt. Acta* **27,** 581.
Carmichael, H. J. and D. F. Walls, 1973, *J. Phys.* **A6,** 1552.

Carmichael, H. J. and D. F. Walls, 1974, *Phys. Rev.* **A8,** 2686.
Carmichael, H. J. and D. F. Walls, 1976, *J. Phys.* **B9,** L 43 and 1199.
Carruthers, P. and K. S. Dy, 1966, *Phys. Rev.* **147,** 214.
Carruthers, P. and M. M. Nieto, 1968, *Rev. Mod. Phys.* **40,** 411.
Carter, W. H., 1980, *J. Opt. Soc. Am.* **70,** 1067.
Carter, W. H. and M. Bertolotti, 1978, *J. Opt. Soc. Am.* **68,** 329.
Carter, W. H. and E. Wolf, 1973, *J. Opt. Soc. Am.* **63,** 1619.
Carter, W. H. and E. Wolf, 1975, *J. Opt. Soc. Am.* **65,** 1067.
Carter, W. H. and E. Wolf, 1977, *J. Opt. Soc. Am.* **67,** 785.
Carter, W. H. and E. Wolf, 1981a, *Opt. Acta* **28,** 227.
Carter, W. H. and E. Wolf, 1981b, *Opt. Acta* **28,** 245.
Carusotto, S., 1974, *Lett. Nuovo Cim.* **10,** 571.
Carusotto, S., 1975, *Phys. Rev.* **A11,** 1629.
Carusotto, S., 1980, *Opt. Acta* **27,** 1567.
Carusotto, S., G. Fornaca and E. Polacco, 1967, *Phys. Rev.* **157,** 1207.
Carusotto, S., G. Fornaca and E. Polacco, 1968, *Phys. Rev.* **165,** 1391.
Carusotto, S. and C. Strati, 1973, *Nuovo Cim.* **15B,** 159.
Caves, C. M., 1981, *Phys. Rev.* **D23,** 1693.
Chand, P., 1979, *Nuovo Cim.* **50B,** 17.
Chandra, N. and H. Prakash, 1969, *Phys. Rev. Lett.* **22,** 1068.
Chandra, N. and H. Prakash, 1970, *Phys. Rev.* **A1,** 1696.
Chang, R. F., R. W. Detenbeck, V. Korenman, C. O. Alley and U. Hochuli, 1967, *Phys. Lett.* **25A,** 272.
Chang, R. F., V. Korenman, C. O. Alley and R. W. Detenbeck, 1969, *Phys. Rev.* **178,** 612.
Chang, R. F., V. Korenman and R. W. Detenbeck, 1968, *Phys. Lett.* **26A,** 417.
Chaturvedi, S., P. Drummond and D. F. Walls, 1977, *J. Phys.* **A10,** L 187.
Chmela, P., 1971, *Acta Univ. Palack. Ol.* **33,** 253.
Chmela, P., 1973, *Czech. J. Phys.* **B23,** 884.
Chmela, P., 1974, *Czech. J. Phys.* **B24,** 1 and 506; *Acta Univ. Palack. Ol.* **45,** 5.
Chmela, P., 1977, *Acta Phys. Pol.* **A52,** 835.
Chmela, P., 1978, *Acta Phys. Pol.* **A53,** 719.
Chmela, P., 1979a, *Opt. Quant. Electr.* **11,** 103.
Chmela, P., 1979b, *Czech. J. Phys.* **B29,** 129.
Chmela, P., 1979c, *Opt. Quant. Electr.* **11,** 287.
Chmela, P., 1979d, *Opt. Appl.* **9,** 223.
Chmela, P., 1979e, *Acta Phys. Pol.* **A55,** 945.
Chmela, P., 1981a, *Czech. J. Phys.* **A31,** 119 (in Czech).
Chmela, P., 1981b, *Czech. J. Phys.* **B31,** 977 and 999.
Chmela, P., R. Horák and J. Peřina, 1981, *Opt. Acta* **28,** 1209.
Chopra, S. and J. P. Dudeja, 1976, *Opt. Acta* **23,** 37.
Chopra, S. and J. P. Dudeja, 1977, *Opt. Comm.* **23,** 51.
Chopra, S. and L. Mandel, 1972, *IEEE J. Quant. Electr.* **QE-8,** 324.
Chopra, S. and L. Mandel, 1973, Coherence and Quantum Optics, eds. L. Mandel and E. Wolf, Plenum Press, New York, p. 805.
Chrostowski, J., 1980, *Opt. Acta* **27,** 1401.
Chrostowski, J. and B. Karczewski, 1977, *Phys. Lett.* **63A,** 239.
Chrostowski, J., A. Zardecki and C. Dalisle, 1981, *Phys. Rev.* **A24,** 345.
Chu, B., 1974, Laser Light Scattering, Acad. Press, New York.
Chung, J. C., J. C. Huang and N. B. Abraham, 1980, *Phys. Rev.* **A22,** 1018.
Churnside, J. H. and C. M. McIntyre, 1978a, *Appl. Opt.* **17,** 2141.
Churnside, J. H. and C. M. McIntyre, 1978b, *Appl. Opt.* **17,** 2148.

Clark, W. G., L. E. Estes and L. M. Narducci, 1970, *Phys. Lett.* **33A,** 517.
Clifford, S. F. and R. J. Hill, 1981, *J. Opt. Soc. Am.* **71,** 112.
Cohen-Tannoudji, C., 1977, Frontiers in Laser Spectroscopy, eds. R. Balian, S. Haroche and S. Liberman, North-Holland, Amsterdam.
Colegrave, R. K. and M. S. Abdalla, 1981, *Opt. Acta* **28,** 495.
Colegrave, R. K. and M. S. Abdalla, 1983, *Opt. Acta* **30,** 849 and 861.
Collet, E. and E. Wolf, 1979, *J. Opt. Soc. Am.* **69,** 942.
Collet, E. and E. Wolf, 1980, *Opt. Comm.* **32,** 27.
Cook, R. J., 1980, *Opt. Comm.* **35,** 347.
Cook, R. J., 1981, *Phys. Rev.* **A23,** 1243.
Cook, R. J., 1982, *Phys. Rev.* **A25,** 1164; **A26,** 2754.
Corti, M. and V. Degiorgio, 1974, *Opt. Comm.* **11,** 1.
Corti, M. and V. Degiorgio, 1976a, *Phys. Rev.* **A14,** 1475.
Corti, M. and V. Degiorgio, 1976b, Recent Advances in Optical Physics, eds. B. Havelka and J. Blabla, Soc. Czech. Math. Phys., Prague, p. 59.
Corti, M., V. Degiorgio and F. T. Arecchi, 1973, *Opt. Comm.* **8,** 329.
Courtens, E., 1972, Laser Handbook, Vol. 2, eds. F. T. Arecchi and E. O. Schulz-Dubois, North-Holland, Amsterdam, p. 1259.
Cresser, J. D., 1983, *Phys. Rep.* **94,** 47.
Cresser, J. D., J. Häger, G. Leuchs, M. Rateike and H. Walther, 1982, Dissipative Systems in Quantum Optics, Topics in Current Physics, Vol. 27, ed. R. Bonifacio (Springer, Berlin), p. 21.
Crosignani, B. and P. Di Porto, 1974, Photon Correlation and Light Beating Spectroscopy, eds. H. Z. Cummins and E. R. Pike, Plenum Press, New York.
Crosignani, B., P. Di Porto and M. Bertolotti, 1975, Statistical Properties of Scattered Light, Academic Press, New York.
Crosignani, B., P. Di Porto, U. Ganiel, S. Solimeno and A. Yariv, 1972, *IEEE J. Quant. Electr.* **QE-8,** 731.
Crosignani, B., P. Di Porto and S. Solimeno, 1968a, *Phys. Lett.* **27A,** 568.
Crosignani, B., P. Di Porto and S. Solimeno, 1968b, *Phys. Lett.* **28A,** 271.
Crosignani, B., P. Di Porto and S. Solimeno, 1969, *Phys. Rev.* **186,** 1342.
Crosignani, B., P. Di Porto and S. Solimeno, 1971, *Phys. Rev.* **D3,** 1729.
Crosignani, B., P. Di Porto and S. Solimeno, 1972, *J. Phys.* **A5,** L 119.
Crosignani, B., P. Di Porto and S. Solimeno, 1980, *Phys. Rev.* **A21,** 594.
Crosignani, B., U. Ganiel, S. Solimeno and P. Di Porto, 1971a, *Phys. Rev. Lett.* **26,** 1130.
Crosignani, B., U. Ganiel, S. Solimeno and P. Di Porto, 1971b, *Phys. Rev.* **A4,** 1570.
Crosignani, B., C. H. Papas and P. Di Porto, 1980, *Opt. Lett.* **5,** 467.
Crosignani, B. and A. Tedeschi, 1976, *Lett. Nuovo Cim.* **17,** 141.
Cummins, H. Z. and E. R. Pike, eds., 1974, Photon Correlation and Light Beating Spectroscopy, Plenum Press, New York.
Cummins, H. Z. and H. L. Swinney, 1970, Progress in Optics, Vol. 8, ed. E. Wolf, North-Holland, Amsterdam, p. 133.
Dagenais, M. and L. Mandel, 1978, *Phys. Rev.* **A18,** 2217.
Daino, B., G. DeMarchis and S. Piazzolla, 1980, *Opt. Acta* **27,** 1151.
Dainty, J. C., ed., 1975, Laser Speckle and Related Phenomena, Springer, Berlin.
Dainty, J. C., 1976, Progress in Optics, Vol. 14, ed. E. Wolf, North-Holland, Amsterdam, p. 1.
Dattoli, G., A. Renieri and F. Romanelli, 1980, *Opt. Comm.* **35,** 245.
Dattoli, G., A. Renieri, F. Romanelli, and R. Bonifacio, 1980, *Opt. Comm.* **34,** 240.
Davidson, F., 1969, *Phys. Rev.* **185,** 446.
Davidson, F., Chung-Min Chao, F. K. Tittel and J. P. Hohimer, 1974, *IEEE J. Quant. Electr.* **QE-10,** 409.
Davidson, F. and A. Gonzales-del-Valle, 1975, *J. Opt. Soc. Am.* **65,** 655.

Davidson, F. and L. Mandel, 1967, *Phys. Lett.* **25A,** 700.
Davidson, F. and L. Mandel, 1968, *Phys. Lett.* **27A,** 579.
Davies, E. B., 1976, Quantum Theory of Open Systems, Academic Press, New York.
Degiorgio, V. and J. B. Lastovka, 1971, *Phys. Rev.* **A4,** 2033.
Delone, N. B. and A. V. Masalov, 1980, *Opt. Quant. Electr.* **12,** 291.
Dembinski, S. T. and A. Kossakowski, 1976, *Z. Phys.* **B24,** 141.
DeSantis, P., R. Grella, D. Paoletti and F. Gori, 1978, *Opt. Acta* **25,** 191.
Dewael, P., 1975, *J. Phys.* **A8,** 1614.
De Wolf, D. A., 1973a, *J. Opt. Soc. Am.* **63,** 171.
De Wolf, D. A., 1973b, *J. Opt. Soc. Am.* **63,** 657.
De Wolf, D. A., 1973c, *J. Opt. Soc. Am.* **63,** 1249.
Dialetis, D., 1967, *J. Math. Phys.* **8,** 1641.
Dialetis, D., 1969a, *J. Phys.* **A2,** 229.
Dialetis, D., 1969b, *J. Opt. Soc. Am.* **59,** 74.
Diament, P. and M. C. Teich, 1969, *J. Opt. Soc. Am.* **59,** 661.
Diament, P. and M. C. Teich, 1970a, *J. Opt. Soc. Am.* **60,** 682.
Diament, P. and M. C. Teich, 1970b, *J. Opt. Soc. Am.* **60,** 1489.
Diament, P. and M. C. Teich, 1971, *Appl. Opt.* **10,** 1664.
Dicke, R. H., 1954, *Phys. Rev.* **93,** 99.
Dicke, R. H., 1964, Quantum Electronics, eds. N. Bloembergen and P. Grivet, Dunod et Cie., Paris, p. 35.
Dirac, P. A. M., 1958, Principles of Quantum Mechanics, 4th ed., Clarendon Press, Oxford.
Dixit, S. N. and P. Lambropoulos, 1980, *Phys. Rev.* **A21,** 168.
DosReis, F. G. and M. P. Sharma, 1982, *Opt. Comm.* **41,** 341.
Drummond, P. D. and C. W. Gardiner, 1980, *J. Phys.* **A13,** 2353.
Drummond, P. D., C. W. Gardiner and D. F. Walls, 1981, *Phys. Rev.* **A24,** 914.
Drummond, P. D., K. J. McNeil and D. F. Walls, 1979, *Opt. Comm.* **28,** 255.
Drummond, P. D., K. J. McNeil and D. F. Walls, 1980a, *Phys. Rev.* **A22,** 1672.
Drummond, P. D., K. J. McNeil and D. F. Walls, 1980b, *Opt. Acta* **27,** 321.
Drummond, P. D., K. J. McNeil and D. F. Walls, 1981, *Opt. Acta* **28,** 211.
Drummond, P. D. and D. F. Walls, 1980, *J. Phys.* **A13,** 725.
Drummond, P. D. and D. F. Walls, 1981, *Phys. Rev.* **A23,** 2563.
Dudeja, J. P. and S. Chopra, 1977, *Phys. Lett.* **64A,** 271.
Durnin, J., C. Reece and L. Mandel, 1981, *J. Opt. Soc. Am.* **71,** 115.
Eberly, J. H. and A. Kujawski, 1967a, *Phys. Lett.* **24A,** 426.
Eberly, J. H. and A. Kujawski, 1967b, *Phys. Rev.* **155,** 10.
Eberly, J. H. and A. Kujawski, 1972, *Acta Phys. Pol.* **A41,** 259.
Eberly, J. H. and N. E. Rehler, 1970, *Phys. Rev.* **A2,** 1607.
Eberly, J. H. and K. Wódkiewicz, 1977, *J. Opt. Soc. Am.* **67,** 1252.
Echtermeyer, B., 1971, *Z. Phys.* **246,** 225.
Einstein, A., 1909, *Phys. Z.* **10,** 185 and 817.
Elbaum, M. and P. Diament, 1976, *Appl. Opt.* **15,** 2268.
Eljutin, S. O., A. I. Mailistov and E. A. Maykin, 1981, *Opt. Spectr.* (USSR) **50,** 354.
Ernst, V., 1969, *Z. Phys.* **229,** 432.
Ernst, V., 1976, *Z. Phys.* **B23,** 103 and 113.
Estes, L. E., J. D. Kuppenheimer and L. M. Narducci, 1970, *Phys. Rev.* **A1,** 710.
Every, I. M., 1975, *J. Phys.* **A8,** L 69.
Fano, U., 1961, *Am. J. Phys.* **29,** 539.
Farina, J. D., L. M. Narducci and E. Collet, 1980, *Opt. Comm.* **32,** 203.
Fercher, A. F. and P. F. Steeger, 1981, *Opt. Acta* **28,** 443.
Ferwerda, H. A., 1978, Inverse Source Problems, ed. H. P. Baltes, Springer, Berlin, p. 13.

Ficek, Z., R. Tanaś and S. Kielich, 1983, *Opt. Comm.* **46**, 23.
Fillmore, G. L., 1969, *Phys. Rev.* **182**, 1384.
Fillmore, G. L. and G. Lachs, 1969, *IEEE Trans. Inform. Theory* **IT-15**, 465.
Fleck, J. A., 1966a, *Phys. Rev.* **149**, 309 and 322.
Fleck, J. A., 1966b, *Phys. Rev.* **152**, 278.
Foerster, Von T. and R. J. Glauber, 1971, *Phys. Rev.* **A3**, 1484.
Forrester, A. T., R. A. Gudmundsen and P. O. Johnson, 1955, *Phys. Rev.* **99**, 1691.
Francon, M., 1966, Diffraction. Coherence in Optics, Pergamon, Oxford.
Francon, M. and S. Slansky, 1965, Cohérence en optique, Centre Nat. Rech. Sci., Paris.
Fray, S., F. A. Johnson, R. Jones, T. P. McLean and E. R. Pike, 1967, *Phys. Rev.* **153**, 357.
Freed, C. and H. A. Haus, 1965, *Phys. Rev. Lett.* **15**, 943.
Freed, C. and H. A. Haus, 1966, *IEEE J. Quant. Electr.* **QE-2**, 190.
Friberg, A. T., 1978a, Coherence and Quantum Optics IV, eds. L. Mandel and E. Wolf, Plenum Press, New York, p. 449.
Friberg, A. T., 1978b, *J. Opt. Soc. Am.* **68**, 1281.
Friberg, A. T., 1979a, *J. Opt. Soc. Am.* **69**, 192.
Friberg, A. T., 1979b, Proc. SPIE, Appl. Opt. Coherence, Vol. 194, ed. W. H. Carter, p. 55 and 71.
Friberg, A. T., 1981a, Optics in Four Dimensions — 1980, eds. M. A. Machado and L. M. Narducci, Am. Inst. Phys., New York, p. 313.
Friberg, A. T., 1981b, *Opt. Acta* **28**, 261.
Frieden, B. R., 1971, Progress in Optics, Vol. 9, ed. E. Wolf, North-Holland, Amsterdam, p. 311.
Funke, J., 1974, *Czech. J. Phys.* **B24**, 245.
Funke, J. and M. Hoffmann, 1976, *Czech. J. Phys.* **B26**, 134.
Furutsu, K., 1972, *J. Opt. Soc. Am.* **62**, 240.
Furutsu, K., 1976, *J. Math. Phys.* **17**, 1252.
Furutsu, K. and Y. Furuhama, 1973, *Opt. Acta* **20**, 707.
Fürth, R., 1928a, *Z. Phys.* **48**, 323.
Fürth, R., 1928b, *Z. Phys.* **50**, 310.
Gambini, R., 1977, *Phys. Rev.* **A15**, 1157.
Gelfand, I. M. and G. E. Shilov, 1964, Generalized Functions, Vol. I, Academic Press, New York.
Georges, A. T., 1981, *Opt. Comm.* **38**, 274.
Germey, K., F. J. Schütte and R. Tiebel, 1981, *Ann. Physik* **38**, 80.
Ghielmetti, F., 1964, *Phys. Lett.* **12**, 210.
Ghielmetti, F., 1976, *Nuovo Cim.* **35B**, 243.
Gibbs, H. M. and R. E. Slusher, 1971, *Appl. Phys. Lett.* **18**, 505.
Gilmore, R., 1975, *Phys. Rev.* **A12**, 1019.
Glauber, R. J., 1963a, *Phys. Rev.* **130**, 2529.
Glauber, R. J., 1963b, *Phys. Rev.* **131**, 2766.
Glauber, R. J., 1963c, *Phys. Rev. Lett.* **10**, 84.
Glauber, R. J., 1964, Quantum Electronics, eds. N. Bloembergen and P. Grivet, Dunod et Cie., Paris, p. 111.
Glauber, R. J., 1965, Quantum Optics and Electronics, eds. C. DeWitt, A. Blandin and C. Cohen-Tannoudji, Gordon and Breach, New York, p. 144.
Glauber, R. J., 1966a, Physics of Quantum Electronics, eds. P. L. Kelley, B. Lax and P. E. Tannenwald, McGraw-Hill, New York, p. 788.
Glauber, R. J., 1966b, *Phys. Lett.* **21**, 650.
Glauber, R. J., 1967, Proc. Symp. Modern Optics, Polytechnic Press, New York, p. 1.
Glauber, R. J., 1969, Quantum Optics, ed. R. J. Glauber, Academic Press, New York.
Glauber, R. J., 1970, Quantum Optics, eds. S. M. Kay and A. Maitland, Academic Press, London, p. 53.

Glauber, R. J., 1972, Laser Handbook, Vol. 1, eds. F. T. Arecchi and E. O. Schulz-Dubois, North-Holland, Amsterdam, p. 1.
Glódź, M., 1978, *Acta Phys. Pol.* **A54,** 213.
Gnutzmann, U., 1969, *Z. Phys.* **222,** 283.
Gnutzmann, U., 1970, *Z. Phys.* **233,** 380.
Golay, M. J. E, 1961, *Proc. IRE* **49,** 958.
Goldberger, M. L., H. W. Lewis and K. M. Watson, 1963, *Phys. Rev.* **132,** 2764.
Goldberger, M. L., H. W. Lewis and K. M. Watson, 1966, *Phys. Rev.* **142,** 25.
Goldberger, M. L. and K. M. Watson, 1964, *Phys. Rev.* **134,** B 919.
Goldberger, M. L. and K. M. Watson, 1965, *Phys. Rev.* **137,** B 1396.
Golubev, Yu. M., 1979, *Opt. Spectr.* (USSR) **46,** 3 and 398.
Gonzáles-Díaz, P. T., 1978, *Opt. Comm.* **26,** 437.
Goodman, J. W., 1975, Laser Speckle and Related Phenomena, ed. J. C. Dainty, Springer, Berlin, p. 9.
Gorbatchev, V. N. and P. V. Zanadvorov, 1980, *Opt. Spectr.* (USSR) **49,** 600.
Gordov, E. P. and S. D. Tvorogov, 1978, The Quantum Theory of Propagation of the Electromagnetic Field, Nauka, Novosibirsk (in Russian).
Gordov, E. P. and S. D. Tvorogov, 1980, *Phys. Rev.* **D22,** 908.
Gori, F., 1980, *Opt. Comm.* **34,** 301.
Gradshteyn, I. S. and I. M. Ryzhik, 1965, Table of Integrals, Series and Products, Academic Press, New York.
Graham, R., 1968a, *Z. Phys.* **210,** 319.
Graham, R., 1968b, *Z. Phys.* **211,** 469.
Graham, R., 1970, Quantum Optics, eds. S. M. Kay and A. Maitland, Academic Press, London, p. 489.
Graham, R., 1973, Springer Tracts in Modern Physics, Vol. 66, ed. G. Höhler, Springer, Berlin, p. 1.
Graham, R., 1974, Progress in Optics, Vol. 12, ed. E. Wolf, North-Holland, Amsterdam, p. 233.
Graham, R. and H. Haken, 1968, *Z. Phys.* **210,** 276.
Graham, R. and A. Schenzle, 1981, *Phys. Rev.* **A23,** 1302.
Greenhow, R. C. and A. J. Schmidt, 1974, Advances in Quantum Electronics, Vol. 2, ed. D. W. Goodwin, Academic Press, London, p. 157.
Griffin, W. G. and P. N. Pusey, 1979, *Phys. Rev. Lett.* **43,** 1100.
Gupta, P. S. and B. K. Mohanty, 1980, *Czech. J. Phys.* **B30,** 1127.
Gupta, P. S. and B. K. Mohanty, 1981, *Opt. Acta* **28,** 521.
Gurvich, A. S., V. Kan, V. I. Tatarskii and V. U. Zavorotnyi, 1979, *Opt. Acta* **26,** 543.
Gurvich, A. S. and V. I. Tatarskii, 1973, *Izv. Vysh. Utscheb. Zav.-Radiophys.* XVI, 434.
Haake, F., 1973, Springer Tracts in Modern Physics, Vol. 66, ed. G. Höhler, Springer, Berlin, p. 98.
Haake, F., J. W. Haus and R. Glauber, 1981, *Phys. Rev.* **A23,** 3255.
Haake, F., J. W. Haus, H. King, G. Schröder and R. Glauber, 1981, *Phys. Rev.* **A23,** 1322.
Haig, N. D. and R. M. Sillitto, 1968, *Phys. Lett.* **28A,** 463.
Haken, H., 1967, Dynamical Processes in Solid State Optics, Part I, eds. R. Kubo and H. Kamimura, W. A. Benjamin, New York, p. 168.
Haken, H., 1970a, Quantum Optics, eds. S. M. Kay and A. Maitland, Academic Press, London p. 201.
Haken, H., 1970b, Handbuch der Physik, Vol. 25/2c, ed. S. Flügge, Springer, Berlin.
Haken, H., 1972, Laser Handbook, Vol. 1, eds. F. T. Arecchi and E. O. Schulz-Dubois, North-Holland, Amsterdam, p. 115.
Haken, H., 1978, Synergetics. An Introduction, 2nd ed., Springer, Berlin.
Haken, H., H. Risken and W. Weidlich, 1967, *Z. Phys.* **206,** 355.
Harwit, M., 1960, *Phys. Rev.* **120,** 1551.
Hassan, S. S., P. D. Drummond and D. F. Walls, 1978, *Opt. Comm.* **27,** 480.

Heitler, W., 1954, The Quantum Theory of Radiation, 3rd ed., Oxford.
Helstrom, C. W., 1972, Progress in Optics, Vol. 10, ed. E. Wolf, North-Holland, Amsterdam, p. 284.
Helstrom, C. W., 1976, Quantum Detection and Estimation Theory, Academic Press, New York.
Helstrom, W., 1979, *IEEE Trans. Inf. Theory* **IT-25**, 69.
Helstrom, C. W., 1981, *Opt. Comm.* **37**, 175.
Hempstead, R. D. and M. Lax, 1967, *Phys. Rev.* **161**, 350.
Herzog, U., 1983, *Opt. Acta* **30**, 639.
Hildred, G. P., 1980, *Opt. Acta* **27**, 1621.
Hildred, G. P. and A. G. Hall, 1978, *J. Phys.* **A11**, L 209.
Hill, R. J. and S. F. Clifford, 1981, *J. Opt. Soc. Am.* **71**, 675.
Hioe, F. T. and S. Singh, 1981, *Phys. Rev.* **A24**, 2050.
Hioe, F. T., S. Singh and L. Mandel, 1979, *Phys. Rev.* **A19**, 2036.
Hirota, O. and S. Ikehara, 1976, *Phys. Lett.*, **57A**, 317.
Ho, T. L., 1969, *J. Opt. Soc. Am.* **59**, 385.
Hoenders, B. J., 1975, *J. Math. Phys.* **16**, 1719.
Hofman, M., 1980, *Acta Univ. Palack. Ol.* **65**, 35.
Holliday, D., 1964, *Phys. Lett.* **8**, 250.
Holliday, D. and M. L. Sage, 1964, *Ann. Phys.* **29**, 125.
Holliday, D. and M. L. Sage, 1965, *Phys. Rev.* **138**, B 485.
Holý, V., 1980, *Phys. Stat. Sol.* (*b*) **101**, 575.
Horák, R., 1971, *Czech. J. Phys.* **B21**, 7.
Horák, R., L. Mišta and J. Peřina, 1971a, *J. Phys.* **A4**, 231.
Horák, R., L. Mišta and J. Peřina, 1971b, *Czech. J. Phys.* **B21**, 614.
Horák, R., L. Mišta and J. Peřina, 1971c, *Phys. Lett.* **35A**, 400.
Hughes, A. J., E. Jakeman, C. J. Oliver and E. R. Pike, 1973, *J. Phys.* **A6**, 1327.
Iwasawa, H., 1976, *Z. Phys.* **B23**, 399.
Jaiswal, A. K. and G. S. Agarwal, 1969, *J. Opt. Soc. Am.* **59**, 1446.
Jaiswal, A. K. and C. L. Mehta, 1969, *Phys. Rev.* **186**, 1355.
Jaiswal, A. K. and C. L. Mehta, 1970, *Phys. Rev.* **A2**, 168.
Jakeman, E., 1970, *J. Phys.* **A3**, 201.
Jakeman, E., 1974, Photon Correlation and Light Beating Spectroscopy, eds. H. Z. Cummins and E. R. Pike, Plenum Press, New York.
Jakeman, E., 1980, *Opt. Acta* **27**, 735.
Jakeman, E., 1981, *Opt. Acta* **28**, 435.
Jakeman, E., C. J. Oliver and E. R. Pike, 1968a, *J. Phys.* **A1**, 406.
Jakeman, E., C. J. Oliver and E. R. Pike, 1968b, *J. Phys.* **A1**, 497.
Jakeman, E., C. J. Oliver and E. R. Pike, 1970, *J. Phys.* **A3**, L 45.
Jakeman, E., C. J. Oliver and E. R. Pike, 1971a, *J. Phys.* **A4**, 827.
Jakeman, E., C. J. Oliver and E. R. Pike, 1971b, *Phys. Lett.* **34A**, 101.
Jakeman, E., C. J. Oliver, E. R. Pike, M. Lax and M. Zwanziger, 1970, *J. Phys.* **A3**, L 52.
Jakeman, E. and E. R. Pike, 1968, *J. Phys.* **A1**, 128.
Jakeman, E. and E. R. Pike, 1969a, *J. Phys.* **A2**, 115.
Jakeman, E. and E. R. Pike, 1969b, *J. Phys.* **A2**, 411.
Jakeman, E. and E. R. Pike, 1971, *J. Phys.* **A4**, L 56.
Jakeman, E., E. R. Pike, G. Parry and B. Saleh, 1976, *Opt. Comm.* **19**, 359.
Jakeman, E., E. R. Pike, P. N. Pusey and J. M. Vaughan, 1977, *J. Phys.* **A10**, L 257.
Jakeman, E., E. R. Pike and S. Swain, 1970, *J. Phys.* **A3**, L 55.
Jakeman, E., E. R. Pike and S. Swain, 1971, *J. Phys.* **A4**, 517.
Jannson, T., 1980, *J. Opt. Soc. Am.* **70**, 1544.
Jannussis, A., N. Patargias, and L. Papaloucas, 1979, *J. Phys. Soc. Jap.* **47**, 1003.
Janossy, L., 1957, *Nuovo Cim.* **6**, 111.

Janossy, L., 1959, *Nuovo Cim.* **12,** 370.

Jansen, R. V. and C. R. Oberman, 1981, *Phys. Rev. Lett.* **46,** 1547.

Javan, A., E. A. Ballik and W. L. Bond, 1962, *J. Opt. Soc. Am.* **52,** 96.

Jodoin, R. and L. Mandel, 1971, *J. Opt. Soc. Am.* **61,** 191.

Johnson, F. A., R. Jones, T. P. McLean and E. R. Pike, 1966, *Phys. Rev. Lett.* **16,** 589.

Johnson, F. A., T. P. McLean and E. R. Pike, 1966, Physics of Quantum Electronics, eds. P. L. Kelley, B. Lax and P. E. Tannenwald, McGraw-Hill, New York.

Jordan, F. T. and F. Ghielmetti, 1964, *Phys. Rev. Lett.* **12,** 607.

Kahn, F. D., 1958, *Opt. Acta* **5,** 93.

Kaminishi, K., R. Roy, R. Short and L. Mandel, 1981, *Phys. Rev.* **A24,** 370.

Kano, Y., 1964a, *J. Phys. Soc. Jap.* **19,** 1555.

Kano, Y., 1964b, *Ann. Phys.* **30,** 127.

Kano, Y., 1965, *J. Math. Phys.* **6,** 1913.

Kano, Y., 1966, *Nuovo Cim.* **43,** 1.

Kano, Y., 1976, *Phys. Lett.* **56A,** 7.

Kano, Y. and E. Wolf, 1962, *Proc. Phys. Soc.* **80,** 1273.

Karczewski, B., 1976, Recent Advances in Optical Physics, eds. B. Havelka and J. Blabla, Soc. Czech. Math. Phys., Prague, p. 53.

Katriel, J. and D. G. Hummer, 1981, *J. Phys.* **A14,** 1211.

Kazantsev, A. P., V. S. Smirnov and V. P. Sokolov, 1980, *Opt. Comm.* **35,** 209.

Kazantsev, A. P., V. S. Smirnov, V. P. Sokolov and A. N. Tumajkin, 1981, *J. Exp. Theor. Phys.* (USSR) **81,** 888.

Keller, E. F., 1965, *Phys. Rev.* **139,** B 202.

Kelley, P. L. and W. H. Kleiner, 1964, *Phys. Rev.* **136,** A 316.

Kiedroń, P., 1980, *Opt. Appl.* **10,** 253.

Kiedroń, P., 1981, *Optik* **59,** 303.

Kielich, S., 1981, Molecular Nonlinear Optics, Nauka, Moscow (in Russian).

Kielich, S., M. Kozierowski and R. Tanaś, 1978, Coherence and Quantum Optics IV, eds. L. Mandel and E. Wolf, Plenum Press, New York, p. 511.

Kikuchi, R. and B. H. Soffer, 1977, *J. Opt. Soc. Am.* **67,** 1656.

Kimble, H. J., M. Dagenais and L. Mandel, 1977, *Phys. Rev. Lett.* **39,** 691.

Kimble, H. J., M. Dagenais and L. Mandel, 1978, *Phys. Rev.* **A18,** 201.

Kimble, H. J. and L. Mandel, 1973, *J. Opt. Soc. Am.* **63,** 1550.

Kimble, H. J. and L. Mandel, 1975, *Opt. Comm.* **14,** 167.

Kimble, H. J. and L. Mandel, 1976, *Phys. Rev.* **A13,** 2123.

Kimble, H. J. and L. Mandel, 1977, *Phys. Rev.* **A15,** 689.

Kitazima, I., 1974, *Opt. Comm.* **10,** 137.

Klauder, J. R., 1960, *Ann. Phys.* **11,** 123.

Klauder, J. R., 1966, *Phys. Rev. Lett.* **16,** 534.

Klauder, J. R., J. McKenna and D. G. Currie, 1965, *J. Math. Phys.* **6,** 734.

Klauder, J. R. and E. C. G. Sudarshan, 1968, Fundamentals of Quantum Optics, W. A. Benjamin, New York.

Kleinman, D. A., 1972, Laser Handbook, Vol. 2, eds. F. T. Arecchi and E. O. Schulz-Dubois, North-Holland, Amsterdam, p. 1229.

Klyackin, V. I., 1975, Statistical Description of Dynamical Systems with Fluctuating Parameters, Nauka, Moscow (in Russian).

Klyackin, V. I. and V. I. Tatarskii, 1973, *Usp. Phys. Nauk* **110,** 499.

Klyshko, A. N., 1980, Photons and Nonlinear Optics, Nauka, Moscow (in Russian).

Kohler, D. and L. Mandel, 1970, *J. Opt. Soc. Am.* **60,** 280.

Kohler, D. and L. Mandel, 1973, *J. Opt. Soc. Am.* **63,** 126.

Koňák, Č., P. Štěpánek, L. Dvořák, Z. Kupka, J. Křepelka and J. Peřina, 1982, *Opt. Acta* **29,** 1105.

Koňák, Č., J. Křepelka and J. Peřina, 1983, *Opt. Acta* **30**, in print.
Korenman, V., 1965, *Phys. Rev. Lett.* **14**, 293.
Korenman, V., 1966, *Ann. Phys.* **39**, 72.
Korenman, V., 1967, *Phys. Rev.* **154**, 1233.
Kovarskii, V. A., 1974, Multiphoton Transitions, Shtiinca, Kishinev (in Russian).
Kozierowski, M., 1981, *Kvant. Electr.* (USSR) **8**, 1157.
Kozierowski, M. and S. Kielich, 1983, *Phys. Lett.* **94A**, 213.
Kozierowski, M. and R. Tanaś, 1977, *Opt. Comm.* **21**, 229.
Kozierowski, M., R. Tanaś and S. Kielich, 1978, Validity of the short-time approximation in the quantum theory of harmonics generation, EKON-78, Univ. Mickiewicz, Poznań, p. 126.
Krasiński, J., S. Chudzyński and W. Majewski, 1974, *Opt. Comm.* **12**, 304.
Krasiński, J., S. Chudzyński and W. Majewski, 1976, Recent Advances in Optical Physics, eds. B. Havelka and J. Blabla, Soc. Czech. Math. Phys., Prague, p. 323.
Kraynov, V. P. and S. S. Todirashku, 1980, *J. Exp. Theor. Phys.* (USSR) **79**, 69.
Krivoshlykov, S. G. and I. N. Sissakian, 1979, *Opt. Quant. Electr.* **11**, 393.
Krivoshlykov, S. G. and I. N. Sissakian, 1980a, *Opt. Quant. Electr.* **12**, 463.
Krivoshlykov, S. G. and I. N. Sissakian, 1980b, *Kvant. Electr.* (USSR) **7**, 553.
Krivoshlykov, S. G. and I. N. Sissakian, 1983, *Kvant. Electr.* **10**, 735.
Kruglik, G. S., 1978, Quantum Statistical Theory of Ring Optical Quantum Generators, Nauka and Tech., Minsk (in Russian).
Kryszewski, S. and J. Chrostowski, 1977, *J. Phys.* **A10**, L 261.
Kujawski, A., 1966, *Nuovo Cim.* **44**, 326.
Kujawski, A., 1968, *Acta Phys. Pol.* **34**, 957.
Kujawski, A., 1969, *Bull. Acad. Pol. Sci.* **17**, 467 and 839.
Kujawski, A. and J. H. Eberly, 1978, Coherence and Quantum Optics IV, eds. L. Mandel and E. Wolf, Plenum Press, New York, p. 989.
Kumar, S. and C. L. Mehta, 1980, *Phys. Rev.* **A21**, 1573.
Kumar, S. and C. L. Mehta, 1981, *Phys. Rev.* **A24**, 1460.
Kumar, S., C. L. Mehta and G. S. Agarwal, 1981, *Opt. Comm.* **39**, 197.
Kuriksha, A. K., 1973, Quantum Optics and Optical Location, Sov. Radio, Moscow (in Russian).
Kühlke, D. and R. Horák, 1979, *Opt. Quant. Electr.* **11**, 485.
Kühlke, D. and R. Horák, 1981, *Physica* 111C, 111.
Lachs, G., 1965, *Phys. Rev.* **138**, B 1012.
Lachs, G., 1967, *J. Appl. Phys.* **38**, 3439.
Lachs, G., 1971, *J. Appl. Phys.* **42**, 602.
Lachs, G. and S. R. Laxpati, 1973, *J. Appl. Phys.* **44**, 3332.
Lachs, G. and D. R. Voltmer, 1976, *J. Appl. Phys.* **47**, 346.
Lambropoulos, P., 1968, *Phys. Rev.* **168**, 1418.
Lambropoulos, P., C. Kikuchi and R. K. Osborn, 1966, *Phys. Rev.* **144**, 1081.
Landau, L. D. and E. M. Lifshitz, 1959, Statistical Physics, 2nd ed., Pergamon, Oxford.
Langer, J. S., 1968, *Phys. Rev.* **167**, 183.
Langer, J. S., 1969, *Phys. Rev.* **184**, 219.
Lax, M., 1967, Dynamical Processes in Solid State Optics, Part I, eds. R. Kubo and H. Kamimura, W. A. Benjamin, New York, p. 195.
Lax, M., 1968a, Statistical Physics, Phase Transitions and Superconductivity, eds. M. Chrétien, E. P. Gross and S. Deser, Gordon and Breach, New York.
Lax, M., 1968b, *Phys. Rev.* **172**, 350.
Lax, M. and W. H. Louisell, 1967, *IEEE J. Quant. Electr.* **QE-3**, 47.
Lax, M. and H. Yuen, 1968, *Phys. Rev.* **172**, 362.
Lax, M. and M. Zwanziger, 1970, *Phys. Rev. Lett.* **24**, 937.
Lax, M. and M. Zwanziger, 1973, *Phys. Rev.* **A7**, 750.

Laxpati, S. R. and G. Lachs, 1972, *J. Appl. Phys.* **43,** 4773.
Leader, J. C., 1981, *J. Opt. Soc. Am.* **71,** 542.
LeCompte, C., G. Mainfray and C. Manus, 1974, *Phys. Rev. Lett.* **32,** 265.
LeCompte, C., G. Mainfray, C. Manus and F. Sanchez, 1975, *Phys. Rev.* **A11,** 1009.
Ledinegg, E., 1967, *Z. Phys.* **205,** 25.
Ledinegg, E. and E. Schachinger, 1983, *Phys. Rev.* **A27,** 2555.
Lehmberg, R. H., 1968, *Phys. Rev.* **167,** 1152.
Lenstra, D., 1982, *Phys. Rev.* **A26,** 3369.
Leuchs, G., M. Rateike and H. Walther, 1979, see D. F. Walls, 1979, *Nature* **280,** 451.
Lipsett, M. S. and L. Mandel, 1963, *Nature* **199,** 553.
Lipsett, M. S. and L. Mandel, 1964, Quantum Electronics, eds. N. Bloembergen and P. Grivet, Dunod and Cie., Paris, p. 1271.
López, A., 1967, *Phys. Lett.* **25A,** 83.
Loudon, R., 1973, The Quantum Theory of Light, Clarendon, Oxford.
Loudon, R., 1980, *Rep. Prog. Phys.* **43,** 913.
Loudon, R., 1981, *Phys. Rev. Lett.* **47,** 815.
Louisell, W. H., 1964, Radiation and Noise in Quantum Electronics, McGraw-Hill, New York.
Louisell, W. H., 1970, Quantum Optics, eds. S. M. Kay and A. Maitland, Academic Press, London, p. 177.
Louisell, W. H., 1973, Quantum Statistical Properties of Radiation, J. Wiley, New York.
Louisell, W. H., A. Yariv and A. E. Siegman, 1961, *Phys. Rev.* **124,** 1646.
Lugiato, L. A., 1981, *Z. Phys.* **B41,** 85.
Lugiato, L. A., P. Mandel, S. T. Dembinski and A. Kossakowski, 1978, *Phys. Rev.* **A18,** 238.
Lugiato, L. A., L. M. Narducci and M. Gronchi, 1977, *Phys. Rev.* **A15,** 1126.
Lukš, A., 1976, *Czech. J. Phys.* **B26,** 1095.
Lyons, J. and G. J. Troup, 1970a, *Phys. Lett.* **31A,** 182.
Lyons, J. and G. J. Troup, 1970b, *Phys. Lett.* **32A,** 352.
Magill, P. J. and R. P. Soni, 1966, *Phys. Rev. Lett.* **16,** 911.
Magyar, G. and L. Mandel, 1963, *Nature* **198,** 255.
Magyar, G. and L. Mandel, 1964, Quantum Electronics, eds. N. Bloembergen and P. Grivet, Dunod et Cie., Paris, p. 1247.
Malkin, I. A. and V. I. Man'ko, 1979, Dynamical Symmetries and Coherent States of Quantum Systems, Nauka, Moscow (in Russian).
Malkin, I. A., V. I. Man'ko and D. A. Trifonov, 1969, *Phys. Lett.* **30A,** 414.
Mandel, L., 1958, *Proc. Phys. Soc.* (Lond.) **72,** 1037.
Mandel, L., 1959, *Proc. Phys. Soc.* (Lond.) **74,** 233.
Mandel, L., 1963a, Progress in Optics, Vol. 2, ed. E. Wolf, North-Holland, Amsterdam, p. 181.
Mandel, L., 1963b, *Proc. Phys. Soc.* **81,** 1104.
Mandel, L., 1963c, *Phys. Lett.* **7,** 117.
Mandel, L., 1964a, *Phys. Lett.* **10,** 166.
Mandel, L., 1964b, *Phys. Rev.* **134,** A 10.
Mandel, L., 1964c, *Phys. Rev.* **136,** B 1221.
Mandel, L., 1965, *Phys. Rev.* **138,** B 753.
Mandel, L., 1966a, *Phys. Rev.* **144,** 1071.
Mandel, L., 1966b, *Phys. Rev.* **152,** 438.
Mandel, L., 1969, *Phys. Rev.* **181,** 75.
Mandel, L., 1976a, Progress in Optics, Vol. 13, ed. E. Wolf, North-Holland, Amsterdam, p. 27.
Mandel, L., 1976b, *J. Opt. Soc. Am.* **66,** 968.
Mandel, L., 1979, *Opt. Lett.* **4,** 205.
Mandel, L., 1980, *J. Opt. Soc. Am.* **70,** 873.
Mandel, L., 1981a, *Opt. Comm.* **36,** 87.

Mandel, L., 1981b, *Opt. Comm.* **39,** 163.
Mandel, L., 1981c, *J. Opt. Soc. Am. Lett.* **71,** 1273.
Mandel, L., 1981d, *Opt. Acta* **28,** 1447.
Mandel, L., 1982a, *Phys. Rev. Lett.* **49,** 136.
Mandel L., 1982b, *Opt. Comm.* **42,** 437.
Mandel, L., 1983, *Phys. Rev.* **A28,** 929.
Mandel, L. and D. Meltzer, 1969, *Phys. Rev.* **188,** 198.
Mandel, L., E. C. G. Sudarshan and E. Wolf, 1964, *Proc. Phys. Soc.* **84,** 435.
Mandel, L. and E. Wolf, 1961, *Phys. Rev.* **124,** 1696.
Mandel, L. and E. Wolf, 1965, *Rev. Mod. Phys.* **37,** 231.
Mandel, L. and E. Wolf, 1966, *Phys. Rev.* **149,** 1033.
Mandel, L. and E. Wolf, 1976, *J. Opt. Soc. Am.* **66,** 529.
Mandel, L. and E. Wolf, 1981, *Opt. Comm.* **36,** 247.
Mandel, P., 1978, Quantum Optics, eds. J. Heldt and J. Czub, Univ. Toruń and Gdańsk, p. 115.
Man'ko, V. I., ed., 1972, Coherent States in Quantum Theory, Mir, Moscow (in Russian).
Marchand, E. W. and E. Wolf, 1972, *Opt. Comm.* **6,** 305.
Marchand, E. W. and E. Wolf, 1974a, *J. Opt. Soc. Am.* **64,** 1219.
Marchand, E. W. and E. Wolf, 1974b, *J. Opt. Soc. Am.* **64,** 1273.
Martienssen, W. and E. Spiller, 1964, *Am. J. Phys.* **32,** 919.
Martienssen, W. and E. Spiller, 1966a, *Phys. Rev. Lett.* **16,** 531.
Martienssen, W. and E. Spiller, 1966b, *Phys. Rev.* **145,** 285.
Martínez-Herrero, R., 1979, *Nuovo Cim.* **54B,** 205.
Martínez-Herrero, R., 1981, *Opt. Acta* **28,** 1151.
Martínez-Herrero, R. and A. Durán, 1981, *Opt. Acta* **28,** 65.
Martínez-Herrero, R. and P. M. Mejías, 1981, *Opt. Comm.* **37,** 234.
May, M., 1977, *J. Phys.* **E10,** 849.
McCall, S. L. and E. L. Hahn, 1967, *Phys. Rev. Lett.* **18,** 908.
McCall, S. L. and E. L. Hahn, 1969, *Phys. Rev.* **183,** 457.
McGill, W. J., 1967, *J. Math. Psych.* **4,** 351.
McLean, T. P. and E. R. Pike, 1965, *Phys. Lett.* **15,** 318.
McMillan, J. L., R. M. Sillitto and W. Sillitto, 1979, *Opt. Acta* **26,** 1125.
McNeil, K. J., P. D. Drummond and D. F. Walls, 1978, *Opt. Comm.* **27,** 292.
McNeil, K. J. and D. F. Walls, 1974, *J. Phys.* **A7,** 617.
McNeil, K. J. and D. F. Walls, 1975a, *J. Phys.* **A8,** 104.
McNeil, K. J. and D. F. Walls, 1975b, *J. Phys.* **A8,** 111.
McNeil, K. J. and D. F. Walls, 1975c, *Phys. Lett.* **51A,** 233.
Mehta, C. L., 1963, *Nuovo Cim.* **28,** 401.
Mehta, C. L., 1966, *Nuovo Cim.* **45,** 280.
Mehta, C. L., 1967, *Phys. Rev. Lett.* **18,** 752.
Mehta, C. L., 1970, Progress in Optics, Vol. 8, ed. E. Wolf, North-Holland, Amsterdam, p. 373.
Mehta, C. L., P. Chand, E. C. G. Sudarshan and R. Vedam, 1967, *Phys. Rev.* **157,** 1198.
Mehta, C. L. and S. Gupta, 1975, *Phys. Rev.* **A11,** 1634.
Mehta, C. L. and A. K. Jaiswal, 1970, *Phys. Rev.* **A2,** 2570.
Mehta, C. L. and L. Mandel, 1967, Electromagnetic Wave Theory, Part 2, Pergamon, Oxford, p. 1069.
Mehta, C. L. and E. C. G. Sudarshan, 1965, *Phys. Rev.* **138,** B 274.
Mehta, C. L. and E. C. G. Sudarshan, 1966, *Phys. Lett.* **22,** 574.
Mehta, C. L. and E. Wolf, 1964, *Phys. Rev.* **134,** A 1143 and 1149.
Mehta, C. L. and E. Wolf, 1967, *Phys. Rev.* **157,** 1183 and 1188; **161,** 1328.
Mehta, C. L., E. Wolf and A. P. Balachandran, 1966, *J. Math. Phys.* **7,** 133.
Meltzer, D., W. Davis and L. Mandel, 1970, *Appl. Phys. Lett.* **17,** 242.

Meltzer, D. and L. Mandel, 1970, *Phys. Rev. Lett.* **25,** 1151.
Meltzer, D. and L. Mandel, 1971, *Phys. Rev.* **A3,** 1763.
Messiah, A., 1961, Quantum Mechanics, Vol. I, J. Wiley, New York.
Messiah, A., 1962, Quantum Mechanics, Vol. II, J. Wiley, New York.
Meystre, P., E. Geneux, A. Quattropani and A. Faist, 1975, *Nuovo Cim.* **25B,** 521.
Mielniczuk, W. J., 1979, *Opt. Acta* **26,** 1115.
Mielniczuk, W. J. and J. Chrostowski, 1981, *Phys. Rev.* **A23,** 1382.
Mikhailov, A. S., 1976, *Phys. Sol. State* (USSR) **18,** 494.
Mikhailov, V. V., 1971, *Phys. Lett.* **34A,** 343.
Milburn, G. J. and D. F. Walls, 1983, *Phys. Rev.* **A27,** 392.
Miller, M. M. and E. A. Mishkin, 1966, *Phys. Rev.* **152,** 1110.
Miller, M. M. and E. A. Mishkin, 1967a, *Phys. Lett.* **24A,** 188.
Miller, M. M. and E. A. Mishkin, 1967b, *Phys. Rev.* **164,** 1610.
Millet, J. and W. Usselio-La-Verna, 1970, *Opt. Comm.* **2,** 12.
Millet, J. and B. Varnier, 1969, *Opt. Comm.* **1,** 211.
Misell, D. L., 1973, *J. Phys.* **D6,** 2200 and 2217.
Mishkin, E. A. and D. F. Walls, 1969, *Phys. Rev.* **185,** 1618.
Mišta, L., 1967, *Phys. Lett.* **25A,** 646.
Mišta, L., 1969, *Czech. J. Phys.* **B19,** 443.
Mišta, L., 1971, *J. Phys.* **A4,** L 73.
Mišta, L., 1973, *Czech. J. Phys.* **B23,** 715.
Mišta, L., 1981, *Acta Univ. Palack. Ol.* **69,** 39.
Mišta, L. and J. Peřina, 1971, *Opt. Comm.* **2,** 441.
Mišta, L. and J. Peřina, 1977a, *Czech. J. Phys.* **B27,** 373.
Mišta, L. and J. Peřina, 1977b, *Acta Phys. Pol.* **A52,** 425.
Mišta, L. and J. Peřina, 1977c, *Czech. J. Phys.* **B27,** 831.
Mišta, L. and J. Peřina, 1978, *Czech. J. Phys.* **B28,** 392.
Mišta, L., J. Peřina and Z. Braunerová, 1973, *Opt. Comm.* **9,** 113.
Mišta, L., J. Peřina and V. Peřinová, 1971, *Phys. Lett.* **35A,** 197.
Mišta, L., V. Peřinová, J. Peřina and Z. Braunerová, 1977, *Acta Phys. Pol.* **A51,** 739.
Mohanty, B. K. and P. S. Gupta, 1981a, *Czech. J. Phys.* **B31,** 275.
Mohanty, B. K. and P. S. Gupta, 1981b, *Czech. J. Phys.* **B31,** 1083.
Mohanty, B. K. and P. S. Gupta, 1981c, *Czech. J. Phys.* **B31,** 857.
Mohr, U., 1981, *Ann. Physik* **38,** 143.
Mohr, U. and H. Paul, 1978, *Ann. Physik* **35,** 461.
Mohr, U. and H. Paul, 1979, *J. Phys.* **A12,** L 43.
Mollow, B. R., 1968a, *Phys. Rev.* **168,** 1896.
Mollow, B. R., 1968b, *Phys. Rev.* **175,** 1555.
Mollow, B. R., 1970, *Phys. Rev.* **A2,** 1477.
Mollow, B. R., 1981, Progress in Optics, Vol. 19, ed. E. Wolf, North-Holland, Amsterdam, p. 1.
Mollow, B. R. and R. J. Glauber, 1967a, *Phys. Rev.* **160,** 1076.
Mollow, B. R. and R. J. Glauber, 1967b, *Phys. Rev.* **160,** 1097.
Morawitz, H., 1965, *Phys. Rev.* **139,** A 1072.
Morawitz, H., 1966, *Z. Phys.* **195,** 20.
Morgan, B. L. and L. Mandel, 1966, *Phys. Rev. Lett.* **16,** 1012.
Morse, P. M. and H. Feshbach, 1953, Methods of Theoretical Physics, Vol. I, McGraw-Hill, Amsterdam.
Mostowski, J. and K. Rzażewski, 1978, *Phys. Lett.* **66A,** 275.
M-Tehrani, M. and L. Mandel, 1977, *Opt. Lett.* **1,** 196.
M-Tehrani, M. and L. Mandel, 1978a, *Phys. Rev.* **A17,** 677.
M-Tehrani, M. and L. Mandel, 1978b, *Phys. Rev.* **A17,** 694.

Narducci, L. M., C. M. Bowden, V. Bluemel, G. P. Garrazana and R. A. Tuft, 1975, *Phys. Rev.* **A11,** 973.
Narducci, L. M., W. W. Eidson, P. Furcinitti and D. C. Eteson, 1977, *Phys. Rev.* **A16,** 1665.
Nath, R., 1979, *Lett. Nuovo Cim.* **24,** 144.
Nayak, N., 1980, *IEEE J. Quant. Electr.* **QE-16,** 843.
Nayak, N., 1982, Quantum theory of third-harmonic generation in a gaseous system-II, preprint, Ind. Inst. Tech., Kharagpur.
Nayak, N. and B. K. Mohanty, 1977, *Phys. Rev.* **A15,** 1173.
Nayak, N. and B. K. Mohanty, 1979, *Phys. Rev.* **A19,** 1204.
Neumann, R. and H. Haug, 1979, *Opt. Comm.* **31,** 267.
Nieto, M. M. and L. M. Simmons, 1978, *Phys. Rev. Lett.* **41,** 207.
Nieto, M. M. and L. M. Simmons, 1979, *Phys. Rev.* **D20,** 1332 and 1342.
Nieto, M. M., L. M. Simmons and V. P. Gutschick, 1981, *Phys. Rev.* **D23,** 927.
Nieto-Vesperinas, M., 1980, *Optik* **56,** 377.
Nieto-Vesperinas, M. and O. Hignette, 1979, *Opt. Pura y Apl.* **12,** 175.
Nikolov, B. and D. A. Trifonov, 1980, *C. R. Acad. Bulg. Sci.* **33,** 309.
Nussenzveig, H. M., 1967, *J. Math. Phys.* **8,** 561.
Nussenzveig, H. M., 1973, Introduction to Quantum Optics, Gordon and Breach, London.
Ohtsubo, J. and T. Asakura, 1977a, *Appl. Opt.* **16,** 1742.
Ohtsubo, J. and T. Asakura, 1977b, *Opt. Lett.* **1,** 98.
Oliver, G. and C. Bendjaballah, 1980, *Phys. Rev.* **A22,** 630.
O'Neill, E. L., 1963, Introduction to Statistical Optics, Addison-Wesley, Reading.
Orszag, M., 1979, *J. Phys.* **A12,** 2205, 2225 and 2233.
Orszag, M., 1981, *Opt. Acta* **28,** 5.
Parry, G., 1981, *Opt. Acta* **28,** 715.
Parry, G., P. N. Pusey, E. Jakeman and J. G. McWhirter, 1978, Coherence and Quantum Optics IV eds. L. Mandel and E. Wolf, Plenum Press, New York, p. 351.
Paul, H., 1964, *Ann. Physik* **14,** 147.
Paul, H., 1966, *Fortschr. Phys.* **14,** 141.
Paul, H., 1967, *Ann. Physik* **19,** 210.
Paul, H., 1969, Lasertheorie, I, II, Akademie, Berlin.
Paul, H., 1973, Nichtlineare Optik, I, II, Akademie, Berlin.
Paul, H., 1974, *Fortschr. Phys.* **22,** 657.
Paul, H., 1976, Recent Advances in Optical Physics, eds. B. Havelka and J. Blabla, Soc. Czech. Math. Phys., Prague, p. 67.
Paul, H., 1980, *Fortschr. Phys.* **28,** 633.
Paul, H., 1981, *Opt. Acta* **28,** 1.
Paul, H., 1982, *Rev. Mod. Phys.* **54,** 1061.
Paul, H. and W. Brunner, 1980, *Opt. Acta* **27,** 263.
Paul, H. and W. Brunner, 1981, *Ann. Physik* **38,** 89.
Paul, H., W. Brunner and G. Richter, 1963, *Ann. Physik* **12,** 325.
Paul, H., U. Mohr and W. Brunner, 1976, *Opt. Comm.* **17,** 145.
Pearl, P. and G. J. Troup, 1968, *Phys. Lett.* **27A,** 560.
Perelomov, A. M., 1977, *Usp. Phys. Nauk* **123,** 23.
Peřina, J., 1965, *Phys. Lett.* **19,** 195.
Peřina, J., 1967a, *Czech. J. Phys.* **B17,** 1086.
Peřina, J., 1967b, *Phys. Lett.* **24A,** 333.
Peřina, J., 1968a, *Acta Univ. Palack. Ol.* **27,** 227.
Peřina, J., 1968b, *Czech. J. Phys.* **B18,** 197.
Peřina, J., 1969a, *Opt. Acta* **16,** 289.
Peřina, J., 1969b, *Czech. J Phys.* **B19,** 151.

Peřina, J., 1970, Quantum Optics, eds. S. M. Kay and A. Maitland, Academic Press, London, p. 513.
Peřina, J., 1972, Coherence of Light, Van Nostrand, London.
Peřina, J., 1972a, *Czech. J. Phys.* **B22,** 1075.
Peřina, J., 1974, Coherence of Light, Mir, Moscow (in Russian).
Peřina, J., 1975, Theory of Coherence, SNTL, Prague (in Czech).
Peřina, J., 1976, *Czech. J. Phys.* **B26,** 140.
Peřina, J., 1977, *Acta Phys. Pol.* **A52,** 559.
Peřina, J., 1979, *Opt. Acta* **26,** 821.
Peřina, J., 1980a, Proc. EKON-78, eds. S. Kielich, F. Kaczmarek, T. Bancewicz, Univ. Mickiewicz, Poznań, p. 153.
Peřina, J., 1980b, Progress in Optics, Vol. 18, ed. E. Wolf, North-Holland, p. 127.
Peřina, J., 1981a, *Opt. Acta* **28,** 325.
Peřina, J., 1981b, *Opt. Acta* **28,** 1529.
Peřina, J. and R. Horák, 1969a, *J. Phys.* **A2,** 702.
Peřina, J. and R. Horák, 1969b, *Opt. Comm.* **1,** 91.
Peřina, J. and R. Horák, 1970, *Czech. J. Phys.* **B20,** 149.
Peřina, J. and R. Horák, 1981, *Opt. Quant. Electr.* **13,** 345.
Peřina, J. and L. Mišta, 1968a, *Phys. Lett.* **27A,** 217.
Peřina, J. and L. Mišta, 1968b, *Czech. J. Phys.* **B18,** 697.
Peřina, J. and L. Mišta, 1969, *Ann. Physik* **22,** 372.
Peřina, J. and L. Mišta, 1974, *Opt. Acta* **21,** 329.
Peřina, J. and V. Peřinová, 1965, *Opt. Acta* **12,** 333.
Peřina, J. and V. Peřinová, 1969, *Opt. Acta* **16,** 309.
Peřina, J. and V. Peřinová, 1971, *Phys. Lett.* **35A,** 283.
Peřina, J. and V. Peřinová, 1972, *Czech. J. Phys.* **B22,** 1085.
Peřina, J. and V. Peřinová, 1975, *Czech. J. Phys.* **B25,** 605.
Peřina, J. and V. Peřinová, 1976a, *Czech. J. Phys.* **B26,** 489.
Peřina, J. and V. Peřinová, 1976b, Recent Advances in Optical Physics, eds. B. Havelka and J. Blabla, Soc. Czech. Math. Phys., Prague, p. 73.
Peřina, J. and V. Peřinová, 1983, *Opt. Acta* **30,** 955.
Peřina, J., V. Peřinová and Z. Braunerová, 1977, *Opt. Appl.* **VII/3,** 79.
Peřina, J., V. Peřinová, P. Diament and M. C. Teich, 1975, *Czech. J. Phys.* **B25,** 483.
Peřina, J., V. Peřinová and R. Horák, 1973a, *Czech. J. Phys.* **B23,** 975.
Peřina, J., V. Peřinová and R. Horák, 1973b, *Czech. J. Phys.* **B23,** 993.
Peřina, J., V. Peřinová and L. Knesel, 1977, *Acta Phys. Pol.* **A51,** 725.
Peřina, J., V. Peřinová, J. Křepelka, A. Lukš, C. Sibilia and M. Bertolotti, 1983, *Opt. Acta* **30,** 959.
Peřina, J., V. Peřinová, G. Lachs and Z. Braunerová, 1973, *Czech. J. Phys.* **B23,** 1008.
Peřina, J., V. Peřinová and L. Mišta, 1971, *Opt. Comm.* **3,** 89.
Peřina, J., V. Peřinová and L. Mišta, 1972, *Opt. Acta* **19,** 579.
Peřina, J., V. Peřinová and L. Mišta, 1974, *Czech. J. Phys.* **B24,** 482.
Peřina, J., V. Peřinová, L. Mišta and R. Horák, 1974, *Czech. J. Phys.* **B24,** 374.
Peřina, J., V. Peřinová, M. C. Teich and P. Diament, 1973, *Phys. Rev.* **A7,** 1732.
Peřina, J., B. E. A. Saleh and M. C. Teich, 1983, Independent photon deletions from quantized boson fields: The quantum analogue of the Burgess variance theorem, *Opt. Comm.*, to be published.
Peřinová, V., 1981, *Opt. Acta* **28,** 747.
Peřinová, V. and J. Peřina, 1978a, *Czech. J. Phys.* **B28,** 306.
Peřinová, V. and J. Peřina, 1978b, *Czech. J. Phys.* **B28,** 1183.
Peřinová, V. and J. Peřina, 1978c, *Czech. J. Phys.* **B28,** 1196.
Peřinová, V. and J. Peřina, 1981, *Opt. Acta* **28,** 769.
Peřinová, V., J. Peřina, M. Bertolotti and C. Sibilia, 1982, *Opt. Acta* **29,** 131.
Peřinová, V., J. Peřina and L. Knesel, 1977, *Czech. J. Phys.* **B27,** 487.

Peřinová, V., J. Peřina, P. Szlachetka and S. Kielich, 1979a, *Acta Phys. Pol.* **A56,** 267.
Peřinová, V., J. Peřina, P. Szlachetka and S. Kielich, 1979b, *Acta Phys. Pol.* **A56,** 275.
Pfleegor, R. L. and L. Mandel, 1967a, *Phys. Lett.* **24A,** 766.
Pfleegor, R. L. and L. Mandel, 1967b, *Phys. Rev.* **159,** 1084.
Pfleegor, R. L. and L. Mandel, 1968, *J. Opt. Soc. Am.* **58,** 946.
Phillips, R. L. and L. C. Andrews, 1981, *J. Opt. Soc. Am.* **71,** 1440.
Phillips, D. T., H. Kleiman and S. P. Davis, 1967, *Phys. Rev.* **153,** 113.
Picard, R. H. and C. R. Willis, 1965, *Phys. Rev.* **139,** A 10.
Picinbono, B., 1967, Proc. Symp. Modern Optics, Polytech. Press, New York, p. 167.
Picinbono, B., 1969, *Phys. Lett.* **29A,** 614.
Picinbono, B., 1971, *Phys. Rev.* **A4,** 2398.
Picinbono, B. and M. Rousseau, 1970, *Phys. Rev.* **A1,** 635.
Picinbono, B. and M. Rousseau, 1977, *Phys. Rev.* **A15,** 1648.
Pieczonková, A., 1982a, *Czech. J. Phys.* **B32,** 831.
Pieczonková, A., 1982b, *Opt. Acta* **29,** 1509.
Pieczonková, A. and J. Peřina, 1981, *Czech. J. Phys.* **B31,** 837.
Pike, E. R., 1969, *Rivista Nuovo Cim.* **1,** 277.
Pike, E. R., 1970, Quantum Optics, eds. S. M. Kay and A. Maitland, Academic Press, London, p. 127.
Pike, E. R. and E. Jakeman, 1974, Advances in Quantum Electronics, Vol. 2, ed. D. W. Goodwin, Academic Press, London, p. 1.
Piovoso, M. J. and L. P. Bolgiano, 1967, *Proc. IEEE* **55,** 1519.
Ponath, H. E. and M. Schubert, 1980, *Ann. Physik* **37,** 109.
Potechin, V. A. and V. N. Tatarinov, 1978, Theory of Coherence of Electromagnetic Field, Svyaz, Moscow (in Russian).
Prucnal, P. R., 1980, *Appl. Opt.* **19,** 3611.
Prucnal, P. R. and M. C. Teich, 1978, *Appl. Opt.* **17,** 3576.
Prucnal, P. R. and M. C. Teich, 1979, *J. Opt. Soc. Am.* **69,** 539.
Purcell, E. M., 1956, *Nature* **178,** 1449.
Pusey, P. N., 1979, *J. Phys.* **A12,** 1805.
Pusey, P. N., D. W. Schaefer and D. E. Koppel, 1974, *J. Phys.* **A7,** 530.
Quattropani, A., P. Schwendimann and H. P. Baltes, 1980, *Opt. Acta* **27,** 135.
Radcliffe, J. M., 1971, *J. Phys.* **A4,** 313.
Radloff, W., 1968, *Phys. Lett.* **27A,** 366.
Radloff, W., 1971, *Ann. Physik* **26,** 178.
Raiford, M. T., 1970, *Phys. Rev.* **A2,** 1541.
Rebka, G. A. and R. V. Pound, 1957, *Nature* **180,** 1035.
Reed, I. S., 1962, *IRE Trans. Inf. Theory* **IT-8,** 194.
Reid, M., K. J. McNeil and D. F. Walls, 1981, *Phys. Rev.* **A24,** 2029.
Reynolds, G. T., K. Spartalian and D. B. Scarl, 1969, *Nuovo Cim.* **61,** 355.
Rezende, S. M. and N. Zagury, 1969, *Phys. Lett.* **29A,** 47 and 616.
Richter, G., W. Brunner and H. Paul, 1964, *Ann. Physik* **14,** 239.
Richter, T., 1981, *Ann. Physik* **38,** 106.
Risken, H., 1965, *Z. Phys.* **186,** 85.
Risken, H., 1966, *Z. Phys.* **191,** 302.
Risken, H., 1968, *Fortschr. Phys.* **16,** 261.
Risken, H., 1970, Progress in Optics, Vol. 8, ed. E. Wolf, North-Holland, Amsterdam, p. 239.
Risken, H. and H. D. Vollmer, 1967, *Z. Phys.* **204,** 240.
Ritze, H. H. and A. Bandilla, 1979, *Opt. Comm.* **30,** 125.
Robl, H. R., 1967, *Phys. Lett.* **24A,** 288.
Robl, H. R., 1968, *Phys. Rev.* **165,** 1426.

Rocca, F., 1967, *J. d. Phys.* **28**, 113.
Rockower, E. B. and N. B. Abraham, 1978, *J. Phys.* **A11**, 1879.
Rockower, E. B., N. B. Abraham and S. R. Smith, 1978, *Phys. Rev.* **A17**, 1100.
Roman, P. and A. S. Marathay, 1963, *Nuovo Cim.* **30**, 1452.
Roman, P. and E. Wolf, 1960, *Nuovo Cim.* **17**, 462 and 477.
Rosenberg, S. and M. C. Teich, 1972, *J. Appl. Phys.* **43**, 1256.
Rosenberg, S. and M. C. Teich, 1973a, *Appl. Opt.* **12**, 2625.
Rosenberg, S. and M. C. Teich, 1973b, *IEEE* **IT-19**L, 807.
Rosenfeld, L., 1958, Niels Bohr and Development of Physics, ed. W. Pauli.
Ross, G. and M. A. Fiddy, 1978, *Opt. Acta* **25**, 205.
Ross, G., M. A. Fiddy, M. Nieto-Vesperinas and M. W. L. Wheeler, 1977, *Optik* **49**, 71.
Ross, G. and M. Nieto-Vesperinas, 1981, *Opt. Acta* **28**, 77.
Rousseau, M., 1969, *C. R. Acad. Sci. Paris* **268**, 1477.
Rousseau, M., 1971, *J. Opt. Soc. Am.* **61**, 1307.
Rowe, D. J. and A. G. Ryman, 1980, *Phys. Rev. Lett.* **45**, 406.
Roy, R. and L. Mandel, 1977, *Opt. Comm.* **23**, 306.
Roy, R. and L. Mandel, 1980, *Opt. Comm.* **34**, 133.
Ruggieri, N. F., D. O. Cummings and G. Lachs, 1972, *J. Appl. Phys.* **43**, 1118.
Rytov, S. M., Yu. A. Kravcov and V. I. Tatarskii, 1978, Introduction to Statistical Radiophysics, Vol. II, Nauka, Moscow (in Russian).
Saleh, B. E. A., 1975a, *J. Appl. Phys.* **46**, 943.
Saleh, B. E. A., 1975b, *Appl. Phys.* **8**, 269.
Saleh, B. E. A., 1978, Photoelectron Statistics, Springer, Berlin.
Saleh, B. E. A. and J. Hendrix, 1975, *J. Phys.* **A8**, 1134.
Saleh, B. E. A. and M. Irshid, 1979, *Opt. Quant. Electr.* **11**, 479.
Saleh, B. E. A., D. Stoler and M. C. Teich, 1983, *Phys. Rev.* **A27**, 360.
Saleh, B. E. A. and M. C. Teich, 1982, *Proc. IEEE* **70**, 229.
Saleh, B. E. A., J. T. Tavolacci and M. C. Teich, 1981, *IEEE J. Quant. Electr.* **QE-17**, 2341.
Sarfatt, J., 1963, *Nuovo Cim.* **27**, 1119.
Sargent, M. and M. O. Scully, 1972, Laser Handbook, Vol. 1, eds. F. T. Arecchi and E. O. Schulz-Dubois, North-Holland, Amsterdam, p. 45.
Sargent, M., M. O. Scully and W. E. Lamb, 1974, Laser Physics, Addison-Wesley, Reading.
Savage, C. M. and D. F. Walls, 1983, *Opt. Acta* **30**, 557.
Saxton, W. O., 1974, *J. Phys.* **D7**, L 63.
Schmidt-Weinmar, N. G., 1978, Inverse Source Problems, ed. H. P. Baltes, Springer, Berlin, p. 83.
Schmidt-Weinmar, N. G., B. Steinle and H. P. Baltes, 1978/79, *Optik* **52**, 205.
Schrödinger, E., 1927, *Naturwiss.* **14**, 644.
Schubert, M., K. E. Süsse, W. Vogel and D. G. Welsch, 1980, *Opt. Quant. Electr.* **12**, 65.
Schubert, M., K. E. Süsse, W. Vogel and D. G. Welsch, 1981, *Opt. Quant. Electr.* **13**, 301.
Schubert, M., K. E. Süsse, W. Vogel, D. G. Welsch and B. Wilhelmi, 1982, *Kvant. Electr.* (USSR) **9**, 495.
Schubert, M. and W. Vogel, 1978, *Phys. Lett.* **68**, 321.
Schubert, M. and W. Vogel, 1981, *Opt. Comm.* **36**, 164.
Schubert, M. and B. Wilhelmi, 1976, Recent Advances in Optical Physics, eds. B. Havelka and J. Blabla, Soc. Czech. Math. Phys., Prague, p. 225.
Schubert, M. and B. Wilhelmi, 1978, Einführung in die nichtlineare Optik, Teil II, Teubner, Leipzig.
Schubert, M. and B. Wilhelmi, 1980, Progress in Optics, Vol. 17, ed. E. Wolf, North-Holland, Amsterdam, p. 163.
Schütte, F. J., K. Germey, R. Tiebel and K. Worlitzer, 1981a, Optical bistability in multimode systems with second and third order dispersive nonlinearity, preprint (see *Ann. Physik* **39**, (1982) 170).

Schütte, F. J., K. Germey, R. Tiebel and K. Worlitzer, 1981b, Dispersive optical bistability of trilinear interacting modes, preprint.
Schütte, F. J., K. Germey, R. Tiebel and K. Worlitzer, 1983, *Opt. Acta* **30**, 465.
Schweber, S. S., 1961, An Introduction to Relativistic Quantum Field Theory, Row, Peterson and Co. Evanston, Ill., Elmsford, New York.
Scudieri, F. and M. Bertolotti, 1974, *J. Opt. Soc. Am.* **64,** 776.
Scudieri, F., M. Bertolotti and R. Bartolino, 1974, *Appl. Opt.* **13,** 181.
Scully, M. O. and W. E. Lamb, 1966, *Phys. Rev. Lett.* **16,** 853.
Scully, M. O. and W. E. Lamb, 1967, *Phys. Rev.* **159,** 208.
Scully, M. O. and W. E. Lamb, 1968, *Phys. Rev.* **166,** 246.
Scully, M. O. and W. E. Lamb, 1969, *Phys. Rev.* **179,** 368.
Scully, M. O. and K. G. Whitney, 1972, Progress in Optics, Vol. 10, ed. E. Wolf, North-Holland, Amsterdam, p. 89.
Sczaniecki, L., 1980, Bistability of multi-photon lasers, Proc. Inter. Conf. Lasers (New Orleans).
Sczaniecki, L., 1982, Quantum theory of multi-photon lasers, preprint; *Opt. Acta* **29,** 69.
Sczaniecki, L. and J. Buchert, 1978, *Opt. Comm.* **27,** 463.
Selloni, A., 1980, Inverse Scattering Problems in Optics, ed. H. Baltes, Springer, Berlin, p. 117.
Selloni, A., P. Schwendimann, A. Quattropani and H. P. Baltes, 1978a, *J. Phys.* **A11,** 1427.
Selloni, A., P. Schwendimann, A. Quattropani and H. P. Baltes, 1978b, *Phys. Rev.* **A18,** 2234.
Semenov, A. A. and T. I. Arsenyan, 1978, Fluctuations of Electromagnetic Waves, Nauka, Moscow (in Russian).
Senitzky, I. R., 1958, *Phys. Rev.* **111,** 3.
Senitzky, I. R., 1962, *Phys. Rev.* **127,** 1638.
Senitzky, I. R., 1967a, *Phys. Rev.* **161,** 165.
Senitzky, I. R., 1967b, *Phys. Rev.* **155,** 1387.
Senitzky, I. R., 1968, *Phys. Rev.* **174,** 1588.
Senitzky, I. R., 1969, *Phys. Rev.* **183,** 1069.
Senitzky, I. R., 1973, *Phys. Rev. Lett.* **31,** 955.
Senitzky, I. R., 1978, Progress in Optics, Vol. 16, ed. E. Wolf, North-Holland, Amsterdam, p. 413.
Seybold, K. and H. Risken, 1974, *Z. Phys.* **267,** 323.
Shapiro, J. H., 1980, *Opt. Lett.* **5,** 351.
Shapiro, J. H., H. P. Yuen and J. A. Machado Mata, 1979, *IEEE Trans. Inf. Theory* **IT-25,** 179.
Shapiro, S. L., 1977, Ultrashort Light Pulses, Springer, Berlin.
Sharma, M. P. and L. M. Brescausin, 1981, *Phys. Rev.* **A23,** 1893.
Shen, Y. R., 1967, *Phys. Rev.* **155,** 921.
Shen, Y. R., 1976, *Rev. Mod. Phys.* **48,** 1.
Shepherd, T. J., 1981, *Opt. Acta* **28,** 567.
Sheremetyev, A. G., 1971, Statistical Theory of Laser Communication, Svyaz, Moscow (in Russian).
Shiga, F. and S. Inamura, 1967, *Phys. Lett.* **25A,** 706.
Short, R. and L. Mandel, 1983, Abstracts 5th Conf. Coher. Quant. Optics, Rochester 1983, p. 221.
Shustov, A. P., 1978, *J. Phys.* **A11,** 1771.
Sibilia, C. and M. Bertolotti, 1981, *Opt. Acta* **28,** 503.
Sillitto, R. M., 1968, *Phys. Lett.* **27A,** 624.
Simaan, H. D., 1975, *J. Phys.* **A8,** 1620.
Simaan, H. D., 1978, *J. Phys.* **A11,** 1799.
Simaan, H. D., 1979, *Opt. Comm.* **31,** 21.
Simaan, H. D. and R. Loudon, 1975, *J. Phys.* **A8,** 539 and 1140.
Simaan, H. D. and R. Loudon, 1978, *J. Phys.* **A11,** 435.
Sinaya, Ya. G. and L. P. Shilnikov, 1981, Strange Attractors, Mir, Moscow (in Russian).
Singh, S., 1981, *Phys. Rev.* **A23,** 837.
Singh, S., 1983, *Opt. Comm.* **44,** 254.

Slusher, R. E., 1974, Progress in Optics, Vol. 12, ed. E. Wolf, North-Holland, Amsterdam, p. 53.
Smirnov, D. F. and A. S. Troshin, 1979, *J. Exp. Theor. Phys.* (USSR) **76,** 1254.
Smirnov, D. F. and A. S. Troshin, 1981, *J. Exp. Theor. Phys.* (USSR) **81,** 1597.
Smith, A. W. and J. A. Armstrong, 1966a, *Phys. Rev. Lett.* **16,** 1169.
Smith, A. W. and J. A. Armstrong, 1966b, *Phys. Lett.* **19,** 650.
Smithers, M. E. and E. Y. C. Lu, 1974, *Phys. Rev.* **A10,** 1874.
Sobolewska, B. and R. Sobolewski, 1978, *Opt. Comm.* **26,** 211.
Solimeno, S., E. Corti and B. Nicoletti, 1970, *J. Opt. Soc. Am.* **60,** 1245.
Solimeno, S., P. Di Porto and B. Crosignani, 1969, *J. Math. Phys.* **10,** 1922.
Sotskii, B. A. and B. I. Glazatchev, 1981, *Opt. Spectr.* (USSR) **50,** 1057.
Spence, J. C. H., 1974, *Opt. Acta* **21,** 835.
Srinivas, M. D., 1978, Coherence and Quantum Optics IV, eds. L. Mandel and E. Wolf, Plenum Press, New York, p. 885.
Srinivas, M. D. and E. B. Davies, 1981, *Opt. Acta* **28,** 981.
Srinivasan, S. K., 1974, *Phys. Lett.* **50A,** 277.
Srinivasan, S. K., 1978, *J. Phys.* **A11,** 2333.
Srinivasan, S. K. and M. Gururajan, 1981, *J. Math. Phys. Sci.* **15,** 297.
Srinivasan, S. K. and S. Sukavanam, 1972, *J. Phys.* **A5,** 682.
Srinivasan, S. K. and S. Sukavanam, 1978, *Phys. Lett.* **66A,** 164.
Srinivasan, S. K. and S. Udayabaskaran, 1979, *Opt. Acta* **26,** 1535.
Steinle, B. and H. P. Baltes, 1977, *J. Opt. Soc. Am.* **67,** 241.
Steinle, B., H. P. Baltes and M. Pabst, 1975, *Phys. Rev.* **A12,** 1519.
Steinle, B., H. P. Baltes and M. Pabst, 1976, *Infrared Phys.* **16,** 25.
Steudel, H., 1981, *Ann. Physik* **38,** 97.
Steyn-Ross, M. L. and D. F. Walls, 1981, *Opt. Acta* **28,** 201.
Stoler, D., 1970, *Phys. Rev.* **D1,** 3217.
Stoler, D., 1971, *Phys. Rev.* **D4,** 1925 and 2309.
Stoler, D., 1972, *Phys. Lett.* **38A,** 433.
Stoler, D., 1974, *Phys. Rev. Lett.* **33,** 1397.
Stoler, D., 1975, *Phys. Rev.* **11,** 3033.
Stoljarov, A. D., 1976, *J. Priclad. Spectr.* (USSR) **25,** 236.
Strohbehn, J. W., 1971, Progress in Optics, Vol. 9, ed. E. Wolf, North-Holland, Amsterdam, p. 73.
Strohbehn, J. W., ed., 1978, Laser Beam Propagation in the Atmosphere, Springer, Berlin.
Sudarshan, E. C. G., 1963a, *Phys. Rev. Lett.* **10,** 277.
Sudarshan, E. C. G., 1963b, Proc. Symp. Optical Masers, J. Wiley, New York, p. 45.
Süsse, K. E., W. Vogel and D. G. Welsch, 1980a, *Phys. Stat. Sol.* (*b*) **99,** 91.
Süsse, K. E., W. Vogel and D. G. Welsch, 1980b, *Opt. Comm.* **33,** 56.
Süsse, K. E., W. Vogel and D. G. Welsch, 1981, *Opt. Comm.* **36,** 135.
Süsse, K. E., W. Vogel, D. G. Welsch and B. Wilhelmi, 1979, *Opt. Comm.* **28,** 389.
Svelto, O., 1974, Progress in Optics, Vol. 12, ed. E. Wolf, North-Holland, Amsterdam, p. 1.
Szlachetka, P., S. Kielich, J. Peřina and V. Peřinová, 1979, *J. Phys.* **A12,** 1921.
Szlachetka, P., S. Kielich, J. Peřina and V. Peřinová, 1980a, *J. Molec. Spectr.* **61,** 281.
Szlachetka, P., S. Kielich, J. Peřina and V. Peřinová, 1980b, *Opt. Acta* **27,** 1609.
Szlachetka, P., S. Kielich, V. Peřinová and J. Peřina, 1980, Proc. EKON-78, eds. S. Kielich, F. Kaczmarek and T. Bancewicz, Univ. Mickiewicz, Poznań, p. 281.
Tänzler, W. and F. J. Schütte, 1981a, *Ann. Physik* **38,** 73.
Tänzler, W. and F. J. Schütte, 1981b, *Opt. Comm.* **37,** 447.
Tatarskii, V. I., 1967, Propagation of Waves in a Turbulent Atmosphere, Nauka, Moscow (in Russian).
Tatarskii, V. I., 1970, Propagation of Short Waves in a Medium with Random Inhomogeneities in Approximation of Markoff Random Process, Inst. Phys. Atmosph., Moscow (in Russian).

Tatarskii, V. I., 1971, *J. Exp. Theor. Phys.* (USSR) **61,** 1822.
Tatarskii, V. I., 1983, *J. Exp. Theor. Phys.* (USSR) **84,** 526.
Tatarskii, V. I. and V. U. Zavorotnyi, 1980, Progress in Optics, Vol. 18, ed. E. Wolf, North-Holland, Amsterdam, p. 204.
Taylor, L. S., 1980, *J. Opt. Soc. Am.* **70,** 1554.
Teich, M. C., 1969, *Appl. Phys. Lett.* **14,** 201.
Teich, M. C., 1977, Topics in Applied Physics, Vol. 19, ed. R. J. Keyes, Springer, Berlin, p. 229.
Teich, M. C., 1981, *Appl. Opt.* **20,** 2457.
Teich, M. C., R. L. Abrams and W. B. Gandrud, 1970, *Opt. Comm.* **2,** 206.
Teich, M. C. and B. I. Cantor, 1978, *IEEE J. Quant. Electr.* **QE-14,** 993.
Teich, M. C. and P. Diament, 1969, *J. Appl. Phys.* **40,** 625.
Teich, M. C., L. Matin and B. I. Cantor, 1978, *J. Opt. Soc. Am.* **68,** 386.
Teich, M. C. and W. J. McGill, 1976, *Phys. Rev. Lett.* **36,** 754.
Teich, M. C., P. R. Prucnal, G. Vannucci, M. E. Breton and W. J. McGill, 1982, Multiplication noise in the human visual system at threshold, preprint; *J. Opt. Soc. Am.* **72,** 419.
Teich, M. C. and S. Rosenberg, 1971, *Opto-Electronics* **3,** 63.
Teich, M. C. and S. Rosenberg, 1973, *Appl. Opt.* **12,** 2216.
Teich, M. C. and B. E. A. Saleh, 1981a, *J. Opt. Soc. Am.* **71,** 771.
Teich, M. C. and B. E. A. Saleh, 1981b, *Phys. Rev.* **A24,** 1651.
Teich, M. C. and B. E. A. Saleh, 1982, *Opt. Lett.* **7,** 365.
Teich, M. C., B. E. A. Saleh and J. Peřina, 1983, Abstracts 5th Conf. Coheren. Quant. Optics, Rochester 1983, p. 128; to be published.
Teich, M. C., B. E. A. Saleh and D. Stoler, 1983, *Opt. Comm.* **46,** 244.
Teich, M. C. and G. Vannucci, 1978, *J. Opt. Soc. Am.* **68,** 1338.
Teich, M. C. and G. J. Wolga, 1966, *Phys. Rev. Lett.* **16,** 625.
Timmermans, J. and R. J. J. Zijlstra, 1977, *Physica* **88A,** 600.
Titulaer, U. M. and R. J. Glauber, 1965, *Phys. Rev.* **140,** B 676.
Titulaer, U. M. and R. J. Glauber, 1966, *Phys. Rev.* **145,** 1041.
Tornau, N. and A. Bach, 1974, *Opt. Comm.* **11,** 46.
Tornau, N. and B. Echtermeyer, 1973, *Ann. Physik* **29,** 289.
Trias, A., 1977, *Phys. Lett.* **61A,** 149.
Trifonov, D. A., 1974, *Phys. Lett.* **48A,** 165.
Trifonov, D. A. and V. N. Ivanov, 1977, *Phys. Lett.* **64A,** 269.
Troup, G. J., 1965a, *Proc. Phys. Soc.* **86,** 39.
Troup, G. J., 1965b, *Phys. Lett.* **17,** 264.
Troup, G. J., 1966, *Nuovo Cim.* **42,** 79.
Troup, G. J., 1967, Optical Coherence Theory, Methuen, London.
Troup, G. J. and J. Lyons, 1969, *Phys. Lett.* **29A,** 705.
Trung, T. V. and F. J. Schütte, 1977, *Ann. Physik* **34,** 262.
Trung, T. V. and F. J. Schütte, 1978, *Ann. Physik* **35,** 216.
Tucker, J. and D. F. Walls, 1969, *Phys. Rev.* **178,** 2036.
Tunkin, V. G. and A. S. Tchirkin, 1970, *J. Exp. Theor. Phys.* (USSR) **58,** 191.
Twiss, R. Q., 1969, *Opt. Acta* **16,** 423.
Twiss, R. Q. and A. G. Little, 1959, *Austr. J. Phys.* **12,** 77.
Twiss, R. Q., A. G. Little and R. Hanbury Brown, 1957, *Nature* **180,** 324.
Vajnshtejn, L. A., V. N. Melechin, S. A. Mishin and E. R. Podoljak, 1981, *J. Exp. Theor. Phys.* (USSR) **81, 2000.**
Van Cittert, P. H., 1934, *Physica* **1,** 201.
Van Cittert, P. H., 1939, *Physica* **6,** 1129.
Vannucci, G. and M. C. Teich, 1981, *J. Opt. Soc. Am.* **71,** 164.
Vinson, J. F., 1971, Optische Kohärenz, Akademie, Berlin.

Voigt, H. and A. Bandilla, 1981, *Ann. Physik* **38,** 137.
Voigt, H., A. Bandilla and H. H. Ritze, 1980, *Z. Phys.* **B36,** 295.
Volterra, V., 1959, Theory of Functionals and of Integral and Integro-Differential Equations, Dover Publ., New York.
Vrbová, M., 1970, *Czech. J. Phys.* **B20,** 959.
Wagner, J., P. Kurowski and W. Martienssen, 1979, *Z. Phys.* **B33,** 391.
Walker, J. G., 1981, *Opt. Acta* **28,** 735.
Walls, D. F., 1970, *Z. Phys.* **237,** 224.
Walls, D. F., 1973, *J. Phys.* **A6,** 496.
Walls, D. F., 1977, *Am. J. Phys.* **45,** 952.
Walls, D. F., 1979, *Nature* **280,** 451.
Walls, D. F., 1980, *J. Phys.* **B13,** 2001.
Walls, D. F. and R. Barakat, 1970, *Phys. Rev.* **A1,** 446.
Walls, D. F., H. J. Carmichael, R. F. Gragg and W. C. Schieve, 1978, *Phys. Rev.* **A18,** 1622.
Walls, D. F., P. D. Drummond, S. S. Hassan and H. J. Carmichael, 1978, Progr. Theor. Phys., Suppl. No 64, 307.
Walls, D. F., P. D. Drummond and K. J. McNeil, 1981, Optical Bistability, eds. C. M. Bowden, M. Ciftan and H. R. Robl, Plenum Press, New York, p. 51.
Walls, D. F., C. V. Kunasz, P. D. Drummond and P. Zoller, 1981, *Phys. Rev.* **A24,** 627.
Walls, D. F. and G. J. Milburn, 1981, Lectures presented at Summer School on Quantum Optics and Experimental General Relativity (Bad Windsheim).
Walls, D. F. and C. T. Tindle, 1972, *J. Phys.* **A5,** 534.
Walls, D. F. and P. Zoller, 1980, *Opt. Comm.* **34,** 260.
Walls, D. F. and P. Zoller, 1981, *Phys. Rev. Lett.* **47,** 709.
Walls, D. F., P. Zoller and M. L. Steÿn-Ross, 1981, *IEEE J. Quant. Electr.* **QE-17,** 380.
Walmsley, I. A. and M. G. Raymer, 1983, *Phys. Rev. Lett.* **50,** 962.
Wang, M. C. and G. E. Uhlenbeck, 1945, *Rev. Mod. Phys.* **17,** 323.
Webber, J. C., 1968, *Phys. Lett.* **27A,** 5.
Webber, J. C., 1969, *Can. J. Phys.* **47,** 363.
Weber, H. P., 1971, *IEEE J. Quant. Electr.* **QE-7,** 189.
Weidlich, W., H. Risken and H. Haken, 1967, *Z. Phys.* **201,** 396.
Welford, W. T., 1977, *Opt. Quant. Electr.* **9,** 269.
Whittaker, E. and G. N. Watson, 1940, A Course of Modern Analysis, 4th ed., Univ. Press. Cambridge.
Wigner, E., 1932, *Phys. Rev.* **40,** 749.
Willis, C. R., 1966, *Phys. Rev.* **147,** 406.
Wódkiewicz, K. and M. S. Zubairy, 1983, *Phys. Rev.* **A27,** 2003.
Wolf, E., 1955, *Proc. Roy. Soc.* **A230,** 246.
Wolf, E., 1957, *Phil. Mag.* (8) **2,** 351.
Wolf, E., 1960, *Proc. Phys. Soc.* (Lond.) **76,** 424.
Wolf, E., 1962, *Proc. Phys. Soc.* (Lond.) **80,** 1269.
Wolf, E., 1963, Proc. Symp. Optical Masers, J. Wiley, New York, p. 29.
Wolf, E., 1964, Quantum Electronics, eds. N. Bloembergen and P. Grivet, Dunod et Cie., Paris, p. 13.
Wolf, E., 1965, *Jap. J. Appl. Phys.* **4,** Suppl. I, p. 1.
Wolf, E., 1966, *Opt. Acta* **13,** 281.
Wolf, E., 1978, *J. Opt. Soc. Am.* **68,** 1597.
Wolf, E., 1981a, *Opt. Comm.* **38,** 3.
Wolf, E., 1981b, Optics in Four Dimensions—1980, eds. M. A. Machado and L. M. Narducci, Am. Inst. Phys., New York, p. 42.
Wolf, E. and G. S. Agarwal, 1969, Polarization, Matière et Rayonnement, Soc. Franc. Phys., Press Univ., p. 541.

Wolf, E. and W. H. Carter, 1975, *Opt. Comm.* **13,** 205.
Wolf, E. and W. H. Carter, 1976, *Opt. Comm.* **16,** 297.
Wolf, E. and C. L. Mehta, 1964, *Phys. Rev. Lett.* **13,** 705.
Yariv, A., 1967, Quantum Electronics and Nonlinear Optics, 1st ed., J. Wiley, New York.
Yariv, A., 1975, Quantum Electronics, 2nd ed., J. Wiley, New York.
Yarunin, V. S., 1978, *Vest. Leningr. Univ.* **3,** 130.
Yen, R. Y., P. Diament and M. C. Teich, 1972, *IEEE Trans. Inf. Theory* **IT-18,** 302.
Yoshikawa, M., N. Suzuki and T. Suzuki, 1981, *Jap. J. Appl. Phys.* **20,** 601.
Yuen, H. P., 1975, *Phys. Lett.* **51A,** 1.
Yuen, H. P., 1976, *Phys. Rev.* **A13,** 2226.
Yuen, H. P. and J. H. Shapiro, 1978a, *IEEE Trans. Inf. Theory* **IT-24,** 657.
Yuen, H. P. and J. H. Shapiro, 1978b, Coherence and Quantum Optics IV, eds. L. Mandel and E. Wolf, Plenum Press, New York, p. 719.
Yuen, H. P. and J. H. Shapiro, 1979, *Opt. Lett.* **4,** 334.
Zardecki, A., 1969, *J. Math. Phys.* **11,** 224.
Zardecki, A., 1971, *Can. J. Phys.* **49,** 1724.
Zardecki, A., 1974, *J. Phys.* **A7,** 2198.
Zardecki, A., 1978, Inverse Source Problems, ed. H. P. Baltes, Springer, Berlin, p. 155.
Zardecki, A., 1981, *Phys. Rev.* **A23,** 1281.
Zardecki, A., J. Bures and C. Delisle, 1972, *Phys. Rev.* **A6,** 1209.
Zardecki, A. and C. Delisle, 1973, *Can. J. Phys.* **51,** 1017.
Zardecki, A. and C. Delisle, 1977, *Opt. Acta* **24,** 241.
Zardecki, A., C. Delisle and J. Bures, 1972, *Opt. Comm.* **5,** 298.
Zardecki, A., C. Delisle and J. Bures, 1973, Coherence and Quantum Optics, eds. L. Mandel and E. Wolf, Plenum Press, New York, p. 259.
Zavorotnyi, V. U. and V. I. Tatarskii, 1973, *J. Exp. Theor. Phys.* (USSR) **64,** 453.
Zernike, F. and J. E. Midwinter, 1973, Applied Nonlinear Optics, J. Wiley, New York.
Zubairy, M. S., 1979, *Phys. Rev.* **A20,** 2464.
Zubairy, M. S. and J. J. Yeh, 1980, *Phys. Rev.* **A21,** 1624.

INDEX

Annihilation of electron–positron pairs, 273, 274
Annihilation operator, 12, 13, 15
Antibunching, 36, 57, 194, 207–209, 215, 218, 224–226, 229, 240, 244, 246, 269, 270–272, 274
Anticorrelation, 36, 214, 225, 229, 246, 274

Bloch states, 107
Brillouin scattering, 157, 158, 234
Bunching, 57–60, 194
Burgess variance theorem, 47

Characteristic function, 17
  antinormal, 81
  normal, 37, 48, 81
  quantum, 17, 81, 96, 97
Characteristic functional, 40
  normal, 40
  quantum, 40
  symmetric, 40
Coherence, 32
  area, 151
  full, 35
  higher-order, 34
  length, 151
  partial, 33
  quantum, 32
  second-order, 34
  time, 53, 62, 64
  volume, 151
Coherence matrix, 55
Coherent γ-emission, 273
Coherent states, 35, 69–78
  atomic, 106
  even, 78
  expansions, 71
  for general potentials, 108
  generalized, 76
  global, 76
  odd, 78
  time development, 77
  two-photon, 103–106
Commutators of the field operators, 17
Completeness condition, 14, 17, 72, 76, 108
Cooperative phenomena, 278
Correlation function, 23
  antinormal, 39
  classical, 24
  fourth-order, 57, 58
  mixed-order, 41
  normal, 24
  properties, 27
  quantum, 23
  second-order, 28
Correlation matrix, 55
Correlation interferometer, 64
Correlation of fluctuations, 58
Correlation spectroscopy, 65, 66
Correlation tensors for blackbody radiation, 112
Creation operator, 12, 13, 15
Cross-spectral density, 29
Cross-spectral purity, 63, 132, 142

Dead-time effect, 66, 67
Degeneracy parameter, 55, 62
Degree of coherence, 31
Degree of polarization, 55
Density matrix, 15, 23
  equation of motion, 172
  Glauber–Sudarshan representation, 78
Diament–Teich description, 196
Dicke states, 107
Dispersion relations, 27
Displacement operator, 69, 73, 74
Distribution of Bose–Einstein, 53
Dynamics of phonon modes, 245
Dynamics of photon modes, 245

Effect of pumping fluctuations, 216

Entropy, 150
Expectation values, 20
  of an operator, 20, 74–75
  of antinormal operator, 80
  of normal operator, 78, 79, 80

Factorial cumulant, 48
Factorial moment, 46, 48, 195, 222, 241
Field, 38
  chaotic, 52, 53
  having no classical analogue, 208
  laser, 119
  strongly coherent, 38
  weakly coherent, 38
  with M degrees of freedom, 54
Filter function, 95
Fluctuations, 185
  fast, 185
  slow, 185
  strong, 185
  weak, 185
Fock space, 14
Fock state, 13
  global, 17
Fokker–Planck equation, 173
  generalized, 174, 187
  solution of, 189
Fokker–Planck equation approach, 167, 172
Free electron lasers, 274
Frequency down-conversion, 153, 216
Frequency up-conversion, 153, 216

Generating function, 45, 195, 222, 240
  antinormal, 193
  normal, 193, 222, 240, 276
Glauber–Sudarshan representation, 78–87, 89
  existence of, 86–87

Hamiltonian for radiation in random medium, 185
Hanbury Brown–Twiss effect, 37, 61–64
Heisenberg equations, 167
Heisenberg picture, 19
Heisenberg–Langevin approach, 167
Heisenberg–Langevin equations, 169, 174, 187
  solution of, 189
Higher harmonic generation, 226
Hyper-Raman scattering, 159, 160, 234, 259

Intensity correlation, 57
Interaction of radiation with atomic system, 174
Interaction of three one-mode boson quantum fields, 219
Interference of independent beams, 101–103

Langevin force, 168
Laser model, 118, 122
  experimental results, 124–128
  ideal, 118–121
  nonlinear, 122–123
  real, 122–128
  statistics, 124
Light, 52
  chaotic, 52, 53, 109
  Gaussian, 84, 109
  non-Gaussian, 63
  partially polarized, 55
  polarized, 56
  pseudothermal, 57, 60
  unpolarized, 56
Log-normal distribution, 179

Master equation, 172, 174
Master equation approach, 167, 172
Maxwell equations, 11, 164
Mercer theorem, 147
Minimum uncertainty wave packets, 72
Mode functions, 11
Mode index, 12
Model for light scattering, 274
Moment problem, 49–52
Multiphoton absorption, 160, 161, 261
Multiphoton emission, 161, 266
Mutual coherence function, 28
Mutual spectral density, 29

Nonlinear optical phenomena, 152
  second-order, 153
  third- and higher-order, 157
Number operator, 12

Optical parametric processes, 205
  degenerate case, 205
  non-degenerate case, 210
  with classical pumping, 205
  with quantum pumping, 219
Ordering, 94–101
  antinormal, 25, 80, 95
  normal, 10, 80, 95
  Ω-, 95
  s-, 94
  symmetric, 39, 95

Parametric amplification, 156
Parametric generation, 153
Phase operators, 88
Phase transition analogies, 278
Photocount distribution 43, 48, 49, 90, 195, 222, 241
  for cosinusoidal modulation, 138
  for exponential modulation, 140
  for square-wave modulation, 138
  for triangular modulation, 138, 139
Photocount statistics, 43, 193
  properties, 47
Photodetection equation, 44
  generalized, 100
Photon antibunching, 209
Photon echo, 165
Photon-number distribution, 46, 90, 277
Photon statistics, 194
  of multiphoton processes, 145–146
Principle of causality, 19
Probability distribution, 49
  for chaotic light, 56
  for laser light, 56
  negative exponential, 53
  of integrated intensity, 49
Propagation of radiation through Gaussian media, 178
  through turbulent atmosphere, 178

Quantum counters, 38
Quantum noise, 207, 214
Quantum statistics, 178, 205
  Diament–Teich description, 196
  in nonlinear media, 205
  in random media, 178
  phenomenological description, 178, 200
  quantum description, 185, 200
  Tatarskii description, 196
Quantum theory of damping, 167
Quasi-probability, 80

Rabi frequency, 270
Radiation,
  Gaussian, 84
  photocount statistics, 114
  of laser, 118–128
  thermal, 16
Raman scattering, 157–159, 234
  quantum description, 252
Rayleigh scattering, 158
Relation of quantum and classical correlation functions, 89–90
Reservoir phonon system, 234
Resonance fluorescence, 269

Schrödinger picture, 19
Second harmonic generation, 153, 154, 226
Second quantization of the electromagnetic field, 11
Self-focusing, 161–162
Self-induced transparency, 164–165
Self-radiation, 191, 192, 196
Shot noise, 198
Speckle phenomenon, 204
Squeezed states, 104, 273
Stationary conditions, 92–94
Subharmonic generation, 154, 226
Sub-Poissonian radiation, 208, 216, 225, 226
  behaviour, 209, 215, 244, 246
  regime, 209
Sum-frequency generation, 153
Superposition of coherent and chaotic fields, 129–151
  accuracy, 147
  characteristic generating function, 131
  entropy, 150, 151
  factorial cumulants, 146
  factorial moments, 140
  integrated intensity probability distribution, 133
  multimode, 131
  one mode, 130
  photocount distribution, 134
Superposition principle, 85
  quantum, 86, 89
Superradiance, 165–166
Synchronization direction, 155

Tatarskii description, 196
Threshold of oscillations, 123
Time development unitary operator, 21
Time ordering, 21
Transient coherent optical effects, 163
Two-beam interference experiment, 33

Vacuum expectation value, 25
Vacuum stability condition, 13
Vector potential operator, 11, 12
  negative frequency part of, 15
  positive frequency part of, 15

Ultradistribution, 87

Wave equations, 29
Wiener – Khintchine theorem, 28, 32
  generalized, 28, 29
Wigner function, 81
Wigner – Weisskopf approximation, 168

Young two-slit experiment, 32
  quantum treatment, 32 – 33